LANDSCAPE

AN INTRODUCTION TO
PHYSICAL GEOGRAPHY

**ADDISON·WESLEY
PUBLISHING COMPANY**

Reading, Massachusetts
Menlo Park, California
London
Amsterdam
Don Mills, Ontario
Sydney

WILLIAM M. MARSH University of Michigan, Flint

JEFF DOZIER University of California, Santa Barbara

LANDSCAPE

AN INTRODUCTION TO
PHYSICAL GEOGRAPHY

Sponsoring Editor: Thomas Robbins
Production Editor: Evelyn Wilde
Text and Cover Designer: Catherine L. Dorin
Art Coordinator: Dick Morton
Illustrators: Oxford Illustrators, Kenneth Wilson, Dick Morton,
          Robert Trevor, Kristin Kramer

PREFACE

Physical geography has held a special place in colleges and universities for many decades. One of the reasons for this is that physical geography courses have been very attractive to undergraduates in the liberal arts and general education programs. The social sciences and teacher training programs, in particular, have found the scope and subject matter of physical geography well suited to their curricula, and many geography departments have grown with the enrollments contributed by these programs.

But physical geography has also grown as a scientific field in the past several decades. Professional geographers, with their diverse interests in matters of environment, land use, and human activities, have added new knowledge and perspectives to the field. Geophysics, hydrology, planning, and other fields have become friendly allies, and physical geography has shared much with each of them. Government programs dealing with environment and resources have brought renewed attention and new research funding to the problems traditionally studied in physical geography. As a whole, a great deal has been learned. New ideas have emerged, some old ones have been discarded, and other old ones have been dusted off and granted new credence. Our purpose in writing this book was to draw together some important lines of thought of the past three or four decades, weld them together with the traditional framework of physical geography, and offer a physical geographer's portrayal of physical geography as it enters the 1980s.

From the wide range of topics and themes embraced by physical geography, we have chosen to focus this book on driving forces and processes in the landscape. Much of the text is, perforce, explanatory in style, with a distinct bent toward fundamentals and the systematic underpinnings of the field. Accordingly, we have written a good deal on concepts about the way the landscape works and changes, emphasizing many that we think are essential to understanding the spatial patterns and trends that are so central to geography.

A conscious effort was made to avoid setting the intellectual scope of this book only on topics that are physical geography by name or tradition; rather, the scope is open to questions and topics that by their very nature belong to physical geography. Therefore, when it was appropriate to do so, we followed topics well beyond the conventional bounds of v physical geography and into the realms of whatever disciplines could help reflect on them. Whether they know it or not, many other fields pursue physical geographic research and make valuable contributions to the field. These fields deserve recognition, and we would be remiss, from an intellectual standpoint, if their findings were not woven into the larger body of geographical knowledge so that the student may gain a sense of how the various fields of the academic community are interrelated.

Many people have contributed to the building of this book, and we owe our sincere thanks to each of them. At the top of the list are those colleagues who provided reviews and advice on all or large parts of the manuscript: Walter A. Schroeder (University of Missouri), Hugh M. Raup (Harvard University), Melvin G. Marcus (Arizona State University), Jerry E. Mueller (New Mexico State University), A. John Arnfield (Ohio State University), Bruce D. Marsh (Johns Hopkins University), Denise Flynn (University of Michigan, Flint), Mark L. Hassett (University of Michigan, Flint), Charles D. Belt, Jr. (Saint Louis University), Eldridge M. Moores (University of California, Davis), Stanley W. Trimble (University of California, Los Angeles), Jay R. Harman (Michigan State University), Orman Granger (University of California, Berkeley), David M. Helgren (University of California, Davis), Donald F. Eschman (University of Michigan), Robert A. Muller (Louisiana State University), Danny Marks (University of California, Santa Barbara), and Charles B. Hunt.

Many people also made contributions in the way of maps, graphics, data, or labor: Waldo Tobler, Peter Van Dusen, Charles Schlinger, Dale Glover, James G. Marsh, T. R. Oke, William Benjey, James J. Parsons, A. L. Washburn, John S. Shelton, Henry Pollack, John B. Carey, Michael Treshow, Kenneth M. Towe, Marion Clawson, Mitchell J. Rycus, Julie Jacobson, and Susan A., Christopher R., and Katherine W. Marsh.

The publication team at Addison-Wesley deserves special recognition: Richard Morton, art coordinator; Thomas Robbins, sponsoring editor; Evelyn Wilde, production editor; and Catherine L. Dorin, designer. To each we extend our heartfelt thanks for unfailing support and for creative guidance in bringing this book to press.

And to our fathers and grandfathers we owe more than we are able to know for our abiding curiosity in the land.

Flint, Michigan   W.M.M.
Washington, D.C.   J.D.
June 1980

CONTENTS

INTRODUCTION

"Physical geography invites you to consider the terrestrial machinery which makes day and night, seed-time and harvest; which lifts the vapor from the sea, forms clouds, and waters the earth; which clothes it with verdure and cheers it with warmth, or covers it with snow."

M. F. Maury, *Physical Geography,* New York: University Publishing Company, 1897.

PHYSICAL GEOGRAPHY: THE FIELD OF STUDY

Physical geography is a small but challenging field which, in the family of natural sciences, is nestled among larger sciences such as geology, botany, and agronomy. It is held together by a set of perspectives at the center of which is the study of the landscape. Although individual geographers would debate this point, most would probably agree that physical geography is concerned with the origin, composition, and spatial expression of the landscape, especially the natural landscape. In other words, we could say that physical geography is concerned with three fundamental questions: How is the landscape formed? Of what is it composed? And what are its dimensional characteristics?

I. C. Russell, one of the explorer-scientists of the U.S. Geological Survey, leading a field party across the Malaspina Glacier, Alaska, in the late 1800s.

Since the nineteenth century, when physical geography emerged as a field of study in American, Canadian, and European colleges and universities, its chief accomplishment has been to describe and to map landscape features over broad regions. Drawing on the studies and surveys of geologists, botanists, anthropologists, human geographers, explorers, and surveyors, physical geographers produced maps and reports on the distribution of soils, land forms, drainage features, vegetation, climate, rocks, and mineral resources.

For certain regions, this information was combined to form large composite maps. By virtue of the correlations that appeared in the distributions of features such as soils, vegetation, and climate, natural regions were identified. Known as physiographic regions, these have become a major focus of study in physical geography throughout most of the twentieth century. Coupled with land-use information, physiographic regions became an important basis for regional geography, the kind of geography most of us were introduced to in grade school, which told about the land and life of the Lapps, the Chinese, and so on.

1

ABOUT THIS BOOK

This book attempts to tell about the landscape by beginning at a more fundamental level than do most other physical geography textbooks. Our main concern is more with the processes that shape the distribution of landscape features than with the distributions themselves. But in order to understand these processes, we must also appreciate the nature of the forces which drive them. Thus we are led to the question of energy, not the popular question about energy the human resource, but the question of energy the driving force of nature and how it relates to streamflow, glaciers, climate, soils, and vegetation.

How do we go about telling the story of the forces in the land and the processes they drive? In geography the storyteller is usually faced with the dilemma of whether to begin at a grandiose scale, encompassing areas the size of continents, or at a more manageable scale, say, the size of a small community or a farm. Although we have tried to use spatial scales that are traditional to geography, we found it necessary in some chapters to focus on very small parcels of space, small at least from the geographer's perspective. The reasons for this are that: (1) most of us who are not earth scientists find it easier to visualize concepts and to relate information to settings about the size of those in which we carry on our daily routines than to those that are much larger; and (2) discussions of many aspects of landscape dynamics are easier to relate, especially from a teacher's standpoint, in a small spatial setting rather than in a broad geographic setting. So in many discussions we will start in small spaces, develop an idea, and then try to apply it to much larger pieces of earth space.

The Main Themes

An introductory textbook must introduce the reader to a great many facts about the field of study. This book is no exception as physical geography texts go, for it utilizes conventional terms and presents many facts about the earth. However, this book goes several steps farther, in that facts and terms are woven into a conceptual fabric that rests on several major themes.

Three themes receive particular emphasis. One, of course, is based on the idea that an appreciation of the driving forces of nature is fundamental to an understanding of process and change in the landscape. One way of addressing energy in this context is with the aid of a simple model called the *energy balance*. This model is based on energy flow to and from a system and the energy taken up in the system for work and storage. A *system* is a group of features linked together by a flow of material or energy. The oceans, rivers, and atmosphere, for instance, are linked by a flow of moisture and thus together constitute a system. Unit I, devoted to a general discussion of the energy balance of the earth's surface, traces the flow of energy through the atmosphere to the land and oceans, then back into the atmosphere, and finally into space beyond. The energy-balance concept is also applied to water (Unit III), soil moisture (Unit IV), vegetation (Unit V), and glaciers and shorelines (Unit VIII).

Two other themes are also set forth. One, called the *stress-threshold concept,* has to do with the relationship between the stress produced by a force such as wind or running water and the resistance earth materials have to it. For each material such as a soil type or a plant species, there is a maximum level of resistance, called a threshold, that can be tolerated without change taking place. If the level of stress surpasses the threshold of resistance, the material gives way. In a soil under the stress of running water, this produces erosion; in vegetation under the stress of wind, this produces damage or death to plants. What makes the landscape interesting in this respect is that not only do the levels of stress vary widely over space and time, but also the thresholds of resistance vary widely among different materials in the landscape. Therefore, the number of combinations of stress and resistance thresholds that produce change in the land is enormous. The stress-threshold concept is particularly helpful in understanding the interplay between plants and environment (Unit V), but it is also useful in understanding rock deformation (Unit VI), mass movement (Unit VII), and soil erosion (Units VII and VIII).

The third main theme of this book concerns the magnitude and frequency of change in the landscape. This has to do with the behavior of landscape processes and which episodes of a process, little or big, frequent or infrequent, do the greatest amount of work in the long run. The *magnitude-and-frequency concept* helps us understand that change in the land is often sporadic and that different combinations of processes can have widely differing influences in the landscape. The magnitude-and-frequency concept is pertinent to many topics, including streamflow, mass movement, and vegetation change. Unit VI is devoted exclusively to this theme.

ENERGY AND PROCESSES ON THE EARTH'S SURFACE

Compared to our neighbor the moon, the earth's surface is very lively. Geologically, the moon's interior is much less active than the earth's interior, because it appears that the moon has lost most of its internal heat supply. As a result, earthquakes, volcanoes, and related geologic processes, which are so active at the earth's surface, have apparently not occurred with much intensity on the moon for hundreds of millions of years. But there are two other conditions that also account for the earth's active surface: (1) the occurrence of large concentrations of heat and mechanical energy in the presence of (2) highly mobile substances, namely, air, water, and living organisms. When these substances are set into motion by this energy, the resultant movements, which we call *processes*, effect change in the earth's surface.

Origin and Destination of Earth Energy

The energy that drives the earth's processes is derived from two sources. The primary source is the sun, which supplies energy only in the form of radiation. Most of this energy is represented by sunlight. The secondary source is the earth's interior, which produces heat and movement in the planet's outer shell. The geosciences have termed these energy sources *exogenous* (external origin) and *endogenous* (internal origin). Figure 1 attempts to show the gross patterns of flow of external and internal energy in and around a cross section of the earth.

As the diagram suggests, the earth's surface is a critical boundary layer for energy; it is here that the energy concentrates, flows, mixes, and changes form. Certain forms of energy penetrate the surface layer and flow through it, whereas other forms are unable to penetrate it at all. Whether the surface layer is permeable or impermeable depends on the form of energy as well as the nature of substances it encounters.

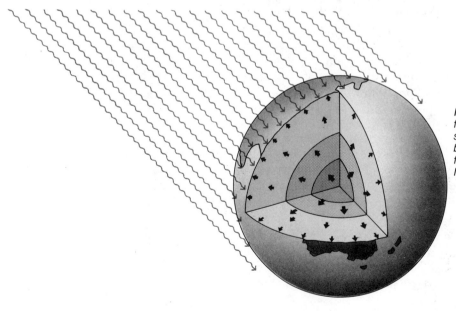

Fig. 1 The gross pattern of energy flow from the endogenous and exogenous sources. The earth's surface is the critical boundary layer toward which energy flows from both sources. (Illustration by William M. Marsh)

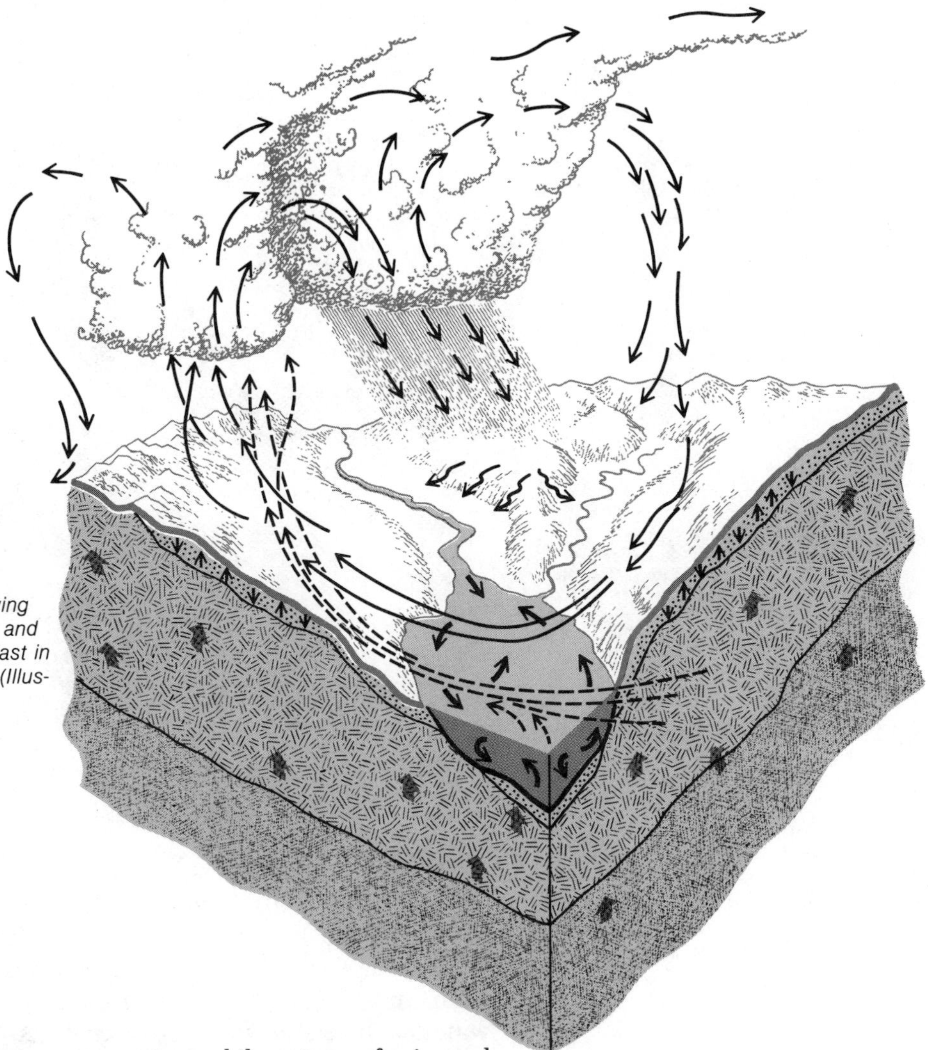

Fig. 2 A schematic illustration showing the nature of heat flow in air, water, and rock. Flow is very slow in rock and fast in air. Water tends to be intermediate. (Illustration by William M. Marsh)

For example, solar radiation can penetrate kilometers of air and meters of seawater, but it is unable to penetrate more than a few millimeters of quartz sand particles on a beach or sand dune.

For energy in the form of heat, the surface layer is a vital regulator of its rate of flow. In moving air, heat flow is rapid, whereas in rock and soil it is very slow, mainly because these substances lack mixing motion. If we could see heat and were to watch it diffuse throughout the surface layer, we might be fascinated by its fast-flowing, almost bursting, motion in air and its slow motion through water (Fig. 2). In contrast, heat flow in rock, soil, and still water (as in the deep oceans) would appear very constrained, with movement limited to small volumes of material over long periods of time.

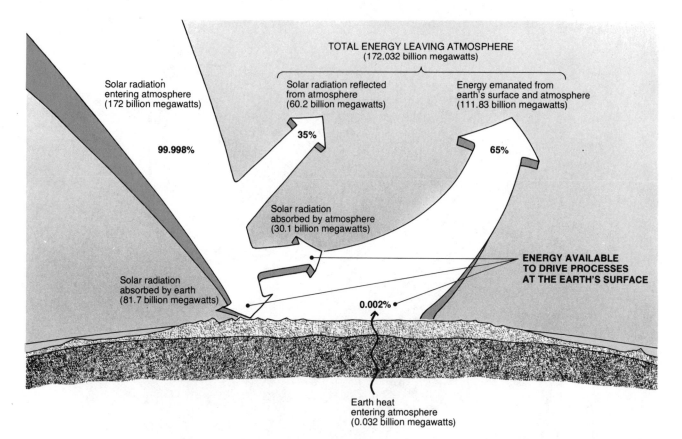

TOTAL ENERGY LEAVING ATMOSPHERE
(172.032 billion megawatts)

Solar radiation
entering atmosphere
(172 billion megawatts)

Solar radiation reflected
from atmosphere
(60.2 billion megawatts)

Energy emanated from
earth's surface and atmosphere
(111.83 billion megawatts)

99.998%

35%

65%

Solar radiation
absorbed by atmosphere
(30.1 billion megawatts)

ENERGY AVAILABLE
TO DRIVE PROCESSES
AT THE EARTH'S SURFACE

Solar radiation
absorbed by earth
(81.7 billion megawatts)

0.002%

Earth heat
entering atmosphere
(0.032 billion megawatts)

Fig. 3 Energy-flow data for the earth as a whole in megawatts of power per year. (Illustration by William M. Marsh)

In all materials, however, heat has the capacity to initiate movement. This movement can actually perform work, i.e., effect physical change on the earth's surface. For instance, heated air generates wind, and as it slides over open water, momentum is transferred to the water surface and waves are formed. As the waves move into shallow water, they rub on the bottom, thereby exerting force on sand and pebbles along the shore. If this force exceeds the strength of the gravitational and molecular forces that hold the sand and pebbles in place, they are moved, work is effected, and the earth's surface is changed.

The total flow of energy toward the earth's surface amounts to an equivalent 172,032,000,000 megawatts of power per year. Except for the small amount that comes from the earth's interior, only 0.002 percent of the total (32,000,000 megawatts), all of this energy emanates from the sun (Fig. 3). But not all of the solar energy that reaches the atmosphere actually gets to the earth's surface; about half of it is either reflected from the atmosphere back into space or absorbed by the atmosphere itself. The total energy available on a yearly basis to drive the processes of the landscape, atmosphere, and oceans is the sum of: (1) the earth heat, at 32 million megawatts; (2) the radiation absorbed by the atmosphere itself, at 30.1 billion megawatts; and (3) the radiation absorbed by the earth's surface, at 81.7 billion megawatts. Thus the skin of the earth is powered by a grand total of 111.8 billion megawatts of energy each year. This is the energy that permits photosynthesis and plant growth, induces winds which generate waves and erode soil, and evaporates water which becomes rainfall on the land and sea.

AN OVERVIEW OF THE EARTH'S LANDSCAPES

What is Landscape?

Before we begin our examination of the energy and processes that change the earth's surface, it would be useful to describe the general character of the major landscapes of the world. The eminent geographer Carl O. Sauer (1889–1975) defined "landscape" as the natural features of an area and the forms superimposed on it by human activities. But what the term "landscape" brings to mind for most of us seems to be an outgrowth of our personal experiences and cultural backgrounds. This point is illustrated by an interesting example based on America's most prominent park planner and the founder of the field of landscape architecture, Frederick Law Olmsted (1822–1903). In the face of rapidly growing and often shabby nineteenth-century industrial cities, Olmsted argued relentlessly for park designs that would bring nature into the cities. But despite his words, the park landscapes he designed, such as Central Park in New York, were not really replicas of the natural landscapes in the regions of the parks. Rather, they tended to be more representative of field and hedgerow landscapes of southern England and the nineteenth-century farm landscape of New England. Olmsted's familiarity with England, New England, and the writings of Romantics such as Ralph Waldo Emerson has been credited with shaping his perception of the American landscape. And so it is with most of us; landscape is what we have learned it to be. To senior Americans it may be a rural landscape comprised of rolling hills, streams, and ponds intermingled with forest and farm fields; to younger Americans it is often the manufactured landscape of suburbia intermingled with patches of tailored nature; to the desert nomad it is vast open space with an irregular cover of small plants; to an Eskimo, it is expansive ice and snow fields in winter and treeless coastal land in summer.

These differences in perception of landscape appear to have had marked influence on human settlement and use of the earth. Historians

Carl O. Sauer (1889–1975), eminent twentieth-century geographer.

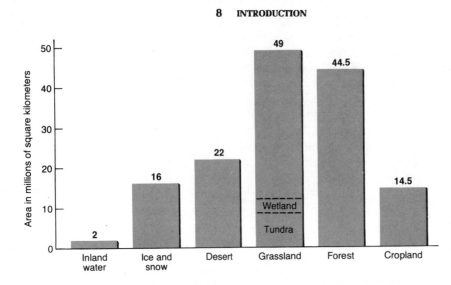

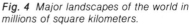

Fig. 4 *Major landscapes of the world in millions of square kilometers.*

tell us they have altered migration trends and even affected the military expansion of empires. But none is very accurate in the impression it gives us of the earth's surface as a whole. Therefore, to set the backdrop and stage for study, it is worthwhile to enumerate briefly the area of the earth's surface presently occupied by the different landscapes.

The Major Classes of Landscapes

The earth's surface has a total area of 510,000,000 km². Approximately 362,000,000 km² are water, principally oceans and seas, and 148,000,000 km² are land. Of the land area, about 2,000,000 km² are covered by inland water in the form of lakes and streams. Approximately 16,000,000 km² are permanently covered by ice and snow; Antarctica and Greenland account for the bulk of this landscape, whereas the remainder is mainly in the insular arctic region of Canada and several major mountainous areas (Fig. 4).

The remainder of the earth's land area can be divided into two broad categories on the basis of the presence or absence of a vegetative cover. Lightly vegetated and nonvegetated lands, principally the deserts, take up 22,000,000 km². The largest of these are the Sahara of Africa and the deserts of central Asia and the Middle East. In Australia desert is the predominant landscape, for more than half of the continent is arid. In the New World the largest deserts are found in the southwestern United States and the adjacent area of northern Mexico and along the Pacific coast of Chile and Peru in South America (Fig. 4).

This leaves 108,000,000 km², about 20 percent of the surface area of the planet, which is covered by vegetation of some kind. Grasslands constitute the largest share of this land, totaling about 49,000,000 km². This figure includes not only the midlatitude grasslands, such as the prairie of North America and the steppe of Russia, but also the tropical savanna, where groves of trees are scattered among expansive areas of grass. It also includes some 8,500,000 km² of tundra, the treeless land north of the Arctic Circle, and around 3,500,000 km² of wetland, located mainly in coastal areas.

Forests occupy 44,500,000 km², of which 20,000,000 km² are tropical forests and 19,000,000 km² are conifer forests. The tropical forests include the vast rainforest of the Amazon Basin, which supports the most diverse assortment of plants and animals on earth. The principal conifer forests are the boreal forests of North America and Eurasia, the western forests of North America, and the subtropical pine forests in areas such as the American South, China, and New Zealand (Fig. 4).

Only 14,500,000 km² remain, and this has been developed into agricultural cropland. Originally, virtually all of this landscape was forest and grassland. In this century cropland has been the most rapidly growing landscape on earth. But a competitor is closing in fast, namely, the urban landscape, which today occupies a total area on the order of several million square kilometers.

THE ENERGY BALANCE AND ITS COMPONENTS AT THE EARTH'S SURFACE

UNIT I

KEY CONCEPTS   OVERVIEW

energy balance
conservation of energy
greenhouse effect
heat transfer
sun angle
albedo
net radiation
ground heat
sensible heat
latent heat
energy gradient
permafrost
convection
roughness length
urban dust dome
flushing capacity
Bowen ratio
heat island
urban climate

The main topic of this unit is the receipt and loss of energy by the earth's surface. The "energy balance" is the conceptual model we use as the framework for tracing the flow of energy through relatively small parcels of the landscape such as your backyard or hometown.

As solar radiation passes into the atmosphere, part of it is absorbed by gases, part is reflected and scattered back into space, and part (45–50 percent) reaches the earth's surface. Most of this energy is absorbed by water, soil, and plants. From these materials it diffuses as heat deeper into the water and soil as well as upward into the atmosphere. The atmosphere and oceans then redistribute much of this heat around the planet. This global-scale redistribution helps to offset the small amounts of solar radiation received in the polar regions and at the same time provides a heat release for the tropical regions. Ultimately, though, virtually all of the energy received by the earth's surface works its way to the top of the atmosphere, where it is released back into space as longwave radiation. Over a year, century, or longer time period, the amount of energy received by the surface of the earth and the lower atmosphere must be equal to the amount released. If this were not the case, the earth's energy balance would change, and the atmosphere would grow warmer or cooler.

Some scientists reason that such energy-balance changes may now be taking place. Although it is difficult to generalize for the planet as a whole, it is clear that the energy balance has been changed measurably over sizable areas as a result of urbanization. Air pollution over cities reduces the annual inflow of solar radiation; however, this energy loss is more than compensated for by the heat emitted from buildings, cars, and hard surfaces such as asphalt and concrete. Coupled with other factors such as slower winds, this results in urban areas' being generally warmer than neighboring rural areas.

In Chapter 1 we examine some basic concepts of the energy balance, including energy flow, the conservation-of-energy principle, and units of energy measurement. Solar radiation and radiation emitted from the earth are taken up in Chapter 2, and heat transfer into the soil, water, and air is described in Chapter 3. Chapter 4 is a discussion of the influence of modern urbanization on the energy balance, including its effects on urban climate.

CHAPTER 1 THE ENERGY BALANCE

INTRODUCTION

Life on earth occupies a very thin zone, effectively only meters deep, on the planet's surface. The atmosphere is a mixture of gases which forms an envelope around the planet, and life occupies the bottom layer of this envelope. As such, the atmosphere is the transitional medium between the zone of the sun and space on one hand and the zone of life on the other. The atmosphere is a vital regulator of both the quantity and quality of solar radiation as well as an important medium in the redistribution of heat across the earth's surface. Fundamental to the field of physical geography is an understanding of the flow of light and heat to and from the earth's surface and lower atmosphere, because many of the important processes which shape the landscape, such as wind, rain, plant growth, and glaciers, are driven by this energy. In order to understand why these processes vary as they do over the earth, we need first to examine the nature of this energy which sets them into motion.

Let us open our discussion of the energy balance on a practical note. "Energy" has become an extremely important word in our daily vocabulary. Indeed, it has become a crisis issue to Americans, Canadians, and Europeans. Although many of us have difficulty understanding some of the technical aspects of energy, we do understand that energy is necessary to make machines work, to raise and lower the temperatures inside buildings, and, in short, to make our whole system go. And whether or not we realize it, we are rapidly developing an appreciation of a concept called the *energy balance*. Are we consuming more energy than we are taking in? Actually, our machines and heating systems do not "consume" energy as such; rather, they change its form to one which is readily given up to the atmosphere and lost to further mechanical processing.

Energy in the form of natural gas is burned to heat millions of homes and buildings. In recent winters in North America the media have carried reports backed by government and industry that we have used gas at a rate faster than it can be produced from the wells. At such times the natural gas energy system is out of balance: More energy is released to the atmosphere as heat than is gained from the earth as gas. How is this possible? It would not be, of course, were it not for large gas reserves that are normally held in storage to make up for use during such shortages. In other words, the energy system that we rely on for heating has a built-in mechanism to allow for short-term imbalances.

What if one very cold winter the reserves were depleted by midwinter, thereby forcing us to rely directly on the gas production from the wellhead? The rate of production would certainly fall behind consumer demand, the gas flow to homes and buildings would decline, and therefore the temperatures inside homes and buildings would fall. Since less heat would be produced, the amount of heat stored in houses would decline, and in turn the amount of heat given up to the atmosphere would decline correspondingly. Soon, the energy outflow (as heat) would fall to a rate equal to that of the energy inflow (as gas), and the system would assume an *energy balance*. In this state, temperatures would hold steady at new, low levels. Figure 1.1 is an idealized portrayal of the natural gas–home heating energy system.

For the earth's surface, the sun is virtually the sole source of energy. This energy heats the surface, and in turn the earth, like all objects, emits this heat as a function of its temperature. For example,

Fig. 1.1 Schematic of natural gas-home heating energy system. The thermographic image (lower right) shows the points of heat loss (in white) in a representative midlatitude house in North America. (Image by Mark L. Hassett)

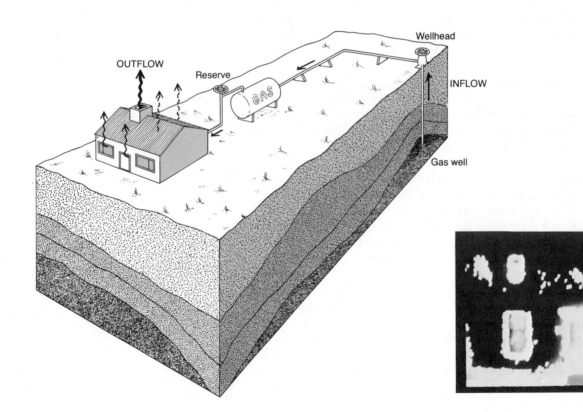

ENERGY INCOMING

ENERGY OUTGOING

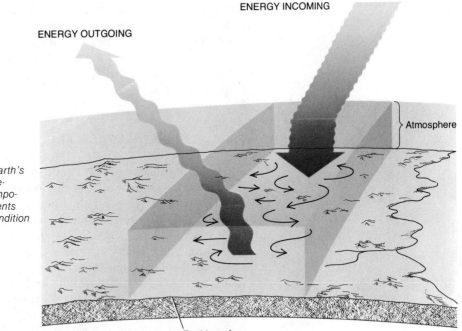

Atmosphere

Earth's surface

Fig. 1.2 The energy balance of the earth's atmosphere and surface can be represented by incoming and outgoing components. Balance of these two components over a period of time constitutes a condition of energy equilibrium.

imagine an unchanging area of earth surface that over a long period of time receives more energy than it emits. The surface temperature rises. But as the temperature rises, more energy must be emitted. On the other hand, if the area receives less energy, its temperature decreases. If the temperature decreases, the energy emitted must decrease. Therefore, if energy is received at a constant rate over a long period of time, a constant temperature develops, and energy must be emitted at a constant rate. The energy received is in equilibrium with the energy emitted, or in *energy balance,* and the temperature is called the *equilibrium surface temperature.* Figure 1.2 shows a generalized model of the energy balance for a section of the earth's surface. Except for the energy sources and directions of the flow, this system is essentially the same as the one in Fig. 1.1.

The energy balance at any point on the earth's surface—say, a square meter of lawn—consists of several components, some incoming and some outgoing relative to the surface. That is, some components bring energy to the surface (+), and others take energy away from the surface (−). Since the *surface* has no thickness and hence no volume, it cannot store energy; therefore, over any period of time, the input and output of energy must balance. The atmosphere, soil, and oceans, on the other hand, do have the capacity to store energy, and daily and seasonal changes in weather and sun position may cause short-term energy imbalances for periods of a few minutes to a few months. For a given moment over the vast surface and atmosphere of the entire earth, however, the energy balance should hold steady, with input equal to output. If this were not the case, the earth would undergo a definite heating or cooling.

PRINCIPLES OF ENERGY BALANCE

Although the energy-balance concept is a variation on a body of theory developed by physicists, it is the ledger of the accountant which provides us with the basic arithmetic needed to put the concept to use. If we substitute the word "budget" for "balance" in energy balance, we have revealed an important clue to understanding this scientific idea. Accordingly, if we imagine money as energy and a financial account as a parcel of earth space, we can conceptualize a system comprised of a basic energy reserve into which and from which energy is continuously flowing. Should inflows exceed outflows, the reserve grows, and the entire system becomes positively balanced; conversely, should outflows exceed inflows, the reserve dwindles, and the system becomes negatively balanced. Further, should the inflow stop altogether while outflow continues, the reserve will ultimately be used up, and the system will fall into complete disorder and effectively cease to exist. Precisely intermediate between these extremes, of course, is a condition of equilibrium, in which the flows are balanced and the system maintains what physicists term a *steady state*. The reserve, though continuously undergoing energy loss and replacement, is unchanging in the total amount of energy from moment to moment.

The operation of any energy system, even one as large as that of the earth, is subject to certain physical principles. *First,* the *conservation-of-energy principle* tells us that there can be no absolute loss of energy in any system. Energy may change form, as from light (a radiant form) to heat, or it may be stored, but eventually the ledger must balance. In other words, all energy that flows into the system must be accounted for in (1) outflow, (2) storage, or (3) work performed. *Second,* because the energy system is *dynamic* (rather than static), it is capable of undergoing spontaneous adjustments to variations in the energy flow. All adjustments represent a trend toward a new state of equilibrium. When the inflow of energy increases and the internal energy reserve grows, the outflow of energy tends to become more rapid. When the outflow adjusts to a level equal to that of the inflow, the system has attained an equilibrium. Authors' Note 1.1 offers a practical example of the possible variations in the balance of the energy system of a greenhouse.

AUTHORS' NOTE 1.1
Energy Balance in the Greenhouse

Let us use a greenhouse to illustrate the variations in the energy balance. On a sunny day, solar radiation is transmitted through the glass roof of a greenhouse and is absorbed by the plants and other objects on the inside. As these materials take on energy, their temperatures rise, and they begin to emit radiation themselves, which in turn heats the air inside the greenhouse. As with a closed car left in the sunshine, this radiation cannot pass out of the greenhouse as easily as the solar radiation passed in. The balance of the energy flow is positive because the greenhouse environment is gaining more energy than it is releasing. One measure of this condition is a steadily rising temperature.

Since extremely high temperatures injure many plants, most greenhouses are designed to maintain an energy equilibrium throughout much of the day. Therefore, as the temperature rises near midday, vents in the roof are opened to facilitate the outflow of hot air. The vents are continually adjusted throughout the day as the inflow of solar radiation varies with the passage of clouds and with the onset of dusk.

The inflow of solar radiation ceases altogether at night, of course, but the greenhouse still continues to release some of the heat gained during the day. The balance of the energy flow is strongly negative because more energy is being lost than gained. Vents are normally closed at night in order to conserve this energy.

Several additional means could be employed to alter the energy balance of the greenhouse. We could paint some of the panels in the glass roof and thus divert the inflow of some energy. Or, conversely, we could change the ventilation system to increase or decrease the outflow of heated air. Or, some of the heat could be drawn off to be used for other purposes, such as heating water or growing plants. In any event, the sum total represented by energy losses, conversion to storage, and work accomplished in the greenhouse cannot exceed the total amount of energy received in the first place.

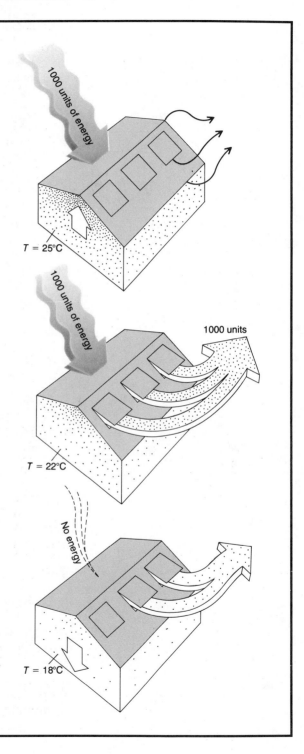

ENERGY, ENERGY UNITS, AND THEIR EXPRESSIONS

Energy is the ability to do work, such as the evaporation of water and the erosion of soil by wind. Work involves the transfer of energy from one body to another or from one energy form to another and is defined as the application of force over distance. Energy is found in various forms in the environment, e.g., radiation in the atmosphere, heat in the soil, or motion in the oceans.

In problems concerning the landscape, we need two types of information about energy: (1) how much energy is held in some material, such as air, water, or soil; and (2) what the rate of energy flow is from, through, or into some material. The rate of energy flow to or through a given area of space, such as a square meter, is referred to as energy flux, radiant-flux density, or irradiance. In this book we shall use the term *energy flux*.

In the case of heat, we can measure both the heat content of a substance and the rate of heat flow in the substance. In the case of radiation, however, we can measure only its rate of flow. Solar radiation passes through the atmosphere so quickly that it makes no sense to measure its content in the atmosphere.

Energy Flow

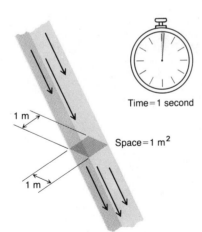

Time = 1 second

Space = 1 m^2

The flow of energy must be expressed in terms of a standard space and time framework. According to the Système Internationale (S.I.), which specifies the preferred units of measurement based on an international consensus of scientists, the most appropriate units are a square meter (abbreviated m^2 or m^{-2}) of space, a second (s) of time, and a joule (J) of energy. Thus, in the case of sunlight coming through the atmosphere, we ask: "How many joules of energy are passing through an area of one square meter in a second of time?" Or, "How many joules are being received by a surface of one square meter per second?" Or, conversely, "How many joules of energy are being released by a surface with an area of one square meter in a second?" Such energy flows are expressed as follows:

$$1500 \text{ J/m}^2 \cdot \text{s}, \quad \text{or} \quad 1500 \text{ J m}^{-2} \text{s}^{-1}.$$

A joule is equal to one unit of force, called a *newton*, applied over a length of one meter. In terms of heat, 4186 joules are needed to raise the temperature of one kilogram of water one degree C when the water is at a temperature of 14.5°C. Since it takes one calorie of heat to raise the temperature of water one degree C (from 14.5° to 15.5°C), a joule is equal to about one-fourth calorie.

Other energy units, as well as the other space and time units, can be used to describe energy flux: calories per square centimeter per minute, watts per square meter per second, and British Thermal Units (BTU) per square foot per hour. A calorie is equal to 4.186 joules; a watt (W), the S.I. unit for power, is equal to 1 joule per second. So 1 $J/m^2 \cdot s$ is equal to 1 W m^2. A BTU is equal to 1054 joules. See Appendix 1 for more complete definitions of energy, work, and force.

Temperature

When energy is added to a substance, its molecules begin to vibrate, heat is generated, and its temperature rises. The amount of temperature rise with the addition of energy varies with the heat capacity of a substance; therefore, the temperature of different materials in the landscape is not directly equal to energy content. However, for materials of the same composition, such as soils of the same particle types and moisture contents or air parcels of comparable makeup, temperature is an accurate indicator of energy content.

Several different temperature scales are in use today. The Fahrenheit scale, which is the most arbitrary one, sets 32 degrees as the freezing temperature of water and 212 degrees as its boiling temperature. The centigrade, or Celsius, scale sets 0 degrees and 100 degrees for these same two temperatures (see Table 1.1). The Kelvin scale is based on *absolute zero*, the state at which there is no molecular vibration in a substance and hence no heat. The unit of the Kelvin scale is the same as that of the Celsius.

Until recently, official atmospheric temperatures in the United States were recorded in Fahrenheit units. Today, however, scientists prefer Celsius units. Thus it often is necessary to convert from one scale to the other. The following formulas can be used for conversion:

$$°C = (F° - 32) / 1.8$$
$$°F = (1.8 \times C°) + 32.$$

The chart next to Table 1.1 provides a graphic comparison of the two temperature scales.

Table 1.1 Key temperatures on the centigrade, Fahrenheit, and Kelvin scales.

	°C	°F	°K
Absolute zero	− 273.15	− 459.67	0
Normal freezing point of H₂0*	0	32	273.15
Normal boiling point of H₂0*	100	212	373.15

*At sea level.

Heat Transfer

Heat transfer, or flow, is at the heart of the earth's energy balance. There are three ways in which heat can be transferred: conduction, convection, and radiation. *Conduction,* the process of heat transfer at the molecular level, takes place when rapidly moving molecules collide with one another. The resultant heat flow is always from a warmer part of a body to a colder part, as is illustrated by the often cited experiment in which a cold iron rod becomes heated by contact with a hot stove. Heat transfer by *convection* takes place when a fluid (either liquid or gas) flows into contact with a solid body or another fluid body of a different temperature. The fluid either gains or loses heat at the contact surface and is then displaced and mixed with cooler or warmer parts of the fluid, thereby effecting heat transfer. Heat transfer by *radiation* involves propagation of energy in electromagnetic waves. This is the only form of heat transfer that does not require a medium such as air or water to carry the energy from place to place.

SUMMARY

The energy balance involves the flow of energy to and from the earth's surface. Incoming radiation from the sun heats the earth's surface, which in turn emits energy into the atmosphere, from which it is ultimately emitted into space. The more energy received, the higher the surface temperature and the higher the rate of emission. Since energy can be neither lost nor gained once it enters the earth system, it must be accounted for in outflows, storage, or work performed. Chapter 2 takes up the flow of radiation to and from the landscape and atmosphere.

CHAPTER 2 THE RADIATION BALANCE

SOLAR AND EARTH RADIATION

The sun produces a massive amount of energy, which it broadcasts into space in the form of radiation. A minute fraction of this radiation (actually 1/2,000,000,000th of the total) is intercepted by earth, where it lights and heats the earth's atmosphere and surface. At the edge of the atmosphere, incoming solar radiation has a strength of 1370 $J/m^2 \cdot s$, or 1.94 cal/cm$^2 \cdot$ min. The flow of this radiation is believed to be very steady, so much so, in fact, that it is referred to as the *solar constant*.

Much, but not all, solar radiation is absorbed by the atmosphere, the land, and the oceans. To maintain an energy balance, the earth must ultimately reradiate this energy back into space. But before it does so, this energy drives such important earth processes as the evaporation of water, the generation of winds and storms, and photosynthesis in plants.

The sun emits a wide variety of radiation types, which physicists have classified according to a scheme called the *electromagnetic spectrum*. Radiant energy travels in waves and is defined on the spectrum in terms of wavelength, the distance from the crest of one wave to the crest of the next. A common unit of measurement is the micrometer (or micron), equal to one millionth of a meter. To facilitate our discussion, we can subdivide the electromagnetic spectrum as shown in Fig. 2.1. The key subdivisions are as follows: (1) very short wavelengths, less than 0.15 micrometer, which includes gamma rays, X-rays, and some ultraviolet radiation; (2) relatively short wavelengths, between 0.15 and 3.0–4.0 micrometers, which is ultraviolet, visible light, and near infrared radiation; (3) relatively long wavelengths, between 3.0–4.0 and 100 micrometers, which is infrared radiation; and (4) very long wavelengths, greater than 100 micrometers, which includes radio waves, television waves, and microwaves. Categories 2 and 3 include most earth and solar radiation and for ease of reference are termed shortwave and longwave, respectively.

The rate at which a body emits radiation is a function of its temperature; the higher the temperature, the higher the rate of emission. More precisely, the rate of emission increases at an increasing rate with temperature; as temperature gets higher, the rate itself gets

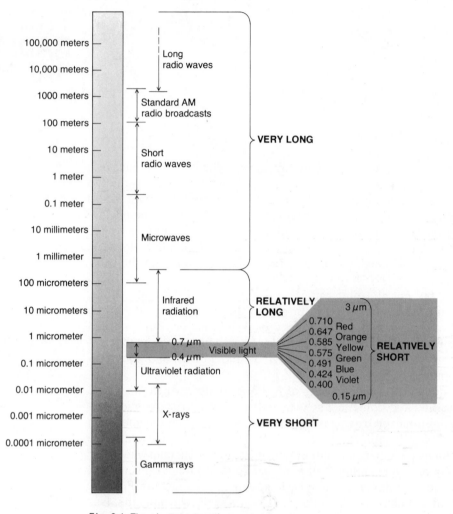

Fig. 2.1 The electromagnetic spectrum, including the terms for the various types of radiation.

higher even faster. Total emission is calculated by the Stefan-Boltzmann equation, which states that the energy (E) radiated from a body increases with the fourth power of its temperature (T) times a constant (σ):

$$E = T^4 \sigma$$

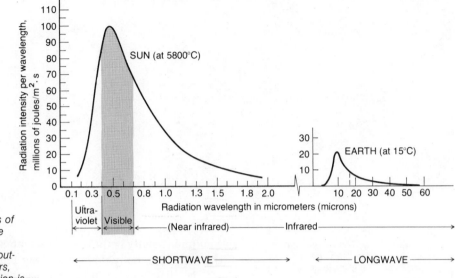

Fig. 2.2 The distribution of intensities of radiation produced by the sun and the earth. The vertical axis is scaled in energy, representing the intensity of output; the horizontal axis, in micrometers, representing wavelength. Solar radiation is concentrated around 0.5 micrometers, between the ultraviolet and infrared wavelengths, whereas earth radiation is entirely infrared.

Since the sun's temperature (5800°C) is so much higher than the earth's (15°C), the sun produces a vastly greater amount of energy than does the earth (Fig. 2.2).

Solar radiation reaches its peak intensity in the visible, or "light," part on the electromagnetic spectrum, whereas radiation emitted by earth reaches its peak intensity in the infrared portion of the spectrum. Thus the sun emits not only more energy than does earth, but also a much greater proportion of shorter wavelengths, These, as well as several other important facts, are illustrated in Fig. 2.2. The left-hand portion represents the energy emitted by the sun and received by the earth's atmosphere; the right-hand portion, the energy emitted by the earth. The region from 0.4 to 0.7 micrometer on the spectrum is the visible, or light, portion.

The bulk of solar radiation is concentrated around 0.48 micrometer; the bulk of earth radiation, at much longer wavelengths, around 10 micrometers. Although it receives energy as relatively shortwave radiation, the earth returns most radiation as relatively long waves in the infrared portion of the spectrum. This important fact is described by Wien's law, which in this context states that the wavelength of

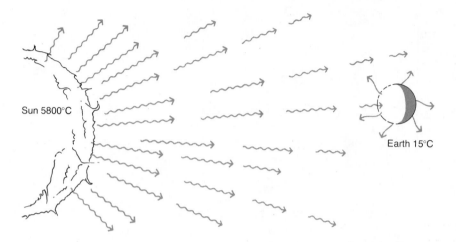

Fig. 2.3 Wien's law: The wavelength of maximum-intensity radiation grows longer as the absolute temperature of the radiation body decreases.

maximum-intensity radiation grows longer as the absolute temperature of the radiating body decreases (see Fig. 2.3).

This difference is an extremely important factor in the earth's energy balance. As indicated in the greenhouse example in Authors' Note 1.1, incoming shortwave radiation passes through the glass, whereas the outgoing longwave radiation produced from the heated surfaces within the greenhouse cannot do so and is thus trapped in the greenhouse atmosphere. This difference in the penetration capacities of short- and longwave radiation is not limited to glass; the atmosphere, which is itself often as transparent or translucent as glass, also induces a similar difference in the behavior of radiation. As a result, the atmosphere produces a greenhouselike effect for the earth, trapping and absorbing longwave radiation. This results in a relatively long residence time in the lower atmosphere for longwave radiation. In fact, energy that took only about nine minutes to travel the 150 million kilometers from sun to earth may take days or weeks to travel upward through only 20 km of atmosphere, though the *rate* of travel is the same for both short- and longwave radiation.

SHORTWAVE RADIATION

Factors Influencing Incoming Solar Radiation

Try to visualize the daily and seasonal variations in the flow of solar radiation to your backyard. Because the sun is the only source of this radiation, it is present only during the daylight hours; radiation from the moon and stars is so minute that it can be ignored. If we measured the total amount of solar radiation getting to your yard, we would find that it is much less than the solar constant. To explain this, we would have to look to three important factors: (1) the daily duration of daylight; (2) the clearness and composition of the atmosphere; and (3) the directness of the angle at which the radiation strikes the landscape.

Length of daylight. The days grow longer in summer and shorter in winter. In the middle of the United States, for example, the longest day of the year exceeds the shortest by about five hours. The summer-winter difference in light and dark is negligible at the equator, but increases poleward until near the poles, it reaches essentially twenty-four hours.

Absorption and reflection by the atmosphere. More important for our backyard observations is the fact that on any day, the atmosphere and its clouds can reduce the amount of incoming shortwave radiation by as much as 75 percent. Reduction begins in the upper atmosphere, where ultraviolet radiation is absorbed by ozone, resulting in a 3-percent decline in the strength of the radiation beam. Farther into the atmosphere, the beam is reduced even more by the scattering of radiation off molecules and dust particles, by reflection from clouds, and by some additional absorption by gases and particles. Infrared radiation, which constitutes a small percentage of solar radiation, is absorbed by molecules of carbon dioxide and water vapor, and virtually none reaches the earth's surface.

   A dirty or cloudy lower atmosphere has an effect similar to that of a thin coat of light-toned paint on a greenhouse; the radiation is either reflected off the paint or scattered by the paint particles. In the atmosphere, scattering occurs when air molecules and particles of dirt intercept certain wavelengths of radiation and randomly disperse them in all directions across the sky. Some of this radiation is diffused toward the earth, and some is backscattered toward space. (The scattering of blue and violet wavelengths, by the way, gives the sky its blue coloration.) Reflection occurs when radiation strikes clouds and the beam is redirected skyward. Together, backscattering and reflection in the atmosphere reduce the beam by an average of 31 percent; on any given day, though, this figure may be as high as 75 percent or as low as 10 percent, depending on sky conditions. Absorption averages 17–18 percent for the atmosphere as a whole.

Sun angle. The angle at which the radiation beam passes through the atmosphere and strikes the earth is very important also. First, if the sun angle is low, the beam slices through the atmosphere obliquely, taking a long route to the surface. This increases the loss by scattering and absorption from what would be expected if the beam traveled

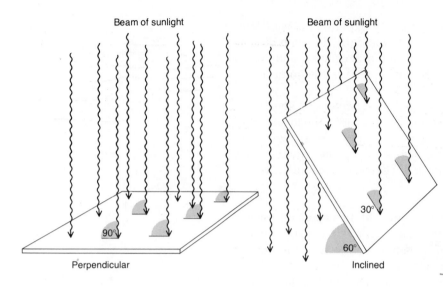

Fig. 2.4 *The influence of sun angle on the intensity of radiation. Where the beam strikes the surface directly, radiation is most intense, because the beam irradiates an area proportional to its cross-sectional area. If the surface is tilted 60°, only half of the beam of radiation strikes the surface, thereby reducing the unit area intensity by 50 percent.*

straight through the atmosphere. Second, sun angle is important because we are interested in the radiation incident on the earth's surface. The more direct the angle, the greater the intensity of bombardment of the surface by radiation.

Consider in Fig. 2.4 a beam of sunlight 1 m^2 shining directly on a surface of 1 m^2, so that the angle of illumination (or sun angle) is 90°. Let us assume that the amount of radiation in the entire beam is 700 W/m^2, or 10,000 cal/m$^2 \cdot$min. The 1-m^2 surface is intercepting the entire 1-m^2 beam. Now, let us tilt the surface 60°, so that the sun angle is only 30°. This reduces the surface area exposed to the beam by one-half. Two facts have not changed: The beam still represents 700 W/m^2, and the area of the surface is still 1 m^2. But one aspect of the situation has changed: The 1-m^2 surface no longer intercepts the entire beam, but only half of it. In order to catch the entire beam, the surface would have to be twice the original area; thus the intensity of radiation on it would be half as great as the 1-m^2 surface. The radiation incident on the surface is equal to the intensity of the radiation beam divided by the surface area which it illuminates:

$$\ast \text{ Incident radiation} = \frac{\text{Intensity of beam}}{\text{Surface area illuminated}} = \frac{700 \text{ W}}{2 \text{ m}^2}$$

$$= 350 \text{ W/m}^2 \text{ at a sun angle of } 30°.$$

Let us apply this principle to the landscape. Instead of tilting a hypothetical surface, however, let us use natural slopes in the land. In the

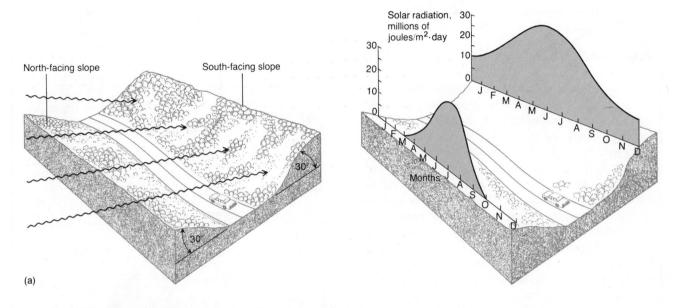

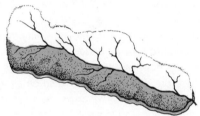

Fig. 2.5 *The influence of slope on sun angle: (a) Total receipt of direct solar radiation may be two times higher on 30° south-facing slopes than on 30° north-facing slopes at latitude 50° N. (Illustrations by William M. Marsh) (b) Differences in* drainage density of the north and south sides of small drainage basins in Niobara County, Wyoming. (From S. A. Schumm, "The Application of Landform Analysis in Studies of Semiarid Erosion," Circular 437, U.S. Geological Survey, 1960.)

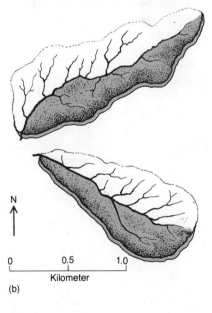

Northern Hemisphere, poleward of the tropics, the sun always shines from the southern part of the sky; therefore, slopes that face south receive more direct radiation than do their north-facing counterparts (Fig. 2.5a). In spring it is not uncommon to find patches of snow on north-facing slopes and barren ground on south-facing slopes. This results from the fact that the sun's radiation heats the south-facing slope while its north-facing counterpart is in shadow much of each day. In late summer a difference in the vigor of herbs sometimes becomes apparent because of soil-moisture differences on these slopes. The south-facing slope has higher rates of soil-moisture evaporation, and as a result, the growth rates of plants such as grasses are markedly slower than are those of the same plants on the north-facing slope. In mountainous areas the plant types may be radically different on north- and south-facing slopes (see Figs. 20.6 and 20.7).

The effect of the sun angle–slope relationship on the landscape can be traced even further. In dry areas the plant cover on south-facing slopes may be so weakened by moisture stress that its capacity to protect the slope against erosion is substantially reduced. Conse-

AUTHORS' NOTE 2.1
Sun Angles

The importance of sun angle can be demonstrated at many geographic scales. At the global scale, the earth's curvature from the equator to the poles strongly influences sun angle and, in turn, surface heating. If the sun's rays strike the surface vertically at the equator, one should be able to observe a progressive poleward decrease in the sun angle, until at the poles themselves, the angle actually closes; that is, there may be no angle at all. Of course, the problem is somewhat more involved than this, and global variations in sun angle are examined in detail in a later chapter.

At the microscale, we can consider the importance of sun angle in building design, e.g., in the size of eaves and the location of windows on the sunny sides of houses. Should one wish to avoid the intensive summer radiation and the heating it produces, one must make the eaves over south-facing walls and windows large enough to provide shade throughout most of the day. In winter, however, the opposite effect is usually desired; therefore, the eaves must also be short enough to permit illumination of windows. To solve this problem, the annual high and low sun angles must be calculated and the eaves and windows designed accordingly. At latitude 45° N (South Dakota or Wisconsin, for example) a house with a south wall 3 m high must have an eave at least 1 m wide to ensure shade from the midday summer sun. This is shown in diagram (b). In winter the sun angle decreases by 47°, and the noon sun is only about 22° above the horizon on New Year's Day. Given this design, one could expect illumination of 2.5 m (83 percent) of this wall by the midday winter sun, as shown in diagram (c). Note that only the uppermost half meter of

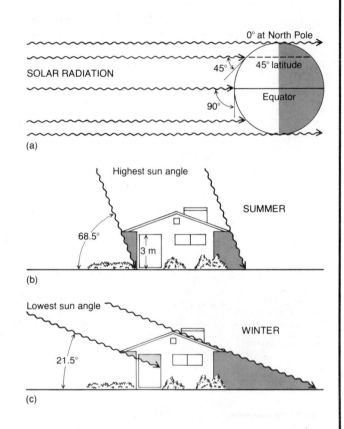

the wall would never receive direct solar radiation at any time. In cities where buildings are tall and close together, many spots rarely receive direct solar radiation. And with the growing reliance on solar collectors as a source of energy, conflicts over "solar space" or rights are developing among landowners in some parts of the United States.

quently, erosion by runoff may be very pronounced on the south-facing slopes, and this is sometimes manifested in the formation of deep gullies. The contrast to the north-facing slope can be striking because the incidence of gullies and related drainage channels is often meas-

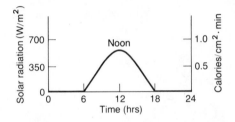

Fig. 2.6 *The ideal curve of incoming shortwave radiation during a one-day period. This smooth curve is often distorted by variations in cloud cover. In addition, it grows higher and wider in summer in response to longer days and higher sun angles. The opposite is true for winter.*

urably less there (Fig. 2.5b). Authors' Note 2.1 offers some additional information on sun angles.

The daily pattern of the amount of incoming shortwave radiation received on a clear day approximates a bell-shaped curve corresponding to the rise and fall in the altitude of the sun through the day (Fig. 2.6). As the day progresses and the sun rises toward its high-noon position, the intensity of incoming radiation increases accordingly; similarly, the intensity of incoming radiation decreases as the sun sinks from noon to dusk.

Surface Reflection

For incoming shortwave radiation reaching the landscape, part is absorbed and part is reflected back into the atmosphere. The capacity of the earth's surface to reflect shortwave radiation is highly variable from place to place, in turn making radiation absorption equally variable.

Surface reflectance is termed *albedo* and is defined as the percentage of the incoming shortwave radiation at ground level that is not absorbed by the surface. We can designate incoming shortwave radiation as S_i and the amount of shortwave radiation reflected as S_o; thus

$$\text{Albedo} = \frac{S_o}{S_i} \times 100.$$

Note that the decimal derived from dividing S_i into S_o is multiplied by 100 to convert it to a percentage.

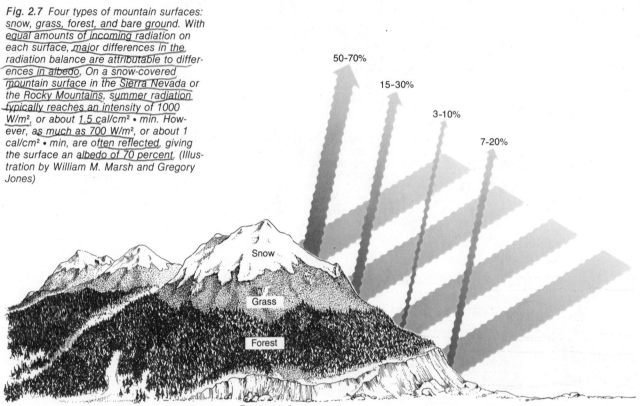

Fig. 2.7 Four types of mountain surfaces: snow, grass, forest, and bare ground. With equal amounts of incoming radiation on each surface, major differences in the radiation balance are attributable to differences in albedo. On a snow-covered mountain surface in the Sierra Nevada or the Rocky Mountains, summer radiation typically reaches an intensity of 1000 W/m², or about 1.5 cal/cm² • min. However, as much as 700 W/m², or about 1 cal/cm² • min, are often reflected, giving the surface an albedo of 70 percent. (Illustration by William M. Marsh and Gregory Jones)

Albedo is influenced by many characteristics of a surface, including texture, water content, and orientation with respect to the sun. Because earth materials are so varied in these respects, some surfaces are good reflectors, and some are better absorbers of shortwave radiation. Generally, dark-toned and rough-textured materials, such as organic soil and asphalt, have lower albedos than do light-toned materials, such as snow and concrete. The actual values range from as high as 90 percent (over fresh snow) to 10 percent or less (over moist, black soil).

Albedo also varies with sun angle. This is especially pronounced on water; at angles above 70° or so, the albedo of water is only 5–10 percent, whereas at angles below 20°, it can be as great as 90 percent. Figure 2.7 shows an example of the wide differences in albedo which can be found in high mountain terrain. With this sort of range, it is easy to appreciate the extreme variations in surface heating which can be produced within small areas on sunny days.

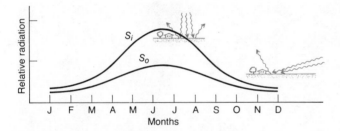

Fig. 2.8 *The relative proportion of incoming (S_i) and reflected (S_o) shortwave radiation over a midlatitude year. The S_i is weak in winter, and as much as 50–70 percent of it may be reflected back into the atmosphere.*

Finally, there are seasonal changes in radiation to consider. At midlatitude sites, such as the midcontinental United States, components of incoming and outgoing shortwave radiation change relative to each other over the seasons. Because of high sun angles in summer, both total incoming radiation and the proportion of it absorbed are usually greater than in any other season of the year. In winter the presence of snow increases the proportion of shortwave radiation reflected, and the incoming shortwave radiation is less also; therefore, the net shortwave radiation is far less in winter (Fig. 2.8).

Net Shortwave Radiation: Summary

As shortwave incoming radiation passes through the atmosphere, part of it is absorbed, part of it is reflected back into space, part of it is scattered across the sky and to the ground, and part of it is directly transmitted to the surface. Clouds, which are made up of water particles, are highly effective in reflecting and scattering the beam. A dirty atmosphere is also effective in scattering solar radiation. The radiation reaching the ground—about 50 percent of the total on the average for the earth—is either reflected or absorbed, depending on the nature of the surface it encounters. The intensity of the energy absorption is finally dependent on the size of the receiving surface, which is a function of sun angle. Thus we come to the critical question: What is the balance after we subtract outgoing shortwave radiation from incoming shortwave radiation? In other words, what is the net shortwave radiation? If it is zero, no incoming shortwave radiation has been absorbed, and the earth has gained no energy. Of course, this is not the case; in fact, net shortwave radiation is strongly positive for the earth as a

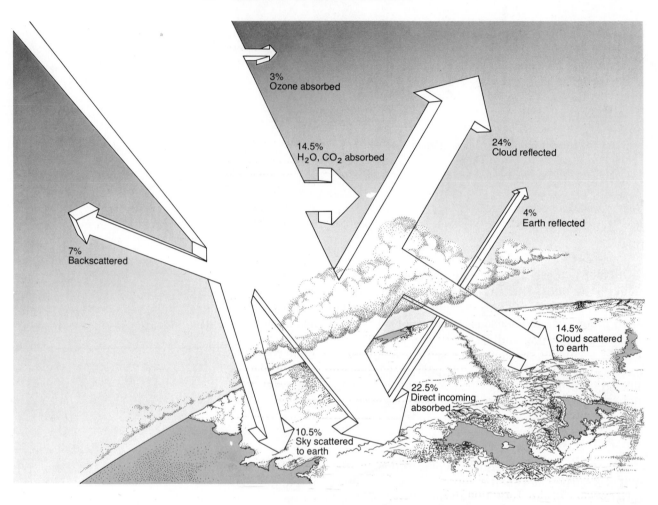

3%
Ozone absorbed

14.5%
H$_2$O, CO$_2$ absorbed

24%
Cloud reflected

4%
Earth reflected

7%
Backscattered

14.5%
Cloud scattered
to earth

22.5%
Direct incoming
absorbed

10.5%
Sky scattered
to earth

Fig. 2.9 The key processes in the breakdown in the atmosphere of incoming solar radiation: scattering, reflection, and absorption. For us, on the earth's surface, the most important values are those of incoming radiation from cloud scattering, sky scattering, and direct beam. The values given here are representative of the earth as a whole; bear in mind that seasonal and geographical variations are quite pronounced. (Illustration by William M. Marsh)

whole, with an average of 47–48 percent of the solar beam absorbed by the earth's surface materials (see Fig. 2.9).

LONGWAVE RADIATION

Barring the minute contributions from geothermal sources, all primary heating of the earth's surface is caused by the absorption of solar radiation. In order to maintain an energy balance, the earth must in turn emit this energy back into space. Using Wien's law, we can reason that this energy must be emitted as longwave radiation, since earth surfaces are much cooler than those of the sun.

According to the Stefan-Boltzmann law, the rate at which longwave radiation is emitted from a surface is a function of its temperature. The capacity of a surface to produce radiant energy is termed *emissivity*. For major materials in the landscape, emissivities range from 0.71 to 0.99 (Table 2.1). The emissivity value itself is a ratio of the total radiant energy produced by a substance to the radiant energy produced by a perfect radiating body, called a *blackbody*, at the same temperature and wavelength. For a grass surface with an emissivity of 0.90 and a temperature of 10°C, for example, the production of outgoing radiation would be equal to 330 W/m², or 0.34 cal/cm² · min.

Outgoing longwave radiation reaches its highest value during the day when the surface temperature is highest (Fig. 2.10). This peak will usually lag two to four hours after the midday peak of incoming shortwave radiation, owing to the time taken to heat the soil. Although both incoming and outgoing shortwave radiation become zero after sunset, outgoing longwave radiation continues all night.

Once in the air, longwave radiation is subject to absorption by water droplets, water vapor, and carbon dioxide, and from these materials it may be reradiated back toward the surface. If the lower atmosphere is highly humid, as it often is when there is a low cloud ceiling, the longwave outflow is slow, and nighttime temperatures will typically remain relatively high. Within this atmospheric greenhouse, longwave radiation is absorbed and then reradiated back to the surface, where it is reabsorbed to produce secondary heating of the ground. Added to this reradiated energy is longwave radiation produced from the absorption of solar radiation in the atmosphere.

Not uncommonly, the energy brought to the surface over a twenty-four-hour period by incoming longwave radiation more than doubles that brought to the earth's surface by incoming shortwave radiation. This is possible because much of the longwave radiation present in the lower atmosphere is recycled several times between atmosphere and earth before it is finally released into space. Thus in calculating the net radiation for all waves, both *outgoing* and *incoming* longwave radiation must be considered.

With each cycle of longwave radiation, the amount of energy decreases as some of it is converted to sensible heat and some of it is lost in skyward radiation. In addition, some longwave radiation is lost directly into space because carbon dioxide and water are simply ineffective in absorbing it. This is particularly so for wavelengths between 8 and 11 micrometers, for which the lower atmosphere presents a

Table 2.1 Emissivities of various surface materials.

SURFACE	EMISSIVITY
Soils	.90–.98
Grass	.90–.95
Crops	.90–.99
Deciduous forest	.97–.98
Coniferous forest	.97–.99
Water	.92–.97
Snow	.82–.99
Ice	.92–.97
Stone	.85–.95
Asphalt	.95
Concrete	.71–.90

(From Sellers (1965) and other sources.)

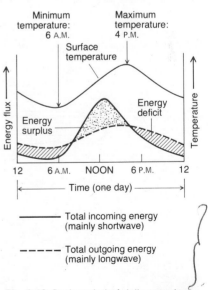

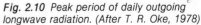

Fig. 2.10 Peak period of daily outgoing longwave radiation. (After T. R. Oke, 1978)

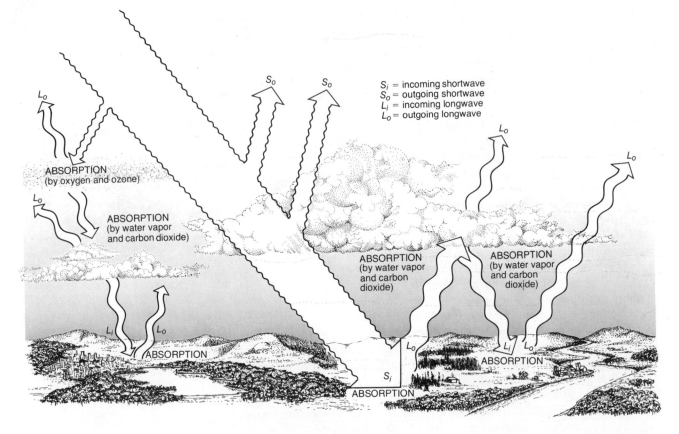

Fig. 2.11 *The general pattern of flow of short- and longwave radiation in the earth's atmosphere. Critical to the flow scheme are the points of radiation absorption. The principal absorber of incoming shortwave radiation (S_i) is the earth's surface; this absorption is the primary source of earth surface heat. From heated surface materials, longwave radiation is emitted, but it is unable to gain ready passage through the atmosphere because water vapor and carbon dioxide absorb it. As their heat content increases, they in turn emit longwave radiation. Some of this is broadcast toward the ground to become incoming longwave (L_i); some, skyward to become outgoing longwave (L_o). The L_i may be absorbed by surface materials which, as their heat content grows, reemit this energy into the atmosphere as L_o.*

After several iterations of this L_i/L_o cycle, the total energy received by the earth's surface as longwave radiation often exceeds that received as shortwave radiation. Reradiated L_i provides the secondary source of earth surface heat. The L_i/L_o cycle represents the heart of the earth's greenhouse effect and is highly variable, depending on local weather conditions. (Illustration by William M. Marsh)

In the figure:
- S_i = incoming shortwave
- S_o = outgoing shortwave
- L_i = incoming longwave
- L_o = outgoing longwave

ABSORPTION (by oxygen and ozone)

ABSORPTION (by water vapor and carbon dioxide)

ABSORPTION (by water vapor and carbon dioxide)

ABSORPTION (by water vapor and carbon dioxide)

ABSORPTION

ABSORPTION

"window" to outgoing longwave radiation. Higher up, however, some of this radiation (between 9 and 10 micrometers) is absorbed by ozone. The pattern of long- and shortwave radiation flow is summarized schematically in Fig. 2.11.

Fig. 2.12 The radiation flow around this cholla cactus can be interpreted from the patterns of snowmelt and shadows. Long-wave radiation from the tree accounts for the snowmelt around the base of the tree. (Photograph by Jeff Dozier)

NET BALANCE OF SHORT- AND LONGWAVE RADIATION

The total incoming and the total outgoing radiation can be compared so as to determine whether the surface is gaining or losing radiative energy. This net radiation is defined by the following equation:

Net radiation = Incoming shortwave – Outgoing shortwave
+ Incoming longwave – Outgoing longwave.

By combining the incoming and outgoing components into two terms, we can simplify this equation to:

Net radiation = Incoming radiation – Outgoing radiation,

which brings us back to the generalized energy-balance scheme illustrated in Fig. 1.2. To bring this concept closer to the visual reality of most of us, we might examine the photograph and diagram in Fig. 2.12.

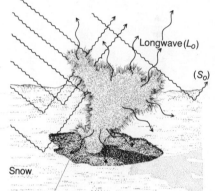

Incoming shortwave (S_i)

Longwave (L_o)

(S_o)

Snow

Absorbed longwave (L_i)

SUMMARY

In this chapter we established that: (1) solar radiation is reduced substantially in quantity as it passes through the atmosphere and strikes the earth's surface; (2) the solar radiation absorbed by the earth varies with location, season, sky conditions, topography, and landscape materials; and (3) radiation absorbed by the earth's surface materials is re-radiated as longwave radiation and that this radiation may heat the atmosphere, which in turn reradiates longwave radiation itself.

In the next chapter we will examine the heat component of the energy balance. This involves the transfer of heat from the surface into the ground, air, and water. With an accounting of both radiation and heat fluxes to and from the earth's surface, we shall be able to compute the complete energy balance.

HEAT AND RADIATION: THE ENERGY BALANCE

CHAPTER 3

FUNDAMENTAL HEAT FLOWS ON THE EARTH'S SURFACE

According to the principle of conservation of energy, we must account for any positive or negative state of net radiation that exists at the earth's surface. If the net radiation is positive—that is, more radiation is gained than lost over a period of time—there is a surplus of energy that must be dissipated, and outflows of energy will develop in response to this condition. These outflows are received by three types of materials: (1) ground (soil, including organic matter, rock, and concrete); (2) water (oceans, lakes, and rivers); and (3) air. If net radiation is negative, there must be additional sources of energy to make up the deficit. This condition produces energy inflows to the surface from ground, water, or air. In either case, an imbalance in radiation always tends toward equilibrium as energy fluxes to or from the surface.

Consider net radiation on a column of soil or water. There are four ways in which excess energy can be used or in which an energy deficit can be made up:

1. *Soil-heat flux.* If the net radiation is positive and the surface heats to a temperature higher than that of the underlying material, heat is transferred downward by conduction (in soil) or convection (in water). If the net radiation is negative, resulting in a surface temperature cooler than that of the underlying soil or water, heat is transferred upward to the surface.

2. *Sensible-heat flux.* Sensible heat is heat energy that can be sensed with a thermometer. Conventional weather bureau temperature readings are a measure of the sensible-heat content of air. Sensible-heat flux involves heat exchange between the surface and the air above it. Heat is transferred by conduction between the surface and the air molecules at the boundary, and groups of these molecules are mixed by convection with other air molecules. Given a positive net radiation, when the surface has reached a temperature greater than that of the overlying air, heat is transferred from the surface into the air. Conversely, if the surface is relatively cool, due

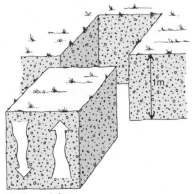

Outflow   Inflow

SOIL-(OR WATER) HEAT FLOW

to a negative net radiation, heat is transferred from the air to the surface. This is represented by the inflow arrow on the diagram.

3. *Latent heat.* Heat released or absorbed when water (or any other substance) changes phase, usually from liquid to vapor, is called latent heat. If the net radiation is positive and sensible heat and water are available on the surface, some sensible heat may be used to evaporate water. With the departure of each molecule of water, heat is taken up and the surface is cooled, just as your skin cools when perspiration evaporates. If, on the other hand, the surface is cooled in response to a negative net radiation, water may condense on the surface and release heat in the process. In this event the heat flow is toward the surface, and for comparable amounts of water, the surface is warmed at a rate equal to the evaporation cooling rate.

4. *Horizontal heat flux.* This is lateral heat flow into or out of the column. In soil this flux is negligibly small, but in water, owing to convective mixing by waves and currents, it is very important. Under a positive net radiation, a column of water may heat and exchange flow with adjacent cooler water. Loss of heat represents an outflow, which in turn represents an inflow to the receiving column of cooler water.

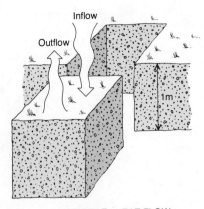

SENSIBLE – HEAT FLOW

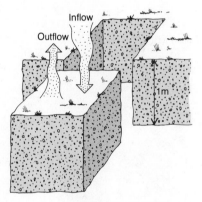

LATENT – HEAT FLOW

THE BASIC ENERGY-BALANCE MODEL

Using the four heat flows described above and the net radiation, we can write a simple equation for the *energy balance* at the surface of a column of soil or water:

Energy balance: Net radiation ± Sensible heat
(for land)   ± Soil-heat flux ± Latent-heat flux = 0.

Energy balance: Net radiation ± Water heat flux
(for water)   ± Sensible-heat flux ± Latent-heat flux
                      ± Horizontal heat flux = 0.

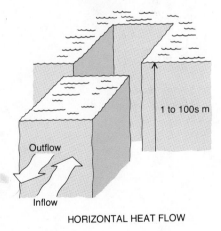

HORIZONTAL HEAT FLOW

The terms in the equations are positive or negative depending on the direction of the flow. *Any flow away* from the surface (i.e., upward into the atmosphere or downward into the soil) is negative, and *any flow toward* the surface (i.e., downward from the atmosphere and upward from the soil) is positive. It is therefore important to remember that the land (or water) surface is always the plane of reference for describing energy flows.

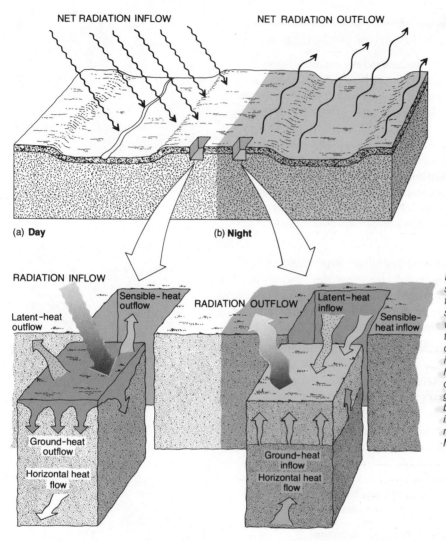

NET RADIATION INFLOW

NET RADIATION OUTFLOW

(a) **Day**

(b) **Night**

RADIATION INFLOW

Sensible-heat outflow

Latent-heat outflow

RADIATION OUTFLOW

Latent-heat inflow

Sensible-heat inflow

Ground-heat outflow

Horizontal heat flow

Ground-heat inflow

Horizontal heat flow

Fig. 3.1 Situation (a) depicts a typical summer midday condition: Surface heating from incoming radiation is sufficiently strong to produce both upward and downward outflows of heat. Situation (b) depicts the nighttime sequal: Longwave radiation outflow produces surface cooling, thereby inducing inflows of sensible heat, latent heat, and soil heat. Latent-heat flow is evidenced by the formation of dew on the ground. Horizontal heat flows, shown in both (a) and (b), relate to these conditions in water and represent current and wave movement. (Illustration by William M. Marsh)

When net radiation is positive and energy is building up on the surface, this energy warms the soil and the air and evaporates the water, as situation (a) in Fig. 3.1 shows. At night or in winter, when net radiation may be negative and the surface is losing energy, heat is transferred upward from the soil below and downward from the air above to the cooler soil surface, as shown in situation (b) in Fig. 3.1. Under certain conditions, the loss of heat from the surface results in the formation of dew, hoar frost, or ground fog.

The two conditions represented in situations (a) and (b) in Fig. 3.1 deserve some thought, for they are the very heart of the energy-balance concept. In contemplating the condition represented in situation (b), bear in mind that although the ground surface is the coolest point in the whole scene and is therefore the recipient of the heat flows, it nonetheless still radiates longwave radiation. This is important, for the model is illogical unless we understand that the heat brought to the surface will always be released to the sky as longwave radiation *so long as the surface temperature is above absolute zero*, which is always the case. Even frozen ground, such as permafrost, radiates longwave radiation, because at a temperature of $-10°C$, it is still $263°$ Kelvin.

These principles of net radiation and heat flow are the keys to the use of solar energy for heating homes. A passive solar furnace can be designed on the basis of the sensible-heat outflow associated with a positive net radiation. Authors' Note 3.1 describes a solar furnace that has been used to heat a midlatitude home over the past several years.

HEAT FLUX INTO SOIL OR WATER

Thermal Properties of Earth Materials

The amount of heat transferred downward into the soil (if net radiation is positive) or upward from the soil (if net radiation is negative) is dependent on the thermal gradient within the soil column and the thermal conductivity of the soil material. *Thermal conductivity* is the rate at which heat energy passes through a 1-m^2 column of a substance having a temperature gradient of $1°K$ per meter (or $1°C/m$). A good insulator, such as organic soil or still air, has a low thermal conductivity, only 1 percent that of ice, for example. The *thermal gradient* is equal to the difference in temperature between two points, divided by the distance separating them; the steeper the gradient, the greater the rate of heat transfer.

Another important thermal property of soil materials is *volumetric heat capacity*, or the amount of heat required to raise the temperature of a unit volume of a substance $1°K$ (or $1°C$). A substance with low volumetric heat capacity (e.g., air) gains a high temperature with the addition of heat, whereas one with a high capacity (e.g., water) gains little temperature with the addition of heat. The volumetric heat capacity is equal to the product of the density and the specific heat of the material. *Specific heat* is the quantity of heat required to raise the temperature of a unit mass of a substance one degree of temperature. A material with a high thermal conductivity and low heat capacity will change temperature easily, whereas a material with opposite properties will not.

AUTHORS' NOTE 3.1
Solar Furnace: An Exercise in Energy Balance

Coal, petroleum, and natural gas are minute by-products of the earth's energy system. Consider that less than one percent of the solar energy received by the earth is utilized by plants in photosynthesis. After plants die or are consumed by animals, their remains are decomposed, and nearly all of the energy represented by the original organic material is released into the atmosphere as heat. Only a tiny fraction of organic debris is stored in special earth environments, where after thousands or millions of years of gradual change, it may be transformed into coal, petroleum, or natural gas. Unfortunately, the rate at which the modern world uses this stored energy far exceeds the rate at which it is produced by nature. The energy system is seriously out of balance; given present use rates, the bulk of the remaining petroleum reserves will be expended in the next several decades. What to do?

Many alternatives are presently being explored, and solar energy is one of the most promising, especially for heating homes and buildings. In addition to being the single greatest supply of energy available to the planet, solar radiation is so widely distributed geographically that it requires less redistribution than do most other forms of energy. However, a number of drawbacks presently prohibit the use of solar energy on more than a small-scale basis. First, low winter intensities of short-wave radiation in the middle and high latitudes necessitate the construction of collector panels larger than one side of most single-family houses. Second, in many of the heavily populated areas of the midlatitudes, notably the central and eastern portions of the United States, northwest Europe, and all of Japan, direct solar radiation is so variable because of cloud cover that auxiliary heating systems or energy-storage facilities are needed. Third, the cost of apparatus with the capacity to store solar energy for more than several days is currently prohibitive.

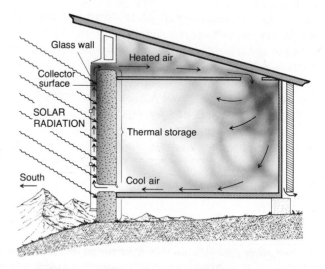

Technological, cost, and environmental drawbacks notwithstanding, research into the uses of solar energy for heating and other purposes is proceeding rapidly, and many experimental projects are now under way. In France, for example, solar scientist Felix Trombe has built a passive solar furnace which has been used to heat an experimental house in the Pyrenees mountains (latitude 43°N) for a number of years. Though simple, the passive design illustrates wise application of basic energy-balance concepts, including net radiation, the greenhouse effect, sun angle, and soil-heat flow.

The basic components of the system are shown in the diagram. The solar collector is the central unit of the furnace, a south-facing concrete slab 4 m high, 12 m long, and 0.33 m thick. The exposed side is painted black and enclosed behind a glass wall. Above and below the collector are air ducts. As the collector is heated with the absorption of the shortwave radiation, sensible heat is transmitted into the adjacent layers of air and into the concrete (analogous to soil heat). When the air between the collector and the glass wall heats, it rises and flows into the house. It is replaced by cool air from the house, which enters the lower duct to the collector surface. While the house is being heated by this process, the concrete slab itself is gaining heat, and, as

occurs in the ground after a sunny day, this heat is slowly reradiated into the house after sunset.

How efficient is this system? In the summer it provides more than enough heat, and the excess must be released to the atmosphere through a duct at the top of the glass wall. In winter, with outdoor temperatures in the Pyrenees falling below freezing, it supplies about 75 percent of the energy needed to heat the house. The additional heat must be supplied by an auxiliary system run, presumably, on conventional energy sources.

Consider the equal volumes of soil and water shown in Fig. 3.2(a). Water has a higher heat capacity than does soil or rock, and if conditions are such that mixing can take place, it has a much higher apparent thermal conductivity as well. If heat is transferred to the surface of both the soil and water, the temperature of the top layer of the soil becomes higher than that of the water, due to the soil's lower volumetric heat capacity, and the soil develops a higher temperature gradient. But the high thermal conductivity of the water—due to mixing—causes more of the heat to be transferred downward. The end result is that the surface temperature of the soil rises greatly, but little heat is transferred downward. At the same time, the overall temperature of the water rises slightly, and most of the heat is transferred downward through a large volume of the material (Fig. 3.2b). This difference in the thermal characteristics of soil and water is the chief reason for the contrasting thermal regimes of climates over continents and those over oceans.

Conduction and the Thermal Gradient

From the soil surface, energy flows mostly by conduction from particle to particle. However, conduction in the soil is not quite as efficient as it would be in a metal. In the soil, conduction is subject to variations related to the density, the chemical composition, and especially the water content. As one might expect, therefore, the denser and wetter the soil, the greater the thermal conductivity. When water is added to

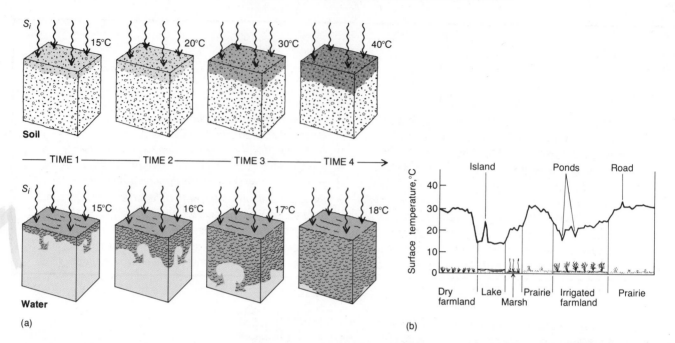

Fig. 3.2 (a) The relative rates of heat flow into water and soil. Note that as time passes, heat is diffused through a much larger volume of water than of soil, because of the mixing motion of water. Also note that for comparable inputs of radiation (S_i), the temperature of water is not raised as high as that of soil. (Illustration by Peter Van Dusen) (b) A profile of surface temperature made on an August afternoon near Brooks, Alberta. Note the sharp contrasts in the temperatures of land and water surfaces. (From R. M. Holmes, "Airborne Measurements of Thermal Discontinuities in the Lowest Layers of the Atmosphere," Ninth Conference on Agricultural Meteorology, Seattle, 1969.)

the soil, conductivity rises as the water displaces air, which has very low conductivity (Fig. 3.3). But this ability to transmit changes in temperature increases only to a certain point, for as the moisture content of the soil increases, the heat capacity of the soil also increases; therefore, additional energy absorption produces smaller temperature rises. Thus the actual diffusion of a temperature change from the surface into the soil reaches a maximum when the moisture content is between 8 and 20 percent and is somewhat less when the moisture content is higher.

Table 3.1 gives some thermal characteristics of various common substances of the landscape. Note the big differences in the thermal conductivities of stirred and still water and air. Further, note the difference that moisture makes in both the volumetric heat capacity and the thermal conductivity in quartz sand, clay, and organic matter.

An outflow of heat from the surface into the soil occurs when temperature decreases with depth into the ground. To determine the rate

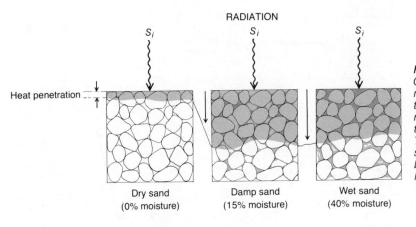

Fig. 3.3 *Thermal conductivity in soil. Given comparable quantities of absorbed radiation (S_i), the soil-heat flow varies with the thermal conductivity and with the volumetric heat capacity; the ratio of these two variables is called thermal diffusivity. The thermal diffusivity of damp sand is significantly greater than that of dry sand, but that for wet sand is actually a little less than for damp sand.*

Table 3.1 Thermal properties of some common earth materials.

SUBSTANCE	VOLUMETRIC HEAT CAPACITY[1]	THERMAL CONDUCTIVITY[2]
Air		
Still (at 10°C)	0.0012	0.025
Turbulent	0.0012	3,500–35,000
Water		
Still (at 4°C)	4.18	0.60
Stirred	4.18	350.00 (approx.)
Ice (at −10°C)	1.93	2.24
Snow (fresh)	0.21	0.08
Sand (quartz)		
Dry	0.9	0.25
15% moisture	1.7	2.0
40% moisture	2.7	2.4
Clay (nonorganic)		
Dry	1.1	0.25
15% moisture	1.6	1.3
40% moisture	3.0	1.8
Organic soil		
Dry	0.2	0.02
15% moisture	0.5	0.04
40% moisture	2.1	0.21
Asphalt	1.5	0.8–1.1
Concrete	1.6	0.9–1.3

[1]Millions of joules needed to raise 1 m³ of a substance 1°K.

[2]Heat flux through a column 1 m² in W/m when the temperature gradient is 1°K per meter.

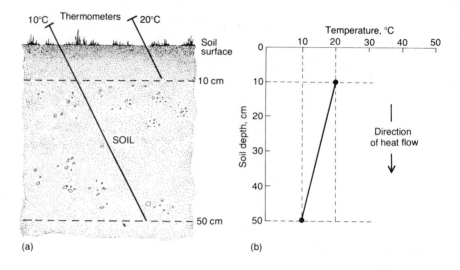

Fig. 3.4 (a) Thermometers (baked-turkey style) inserted to depths of 10 cm and 50 cm in a section of soil record a temperature of 20°C and 10°C, respectively. (b) Graphic representation of these measurements. The rate of outflow (downflow) of heat can be calculated from these data and the thermal conductivity of the soil.

Fig. 3.5 Common energy gradients of the atmosphere. Energy flow is always from the high value to the low value. The rate of flow depends on the steepness of the gradient, i.e., the difference between the high and low values, divided by the distance over which this difference occurs. Flow may be up, down, or lateral.

of flow, one needs to know the thermal conductivity of the soil and the soil temperature at two depths, as shown in Fig. 3.4(a). For example, let us suppose that a clay soil has a thermal conductivity of 1.0 W/m (-25 W/m^2) and temperatures of 20°C and 10°C at depths of 10 cm and 50 cm. The computed rate of soil-heat flow is -25 W/m^2, or -0.036 cal/cm^2 · min (Fig. 3.4b). (The number is negative because the flow is away from the surface.)

It is important to remember that energy flow is always from an area of greater energy (shown by high temperature, greater pressure, or higher elevation) to an area of lower energy. The rate of flow is proportional to the energy gradient, which can be shown graphically by a curve representing the change in energy over distance through a substance. The flow of water down a steep slope is faster than down a gentle slope; soil heat flows faster when there is a larger temperature

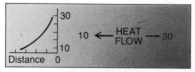

Temperature, °C

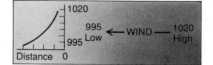

Air pressure (millibars)

gradient. In short, the steeper the gradient of any energy form, the more rapid the energy flow (see Fig. 3.5).

Seasonal and Daily Variations in Soil Heat

The *net annual* soil-heat flow in and out of the soil at any place must equal zero; otherwise, the soil would heat up or cool down. Annual and longer-term imbalances do occur, but they are difficult to detect because they usually involve a long period of time, and for most places we do not have many years of records on soil temperatures. Imbalances over shorter periods have been documented, however, and they correlate with daily temperature changes, weekly weather changes, and seasonal changes in the availability of heat at the earth's surface (Fig. 3.6). The seasonal changes give rise to distinctive gradients, or *temperature profiles*, in the upper two meters or so of the soil. A representative set of soil-temperature profiles for a midlatitude location is

Fig. 3.6 Air-temperature variation at 100 cm above the ground over 4.5 days (top graph) and the corresponding ground temperatures at depths of 10 cm, 50 cm, and 100 cm (lower graphs). Note that the warming trend in the upper graph also appears in the ground-temperature graphs, *but is less pronounced at lower levels. Also note that the diurnal variation in air temperature appears at 10 cm and 50 cm, but the variance is smaller and is offset by as much as twelve hours at 50 cm. Data are from a grass-covered site in the Midwest.*

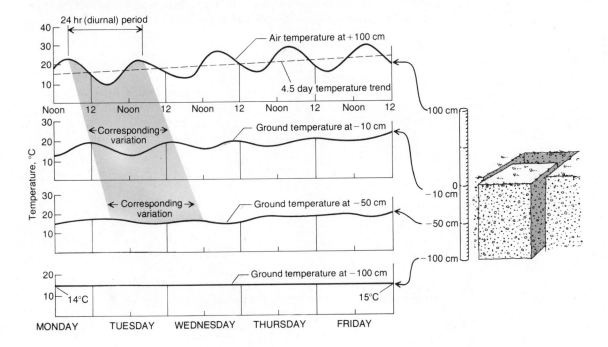

given in Fig. 3.7. Similar profiles also appear in miniature on a *daily* basis in the upper 10 to 20 cm of soil and are superimposed on the seasonal pattern. That is, the soil-temperature profiles near the surface typically flop back and forth from day to night, as the enlargements in Fig. 3.7 illustrate. The surface is coolest in early morning and warmest in early afternoon.

Because of the time needed for heat transfer into and from the soil, the coolest and warmest temperatures usually lag behind the maximum and minimum air temperatures. The deeper the level, the longer the temperature lag; at a depth of 3 m, it may be as long as two months. Thus in midlatitude locations in the Northern Hemisphere the warmest soil temperature at this depth is not often reached until the end of September. Daily temperature changes do not occur much below 20 cm, called the *diurnal damping depth*, and in most places seasonal change cannot be detected much below 3 meters. Obviously, these principles of soil-heat flow can have some important practical applications, and one is illustrated in Authors' Note 3.2.

Air temperature changes little in the tropics over a year, and for this reason soil heat varies little there from season to season. In fact, average air temperatures in the tropics vary less over the year than they do from day to night. Consequently, the soil temperatures vary more diurnally than they do seasonally.

Perhaps nowhere is the seasonal soil-heat flow more vividly expressed than in the permafrost regions of the Arctic and subarctic. Permafrost is ground in such a low heat state that it is continuously frozen, but usually only at depths greater than 1 to 3 meters. The upper

Fig. 3.7 Soil-temperature profiles for a typical winter, spring, summer, and fall day. In spring and fall the heat flux is both upward and downward, indicating that the subsoil is warmer than the layers above and below it. In winter the flux is upward, and in summer it is usually downward.

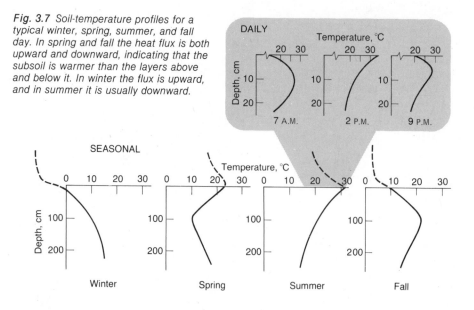

zone, called the *active layer*, freezes and thaws seasonally, in much the same fashion as do the upper 10 cm to 20 cm of midlatitude soils. In response to the outflow of soil heat in the fall, frost is formed at the surface and ultimately expands across this zone toward the frozen subsoil. In spring and summer the trend reverses, with a downflow of heat from the surface. As the thaw line (marked by a temperature of 0°C) progresses downward (Fig. 3.6), the temperature profile steepens, reflecting a growing departure of soil-surface temperatures from that of the permafrost around – 10°C. With the penetration of new frost from the surface in fall (Fig. 3.7), the temperature profile becomes spoon-shaped and straightens somewhat; in ensuing months of winter, when surface temperatures fall far below 0°C, the gradient is fully reversed from that of summer (Fig. 3.8). Paradoxically, the permafrost is the primary source of soil heat in winter. Downward temperature increases as great as 4° to 5°C per meter leave little doubt about this fact.

Fig. 3.8 Map showing the distribution of permafrost and ground frost in the Northern Hemisphere, with seasonal variations in soil heat in a permafrost environment as shown in the diagrams and graphs. Summer heat flow is downward, producing thawing of the active layer. In fall the surface freezes, and the active layer loses heat to the frozen ground above and below it. By winter the active layer is frozen out (except for some pockets), and the surface is so cold that the heat flow, though slight, is upward from the permafrost. (Map after R. U. Cooke and J. C. Doornkamp (1974), based on Corte (1969), MacKay (1972), Brown (1967), and others.)

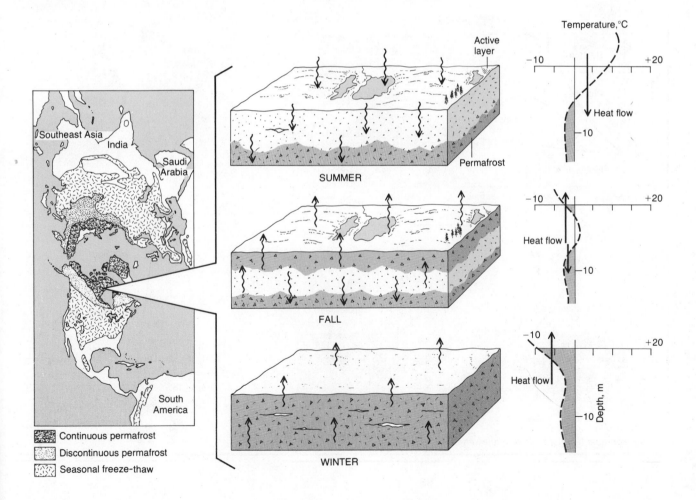

Active layer

Permafrost

SUMMER

FALL

WINTER

Temperature,°C

Heat flow

Heat flow

Heat flow

Depth, m

Southeast Asia
India
Saudi Arabia

South America

Continuous permafrost
Discontinuous permafrost
Seasonal freeze-thaw

AUTHORS' NOTE 3.2
Soil-Heat Concepts Applied to Earthen Houses

The thermal objective in the design and construction of human dwellings is to reduce the variation in the daily and seasonal supply of natural heat. This entails raising winter temperature or lowering summer temperature or both, depending on the particular climate and season. From an energy-balance standpoint, most buildings, especially those constructed of earthen materials such as sod, stone, or clay brick, can be considered as extensions of the soil environment. Such buildings should possess thermal characteristics similar to those of soil, and in certain climates this gives them a clear advantage over buildings constructed of wood or metal, for example, in terms of both energy economy and living comfort.

Recent history is rich in examples of the mismatch between building design and materials on one hand and climate on the other. One of the most notable examples was the adoption of the wood frame house by the

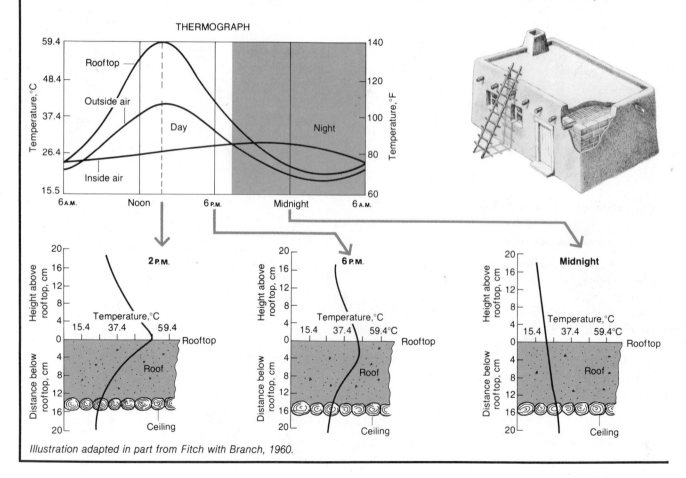

Illustration adapted in part from Fitch with Branch, 1960.

Eskimos on the northern coast of Alaska. In addition to calling for lumber in a treeless environment, these houses proved to be a very poor second, thermally, to the Eskimos' smaller, traditional houses built of rock, sod, and animal skins.

Though no longer extensively used by the Indian in the American Southwest, the adobe, or mud masonry, house is still the principal house type of desert people throughout the world. As shown in the diagram, the thick walls and closed design of the adobe house may seem inappropriate for the desert climate. However, on closer examination we find that it is an especially good adaptation to the daily thermal regime of the desert, because of the low thermal conductivity of the material used for the roof and walls. This material usually consists of a mixture of clay, sand, and some organic matter, such as straw or sticks, all of which have relatively low thermal conductivities when dried. Thus if the roof and walls are thick enough, the penetration of each day's heat will not reach the interior of the structure. Therefore, the temperature of inside air should remain relatively low, especially if circulation with outside air can be minimized. The thermograph illustrates just how effective the adobe house can be in this regard; the midafternoon difference in temperature between the rooftop and inside air is 33°C (60°F).

What happens to the heat absorbed by the surface of the roof? As the profiles of the soil (roof) heat show, most of it penetrates only 5 to 7 cm and then flows back to the surface, where it is released into the outside air. Some heat, however, flows through the roof, which accounts for the late-afternoon and early-evening rise in indoor temperature. On balance, the house is the coolest place during the day and the warmest place at night. Anyone for sleeping out?

HEAT FLOW INTO THE ATMOSPHERE

Energy Transfer as Sensible Heat and as Latent Heat

There are two forms of heat-energy flow from the ground level into the overlying atmosphere: as either sensible heat or latent heat. *Sensible-heat* transfer takes place when the temperature of a soil or water surface is different from that of the overlying air. Heat is transferred by conduction between the lowermost air molecules and the soil or water.

Latent-heat transfer takes place when water changes phase. Water can exist on earth in three phases: solid (ice), liquid (water), or gas (water vapor). A change from solid to liquid (*melting* or *thawing*) requires an energy input of 0.334×10^6 joules per kilogram, or 80 calories per gram. (The scientific notation 10^6 stands for 1,000,000; therefore, $0.334 \times 10^6 = 334,000$.) A phase change from liquid to vapor (evaporation) at a temperature of 0°C requires 2.5×10^6 joules per kilogram, or 597 calories per gram. At higher temperatures, the value

Fig. 3.9 The vapor layer over water may form minute droplets, indicating an upward flow of latent heat into the drier air above. (Photograph by Charles Schlinger)

is lower; for example, at 100°C, it is 2.3×10^6 J/kg (540 cal/g). Change directly from ice to vapor (*sublimation*) requires 2.834×10^6 J/kg (676 cal/g), the sum of the melting and vaporization values. The reverse changes—vapor to liquid (*condensation*), liquid to solid (*freezing or fusion*), or vapor to solid (also called *sublimation*)—result in the release of equivalent quantities of heat.

When ice or liquid water at the earth's surface changes into water vapor, the vapor passes into the atmosphere, thereby producing a flow of energy into the atmosphere (Fig. 3.9). Conversely, when water vapor in the atmosphere condenses on the surface to form dew or frost, heat energy is transferred from the atmosphere to the surface. In either case, a humidity gradient must exist in order for a latent-heat flow to take place.

From the standpoint of the energy balance at ground level, the most important latent-heat flow is produced by the combination of evaporation from soil and open water and transpiration from plants. From a land parcel of one acre, about $250,000 \times 10^6$ joules $(60,000 \times 10^6$ calories) of heat are pushed into the atmosphere for every 2.5 cm (1 inch) of water vaporized. In the eastern United States, this amount of latent heat leaves each acre on the average of once every four days in the summer. The driving force for this enormous energy flow is generated by the absorption of solar energy on the soil surface. Thus it is easy to see that latent-heat flow is of paramount importance as a mechanism for disposing of sensible heat from the surface and in turn for controlling surface temperature.

The flow of latent heat into the atmosphere is controlled by many conditions, two of the most critical ones being wind speed and the amount of vapor (humidity) in the air. The vapor content of air is expressed in various ways (discussed in detail in Unit II). What is necessary to point out here is that the capacity of air to hold vapor is related to its temperature and that this capacity increases rapidly with higher temperature (Fig. 3.10). For example, a temperature increase from 10°C to 15°C increases holding capacity by only about 2 grams per cubic meter of air (from 8 to 10 grams), whereas a temperature increase from 35°C to 40°C increases holding capacity by about 13 grams per cubic meter of air (from 37 grams to 50 grams). Colder air is therefore generally drier, and evaporation rates in winter from open water are surprisingly high.

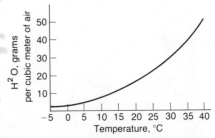

Fig. 3.10 *The maximum amount of water vapor that can be held at a given temperature. Note that the vapor-holding capacity increases at an increasing rate with temperature.*

The Role of Wind in the Energy Balance

The flux of sensible and latent heat from ground to air is heavily dependent on the wind speed near the ground. When wind is light, the thin layer of calm over the ground, called the *laminar sublayer* (so called because the air in it moves only parallel to the surface), is thicker. Since heat crosses this layer by conduction, the transfer is exceedingly slow, and heat and vapor build up in it. This buildup, in turn, reduces the energy gradients at the interface between ground and air, and transfer declines correspondingly. Figure 3.11 portrays this schematically in a sequence of illustrations representing a segment of time.

If stronger wind is brought to the surface, a large part of the energy-saturated surface layer can be swept away and replaced by cooler, drier air. As a result, the energy gradients are reestablished; if the wind continues, the gradients can be maintained, thereby ensuring a relatively high rate of latent- and sensible-heat transfer. The actual transfer rates achieved, of course, depend on the wind speed as well as the temperature and humidity differences between the air and surface.

The action of the wind in the transfer of energy over small distances—say, on the order of 1 to 15 km—is referred to as *convectional mixing*. This includes not only vertical movement, but also certain types of motion related to horizontal movement near the surface. Convectional mixing is produced in two ways. The first is by wind movement in response to density differences in the atmosphere resulting from variations in the heating of air. This is called *free convection*, or *thermal turbulence*, and is associated with vertical movement of unstable parcels of air both near the ground and at higher elevation (as in a thun-

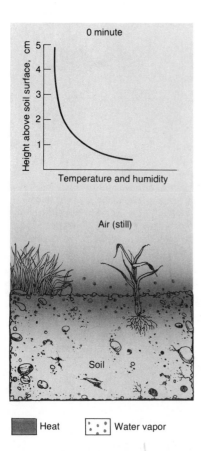

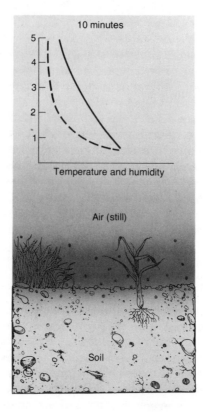

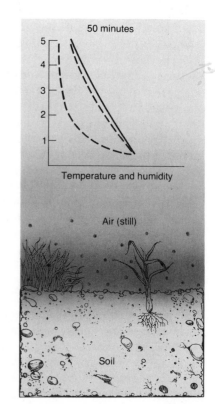

Heat | Water vapor

Fig. 3.11 *The flow of sensible and latent heat into the laminar boundary layer of still air. Initially (0 minute), the energy gradient between the soil surface and air is steep, and the rate of heat flow is relatively high in the first 10 minutes, after which the rate is very slow, owing to the weakness of the gradient over the surface. At 50 minutes, the heat and vapor content of the air have increased only slightly.*

derstorm). The second is by movement of wind over rough surfaces in the landscape. This is called *mechanical*, or *forced*, *convection* and is characterized by a swirling flow similar to that of river water moving through rapids. Both types of convection are essential mechanisms in transferring sensible and latent heat between the surface and the atmosphere. Let us examine them in greater detail.

Free convection. Any body immersed in a fluid (liquid or gas) is accelerated upward if its density is less than that of the fluid or downward if its density is greater than that of the fluid. This principle was first described by the Greek mathematician Archimedes more than 2000 years ago. This principle explains the rise of bubbles in beer and the sinking of most bars of soap in water. To apply this principle to the atmosphere, we need only imagine a body of gas containing some pockets with densities markedly lower than that of the body as a whole. Such bubbles of air should rise through the atmosphere because they are gravitationally unstable.

Parcels of light air form over the land as a result of intensive out-flows of sensible heat from the ground. As the air is heated, it expands and becomes buoyant, eventually breaking free of the surface. As it ascends, an upward-flowing wind, called a *thermal*, is generated (Fig. 3.12). This is a mild form of free convection; a thunderstorm, on the other hand, represents one of the most intensive forms of convection, but it too is usually initiated by thermals.

Since the atmosphere heats from the bottom up and rising air cools due to expansion, air near the surface is usually warmer than air aloft. The cooling of air due to expansion is called *adiabatic cooling* and is discussed in detail in Unit II. Here it is necessary to note only that air rising through the atmosphere cools at a rate of 0.0098°C per meter (the *dry adiabatic lapse rate*), as long as no condensation is occurring. As air descends, it warms at the same rate.

Whether air near the surface is *unstable* or *stable* depends on whether the air tends to rise on its own or to stay on the ground. If there are pockets of air that are warmer (and hence lighter) than the surrounding air, these pockets will rise, and we would say that the con-dition is unstable and that there is free convection. Whether or not this will be the case depends on the temperature profile of the air above the parcel. Since rising air cools adiabatically, instability can exist only if the rising air remains warmer than the surrounding air, which can be the case only if the temperature gradient in the overlying atmosphere is colder than 0.0098°C per meter. Where cold air overlies warm sur-face air, free convection is enhanced, but where the air aloft is only a little colder than the surface layer, it is subdued because as the rising parcel of air cools, it quickly becomes colder than the surrounding air and rises no further.

In addition to instability caused by surface heating, condensation may also contribute to *instability* and convection. In the case of a rising

Fig. 3.12 The development and ascent of a heated parcel of air. The vertically flow-ing air, called a thermal, is one of the most prevalent forms of free convection. (After Woodward, 1960)

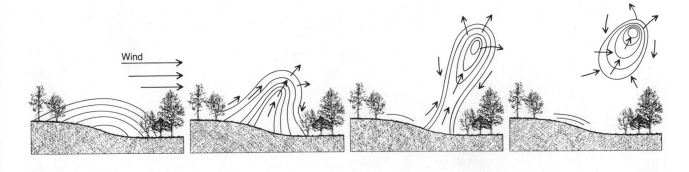

AUTHORS' NOTE 3.3

Mixing the Atmosphere Near the Ground

Atmospheric mixing related to instability within several hundred meters of the ground is common in most locales, though it may be so subtle as to pass unnoticed by most of us. It is related to variations in surface heating that arise from differences in surface materials or exposure to solar radiation. For example, land usually heats faster than water, and on clear summer days, the afternoon air temperature over the land may be 5°–10°C higher than that over adjacent water. Since the pressure of unconfined air tends to decrease as it is heated, air pressure over the land decreases during the day; that over the water changes little, remaining relatively high. Thus across the coastline a pressure gradient develops that induces an onshore wind, or sea breeze, as shown in diagram (a).

As the land cools in the evening, both the pressure differential and the breeze subside. However, if the atmosphere is clear, allowing heat to escape rapidly from the land at night, the temperatures and pressure pattern may actually reverse from that of the day. Although the pressure differential is usually small, it is often sufficient to generate a noticeable land breeze several hours after dark.

Another example of atmospheric mixing near the ground is found in areas of rough terrain, where variations in sun angle produce strong differences in slope heating. In east-west–trending mountain valleys, shadows envelop the low-exposure slope in late afternoon. As a result, this slope may develop a layer of cool surface air, which, owing to its high density, slides downslope in the evening to occupy the valley bottom. The warmer air on the valley floor is displaced upward and settles over the cool air layer. The resultant temperature profile, shown in diagram (b), upward from the valley floor is reversed from that of the day. Because of the thermal structure, this condition is referred to as a temperature inversion, and it is extremely stable at ground level because that is where the densest air rests.

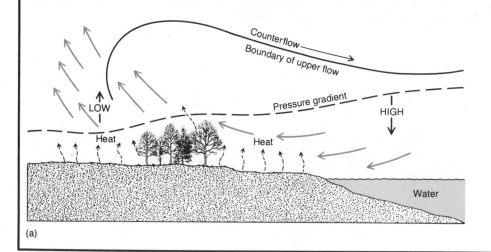

(a)

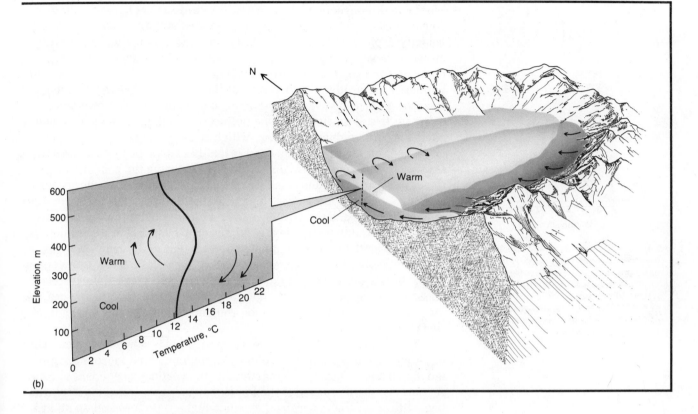

(b)

parcel of air that ordinarily would be stable at some height were it not for the occurrence of condensation, the latent heat released raises the air temperature and induces instability. In essence, the latent heat of vaporization supplies an extra charge of energy that drives the parcel even higher. This is typically the case in most rainstorms. Finally, to bring this discussion full circle, as the rain falls to the ground, water is resupplied to the soil, where it may subsequently vaporize and serve as the vehicle for further heat transfer from the surface to high in the atmosphere.

Although free convection is probably best expressed in its most dramatic form, the thunderstorm, it is important from our standpoint to remember that this process is usually initiated at or near the earth's surface, that it draws energy off the surface, and that it occurs frequently at the local scale in the lower 10–100 m or so of the atmosphere. At this scale, convection can often be related to local variations in surface heating, topography, and even land use. Authors' Note 3.3 presents two common examples of local atmospheric mixing related to free convection.

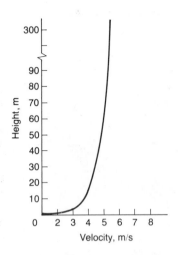

Fig. 3.13 The change in wind velocity with height above the ground. Note that the graph line, or velocity profile, changes most rapidly just above the surface. From 20 to 300 m, velocity increases at a lower rate, and in the zone above an elevation of 300 m, it generally holds steady with increasing elevation.

Mechanical convection. The second kind of convection—*mechanical convection*—results when air flows over a rough surface. Wind velocity typically increases rapidly with height in the first several meters above ground. Above 20 m or so, the rate decreases and changes little for the next several hundred meters (Fig. 3.13). At the bottom of the profile, it is usually assumed that wind velocity vanishes to zero at some near surface height, called the *roughness length,* or *effective frictional surface.* This defines the ceiling of the envelope of calm air, the laminar sublayer, which covers a surface. It is a very important factor in sensible- and latent-heat flux because heat must first cross this zone by conduction before entering the zone of convective flow. If the roughness length is great, as over a city, energy flow into the atmosphere is slow, and heat builds up in the air over the ground. If it is small, however, as over a plowed field, the laminar sublayer is thin and therefore less of a barrier to heat flow from the soil. In a forested landscape, the laminar sublayer is usually marked by high humidity, reflecting the absence of wind flushing of the water vapor which builds up under the canopy. The roughness length varies from about 0.01 cm over smooth snow or mud flats to several meters in forests (Fig. 3.14). Table 3.2 gives roughness lengths for a variety of surfaces.

The roughness length affects mechanical convection because large lengths generally lead to more turbulence (and more mixing) in the air above the surface. In particular, the vertical component of convection may be increased if the landscape is highly irregular in elevation, with some objects protruding much higher into the air than others. On the other hand, a small roughness length leads to a thin but undisturbed laminar sublayer and reduced mixing in the air above the surface. If a surface combines topographic irregularity with a small roughness length on individual objects, energy flux due to mechanical

Table 3.2 Roughness lengths for various surfaces.

TYPE OF SURFACE	ROUGHNESS LENGTH	SOURCE
Fir forest	5.55 – 2.8 m*	Baumgartner (1956)
Citrus orchard	3.35 – 2.0 m*	Kepner *et al.* (1942)
Large city (Tokyo)	1.65 m	Yamamoto and Shimanuki (1964)
Corn	1.3 – 0.72 m*	Wright and Lemon (1962)
Grass	0.15 – 0.08 m*	Deacon (1953)
Smooth desert	0.03 cm	Deacon (1953)
Dry lake bed	0.003 cm	Vehrencamp (1951)
Smooth mud flats	0.001 cm	Deacon (1953)

*The two values represent low- and high-velocity winds.

FOREST SURFACE

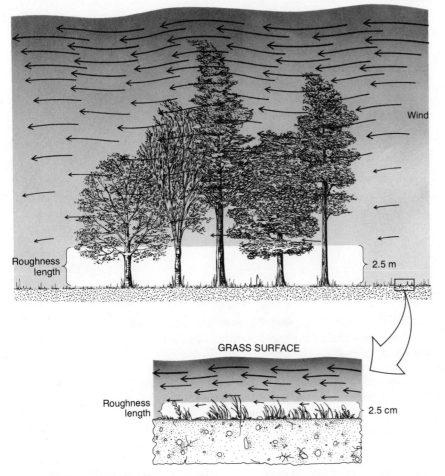

Wind

Roughness
length

2.5 m

GRASS SURFACE

Roughness
length

2.5 cm

Fig. 3.14 Conceptualization of roughness length, or the depth of the zone of calm air which envelopes a surface. As surface roughness decreases, so too does roughness length. The ceiling of the calm layer (at 2.5 m and 2.5 cm, respectively) marks the bottom of the zone of wind turbulence. (Illustration by William M. Marsh and Peter Van Dusen)

convection should be relatively high from the taller objects, particularly if the wind is fast.

The aerodynamic behavior of fast wind over such a surface is illustrated reasonably well by airflow on a windy day in a city. Flow is irregular, or gusty, with eddies (whirls) developing off the corners of buildings. The eddies of the strongest gusts typically violate the roughness length, sweeping heat and vapor off building surfaces. At street

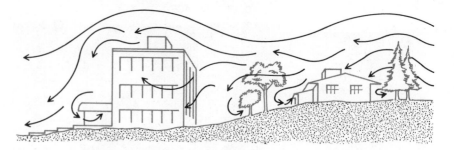

Fig. 3.15 Wind flow over an irregular surface. Note the eddy mixing on the lee side of obstacles such as buildings.

level, however, heat exchange is light; Fig. 3.15 illustrates what this pattern would look like if we could see wind. It is important in the context of surface-energy transfer to underscore the vertical component of flow in mechanical convection. A detailed discussion of airflow over rough terrain is presented in Chapter 30.

SUMMARY ON ENERGY BALANCE

Net radiation, soil- (or water-) heat flux, sensible-heat flux, and latent-heat flux are the major components of the atmospheric energy balance and are related by the equation:

> Incoming shortwave − Outgoing shortwave
> + Incoming longwave − Outgoing longwave
> ± Soil-heat flux ± Sensible-heat flux
> ± Latent-heat flux
> = 0.

In a still atmosphere, energy flows essentially up and down. Shortwave radiation from the sun enters the atmosphere, much of it passing through and striking the earth's surface, where it is absorbed or reflected. Absorbed radiation heats the surface, producing temperature gradients downward into the cooler soil or water and upward into the cooler air.

Heat flows in both directions as long as the temperature of this surface is greater than that of the adjacent air and soil or water at depth. The rate of downward heat flow in soil depends mainly on its water content and particle composition. The rate of heat flow in water depends on the mixing motion of waves and currents.

Heat flows into the air in two forms—sensible and latent. Both sensible and latent heat enter the atmosphere by conduction and move through the atmosphere by convection, created by atmospheric insta-

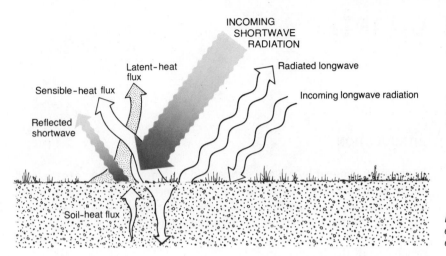

Fig. 3.16 *The major components of the energy balance at the surface of the earth.*

bility and wind. Greater instability and higher wind velocity lead to greater rates of transfer of sensible and latent heat.

Energy is lost from the surface as reflected shortwave radiation and as reradiated flows of longwave radiation, sensible heat, latent heat, and soil heat. The soil heat ultimately flows back to the surface when the soil surface cools, as at night or in winter, when the soil-temperature gradient reverses. The net soil-heat flow over a long period of time equals zero. However, net sensible- and latent-heat flows need not be zero, because heat transferred into the atmosphere can be moved to another place, a topic taken up in Chapter 6. Figure 3.16 illustrates the heat and radiation components of the energy balance and the direction of flows.

THE INFLUENCE OF URBANIZATION ON THE ENERGY BALANCE

CHAPTER 4

INTRODUCTION

In physical geography we are concerned with not only the natural landscape as it is shaped by natural process, but also the ways in which human activities shape the landscape. Accordingly, it is important to ask what influences people have on the energy balance, because it is the energy on the earth's surface that drives many of the essential processes of the landscape.

In recent years the question of the impact of human land use on the atmosphere has taken on major proportions. Cities consist of the most intensive forms of land use yet devised, and it is not surprising that some of the most drastic alterations of local climate, or microclimate, have traditionally been associated with them. In the past several decades urbanized areas have undergone a geographical explosion and now cover such large areas that it is no longer appropriate to call them "localized environments" (Fig. 4.1). The greater metropolitan areas of cities such as New York, London, and Los Angeles occupy

Fig. 4.1 (a) *Urbanized areas in the coterminus United States. The black areas show the 1960 pattern; the gray areas, the projected pattern of urbanization in the year 2000. (b) Trends in the number of cities, urban population, and the area of cities between 1800 and 2000. (From Marion Clawson,* America's Land and Its Uses, *Baltimore: The Johns Hopkins University Press, 1972. Copyright © The Johns Hopkins Press and Resources for the Future, Inc. Reprinted by permission.)*

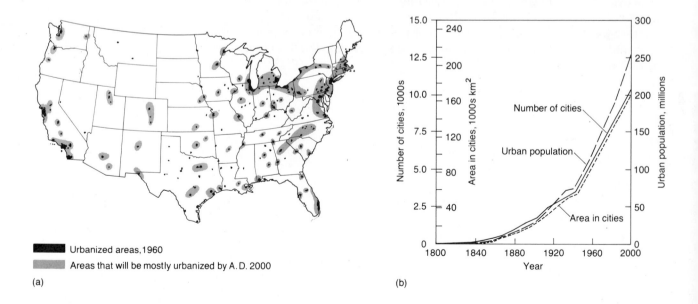

Urbanized areas, 1960

Areas that will be mostly urbanized by A.D. 2000

(a)

(b)

areas of 1000 square kilometers or more, for example. Scientific studies on urban climate are revealing some interesting trends. For example, solar radiation is generally less intensive in cities. However, temperature may be as much as 10–15°C higher in cities than in the surrounding countryside. Wind at ground level is less strong in cities, due to the increased roughness length produced by closely spaced buildings. And fog is characteristically much more common in urban settings than elsewhere. As for other components of climate, such as humidity, snowfall, and rainfall, the picture is not altogether clear whether urban areas are responsible for inducing marked change. Variation in climate associated with cities can be readily appreciated when examined in the context of the energy balance. Based on our earlier discussion, we can now identify several changes in the forms and relative quantities of radiation and heat which are brought about by urbanization.

DIRTY AIR AND THE FLOW OF ENERGY

Alterations in the Flow of Radiation

Large quantities of gases and minute particles of foreign matter are expelled into the atmosphere from cities. If the rate of expulsion exceeds the rate of removal by airflow through the city, these materials build up in the air. Thus for a given volume of air, the mass balance of pollutants is a product of the flushing capacity of the atmosphere as well as the rate of production of airborne materials by human activities. Most of the time airflow in cities is insufficient to remove the contaminated air. This affects not only the smell and taste of the air, but also its visibility. On sunny days the atmosphere appears discolored and hazy, and the sun itself often has a certain dullness about it. Much of this is due to airborne particulates which are usually found in concentrations ten times greater over cities than over rural lands. Studies show that in the lower 1000 m of atmosphere above cities, the intensity of solar radiation may be reduced by more than 50 percent. A large part of this loss is attributable to upward scattering of shortwave radiation from the dome of particulates which blankets the city (Fig. 4.2).

In addition to retarding the entry of incoming shortwave radiation, the dust dome and associated gaseous pollutants decrease the rate of release of longwave radiation from the city. This, of course, is a greenhouse effect, and though it exists in an undisturbed (i.e., natural) atmosphere, it is greatly increased over cities. In the city this involves not only atmospheric absorption and reradiation of longwave radiation, but also some backscattering of longwave radiation off the ceiling formed by pollutants. As a result, heat is retained in the city atmo-

Boundary of urban atmosphere   INCOMING SHORTWAVE RADIATION

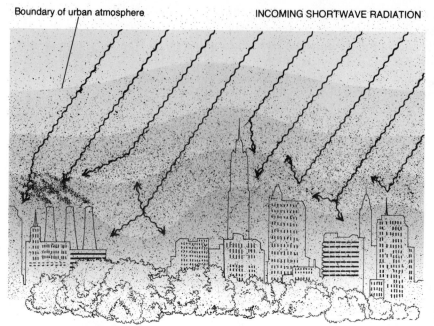

Fig. 4.2 Backscattering of incoming short-wave radiation by the urban dust dome.

sphere for a longer time than is the case in a cleaner atmosphere, thereby inducing higher air temperatures.

The Heat Island Effect

To most of us, the process of urbanization is most vividly expressed in changing patterns of land use and land cover. This change also has important energy-balance implications. Locally, albedos may be reduced with the construction of dark roofs and streets, but overall, urban surfaces have a combined albedo comparable to that of rural areas. On the other hand, the change to urban land uses may greatly influence the latent-heat flux, because hard surface materials such as asphalt, concrete, and brick nearly eliminate the natural moisture flow from the soil to the air. As a result, the ratio between sensible-heat flux and latent-heat flux, called the Bowen ratio, is increased rather significantly. With a greater proportion of energy converted to sensible heat, the ground-level air temperatures in cities are usually higher than those in the neighboring countryside.

Another important land-cover factor in the energy balance of cities is the thermal properties of hard surface materials. Due to the low volumetric heat capacities of asphalt, concrete, and roof materials, their temperatures tend to rise more rapidly and to reach a higher level with the absorption of radiation than do plant and soil

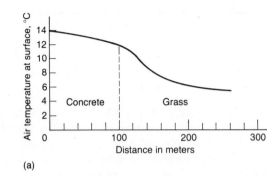

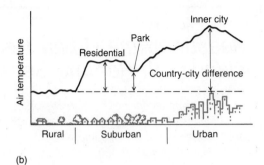

(a)

(b)

materials. Measurements by Rudolph Geiger, a pioneer in the study of microclimate, showed that in early summer the temperature of asphalt would often double that of the grass it replaced. Coupled with the high Bowen ratio of dry materials such as asphalt, it is understandable why the air above hard surfaces tends to heat rapidly (Fig. 4.3a). One can verify the remarkable difference in the heating of hard surfaces and vegetated soil surfaces by touching, say, an asphalt parking lot and an adjacent lawn during a sunny afternoon in any city. Often just walking on them is sufficient to detect the difference. It is little wonder that foot-patrol cops and mail carriers have traditionally lamented hot, swollen feet. And from the standpoint of urban planning, this speaks well for the incorporation of parks, greenbelts, and water features into neighborhoods (Fig. 4.3b).

The high emittance of heat from hard surfaces raises the overall air temperature of cities, creating a warm spot, or heat island, in the

Fig. 4.3 The effect of surface on air temperature: (a) Air temperature over adjacent grass and concrete surfaces. (Data from Aronin, 1943) (b) The nature of air-temperature change associated with rural, suburban, and urban landscapes. (After T. R. Oke, 1976)

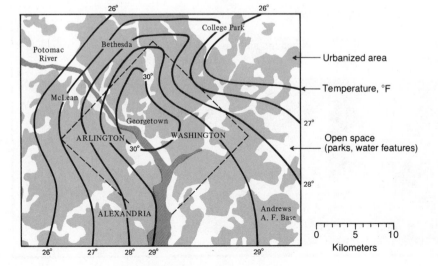

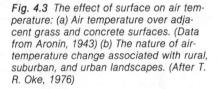

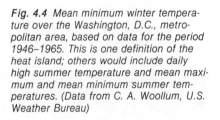

Fig. 4.4 Mean minimum winter temperature over the Washington, D.C., metropolitan area, based on data for the period 1946–1965. This is one definition of the heat island; others would include daily high summer temperature and mean maximum and mean minimum summer temperatures. (Data from C. A. Woollum, U.S. Weather Bureau)

AUTHORS' NOTE 4.1

Measuring the Urban Heat Island with Remote Sensors

Remote-sensing techniques provide one of the most useful means of gaining geographical data about the earth's surface. As the name implies, remote sensing involves measuring something from a distance, that is, without actually touching it with hand or instrument. For large portions of the earth's surface, this is done with cameras or other sensing devices carried by aircraft or satellites.

Aerial photographs are the most widely used form of remotely sensed imagery, and geographers, geologists, foresters, and planners, for example, use them as a source of information on drainage, vegetation, land use, pollution, and so on. In the past few decades the capacity of remote sensing has been extended far beyond that of photographs, and we now possess the means to detect things that cannot be "seen" by a camera. For instance, from altitudes of several thousand meters, radiation sensors can accurately record variations in the longwave radiation emitted from the landscape. One type of infrared sensor that has proved highly successful is the airborne scanner. As an aircraft

flies in a straight-line route over the land, this device scans back and forth across the flight path. Along each scan line, the intensities of the infrared emissions are recorded on electromagnetic tape. Arranged in successive order, the scan lines form a maplike image, similar to a television image, that shows variations in surface heat.

The print shown here is a thermal infrared image in the vicinity of Ann Arbor, Michigan, taken on a December night. The lighter tones indicate areas which are several degrees C warmer than the darker-toned areas. Not only are the surfaces of the inner city warmer than those in the suburban and rural fringe, but in addition many water features are warmer than the adjacent land surfaces.

S. I. Outcalt of the University of Michigan has demonstrated that the temperature difference in summer is due primarily to the decreased latent-heat exchange in the central city. The cooler area around the center is an older residential area, with large trees; it is in fact cooler than the newer suburbs.

Infrared image (10.4–12.5 μm) wavelengths, urbanized area near Ann Arbor, Michigan, 10 P.M., December 23, 1975. (Courtesy of the Environmental Research Institute of Michigan)

Inner urban—relatively warm

Vegetated surface—relatively cool

River—warmest surface in the scene

land (Fig. 4.4). According to T. R. Oke of the University of British Columbia, heat islands usually drop off sharply where the city gives way to the rural landscape, forming a "cliff" in the temperature profile (Fig. 4.3b). The magnitude of the heat island—defined as the difference between urban and rural temperatures—increases with city size, based on population.

For any city, the magnitude of the heat island fluctuates on a daily basis. Usually it is strongest in the evening because the hard surface materials continue to radiate heat long after sunset. This trend is shown in the temperature graphs in Fig. 4.5, which compare an inner-city location with a suburban location in Vienna, Austria, over a twenty-four-hour period in winter and summer. If this graph were extended to include a rural location, the difference would be even greater. Authors' Note 4.1 discusses a technique for measuring the urban heat island.

The Not-So-Windy City

Wind is a critical factor in the energy balance and the climate of cities because it is the principal disperser of dirty, heated air. Cities influence wind in two ways. First, and most important, they increase the roughness of landscape, thereby raising the roughness length. As wind moves from a rural area toward the inner city, its velocity profile (see Fig. 3.13) is displaced upward when it encounters taller, more closely spaced buildings. This is a response to the stronger frictional, or drag, resistance posed by the irregular terrain formed by the mass of buildings and other large urban structures. From his studies of urban climate in the 1950s and the 1960s, Helmut E. Landsberg of the University of Maryland found that average wind speed was 20 to 30 percent lower and that extreme gusts were 10 to 20 percent slower over cities than over the nearby countryside. Consequently, the mixing and removal of the surface air is less efficient in cities than in rural landscapes. In other words, surface air tends to linger over urban areas, allowing more time for it to be altered by surface heat and human activities (Fig. 4.6).

Convection related to the urban heat island is the second influence on wind, but it appears to have little influence on the urban energy balance. Measurements and observations have shown that during periods of calm atmosphere, a gentle inflow of surface air from the suburbs to the inner city may develop. It is especially likely to occur if the calm is associated with a strong heat island over the city. The heat island produces a pocket of somewhat unstable low-pressure air, which generates internal upflow and lateral inflow along the surface. Some studies conducted over New York City have recorded not only a marked upward flow over heavily built-up Manhattan Island, but also a down-

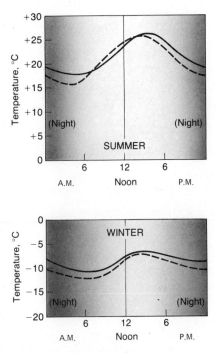

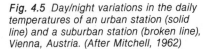

Fig. 4.5 Day/night variations in the daily temperatures of an urban station (solid line) and a suburban station (broken line), Vienna, Austria. (After Mitchell, 1962)

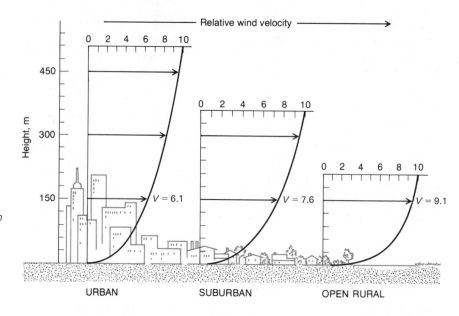

Fig. 4.6 Wind-velocity profiles over urban and rural surfaces. Note that the rate of velocity increase over the city is much lower than it is over the suburban and rural surfaces. (Adapted in part from Davenport, 1965)

ward flow over the nearby Hudson and East rivers. This circulation pattern is consistent with the temperature and pressure regimes which would be expected for surfaces with energy characteristics as different as these two.

The influence of convectional circulation generated by the urban heat island on the distribution of pollutants, diffusion of the dust dome, and dissipation of the heat island itself has not been well documented. Some researchers claim that the upflow from the heat island over large cities is strong enough to help break up the dust dome. Relative to the influence of regional winds in this capacity, airflow brought on by the heat island would appear to be very slight indeed. Does the heat island instability induce precipitation? Again, the evidence is inconclusive. In some instances it appears that it may promote greater convectional rainfall in and around the urban region, but this is often masked by the broader regional weather patterns and events. A recent study in the Detroit metropolitan region, however, reveals that precipitation in some urbanized areas is significantly greater than in the outlying suburban and rural areas (Fig. 4.7). With respect to cloud cover and fog,

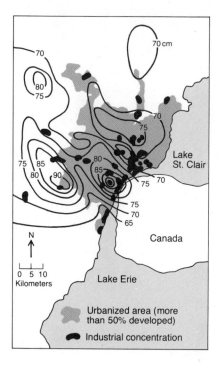

Fig. 4.7 Average annual precipitation (cm) in the Detroit metropolitan region, based on thirteen years of records from fifty stations. (Adapted from data and maps by the Southeast Michigan Council of Governments, Detroit, Michigan, 1976)

Table 4.1 Energy contributions to urban climate from artificial sources.

CITY	ENERGY FROM ARTIFICIAL SOURCES AS A PERCENTAGE OF TOTAL ANNUAL SOLAR RADIATION	SOURCE
Berlin, Germany	33	Kratzer, 1956
Vienna, Austria	17–25	Kratzer, 1956
Sheffield, England	33	Garnett and Bach, 1965
Manhattan (New York City)	33 (approx)	Bornstein, 1968

the data leave no doubt that both are more frequent in most urban areas, especially large ones. Although this may be related to heat island circulation, it is probably related more directly to the abundance of minute airborne particles, which serve as condensation nuclei for moisture droplets. The concentration of condensation nuclei over medium-size cities is typically five to seven times greater than that over rural areas.

Heat from Artificial Sources

Finally, we must consider the influence on the urban energy balance of heat generated by artificial means, namely, internal combustion within automobiles, factories, homes, and institutional establishments. The heat from these sources is expelled into the air around them, and if wind is absent or light, this air can be heated rather quickly. The greatest influence from artificial sources is, as one would expect, most pronounced in northern cities during winter. Calculations of the actual quantities of energy involved have been made for various cities. The examples in Table 4.1 express these types of energy contributions as a percentage of the total annual energy received as solar radiation. In the case of Manhattan in New York City, it was also reported that in winter, heat from combustion was 2.5 times (250 percent) greater than that of the solar energy reaching ground level. In summer, on the other hand, this factor dropped to one-sixth (17 percent).

Summary

Modern urbanization has a pronounced influence on most components of the energy balance. Generally, the larger the city and the denser its development, the greater the magnitude of change. Many factors contribute to this change, and much remains to be learned about them.

Table 4.2 Summary of changes in the components of the urban energy balance.

COMPONENT	CHANGE	CAUSE
Incoming shortwave radiation	Decrease	Dust dome backscattering
Outgoing longwave radiation	Decrease	Dust dome greenhouse effect
Sensible heat	Increase	Low thermal capacities of hard surface materials; heat from artificial sources
Latent heat	Decrease	Hard surfaces over soil; sparse plant cover*
Mechanical turbulence (ground level)	Decrease	Large buildings and other structures
Convectional turbulence	Increase	Heat island

*Cities in arid climates may be an exception with regard to plant cover. Here shrubs, shade trees, and lawns are maintained, with the aid of irrigation, in densities greater than those present under natural conditions.

Table 4.2 provides a qualitative summary of energy-balance changes and the related urban cause(s).

These changes in the energy balance have produced a measurable climatic change in most areas which have undergone urbanization. "Climate" is defined as a generalization of the weather conditions of an area. For large urban areas, the magnitude of the change in weather conditions, when compared to general weather conditions of the nearby countryside, is sufficiently strong to warrant the special designation "urban climate" or "climate of cities." Table 4.3 is a summary of the climatic changes associated with urbanization. The data were compiled by several scientists, based on studies conducted mainly in North America and Europe. The figure given for each climatic element represents the general variation in that element between the city and its neighboring countryside.

URBANIZATION AND THE GLOBAL ENERGY BALANCE

Finally, it is necessary to underscore that much remains to be learned about the influence of urbanization on energy balance and climate. Virtually all findings to date are based on research in midlatitude European, North American, and Japanese cities, which are characterized by industrialization, large buildings, an abundance of hard surfaces and automobiles, and a high rate of heat output from artificial sources. In contrast, cities in nonindustrialized Asia and Africa are

Table 4.3 Climate changes identified with urbanization.

ELEMENT	COMPARISON WITH RURAL ENVIRONS
Temperature	
Annual mean	.55–.83°C higher
Winter minima	1.1–1.7°C higher
Relative humidity	
Annual mean	6% lower
Winter	2% lower
Summer	8% lower
Dust particles	10 times more
Cloudiness	
Clouds	5–10% more
Fog, winter	100% more
Fog, summer	30% more
Radiation	
Total on horizontal surface	15–20% less
Ultraviolet, winter	30% less
Ultraviolet, summer	5% less
Wind speed	
Annual mean	20–30% lower
Extreme gusts	10–20% lower
Calms	5–20% more
Precipitation	
Amounts	5–10% more
Days with < 0.5 cm	10% more

Based mainly on Helmut E. Landsberg, "City Air—Better or Worse," in *Symposium: Air Over Cities*, Cincinnati, Ohio, U.S. Public Health Service, Taft Sanitary Engineering Center Technical Report A62–5, 1962.

less built up, have fewer automobiles and factories, generate less artificial heat, have less area covered by hard surface materials, and are generally smaller in size. Whether the influence of these cities on climate is correspondingly less than that of Western cities is not known. But one would tend to think so.

Particulates versus Carbon Dioxide

What is the influence of urbanization on the climate over large geographical regions? Unfortunately, the answer to this question is largely speculative at the present time. Some scientists, such as Basil Mason of the British Meteorological Office, argue that humans have had little influence on climate. Other scientists, however, have identified what may be some significant trends. For instance, climatologist Reid Bryson of the University of Wisconsin has been able to show data suggesting a cooling trend in global climate of about one-half degree C over the past forty years. Bryson argues that the global energy balance has become slightly negative since 1940 as a result of a decrease in incoming radiation. He thinks that this can be attributed mainly to the massive production and circumglobal dispersal of airborne particulates from urban areas and world population in general. As with the dust dome over a city, these particulates increase backscattering of incoming shortwave radiation, reducing the total intensity of incoming radiation in the lower atmosphere and at the earth's surface. However, the picture is complicated by the large increase in atmospheric carbon dioxide that has occurred in the past 100 years. This gas has increased 11–12 percent since about 1850, and were it not for the increase in particulates, it seems almost certain that an increase of this magnitude would have produced a slight rise in global temperature. Could it be that the cooling effect of particulates and the heating effect of carbon dioxide are offsetting each other? Maybe so, but it seems improbable that such a balance can last for long. In fact, one team of scientists recently identified a heating trend in the Southern Hemisphere which they attribute to the increase in carbon dioxide. Although the picture is further complicated by widespread changes in albedo, surface thermal properties, and artificial heat production, the evidence indicates that humans *can* measurably alter regional climate, and this is sufficient to warrant serious concern over future land use and population trends. Worldwide climate change is examined further in Chapter 9.

UNIT I SUMMARY

- The principal components in the energy balance of the earth's surface and the lower atmosphere are incoming solar radiation and outgoing earth radiation. For the earth as a whole or for any geographic parcel of it, these components must be equal over any period of time in order to maintain an energy balance.

- Solar radiation is relatively shortwave, concentrated between 0.15 and 3.0 micrometers, whereas earth radiation is relatively longwave, concentrated between 3 and 20 micrometers. Earth radiation is generated from land and water surfaces heated by the absorption of solar radiation. Longwave radiation is readily absorbed by water vapor and carbon dioxide in the atmosphere. Thus the atmosphere serves to detain outgoing earth radiation, and in this regard the atmosphere can be likened to a greenhouse.

- About half of the solar radiation entering the atmosphere is absorbed by the earth's surface on a worldwide basis. This energy heats surface materials, and they in turn produce outflows of energy in the form of longwave radiation, sensible heat, latent heat, and soil heat.

- Based on the conservation-of-energy principle, energy can be neither created nor destroyed; therefore, energy received by a surface must be accounted for in outflows of radiation, heat, or storage on the surface.

- Appreciable variations in the energy balance are detectable in the landscape, even over small areas. Many factors contribute to such variations: cloud cover, sun angle, slope exposure, and albedo influence radiation receipt; surface materials influence heat flux into the air and soil in as much as they control conductivity; and wind speed and turbulence are important regulators of heat loss to the atmosphere.

- The influence of cities on the energy balance is pronounced in middle latitudes. Radiation receipt is lower in urban areas because of increased backscattering from the dirty atmosphere. Air temperatures, however, tend to be higher because of the thermal properties of artificial surface materials, lower rates of latent-heat flux, heat production from artificial sources, and lower rates of wind flushing.

- The climate of cities is warmer and less illuminated by solar radiation than the climate of the countryside. In addition, cities tend to have more fog, but whether the precipitation regime is different for cities is uncertain. The trends toward increased urban population and world population in general indicate that urban climate will probably intensify and spread over larger areas in the future.

FURTHER READING

Bach, Wilfred, *Atmospheric Pollution*, New York: McGraw-Hill, 1972. *A discussion of various types and effects of modern air pollution.*

Geiger, Rudolph, *Climate Near the Ground*, 4th ed., Cambridge, Mass.: Harvard University Press, 1965, 611 pages. *First published in 1927, this is the classic on climate at ground level.*

Landsberg, Helmut E., "The Climate of Towns," in *Man's Role in Changing the Face of the Earth*, Chicago: University of Chicago Press, 1956, pp. 584–606. *Article written before urban climate was popular research topic; appears in volume important to practically all fields of geography.*

Mather, John R., *Climatology Fundamentals and Applications*, New York: McGraw-Hill, 1974, 412 pages. *Application of climatology, including the energy balance, to soils, water, agriculture, architecture, industry, and related topics.*

Neiburger, Morris, James G. Edinger, and William D. Bonner, *Understanding Our Atmospheric Environment*, San Francisco: W. H. Freeman, 1973, 293 pages. *A good description of the basic principles and process of weather and climate.*

Oke, T. R., *Boundary Layer Climates*, London: Methuen, New York: Halsted Press, 1978, 372 pages. *An excellent compilation of modern findings on the nature of the atmosphere near the ground; a must for the physical geographer's library.*

Sellers, William D., *Physical Climatology*, Chicago: University of Chicago Press, 1965. *A comprehensive description of energy-balance principles and methods of calculating components.*

REFERENCES AND BIBLIOGRAPHY

Aronin, J., *Climate and Agriculture*, New York: Reinhold, 1943.

Baumgartner, A. "Das Eindringen des Lichtes in den Boden," *Forstwiss. Zentr.* **72** (1956): 172–184.

Black, R. F., "Permafrost—A Review," *Bulletin of the Geological Society of America* **65** (1954): 839–855.

Bornstein, R. D., "Observations of the Urban Heat Island Effect in New York City," *Journal of Applied Meteorology* **7** (1968): 575.

Brown, R. J. E., "Comparison of Permafrost Conditions in Canada and the U.S.S.R.," *Polar Record* **13** (1967): 741–751.

Buffo, J., L. J. Fritschen, and J. L. Murphy, "Direct Solar Radiation on Various Slopes From 0 to 60 Degrees North Latitude," U.S.D.A. Forest Service Research Paper, PNW-142, 1972.

Cooke, R., and J. C. Dornkamp, *Geomorphology in Environmental Management: An Introduction*, Oxford: Clarendon Press, Oxford University Press, 1974.

Corte, A. E., "Geocryology and Engineering," *Reviews in Engineering Geology* **2** (1969): 119–185.

Davenport, A. G., "The Relationship of Wind Structure to Wind Loading," National Physical Laboratory, Symposium No. 16, Wind Effects on Buildings and Structures, London: Her Majesty's Stationery Office, 1965.

Deacon, E. L., "Vertical Profiles of Mean Wind in the Surface Layers of the Atmosphere," *Geophysical Mem. No. 91*, London: Meteorol. Office, Air Ministry, 1953.

Fitch, J. M., with D. P. Branch, "Primitive Architecture," *Scientific American* **203** (1960): 133–144.

Garnett, A., and W. Bach, "An Estimation of the Ratio of Artificial Heat Generation to Natural Radiation Heat in Sheffield," *Monthly Weather Review* **93** (1965): 383.

Givoni, B., *Man, Climate and Architecture*, London: Elsevier Architectural Sciences Series (Elsevier Publishing Company), 1969.

Kepner, R. A., L. M. K. Boelter, and F. A. Brooks, "Nocturnal Wind Velocity, Eddy Stability and Eddy Diffusion Above a Citrus Orchard," *Transactions of the American Geophysical Union* **23** (1942): 237–249.

Kratzer, P., *Das Stadtklima*, Braumschweig: Frederich Vieweg and Sohn, 1956.

Landsberg, H. E., *Weather and Health, An Introduction to Biometeorology*, Anchor Science Study Series (Doubleday), 1969.

Mackay, J. R., "The World of Underground Ice," *Annals of the Association of American Geographers* **62** (1972): 1–22.

Mitchell, J. M., Jr., "The Thermal Climate of Cities," in *Symposium: Air Over Cities*, Cincinnati, Ohio: U.S. Public Health Service, Taft Sanitary Engineering Center, Technical Report A62-5, 1962.

Monin, A. S., and A. M. Obukhov, "Principal Law of Turbulent Mixing in the Air Layer near the Ground," *USSR Akad. Nauk. Geophys. Inst.* (1954).

Nicholas, F. W., "The Changing Form of the Urban Heat Island of Metropolitan Washington," *Tech. Paper*, American Congress of Surveying and Mapping, Annual Meeting, March 7–12, Washington, D.C., 1971.

Nunez, M., and T. R. Oke, "The Energy Balance of an Urban Canyon," *Journal of Applied Meteorology* **16**, 1 (1977): 11–19.

Oke, T. R., "City Size and the Urban Heat Island," *Atmospheric Environment*, vol. 7, New York: Pergamon Press, 1973, 769–779.

_____,"Inadvertent Modification of the City Atmosphere and the Prospects for Planned Urban Climates," *Proceedings Symposium on Meteorology Related to Urban and Regional Land-Use Planning*, Asheville, N.C.: World Meteorological Organization, Geneva, 1976.

_____, *Boundary Layer Climates*, London: Methuen, New York: Halsted, 1978.

Olgyay, V., *Design with Climate: Bioclimatic Approach to Architectural Regionalism*, Princeton, N.J.: Princeton University Press, 1973.

Peterson, J. T., *The Climate of Cities: A Survey of Recent Literature*, Raleigh, N.C.: U.S. Department of Health, Education and Welfare, 1969.

Vehrencamp, J. E. "Experimental Investigation of Heat Transfer at an Air-Earth Interface," *Transactions of the American Geophysical Union* **34** (1953): 22–30.

Woodward, B., *Cumulus Dynamics*, New York, Pergamon Press, 1960.

Woollum, C. A., and N. Canfield, "Washington Metropolitan Area Precipitation and Temperature Patterns," *U.S. Weather Bureau Tech. Memo.*, 1968.

Wright, J. L., and E. R. Lemon, "Estimation of Turbulent Exchange Within a Corn Crop Canopy at Ellis Hollow," *Interim Report 62-7*, Ithaca, New York State College of Agriculture, 1962.

Yamamoto, G., and A. Shimanuki, "Profiles of Wind and Temperature in the Lowest 250 Meters in Tokyo," *Geophysics*, Tohoku Univ., Science Report Series 5: 15 (1964): 111–114.

CLIMATIC PROCESSES AND REGIONAL CLIMATE

UNIT II

KEY CONCEPTS   OVERVIEW

In Unit I we examined the components of the energy balance at an arbitrary location on the earth's surface, limiting our discussion to small parcels of space and the flows of energy to and from it. In this unit we look at the spatial component of the energy balance and examine the variations in heat, light, and related forms of energy over the planet. This variation is at the heart of the geographer's interest in the energy balance, because it is the basis for understanding the world's climates and many other geographical characteristics of the planet, such as vegetation, soils, and the availability of water.

When we look at the earth as a whole, the components of the energy balance—net radiation, surface temperature, rates of sensible- and latent-heat exchange, and soil-heat flow—vary with both location on the earth and season at a given location. There are two major reasons for this variation.

The first is related to differences in sun angle. Due to differences in sun angle with latitude and to sun-angle variations with season, shortwave radiation coming into the atmosphere is greatest, on an annual basis, at the equator and decreases toward the poles. In addition, for any location, shortwave radiation is greater during summer.

The second is related to the thermal differences between land and water. The greater heat capacity and thermal conductivity of water, when compared to land, causes the surface temperatures of a water surface to vary less than those of a land surface.

These variations in surface temperature, on both a global and a regional scale, cause air to move horizontally and vertically within the atmosphere. This moving air carries with it both heat and water vapor and thus transfers energy from one location to another. But because this air circulates over a rotating earth and within a rotating atmosphere, deviations, called the *Coriolis effect*, occur in the circulation pattern.

Winds not only transfer heat within the atmosphere, but also cause the oceans to circulate. As winds blow over open water, they tend to drag the surface water with them, setting up currents. Thus there are two principal ways in which energy is transferred around the globe: (1) the movement of air containing heat and moisture, resulting in sensible-heat exchange when warm air moves to cold regions or cold air moves to warm regions, as well as latent-heat exchange when evaporation takes place in one location and the water vapor condenses out at another; and (2) the transfer of heat by ocean currents.

In this unit we shall examine in detail these transfers of energy over the earth's surface. First, we shall consider the size and shape of the earth (Chapter 5), then the seasonal and latitudinal variations in incoming shortwave radiation, the basic energy source which drives the atmospheric and oceanic circulation (Chapter 6). Next, we shall describe and explain the circulation of the atmosphere and oceans (Chapter 7). Following this is a discussion of precipitation processes (Chapter 8). Chapter 9 covers regional climates and climatic change.

THE EARTH'S SIZE, SHAPE, AND MOTION

CHAPTER 5

INTRODUCTION

In Unit I we examined the flow of energy to and from a small area on the earth's surface. In this unit we examine the geographic distribution of energy on the earth as the basis for understanding the climates of the world. But before we can properly describe the distribution of energy on the earth and relate this energy to climate, it is necessary to review some concepts about the size, shape, and motion of the earth.

THE SIZE AND SHAPE OF THE EARTH

Roughly spherical in shape, the earth spins on its axis while traveling through space around the sun. The climatic effect of this rotation and orbit is covered in Chapter 6; here we consider only those effects on shape and location. The centrifugal effect of the earth's rotation causes a slight equatorial bulge, which represents the principal departure of the earth's shape from the true sphere. The term we give to the true shape of the earth is geoid, and the science of geodesy is devoted to the determination of the nature of the geoid.

The two ends of the earth's axis give us two points on the earth's surface which are located in a nonarbitrary way: the North Pole and South Pole. We can define the center of the earth as the midpoint of the axis, half way between the two poles. An imaginary plane, constructed perpendicular to the earth's axis at the center, defines the equator, where it intersects the surface. The equator is thus midway between the poles.

The shape of the equator is very close to a true circle, with a radius of 6378 km (3963 miles). However, an imaginary plane perpendicular to the equator and passing through the center of the earth and the poles defines an approximate ellipse, not a circle, where it intersects the surface (Fig. 5.1). That is, the shape of the earth's cross-section through the poles is elliptical. The polar radius of the earth —the distance from the center to either pole—is 6357 km (3950 miles). As a first approximation, then, the earth has the shape of an ellipsoid. In the construction of large-scale maps, map makers must take this departure from the spherical shape into account.

78

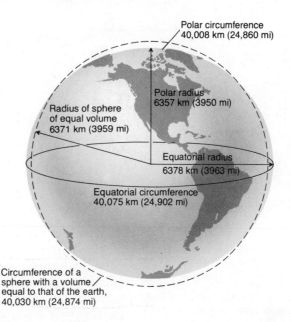

Fig. 5.1 Important dimensions of the earth.

For climatic considerations, however, a further approximation may be used. We will consider the earth as a true sphere in the chapters that follow. We choose as the size a sphere having a volume equal to that of the true volume of the earth ($1.083 \times 10^{12}\,km^3$). The radius of this sphere is 6371 km (3959 miles), and its relations with the equatorial and polar radii are shown in Fig. 5.1. The surface area of the earth is 510 million km^2 (197 million mi^2).

LATITUDE AND LONGITUDE

We measure locations on the earth with the aid of a grid network consisting of lines running east-west and north-south. This global coordinate system is constructed according to the principle of angular rotation around a circle. Let us briefly describe the system by showing how the lines are drawn (Fig. 5.2).

To construct the east-west lines, which are called *parallels*, the first step is to bisect the globe from pole to pole. Next, a protractor is placed on the plane of bisection, with the base of the protractor aligned with the equator. Starting at the equator, angles northward to the pole are ticked off; the procedure is repeated toward the South Pole (Fig. 5.2a). The angles are then numbered, beginning with 0 degrees at the equator and ending with 90 degrees at each of the poles. Finally, the parallels themselves are drawn by rotating each angle entirely around

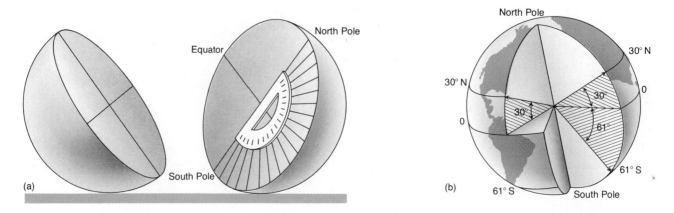

Fig. 5.2 Constructing lines of latitude and longitude: (a) If the earth is bisected along the polar axis, angles can be measured northward and southward from the equator with the aid of a protractor. (b) The latitude of any location represents an angle constructed through the equator, the center of the earth, and the location on the earth's surface.

latitude ⟹ east west
longtitude ⟹ north south.

the earth so that the ray of the angle inscribes a line into the surface of the globe. Thus the *latitude* of any location on the earth's surface represents an angle between the equatorial plane and a line drawn from the center of the earth to the location (Fig. 5.2b).

Each parallel encircles part or all of the earth, running parallel to the equator. Only one parallel, the equator, traces the full circumference of the earth. Therefore, it qualifies as a *great circle*, defined as the perimeter of any plane which passes through the center of the earth. All of the other parallels are *small circles*, because the planes they define do not pass through the center of the earth, and hence their perimeters represent less than the earth's full circumference.

The north-south lines, called *meridians,* are constructed in the same fashion as are the parallels (Fig. 5.3). The earth is bisected along the equatorial plane, and angles are measured around the perimeter. However, there is a problem as to where to start the system because there is no convenient point at which to place zero. International agreement has specified an arbitrary starting point that coincides with the Royal Observatory at Greenwich, England (Fig. 5.3a). A north-south line drawn through this point to the North Pole and the South Pole is the *Greenwich* (or prime) *Meridian*, and it is labeled 0 degrees longitude. From the Greenwich Meridian all meridians westward to 180 degrees are designated west longitude, and those eastward to 180 degrees are designated east longitude.

As half of a great circle, every meridian connects the North and South poles. Since the 0° meridian does not completely encircle the earth, but only half of it, longitudes can vary from 0° to 180° east and west. The line at 180° E coincides with the one at 180° W, both being half way around the globe, in opposite directions, from the Greenwich Meridian; the direction indication for 180° is thus omitted (Fig. 5.3c).

Thus locations for both longitude and latitude are always given as portions of a circle and are usually measured in degrees, minutes, and

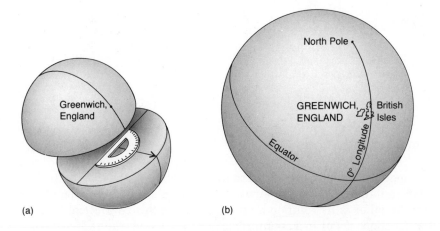

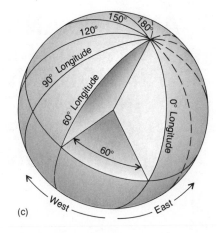

Fig. 5.3 *Constructing meridians: (a) Meridians are constructed on the perimeter of the equatorial plane, with the protractor fixed on the earth's center. (b) Since there is no geometrically convenient place to begin, the 0 meridian is fixed on the Royal Observatory, Greenwich, England, and is called the Greenwich Meridian, or the prime meridian. (c) Every meridian is half of a great circle. Longitude is given as degrees east or west of the Greenwich Meridian, which is 0° longitude.*

seconds (symbolized °, ′, ″). There are 360 degrees in a complete circle; each degree is divided into 60 minutes; and each minute is divided into 60 seconds. In this book our considerations of angles are generally not precise enough to justify their measurement to the second, so we will usually use just degrees and minutes.

The distance represented on a sphere by a degree of latitude is always the same. On the earth this distance is 111 km (69 miles), although if we consider the true, ellipsoidal shape of the earth, there is some variation, about 1 km between the equator and the poles. All meridians, on the other hand, converge at the poles; so the *length* (distance) represented by a degree of longitude varies from 111 km at the equator, to 96 km at 30° latitude, to 56 km at 60° latitude, to 0 km at 90° latitude.

SUMMARY

The earth's shape deviates slightly from that of a true sphere, but for most geographical problems it can be treated as a sphere. Locations on earth are measured according to the global coordinate system, which consists of a network of intersecting lines called parallels and meridians. Parallels are east-west–running lines which parallel the equator. Except for the equator itself, all parallels encircle less than a full circumference of the earth and therefore qualify as small circles. The equator qualifies as a great circle because it encircles the full circumference of the earth, marking the perimeter of a plane that passes through the center of the earth. Meridians are north-south lines which run from pole to pole. Each meridian forms half of a great circle. Locations on earth are measured in degrees, minutes, and seconds longitude (east or west of the Greenwich Meridian) and latitude (north or south of the equator).

ANNUAL AND LATITUDINAL VARIATIONS IN INCOMING SHORTWAVE RADIATION

CHAPTER 6

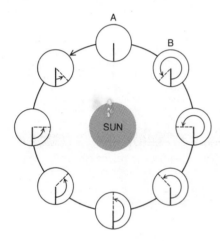

Fig. 6.1 The earth's revolution around the sun and rotation on its axis. At point A the heavy radial line represents a point directly "beneath" the sun. As the earth moves in its orbit to point B, one true rotation takes place, but because of the travel in the orbit, point B is not yet beneath the sun. To get beneath the sun, an additional rotation, to the position indicated by the dotted line, is necessary. As the earth makes a complete revolution in its orbit, the number of apparent rotations (with respect to the sun) is one less than the number of true rotations. The average apparent rotation period is called the solar day and is equal to 24 hours; the true rotation period is called the sidereal day and is equal to 23 hours, 56 minutes, 4.09 seconds. (Adapted from Horace R. Byers, General Meteorology, 4th ed., New York: McGraw-Hill, 1974. Used by permission.)

EARTH-SUN RELATIONS

Two motions of the earth in space are important in the distribution of solar radiation. One is the earth's revolution in an orbit around the sun; the other is its rotation on its axis during this revolution. Both of these motions occur in a counterclockwise direction when viewed from the North Pole. That is, a point on the surface of the earth rotates from west to east, and the path of the earth in its orbit is in the same direction.

The period of revolution is called the *year* and is equal to 365.242 solar days. A *solar day* is the apparent rotation period with respect to the sun. Because the earth's orbit is slightly elliptical, rather than circular, the length of the solar day is not constant, but averages twenty-four hours. This apparent rotation speed is slightly slower (that is, slightly longer rotation period) than the true rotation speed. The reason for this is shown in Fig. 6.1.

As the earth moves in its elliptical orbit, its distance from the sun varies (Fig. 6.2). The mean distance from the earth to the sun is 149.6 million kilometers (93 million miles). *Perihelion*, the closest distance, is 147 million kilometers (91.5 million miles) and occurs on January 3. *Aphelion* (the farthest distance) is 152 million kilometers (94.5 million miles) and occurs on July 5.

THE SEASONS

Figure 6.2 shows that the closest position of the earth to the sun does not coincide with the Northern Hemisphere's summer. Rather, winter in the Northern Hemisphere coincides with perihelion; summer, with aphelion. In fact, the variation in distance between the earth and sun, due to the elliptical nature of the earth's orbit, has only a minor effect (about ± 3.5 percent) on the seasonal variations in the receipt of solar energy. The cause of the seasons is entirely different.

The true explanation of the seasons begins with Fig. 6.3. As the earth revolves around the sun, its axis is not perpendicular to the plane of revolution, called *plane of the ecliptic*, but is instead inclined at an angle of 66 degrees 33 minutes, or 23°27′, off the vertical. Since the orientation of the axis remains constant throughout revolution, during one portion of the orbit the Southern Hemisphere is pointed toward the sun, and during another portion the Northern Hemisphere is pointed toward the sun. When one hemisphere is pointed toward the sun, the sun angle in that hemisphere is greater (more toward vertical) than it is at comparable latitudes in the other hemisphere. In an earlier chapter we explained how the angle and orientation of a slope can affect the receipt of solar energy at a local scale (see Fig. 2.5), and now we see a similar effect on a global scale.

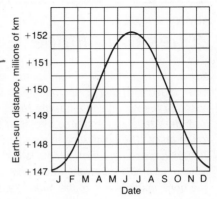

Fig. 6.2 Distance between the earth and the sun, in millions of kilometers. (Data from U.S. Nautical Almanac, *1976*)

SUN ANGLE

Sun Declination, Solstices, and Equinoxes

Let us examine in detail seasonal and latitudinal variations in sun angle. At any moment in the earth's revolution about the sun, there is always one, and only one, latitude at which the noon sun angle is vertical to the earth's surface, i.e., directly overhead to someone standing there. This latitude is called the *declination* of the sun. In the course of one complete revolution of the earth, the declination of the sun migrates back and forth across the tropics as the attitude of the earth's axis changes with respect to the sun. Its northernmost position is 23°27′ N (the Tropic of Cancer) on June 22; in the Northern Hemisphere this date is called the *summer solstice*. The southernmost position is 23°27′ S (the Tropic of Capricorn) on December 22, the *winter solstice* for the Northern Hemisphere. At two dates midway between the sol-

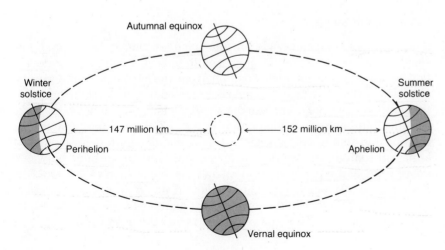

Fig. 6.3 The revolution of the earth and the seasons. Note that the angle of inclination of the earth's axis is the same in all seasons. (After Byers, 1974)

AUTHORS' NOTE 6.1
Zones of Latitude

A note on the definition of the various zones of latitude may be useful at this point. The high latitudes are comprised of the Arctic and Antarctic zones, which extend from the Arctic and Antarctic circles at 66°27' N and S to the poles. The polar zone is the upper high latitudes, 75–90° latitude. The term middle latitude refers to the zones intermediate between the poles and the equator. The north-south extent of the middle latitudes is somewhat arbitrary, but 35–55° is usually given for it. The subarctic zone lies between 55° and the Arctic and Antarctic circles; only in the Northern Hemisphere do large land masses lie in this zone.

Used correctly, the term tropics refers to the Tropic of Capricorn and the Tropic of Cancer. It follows that the zone between 23.5° south latitude and 23.5° north latitude should be the intertropical zone, and indeed many scientists do follow this convention. However, the term intertropical has declined in usage, and today tropics or tropical zone seems to be the preferred term for this zone. The equatorial zone is the middle belt of the earth, extending 10° latitude or so north and south of the equator. The subtropical zone lies between 23.5° and 35° latitude.

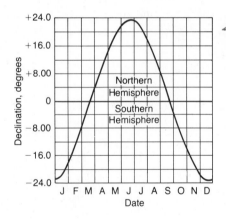

Fig. 6.4 Declination of the sun, in degrees. *(Data from* U.S. Nautical Almanac, *1976)*

stices, the sun is vertical at the equator. These dates are the *equinoxes*, the *vernal equinox* occurring on March 21 and the *autumnal equinox* on September 22 in the Northern Hemisphere. These four dates may vary by a day in either direction because our calendar has to be adjusted every four years to bring it back in phase with the period of the earth's revolution. Figure 6.4 shows graphically the declination of the sun for the entire year. The dates of the solstices and equinoxes are opposite for the Southern Hemisphere.

Calculation of Sun Angle

Calculation of noon sun angle for any given latitude and declination is simplified by considering the *solar zenith angle* (Z), the angular deviation of the sun's noon high position from the vertical. If the sun is directly overhead, the zenith angle is 0°; when the sun is at the horizon, the zenith angle is 90°. The *sun angle*, which is the angle between noon sun and the horizon, is thus equal to 90° − Z. Figure 6.5 depicts the relationship among zenith angle, latitude, and declination for a variety of possible situations: Zenith angle is equal to the difference between the latitude of the place in question and the declination of the sun. For the location 22° north latitude, the zenith angle is 30° on March 1 (equal to the 22° of latitude north of the equator plus the 8° of latitude of the sun south of the equator, as read from the graph in Fig. 6.4).

Global Variations in Sun Angle

Consider now sun-angle variation with season and latitude. At the equator, the zenith angle is never greater than 23°27′, because the declination of the sun is never farther away than the Tropic of Cancer (23°27′ N) and Tropic of Capricorn (23°27′ S). Thus the sun angle at the equator varies from 90° to 66°33′ over the year.

In the midlatitudes the variation in the sun angle is twice as great as at the equator. At 40° N, for example, the zenith angle is 16°33′ on June 22 and 63°27′ on December 22, a difference of 46°54′. At the Arctic Circle (66°27′ N), the zenith angle is 43°06′ on June 22 and 90° on December 22. Consequently, the noon sun on December 22 is on the horizon and the sun angle is 0°.

North of the Arctic Circle the noon sun is below the horizon on December 22, and at the North Pole itself, the sun angle is − 23°27′ on this date, or effectively 0°. The highest sun annual angle at the poles is + 23°27′, which occurs on June 22 in the north and December 22 in the south. The effective variation in sun angle at the poles—assuming that any negative angle is 0—is therefore 23°27′ annually. Thus in considering the world as a whole, the annual variation in sun angle is greatest for the midlatitudes, and, accordingly, we would expect the climates there to be more seasonal in character than those in the polar and equatorial zones.

Fig. 6.5 The relationship among zenith angle, latitude, and sun angle on June 22. Note that the sun is in the southern sky for the observer at 45°N, but in the northern sky for the observer at 45°S. (Illustration by William M. Marsh)

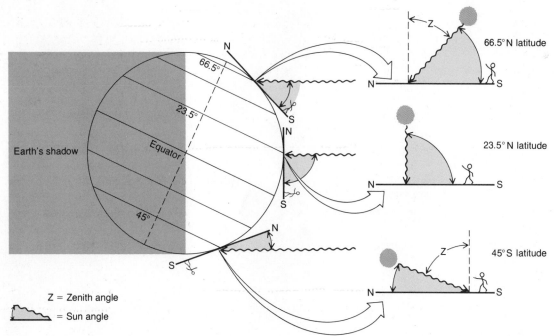

Z = Zenith angle

= Sun angle

VARIATIONS IN THE LENGTH OF THE DAY

In addition to variations in sun angle, variations in day length also are related to latitude and solar declination. Figure 6.6 shows the portions of the earth illuminated by the sun at three separate positions: summer

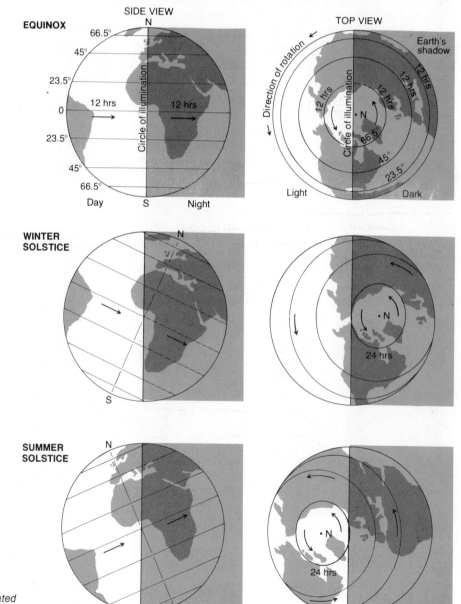

Fig. 6.6 Portions of the earth illuminated by the sun at the solstices and equinoxes.

solstice, winter solstice, and equinox. Note that the *circle of illumination*, which separates the lighted portion from the unlighted portion, is a great circle, which always divides the earth in half. The equator is also a great circle, and since two great circles always bisect each other, the equator is always half in the light and half in the dark, regardless of the date and declination. Thus any point on the equator will be twelve hours in light and twelve hours in darkness as the earth makes a complete rotation. But all of the other parallels of latitude are small circles. Therefore, except when the earth is in one of the equinox positions, the circle of illumination intersects these latitudes unequally, and a point on the surface of the earth will not spend equal times in light and darkness as the earth rotates. Furthermore, on June 22 all latitudes north of the Arctic Circle are in the sunlight for twenty-four hours, and all latitudes south of the Antarctic Circle are in darkness for twenty-four hours. The situation is reversed on December 22.

GLOBAL VARIATIONS IN SOLAR RADIATION

In the absence of cloud cover, the amount of incoming solar radiation possible at any given latitude on the earth's surface is a product of sun angle and duration of daylight. Figure 6.7 shows the global variations

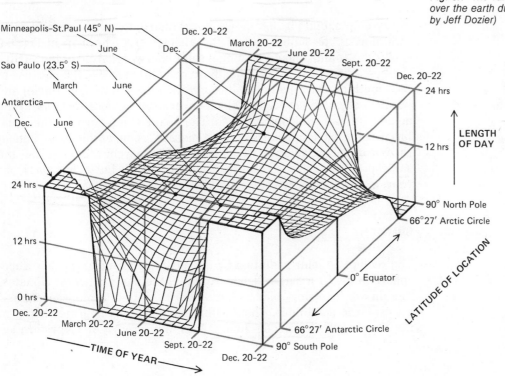

Fig. 6.7 The variation in daylight hours over the earth during a year. (Illustration by Jeff Dozier)

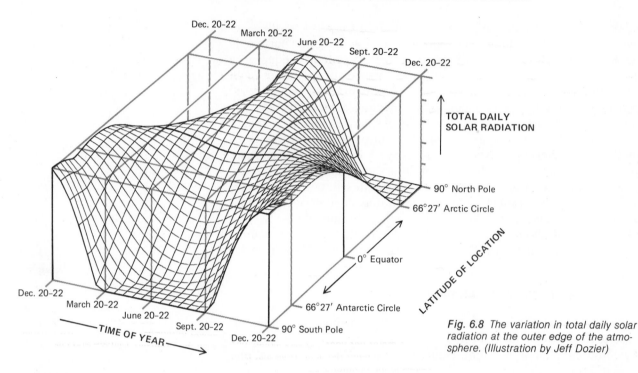

Fig. 6.8 *The variation in total daily solar radiation at the outer edge of the atmosphere. (Illustration by Jeff Dozier)*

in day length, and Fig. 6.8 shows global distributions of incoming solar radiation at the outer edge of the atmosphere. The map in Fig. 6.9 shows the actual amount of annual solar energy received at the earth's surface. The values differ from those calculated in Fig. 6.8 because of the length of the travel path through the atmosphere (sun angle) and because of cloud cover. The cloud-cover effect is most evident in the equatorial and subtropical zones. The heavy cloud cover of the equatorial zone, particularly over land masses, reduces the total annual receipt of solar radiation by as much as 50 percent. In contrast, the subtropical deserts, which are cloudless for all but a few days per year, experience very little reduction in solar radiation.

SUMMARY

The receipt of solar radiation by the earth's surface and lower atmosphere is the first major consideration in the study of world climates. The flux of incoming solar radiation varies with latitude, season, and atmospheric conditions. The declination of the sun migrates between the Tropic of Cancer on June 22 to the Tropic of Capricorn on

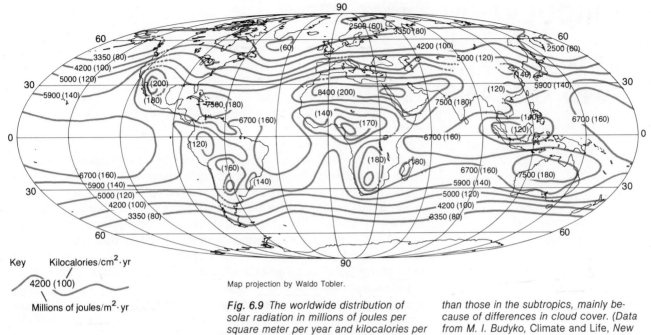

Key   Kilocalories/cm^2·yr

4200 (100)

Millions of joules/m^2·yr

Map projection by Waldo Tobler.

Fig. 6.9 The worldwide distribution of solar radiation in millions of joules per square meter per year and kilocalories per square centimeter per year. Note that the values in the equatorial zone are lower than those in the subtropics, mainly because of differences in cloud cover. (Data from M. I. Budyko, Climate and Life, New York: Academic Press, 1974. Used by permission.)

December 22 in response to the earth's annual revolution about the sun. As the declination changes, so change the sun angles and duration of daylight for every location of earth. Coupled with the influence of cloud cover, a global pattern of ground-level solar radiation emerges, in which the highest values occur not in the equatorial zone, but in the subtropical deserts.

CIRCULATION IN THE ATMOSPHERE AND OCEANS

CHAPTER 7

INTRODUCTION

In order to maintain an energy balance, the earth must release as much radiant energy as it receives. Recall that shortwave is the principal form of radiation coming into the atmosphere and that longwave is the principal form of radiation leaving the atmosphere. In Fig. 7.1, curve I represents the total radiation received by the earth and its atmosphere; curve II, the amount of longwave radiation and reflected shortwave radiation leaving the atmosphere. The graph shows that an area of radiation surplus exists at latitudes lower than about 38° in both hemispheres, whereas a radiation deficit exists poleward of these latitudes. If there were no energy exchanged between the zones of energy surplus and deficit, the tropics would grow hotter, ultimately reaching some equilibrium temperature much higher than those in the area today. The polar areas, on the other hand, would become colder and colder, eventually stabilizing at some lower temperature. But because of the oceans of air and water on the planet, energy exchange does take place between the equatorial and polar zones. The redistribution of energy by the atmosphere and the oceans is possible because these bodies are continuously moving or circulating over the planet. This chapter examines atmospheric circulation, and we begin with a brief examination of the physical properties of the atmosphere.

THE NATURE OF AIR AND SOME FACTS ABOUT THE ATMOSPHERE

Air Pressure

Air is a mixture of gases, and the atmosphere is composed of gas molecules held near the earth's surface by gravity. As in any fluid, the deeper one is in the atmosphere, the greater the pressure from the overlying fluid. For this reason, air pressure is greatest at sea level and decreases with altitude. When you gain altitude rapidly in a car or an airplane, the unpleasant feeling in your ears is caused by the pressure differential between the environment and the air in your eustachian tubes. This is relieved when your ears "pop." Similarly, when you dive to the bottom of a swimming pool, you can feel the increased pressure from the increased depth of the overlying water. The major differences between the atmosphere and the swimming pool are explained by two

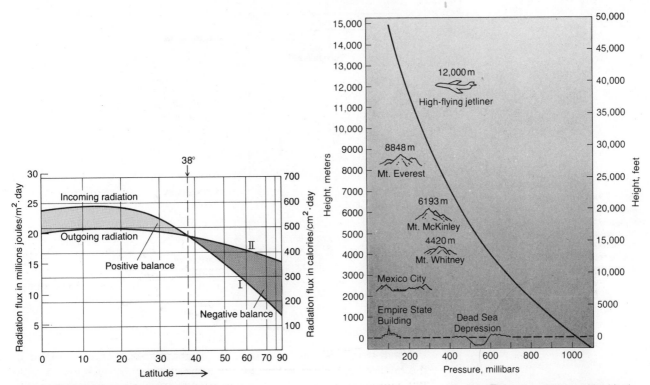

Fig. 7.1 Distribution with latitude of absorbed and outgoing radiation. Curve I shows the solar radiation absorbed by the earth and atmosphere; curve II, short- and longwave radiation leaving the atmosphere. The scaling of the latitude axis is representative of the areas on the earth's surface.

Fig. 7.2 Pressure variations with altitude in a "standard" atmosphere. In the polar areas, pressures for a given altitude are slightly lower than indicated by the graph; near the equator, they are slightly higher.

facts: (1) since water is so much denser than air, greater pressure changes in the pool take place over a relatively small depth; and (2) since air is compressible, the density is not constant throughout the atmosphere.

Air pressure is normally measured in millibars (abbreviated mb). Pressure is defined as a force per unit area. In the Système Internationale we measure this force in *newtons*; a newton is the force necessary to accelerate a 1-kg mass 1 m/sec². But since a newton per square meter is a very small pressure, in the physical sciences pressure is often measured in *bars*; a bar is 100,000 newtons per square meter. A millibar is 0.001 bar, or 100 newtons per square meter. Normal sea-level pressure is 1013.25 mb, or just a little more than a bar. At 5500 m (18,000 feet) pressure is only about 500 mb; thus half of the atmosphere is below this altitude. Figure 7.2 shows variations of pressure with altitude.

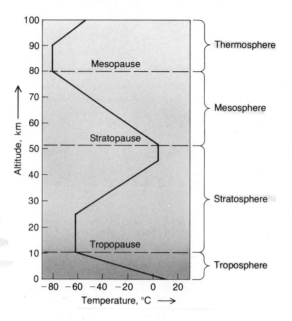

Fig. 7.3 Temperature profile of the atmosphere, with subdivisions. This change in temperature with altitude is considered to be typical of temperature conduction in the middle latitudes and is called the U.S. Standard Atmosphere. The boundaries between the subdivisions are termed pauses, *meaning change.*

Thermal Structure

Upward from the earth's surface, the atmosphere can be subdivided on the basis of the distribution of temperatures within it. A very simple division of the atmosphere is pictured in Fig. 7.3. The *troposphere* extends from the surface to an average altitude of about 12 km, ranging from 7 km over the poles to 17 km over the equator. It is characterized by temperatures decreasing with altitude at a rate of about 6.5°C per km and is the most dynamic part of the atmosphere. Above the troposphere is the *stratosphere*, which extends to an altitude of about 30 km. In the stratosphere temperatures hold roughly constant with altitude or may decrease or increase slightly. In contrast to the troposphere, little vertical mixing takes place in the stratosphere. At the top of the stratosphere, temperatures are almost as high as they are at the earth's surface. This is because of the absorption of ultraviolet radiation by the ozone which is concentrated there. The *mesosphere* extends from 30 km to about 90 km, and beyond it is the thermosphere.

Gaseous Composition

Air is a mixture of many gases, of which nitrogen and oxygen make up over 99 percent by volume (Table 7.1). Most of the remaining gases constitute a small fraction of 1 percent of the atmosphere and are referred to as minor constituents. Despite their low quantities, however, some of

Table 7.1 Composition of the atmosphere.

GAS	% BY VOLUME IN TROPOSPHERE	NOTES
Nitrogen (N_2)	78.084	
Oxygen (O_2)	20.946	Has developed with the evolution of plant life in past billion years.
Argon (A)	0.934	
Carbon dioxide (CO_2)	0.033	Only 0.029 in nineteenth century; absorbs longwave radiation in the 1–5 and 12–14 micrometer range.
Neon (Ne)	0.00182	
Helium (He)	0.000524	
Methane (CH_4)	0.00016	
Krypton (Kr)	0.00014	
Hydrogen (H_2)	0.00005	
Nitrous oxide (N_2O)	0.000035	Absorbs radiation above one micrometer.
Important Variable Gases		
Water vapor (H_2O)	0–4	Absorbs radiation in the 0.85–6.5 micrometer range and the range longer than 18 micrometers.
Ozone (O_3)	0–.000007	Absorbs ultraviolet radiation in upper atmosphere.
	at ground level (0.00001 –0.00002 in stratosphere and mesophere)	

the minor constituents are nevertheless important. In the mesophere, ozone strongly absorbs ultraviolet radiation in the 0.1–0.3 micrometer range, shielding us from potentially lethal exposures of radiation. Carbon dioxide is a good absorber of the longwave radiation emitted from the earth's surface in the 12–14 micrometer range and, along with water vapor, performs the important function of slowing the outflow of energy from the atmosphere. Water vapor, which varies from 0 to 4 percent in the lower troposphere, absorbs radiation in wavelengths ranging from 0.85 to 6.5 micrometers and above 18 micrometers. The effect of water vapor on the behavior of air as a gas will be discussed in a later section.

ATMOSPHERIC CIRCULATION

Hypothetical Circulation on a Uniform, Nonrotating Earth

If the planet did not rotate and were uniformly covered with one material, say, barren rock, the atmosphere would circulate in response to the latitudinal differences in the heat generated from incoming solar radiation. The overall circulation pattern should be very simple compared to that of a rotating planet. Air would be heated most intensively in the equatorial regions, where it would become unstable and would rise. A low-pressure area would thus develop along the equator. In contrast, high pressure would develop over the poles, where heating would be weakest. Between the equator and the poles there would be a pressure gradient, and a gigantic convectional system would form, in which the equatorial low would be fed with air from the polar highs. At the surface, air would flow toward the equator; aloft, it would flow toward the poles. That is, in the Northern Hemisphere the winds at the surface would blow from the north. This circulation is shown in Fig. 7.4.

Rotation: The Coriolis Effect

Let us examine the consequences of rotation by first considering a much simpler system—a flat, rotating disk. Authors' Note 7.1 describes a dart game played on such a disk, demonstrating that when the direc-

Fig. 7.4 Atmospheric-circulation pattern that would develop on a nonrotating planet. The equatorial belt would heat intensively and would produce low pressure, which would in turn set into motion a gigantic convection system. Each side of the system would span one hemisphere.

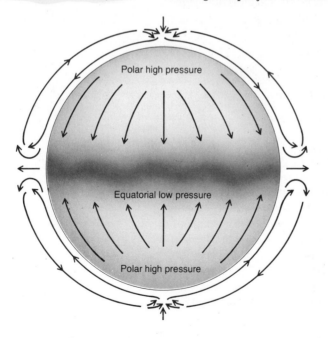

Polar high pressure

Equatorial low pressure

Polar high pressure

AUTHORS' NOTE 7.1
Coriolis Games

To demonstrate the Coriolis effect, imagine that you are playing a game of darts on a large disk rotating in a counterclockwise direction. All points on the disk have the same angular velocity (*i.e., the same number of revolutions per minute*), but those farther from the center have a larger actual velocity, as they travel a greater distance in the same amount of time. In our analysis of the dart game, two facts must be considered: (1) the board is moving in a circular path and will continue to do so after you release the dart; (2) the dart is moving even before you release it and will retain this component of its velocity after you throw it, although it will travel in a straight line. On a disk that rotates counterclockwise, you will always miss the board to the right, regardless of your position relative to the board. The diagram demonstrates this for three separate orientations. The disk has a radius of 15 m and an angular

velocity of 5 revolutions per minute (30 degrees per second). The dart is assumed to travel 10 m/sec. Letter designations are as follows:

R: rotation of the edge of the disk in one-half second

T: movement of the thrower in the one-half second after the dart is released

B: movement of the dart board in the one-half second after the dart is released

P: true path of the dart in one-half second as seen by an observer removed from the rotating disk

A: apparent path of the dart in one-half second as seen by an observer on the rotating disk.

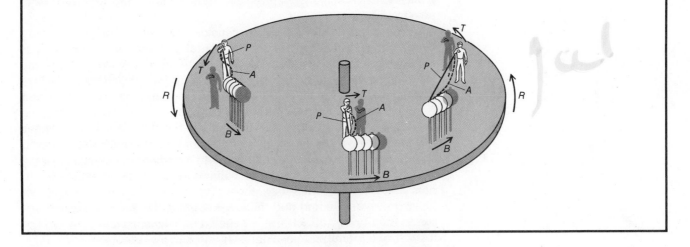

tion of rotation is *counterclockwise*, the apparent motion of an object, in reference to a coordinate system located on the disk itself, is always to the *right* of the force causing the motion. Similarly, if the rotation of the disk is *clockwise*, the apparent motion is always to the *left* of the force causing the motion. This effect is called the *Coriolis effect* (after Georges de Coriolis) and is independent of the direction of relative motion of the object. The magnitude of the Coriolis effect depends only

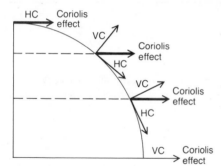

HC=Horizontal component of the Coriolis effect
VC=Vertical component of the Coriolis effect

Fig. 7.5 The Coriolis effect (heavy arrow) is the product of its horizontal and vertical components everywhere except at the poles, where it is equal to the horizontal component, and at the equator, where it is equal to the vertical component.

on the angular velocity of rotation and the relative speed of motion of the moving object.

When viewed above the North Pole, the earth rotates counterclockwise; when viewed from above the South Pole, the earth rotates clockwise. This explains why the Coriolis effect is to the right in the Northern Hemisphere and to the left in the Southern Hemisphere. But what happens at the equator? Do we find a sudden switch in directions? No, instead we find that there is no Coriolis effect at the equator. From the poles, where it is strongest, the Coriolis effect decreases with decreasing latitude, becoming zero at the equator.

The variation of the Coriolis effect with latitude results from the fact that the effects of rotation always operate on a plane that is perpendicular to the axis of rotation, that is, the earth's axis. We also know that a force operating in any direction can be divided into two forces acting at right angles to each other. Thus the force, which is perpendicular to the earth's axis, can be treated instead as a composite of a force horizontal to the earth's surface and another vertical to the earth's surface, as illustrated in Fig. 7.5. When we do this, we find that at the poles, the vertical component is zero, with the horizontal component accounting for the entire effect. As we move toward the equator, the magnitude of the horizontal component decreases and that of the vertical component increases; at the equator, the horizontal component is zero.

You might think that when considering motions on our rotating earth, we ought to consider the horizontal and vertical components of the Coriolis effect. However, the Coriolis effect is very weak, due to the relatively slow angular velocity of the earth (360 degrees in 23 hours 56 minutes 4.09 seconds, or 7.29×10^{-5} radians per second). When we drive down the street in the midlatitude Northern Hemisphere, we do not feel our car being pulled violently to the right. We need consider the Coriolis effect only when we examine motions over long paths in situations where frictional resistance to a change in direction is very weak. Such motions include atmospheric circulation, oceanic circulation, and long-range artillery shots. Even when we look at these motions, however, we find that the force of gravity is so large in comparison to the Coriolis effect that we can ignore the vertical component of the Coriolis effect and consider only the horizontal component.

Thus when we use the term "Coriolis effect," we mean only its horizontal component, described by the following two rules:

1. With respect to the direction of motion, it acts to the right in the Northern Hemisphere and acts to the left in the Southern Hemisphere.

2. It is strongest at the poles and decreases toward the equator, where it is zero.

Air Movement and Pressure Gradients

Whenever two places on the earth's surface or at equal elevations in the atmosphere have different air pressures, a *pressure gradient* exists. A pressure gradient creates a force in the direction of the gradient, and this will cause air to move in that direction. Pressure gradients result from a variety of causes. At any geographic scale, from local to global, heating of the surface and of the lower atmosphere often leads to areas of low pressure, whereas cooling of the surface and lower atmosphere usually leads to areas of high pressure. The reason for the low pressure is that as air is heated, it decreases in density and becomes buoyant, displacing air in the column of atmosphere above it. The reason for high pressure is basically the opposite; the air grows denser as it cools, and the column of atmosphere grows heavier. At scales in between the local and global, what we call the *synoptic* scale, high- and low-pressure areas can also result from complications of fluid flow in the atmosphere.

On a local scale, the distances traveled by air are very short, and we can ignore Coriolis effects. An example of local-scale circulation is the land-sea breeze (see Authors' Note 3.3). In coastal areas the land is typically colder than the water at night, but warmer during the day. During the day, then, a low-pressure area forms over the land, and a "sea breeze" results, the wind blowing along the pressure gradient from water to land. At higher elevations, where cooling of the rising air over the land has taken place, the pressure gradient may be reversed, and the air will flow out over the water. At night the situation is reversed (Fig. 7.6).

Fig. 7.6 Day/night differences in airflow in response to changes in pressure over a land area. Note that because of the vertical component of airflow, the pressure that develops aloft is always opposite that near the ground.

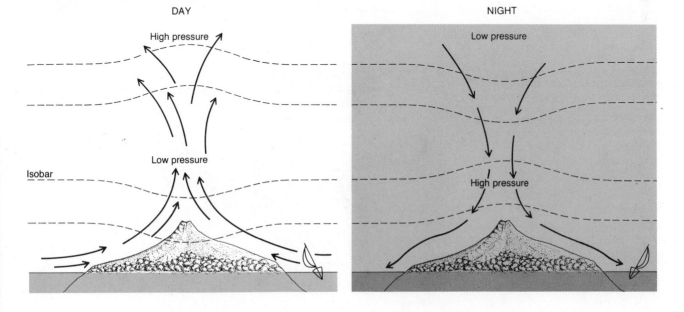

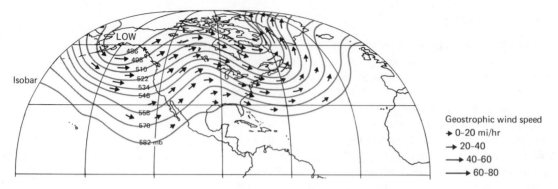

Geostrophic wind speed
→ 0–20 mi/hr
→ 20–40
→ 40–60
→ 60–80

Map projection by Waldo Tobler.

Fig. 7.7 *A geostrophic wind on December 19, 1979, at an altitude of about 5500 m, where pressure is approximately half that at sea level. Note that the wind travels parallel to the isobars. (After U.S. National Weather Service map)*

At larger scales, we must consider the Coriolis effect on the airflow. In the Northern Hemisphere the wind direction will be to the right of the direction of the pressure gradient; in the Southern Hemisphere, to the left. For any given latitude (except at the equator, where there is no Coriolis effect), there is a theoretical limiting wind velocity, even in the absence of friction. This occurs because the magnitude of the Coriolis effect increases with wind velocity; in the absence of friction, therefore, the wind velocity will increase until the Coriolis effect exactly balances the pressure gradient. At this point the wind will be blowing parallel to the isobars (Fig. 7.7). Such winds are called *geostrophic winds* and have the singular distinction of always flowing with the isobars.

Fig. 7.8 *Wind-direction change with elevation above the ground in the Southern Hemisphere. As a result of the Coriolis effect and the lack of friction with the ground, wind shifts counterclockwise with elevation. The opposite shift can be expected in the Northern Hemisphere. (Illustration by William M. Marsh)*

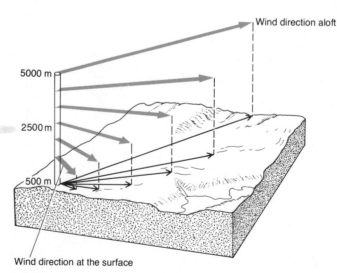

Wind direction aloft

5000 m

2500 m

500 m

Wind direction at the surface

SOUTHERN HEMISPHERE

In the earth's atmosphere geostrophic winds occur only at latitudes above 30° and at elevations greater than about 500 meters. Nearer the equator, the Coriolis effect is so slight that theoretical wind velocities are exceedingly high. At elevations lower than about 500 meters at all latitudes, friction between air and the earth's surface prevents wind from attaining geostrophic velocities. The wind direction nevertheless stabilizes, but generally not in the geostrophic direction. Instead, the direction varies from perpendicular to the pressure gradient to parallel to it, becoming more parallel with height above the surface. Thus in the Northern Hemisphere wind direction changes in the clockwise direction as we move up through the lower portion of the atmosphere. In the Southern Hemisphere the change is counterclockwise (Fig. 7.8).

The General Circulation

Figure 7.9 is a schematic portrayal of the surface circulation of the atmosphere at the earth's surface. Note the low-pressure zone along the equator. This is called the *intertropical convergence zone* (ITC),

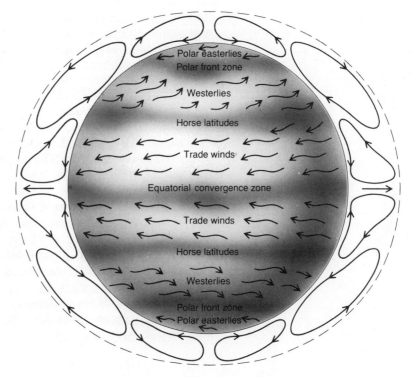

Fig. 7.9 Idealized circulation of the atmosphere at the earth's surface, showing the principal areas of pressure and belts of winds.

and it results from surface overheating caused by intensive incoming solar radiation. The variable winds in the ITC are often calm and frequently caused grief during the days of ocean sailing, when ships attempting to cross the equator would be becalmed. This region was also called the *doldrums*, and this phrase is sometimes used to portray a weary, depressed state. The English poet Samuel Taylor Coleridge described the tragic becalming of a ship in this area in "The Rime of the Ancient Mariner":

Down dropped the breeze, the sails dropped down,
'Twas sad as sad could be,
And we did speak only to break
The silence of the sea!

All in a hot and copper sky,
The bloody sun, at noon,
Right up above the mast did stand,
No bigger than the Moon.

Day after day, day after day,
We stuck, nor breath nor motion;
As idle as a painted ship
Upon a painted ocean.

Water, water, everywhere
And all the boards did shrink;
Water, water, everywhere,
Nor any drop to drink.

Poleward of the ITC in both hemispheres are the *trade winds*. In the Northern Hemisphere they blow from the northeast to the southwest and are called the northeast trades. In the Southern Hemisphere they are called the southeast trades. The Coriolis effect causes the directions to deviate from true north (in the Northern Hemisphere) and true south (in the Southern Hemisphere).

Aloft, at an elevation of about 12 kilometers, the air converging on the ITC after moving up the chimneys of the convectional cells fans out and moves back toward the poles. But it never reaches them, because at latitudes 25°–30° north and south, in the zone called the *horse latitudes*, this air descends, forming large, subtropical high-pressure cells. The reason the air aloft does not reach the poles is because of the *conservation of angular momentum*, described in Authors' Note 7.2. This pattern of cyclical circulation in the tropics, with the trade winds at the surface bringing air to the equator and the antitrades aloft bringing air back to the subtropical high-pressure cells, is called the *Hadley*

AUTHORS' NOTE 7.2

The Conservation of Angular Momentum and Atmospheric Circulation

Just as a body in motion remains in motion unless a force is applied to it, a rotating body tends to keep rotating unless a torque is applied to it. More specifically, the angular momentum of a mass rotating around an axis will not change unless a torque is applied. Mathematically, the angular momentum is defined as:

Mass × Velocity × Radius of rotation.

If the mass is constant in a given case, the products of velocity and radius of rotations must remain constant; therefore, if one increases, the other must decrease. When you were a child, you probably applied this principle many times. If, for example, you wound yourself up by twisting a swing and then unwound, you found that you could control your angular velocity by either extending your legs to slow down or compressing them into a ball to speed up. Other examples of this principle are:

1. A diver or a trampolinist tucks into a ball in order to more easily turn a flip;

2. An ice skater folds his or her arms in order to spin more rapidly;

3. A ball whirled around one's head on a string will increase in velocity if the length of the string is shortened and will decrease in velocity if the string is lengthened.

In all of these examples, the rotating mass remains constant, but since the radius of rotation changes, velocity must change.

How does this apply to the atmosphere? Consider an air particle that is stationary over the equator at an altitude of 10 km. By "stationary" we mean that the particle is traveling in a circle at the same angular velocity as the earth, so that it remains above the same point on the ground. The particle's radius of rotation is 6381 km (assuming a spherical earth), and its angular velocity is 7.29×10^{-5} radians per second, so its actual velocity is 465 meters per second. Now move the particle to the north, maintaining its altitude at 10 km above the surface. If no torque is exerted, the air particle must maintain its angular momentum. Since the farther north it goes, the closer it gets to the axis of rotation, the faster its velocity must become. However, the earth underneath it is moving more slowly, because its angular velocity is still 7.29×10^{-5} radians per second, but its radius of rotation is smaller. To consider an extreme example, at latitude 45° N, a point on the earth's surface has a velocity of 328 m/sec. The particle, on the other hand, must now have a velocity of 658 m/sec, 330 m/sec faster than the surface beneath it. Such extreme velocities simply do not occur in the atmosphere.

We now see why the atmospheric circulation pictured in Fig. 7.4 is impossible. The circulation system breaks down because of the huge zonal component in the motion, and the air moving northward above the N.E. trades (and southward above the S.E. trades) descends around latitudes 25° to 30°.

cells, after George C. Hadley, the British physicist who formulated the first theoretical explanation of the trade winds in 1735. At latitudes 50°–60° north and south, zones of low pressure exist, where the prevailing *westerlies* converge. High pressure exists over the poles, and the resulting winds, blowing away from the poles, are the *polar easterlies*.

Thermal Effects of Land and Water on the General Circulation

The thermal differences of land and water affect the circulation of the atmosphere on a global scale in much the same way as they do on a local scale. Much of the difference is related to the depth to which heat can penetrate land and water. Recall that heat moves into the land by conduction; as a result, only the upper several meters of the land exchange much heat with the atmosphere on an annual basis. Moreover, land materials generally have low heat capacities and in turn produce strong thermal gradients into the atmosphere. For these reasons, the heat taken in by the soil in summer is quickly lost in the ensuing cool season. In contrast, heat moves into the ocean by convection and is carried to much greater depths than it is on land. In addition, water has a high heat capacity and, since it is usually cooler than the air, fails to develop upward thermal gradients into the overlying atmosphere. On balance, then, the ocean gives up its heat much more slowly than the land does. Therefore, land areas in winter are typically much colder than the oceans, whereas in the summer they are much warmer. For example, the North American cities of San Francisco and St. Louis are at approximately the same latitude and would, under the same atmospheric conditions, receive about the same amount of solar radiation. However, San Francisco is near the ocean and is downwind from it, whereas St. Louis is inland. The coolest month in St. Louis is January, with a mean temperature of 0°C (32°F), and the warmest is July, with a mean temperature of 26°C (79°F). In contrast, January in San Francisco averages 10°C) (50°F), and the warmest month is September, with a mean temperature of 17°C (62°F).

Such differences in heat storage result in pressure differences in the atmosphere. The large land masses of the world, particularly North America and Eurasia, typically have very high pressures in the winter and low pressures in the summer. This difference is particularly well developed over Eurasia, because of its very large size. The resulting winds also reverse seasonally, and in Asia these are called the *monsoon* (from the Arabic word *mausim*, meaning "seasonal wind"). During the summer, warm, moist winds, called the summer monsoon, blow from the Indian Ocean and Arabian Sea toward the land mass

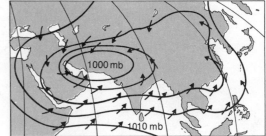

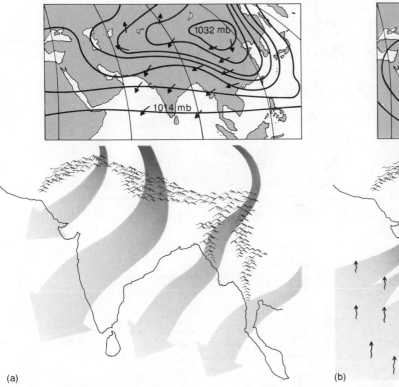

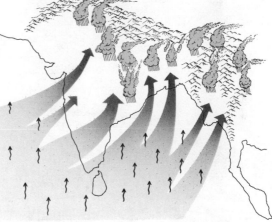

(a)

(b)

Fig. 7.10 Monsoon circulation over south Asia in winter (a) and summer (b). The summer monsoon develops in response to low pressure over land (see inset) and gains large amounts of vapor as it passes over the Indian Ocean and Arabian Sea. In winter the flow reverses as high pressure forms over central Asia. (Illustration by William M. Marsh)

(Fig. 7.10). In the winter, the strong high-pressure system that develops over Asia causes the winds to blow from the land mass to the ocean, and this is called the winter monsoon.

During the period of Portuguese trade with India, the voyages from Europe to the Indian peninsula were timed so that the ships crossed the Arabian Sea from Africa to India during the summer monsoon and then waited in India until the onset of the winter monsoon. The first fleet to sail to India, commanded by Vasco da Gama in 1497–1498, did not do this, but instead crossed to India in May, at the beginning of the summer monsoon; the voyage from Malindi on the coast of Africa required

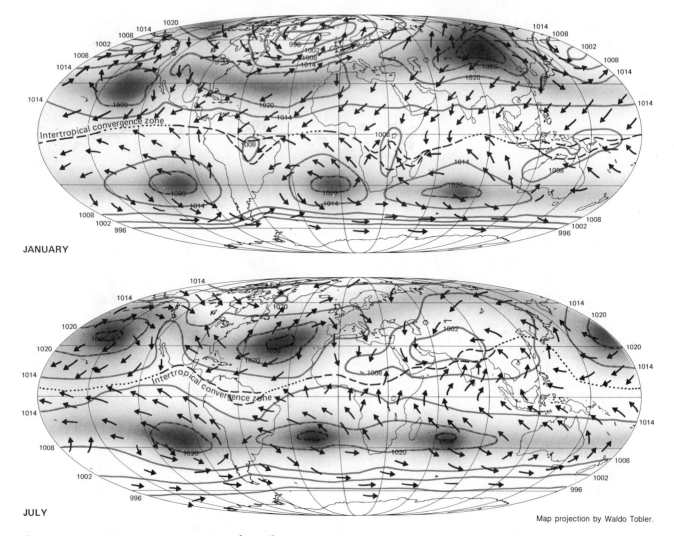

JANUARY

JULY

Map projection by Waldo Tobler.

Fig. 7.11 General distribution of air pressure (in millibars) and atmospheric circulation at a global scale in January and July.

less than a month. The return voyage at the end of August, with the summer monsoon still in progress, required three months (Fig. 7.11).

OCEANIC CIRCULATION

The Effect of Winds

As they blow over a water surface, winds create waves. These waves cover a spectrum of sizes from very long wavelengths (in strong winds) to ripples, all of them superimposed on one another and moving at different velocities. In a true wave only the wave form moves, and there is no net forward motion of the water. However, as a wave increases in size, its crest comes into contact with increasingly stronger winds. At some point the crest may become oversteepened and break, creating white water and resulting in some net forward motion. In this way, winds blowing over open water impart some forward motion to the surface, and the molecules of the surface water drag along those underneath, thereby producing currents.

Currents assume the direction of the prevailing winds. On a uniform earth covered entirely by ocean, we would find east-to-west currents in the trade-wind belts, west-to-east currents in the regions of the prevailing westerlies, and east-to-west currents in the polar regions.

The Effect of Continents

The ideal current pattern on a uniform, ocean-covered earth, however, is disrupted by the earth's land masses. At either side of the ocean basins, continents deflect the ocean currents, resulting in a series of return currents. A circular pattern of currents, called a gyre, thus develops in adjacent wind belts.

As shown in Fig. 7.12, we can generalize a hypothetical ocean as a ellipse surrounded by land. The largest currents are the subtropical gyres; these are formed when the east-to-west currents in the trade-wind zones abut against the east coast of the continents and are deflected poleward, where they return across the ocean in the westerlies. On either side of the equator, narrow equatorial gyres form from the portion of the currents deflected into the doldrums. Poleward of the subtropical gyres are the subpolar gyres. The currents that flow toward the poles carry warm water to higher latitudes, whereas those that flow toward the equator carry cool water to lower latitudes.

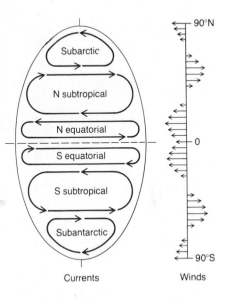

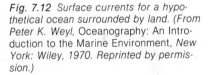

Fig. 7.12 Surface currents for a hypothetical ocean surrounded by land. (From Peter K. Weyl, Oceanography: An Introduction to the Marine Environment, New York: Wiley, 1970. Reprinted by permission.)

The Actual Circulation

Figure 7.13 shows the actual circulation of the oceans. This pattern is remarkably like that of our hypothetical circulation except in the area north of Antarctica, where the ocean basin is not bounded by continents, and the West-Wind Drift circles the entire earth. This difference is extremely important in large-scale energy transfer, because only currents that are deflected off continents are able to exchange much energy with waters north or south of their place of origin.

The Gulf Stream is an example of a warm current that flows into the cold North Atlantic. Though the temperature of the Gulf Stream is usually only several degrees warmer than that of the water of the Mid- and North Atlantic, the additional heat is sufficient to increase greatly

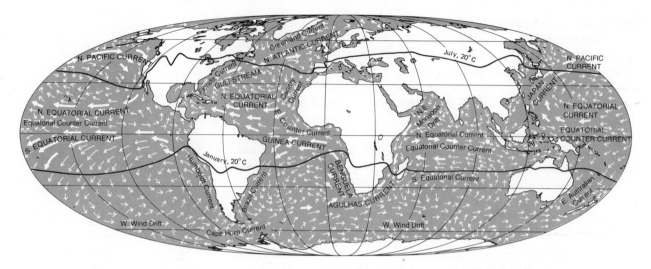

Map projection by Waldo Tobler.

Fig. 7.13 The actual circulation of the oceans. Major currents are shown with heavy arrows. (After maps in Dietrich, 1963, and Defant, 1961)

the sensible- and latent-heat flux into the atmosphere (Fig. 7.14). As a result, the atmosphere over the North Atlantic and the climate of Northwest Europe are somewhat warmer and wetter than they would be otherwise. The opposite tends to be true where cold currents, such as the Humboldt Current along the Pacific Coast of South America, move against a warm coastline. The coastal climate is drier and cooler because of the heat lost to the current and the air over it.

Fig. 7.14 Latent-heat flux from the ocean into the atmosphere for the North Atlantic and Pacific southwest of North America in December. The Gulf Stream and the prevailing surface winds are shown by the arrows. The large energy flux off the East Coast is related to the warm water brought northward by the Gulf Stream and the cold, dry condition of the air flowing off North America. (Adapted from A. H. Perry and J. M. Walker, The Ocean-Atmosphere System, *New York: Longman, 1977. Used by permission.*)

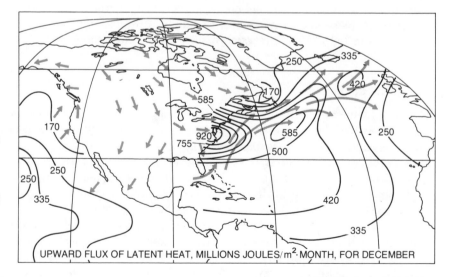

UPWARD FLUX OF LATENT HEAT, MILLIONS JOULES/m^2·MONTH, FOR DECEMBER

Map projection by Waldo Tobler.

SUMMARY

The transfer of energy by the atmosphere and oceans is critical to the earth's energy balance. Without a poleward transfer of energy from the low latitudes, the tropics (or intertropics) would be warmer and the high latitudes colder than they are. The general circulation of the atmosphere is a response to latitudinal differences in solar heating of the earth's surface, the distribution of land and water, the mechanics of the atmosphere's fluid flow, and the Coriolis effect. The Coriolis effect causes winds to deviate from a direct path along the pressure gradient. This explains both the great curved paths that winds take in moving from high- to low-pressure zones and the presence of the prevailing easterly and westerly winds between the major zones of pressure. Airflow aloft tends to counterbalance flow near the surface; in the tropics this takes the form of Hadley cells. Oceanic circulation generally follows that of the prevailing winds, except that the currents are deflected where they meet land masses. The result is the formation of gyres in the ocean basins, with warm currents on one side and cold currents on the other.

PRECIPITATION CHAPTER 8
PROCESSES

INTRODUCTION

Whenever we examine the climate of a place on the earth, precipitation is a very important consideration. How much does it rain? Does it snow? When does it rain? Is there a rainy season? Is rainfall characterized by drizzles or torrential downpours? These questions can be answered by scrutinizing the climatic data recorded at a weather station. But the reasons for the particular precipitation characteristics of a place are more difficult to discover. In this chapter we take up the question of why it rains and snows, examining the atmospheric processes that cause condensation of water vapor.

Precipitation is that portion of the hydrologic cycle that returns water from the atmosphere to the earth's surface. Precipitation may occur in liquid form—rain—or as hail or in a myriad of other frozen forms—collectively called snow. Apart from its effect on human comfort, agriculture, streams, etc., precipitation is important because it is one of the ways in which energy is transferred in the earth's atmosphere. Evaporation, the transfer of water from the liquid to vapor form, requires energy; condensation, the reverse process, gives up energy. Thus when evaporation occurs at one location and the water vapor is condensed elsewhere, a transfer of energy results. This chapter examines some characteristics of water vapor in the atmosphere and explains why and how precipitation occurs.

WATER VAPOR IN THE ATMOSPHERE

As was noted in Table 7.1, water vapor comprises from 0 to 4 percent of the gas in the troposphere, and it is by far the most variable component. There are several ways to describe the amount of water vapor in a parcel of air, and we shall consider these in some detail. First, however, let us understand one important fact: The amount of water vapor contained in air when it is saturated is dependent on air temperature. The warmer the air, the greater the amount of water vapor that can be held; the colder the air, the smaller the maximum water content can be.

Expressions for Water Vapor

The amount of water vapor in a parcel of air can be considered in relation either to a given volume of space or to a given mass of air molecules. Because the density of air (which is a product of the number of molecules per cubic meter of air) changes with height, we must be familiar with both of these types of measures.

Vapor pressure and absolute humidity. The amount of water vapor in a given volume of space can be expressed in terms of *vapor pressure,* or *absolute humidity.* Vapor pressure—the pressure exerted by the weight of the water vapor molecules—is expressed in pressure units, e.g., as millibars (mb) or as newtons/m^2 (also called the pascal, or Pa). We know from the gas laws of physics that in a given volume of air, pressure is dependent on the number of molecules and their temperature. We also know that the total pressure exerted by a mixture of gases is simply the sum of the *partial* pressures exerted by individual gases, because gases in mixture behave independently of one another. The *saturation vapor pressure* is the maximum value the vapor pressure can attain at a given temperature. It is dependent on temperature, as Fig. 8.1 shows, *but not on total pressure.*

The same principle is true of absolute humidity, which is a measure of the mass of water vapor in a given volume of space, regardless of the mass of air occupying the same volume. Absolute humidity is commonly expressed in grams per cubic meter (g/m^3). Figure 8.2 shows the *saturation absolute humidity.* It too is dependent on temperature, but not on total pressure.

Mixing ratio and specific humidity. The other major way to express the amount of water vapor is in relationship to the amount of air occupying the same volume. The mass of water vapor per unit mass of dry air is called the *mixing ratio.* The mass of water vapor per unit mass of moist air (dry air plus water vapor) is called the *specific humidity.* These expressions are typically given in grams per kilogram (g/kg) and are generally so close in value that we often consider them to be approximately equal.

Figure 8.3 shows the *saturation mixing ratio* as a function of *both temperature and pressure.* The saturation mixing ratio is related to pressure as well as to temperature, because the *saturation vapor pressure* is dependent only on temperature. For example, consider a box filled with air and water vapor at saturation at a given temperature. If we add dry air to the box while maintaining the temperature, the air remains at saturation. But the total pressure increases because there is more gas inside the box, and the mixing ratio (which must be the

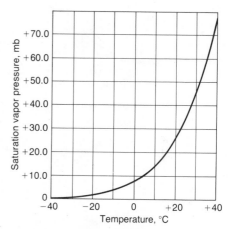

Fig. 8.1 The relationship between temperature and saturation vapor pressure.

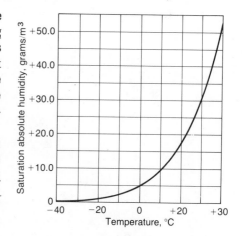

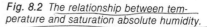

Fig. 8.2 The relationship between temperature and saturation absolute humidity.

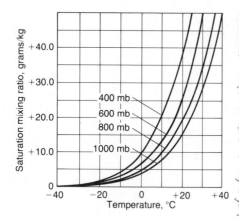

Fig. 8.3 *The saturation mixing ratio as a function of temperature and pressure.*

saturation mixing ratio because the air in the box is saturated) decreases because we have the same mass of water vapor mixed with a greater mass of dry air.

We explore the consequences of these facts in subsequent sections. At this point let us note one other fact that will help to explain why it is useful to have all of these different ways of measuring water vapor content. Consider this question: If we have a parcel of nonsaturated air and lift it to a higher elevation, what happens to its actual vapor pressure and to its actual mixing ratio? The answer is that the vapor pressure (and the absolute humidity) *decrease*, and the *mixing ratio remains the same*. For example, suppose that the air parcel is enclosed in a balloon, so that neither air nor water vapor can escape. Clearly the mixing ratio must remain the same, as long as condensation does not occur. On the other hand, at higher elevation, where pressure is lower, the balloon will expand. The partial pressure due to the water vapor will decrease, as will the absolute humidity, because we have the same mass of water vapor contained in a bigger volume. The measures of the actual water vapor content can be summarized as in Table 8.1.

Relative humidity. Relative humidity, the most commonly used measure of water vapor content of air, expresses vapor content as a percentage of the amount of vapor held by a parcel of saturated air. Any of the four measures of vapor can be used to compute relative humidity. For example,

$$\text{Relative humidity} = \frac{\text{Vapor pressure}}{\text{Saturation vapor pressure}} \times 100$$
$$\text{(at a given temperature)} \quad \text{(at same temperature)}$$

(Multiplying by 100 converts it to a percentage.)

For a numerical example, consider air at 25°C with a vapor pressure of 10 mb. We note from Fig. 8.1 that the saturation vapor pressure

Table 8.1 Measures of water vapor content.

EXPRESSION	TYPICAL UNITS	SATURATION VALUE DEPENDS ON:	AS AIR PARCEL MOVES TO DIFFERENT PRESSURES, THE VALUE:
Vapor pressure	mb	Temperature	Changes
Absolute humidity	g/m³	Temperature	Changes
Mixing ratio	g/kg	Temperature/pressure	Remains constant
Specific humidity	g/kg	Temperature/pressure	Remains constant

at 25°C is 31.7 mb. The relative humidity is therefore

$$\frac{10}{31.7} \times 100 = 32\%.$$

If the air were to cool to 15°C, the relative humidity would rise to

$$\frac{10}{17.0} \times 100 = 59\%.$$

If the air were to cool to 7°C, it would be saturated. This temperature is called the *dew point* of the air and depends on its vapor pressure.

Global distribution of water vapor. Another measure of the vapor content of the atmosphere is the *precipitable water vapor*—the depth of liquid water that would result if all of the water vapor in a column of atmosphere were to condense and fall to earth. Average precipitable water vapor content for the entire atmosphere is 25 mm (1 inch), but there are large, important regional variations (Fig. 8.4). Three latitudinal zones stand out. (1) In a belt extending 30 degrees north and south of the equator, 61.8 percent of the total vapor content of the atmosphere is held. This is explained on the basis of the large volume of air in this zone, its high temperature, and the availability of large areas of ocean. (2) The belt between 30 and 60 degrees in both hemispheres contains 29.2 percent of the atmosphere's moisture. Note, however, that in the Southern Hemisphere this belt contains more vapor than its counterpart in the Northern Hemisphere. This is due to the larger proportion of water area in this belt in the Southern Hemisphere. (3) In the polar latitudes the vapor content of the atmosphere is low in both

Fig. 8.4 Precipitable water vapor in the atmosphere by latitude. Note the differences between the Northern and Southern hemispheres in the 30°–60° and the 60°–90° latitude ranges.

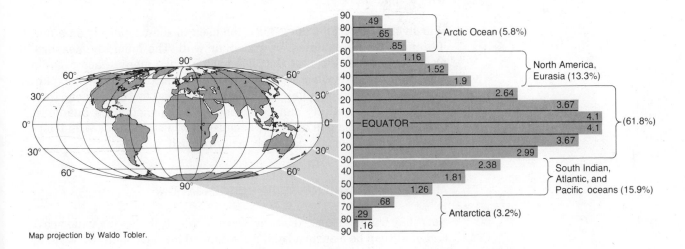

Map projection by Waldo Tobler.

hemispheres, totaling only 9 percent. But the content between 60 and 90 degrees north latitude is appreciably greater than it is between 60 and 90 degrees south latitude. This difference is due to the contrasts in land and water areas, represented by the Arctic and North Atlantic oceans in the north and the continent of Antarctica in the south.

How to Measure Humidity

The humidity of the air can be measured in a number of ways, based on different physical principles or different effects of water vapor in air. Two of the most common devices for making such measurements are the psychrometer and the hair hygrometer.

Psychrometer. Because evaporation of water requires that the latent heat of vaporization be given up, an evaporating surface will be cooler than a dry surface if other environmental conditions are equal. The amount of cooling will depend on the vapor content of the receiving air; thus the loss of temperature can be taken as an indicator of atmospheric humidity. Based on this principle, we can build a psychrometer using two thermometers—one with a saturated gauze attached to the bulb and one without.

If we now cause air to circulate past the thermometers, the dry-bulb thermometer will record the air temperature, whereas the wet-bulb thermometer will lose temperature, eventually stabilizing at some lower reading. The humidity value is computed on the basis of the air (dry-bulb) temperature and the *difference* between the dry- and wet-bulb temperatures (Fig. 8.5). Two conventional methods of ventilating the thermometers are by whirling them in the air (a "sling" psychrometer is pictured in Fig. 8.5) or by drawing an air current past them with a small fan.

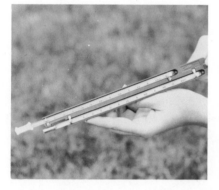

Fig. 8.5 Sling psychrometer. The wet-bulb thermometer is covered with the gauze sock.

Hair hygrometer. This device is often used to show humidity on a dial in an instrument mounted on a desk or wall. The humidity measurement depends on the fact that human hair, like many other organic materials, expands when wet and contracts when dry. A typical expansion value for a human hair is 2.5 percent from 0 to 100 percent relative humidity. Humidity can be measured by connecting a bundle of hairs to a dial on a pen arm. But because hair also expands with temperature, measurements cannot be very precise.

AIR MOVEMENT AND MOISTURE CONDENSATION

Now that we have described the ways in which the water vapor content of air can be measured and have defined the dew point, we want to examine how condensation takes place. Condensation results when air is cooled below its dew point. If the moisture falls to the surface, we call it *precipitation*. The simplest way air can be cooled is for it to blow over a colder surface. This process is known as *advection*, and it is very common in coastal areas. During the winter in the Great Lakes region, for example, air blowing from the relatively warm lake surface (at about 1°C) over the cold land surface (at about −5°C) causes higher amounts of snowfall on the east sides of the lakes. Coastal fog in California often originates in a similar fashion when warm, moist air blows over cold currents near the shore (Fig. 8.6).

But the cooling processes that induce most precipitation involve rising air, usually some form of *convection*. Air may rise spontaneously because it is warmer and lighter than surrounding air, or it may be forced aloft as an air mass moves over a mountain range or a colder, denser air mass. In all such cases the rising air undergoes pressure and temperature changes which are important to our understanding of precipitation and energy transfer in the atmosphere.

Fig. 8.6 *Coastal fog produced from advective cooling of onshore winds. (Photograph by Jeff Dozier; illustration by William M. Marsh)*

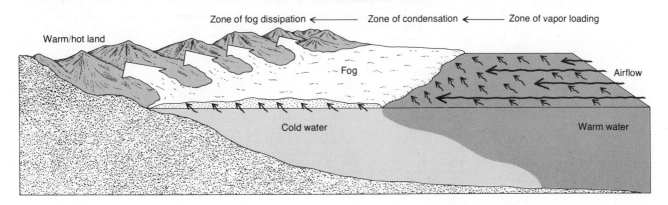

The Dry Adiabatic Lapse Rate

When air rises, it expands because of the decreased pressure at higher altitude. This expansion involves work, so energy (heat) is removed from the air, thereby reducing its temperature. Likewise, when air is forced to descend, it is compressed; the work done is converted into heat, and the temperature of the air rises.

You can note some of these effects in your own experiences with air. If you rapidly pump up a bicycle tire, for example, the pump becomes hot. If air is let out of a tire, it cools when it comes out, due to the lower pressure. This principle is used in snow-making machines. A mixture of water and compressed air is sprayed over a hillside. When it is released from the nozzle, the air cools dramatically, and the water vapor in it condenses to snowflakes.

The rate of cooling in rising air, called the dry adiabatic lapse rate, is − 0.0098° C/m, or − 0.98°C/100 m, or approximately − 1°C per 100 m. We use the minus sign to indicate that air cools as its height increases. For example, a parcel of air at 20°C at sea level will cool to 13°C if raised to 700 m. If the air is brought back down again, it will warm to 20°C. Note that these temperature changes are a result solely of expansion and compression, *not* of interchange with air of a different temperature.

The Saturated Adiabatic Lapse Rate

If rising air cools to its dew point, it will be saturated. Any further increase in elevation will cause condensation to occur, and the condensing water will give off heat. This heat will *partially* offset the temperature decrease due to adiabatic cooling; the saturated air will still cool as it rises, but at a lower rate. The rate of cooling will depend on the amount of condensation taking place, which is dependent on the temperature and pressure of the air (Fig. 8.3). Table 8.2 gives the range of

Table 8.2 Saturated adiabatic lapse rate (°C/100 m).

TEMPERATURE (°C)	PRESSURE (MB)			
	1000	850	700	500
40	− 0.30	− 0.29	− 0.27	
20	− 0.43	− 0.40	− 0.37	− 0.32
0	− 0.65	− 0.61	− 0.57	− 0.51
− 20	− 0.86	− 0.84	− 0.81	− 0.76
− 40	− 0.95	− 0.95	− 0.94	− 0.93

Note: At very cold temperatures, the saturated adiabatic lapse rate is almost equal to the dry adiabatic lapse rate of − 0.98°C/100 m. At higher temperatures, the difference is much greater because more water is condensing.

From *Understanding Our Atmospheric Environment* by Morris Neiburger, James G. Edinger, and William D. Bonner. W.H. Freeman and Company. Copyright © 1973.

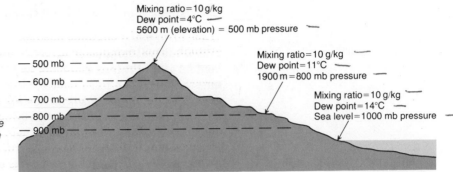

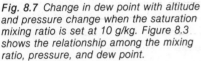

Fig. 8.7 *Change in dew point with altitude and pressure change when the saturation mixing ratio is set at 10 g/kg. Figure 8.3 shows the relationship among the mixing ratio, pressure, and dew point.*

values for the saturated adiabatic lapse rate; an average value is $-0.6°C/100$ m. Note that the saturated adiabatic lapse rate applies only to air that is saturated. For purposes of calculating temperature change with elevation change, air at 80 percent relative humidity is considered "dry" until it reaches its dew point.

Because most of the water vapor that condenses will fall from the air as precipitation, the saturated adiabatic lapse rate, unlike the dry, is generally not reversible. Therefore, air that is lifted beyond its elevation of saturation and releases its water vapor will be *warmer* if it is returned to its original elevation. Remember, this happens because the heat that was released in condensation adds a thermal surcharge to the rising air, which offsets the full effect of the dry adiabatic cooling rate.

Changes of Dew Point with Elevation

We previously explained that the actual mixing ratio of water vapor and air remains the same as air rises, as long as condensation does not occur. Figure 8.3 showed that for a given mixing ratio, the dew point is lower when pressure is lower (i.e., at higher elevation). For example, if the water vapor mixing ratio is 10 g/kg, the dew point at 1000 mb (sea level) is about 14°C. At 800 mb (about 1900 m) the dew point of air at the same mixing ratio is 11°C, and at 500 mb (5600 m) the dew point is 4°C. These values show that the dew point *decreases* with elevation, at an average value of about $-0.2°C/100$ m (Fig. 8.7).

Therefore, if we know the temperature and water vapor content of air at a given elevation, and if we want to calculate the height to which this air would have to rise before condensation occurred, we would need to use not only the dry adiabatic lapse rate, but also the fact that the dew point decreases with height.

Calculations with Skew *T* Log *P* Graphs

In order to make weather forecasts, a meteorologist must often calculate what will happen to air of known temperature and water vapor content if it is raised to a given elevation. Although some parts of this question are easy to calculate, others are more difficult. Because the dry adiabatic lapse rate is a constant and the dew point lapse rate does not vary very much, calculation of initial condensation height can be done algebraically. Beyond that height, however, our elementary algebraic methods fail, because that saturated adiabatic lapse rate is not constant, but changes with elevation.

Such problems can be solved more easily by graphical methods. A special kind of paper for solving such problems is "skew *T* log *P*" paper (Fig. 8.8). The "log *P*" refers to the logarithmic pressure scale on the left. The "skew *T*" refers to the fact that the temperature axis is not perpendicular to the elevation/pressure axis, but rather oblique. The reason for this is so that the dry adiabats (lines slanting upward to the left that show the rate of cooling of nonsaturated air) intersect the isotherms at approximate right angles.

Let us illustrate the solution to a given problem. Assume that we start out at 1000 mb (sea level) with air at 25°C and a mixing ratio of 10 g/kg. We note immediately that the relative humidity is 50 percent (the air could hold 20 g/kg) and that the dew point is 14°C. The "location" of our air parcel is at the intersection of the 25°C isotherm and the 1000-mb isobar, indicated by point *A* on the detail graph in Fig. 8.8. Suppose now that the air rises. At what elevation does condensation begin? We follow the dry adiabat upward to the left until it intersects the 10 g/kg mixing ratio line, at the 850-mb level (just below 1500 m). This is point *B* on the graph. The temperature of the air parcel is now between 11°C and 12°C, and the air is saturated. Note that the surface dew point of 14°C was reached at an elevation of 1100 m, but the air was not saturated there.

Suppose that this saturated air is lifted further, to an elevation of 3500 m (about 660 mb). What will happen? To calculate the resulting temperature changes, we interpolate between the 15° and 20° saturated adiabats, the dashed lines, as shown. Note that the temperature is decreasing, but is warmer than the 25° dry adiabat. Point *C* indicates our arrival at 3500 m. The temperature is now 0°C; under dry adiabatic conditions, if the air had risen to 3500 m, its temperature would be about −9.3°C {25 − (3500 × 0.0098)}. The air is still saturated, but its mixing ratio is now about 6 g/kg. Thus 4 g/kg condensed between 1500 m and 3500 m.

Exploring further, let us assume that almost all of this condensate precipitated. What happens now if the air descends? We follow its descent along the appropriate dry adiabat (35° in this case), as indi-

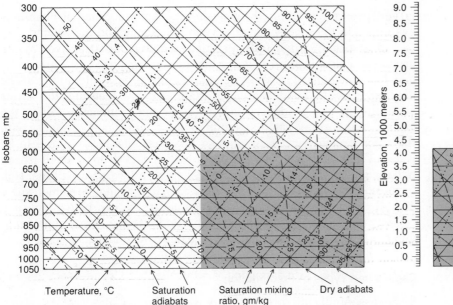

Temperature, °C   Saturation adiabats   Saturation mixing ratio, gm/kg   Dry adiabats

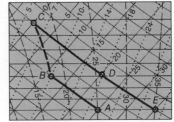

cated. Note that we cross the 20° isotherm at point *D*, which is just above our original level of condensation. At this point the mixing ratio is still 6 g/kg; the temperature is 20°C, as compared to the condensation temperature of 11°C. The extra heat came from the 4 g/kg which condensed. As the air continues to descend, it arrives back at the 1000-mb level, with a temperature of 35°C (point *E*).

Fig. 8.8 Graph, called the skew T, *log* P *diagram, used to find the elevation at which condensation occurs in an air mass of known temperature and vapor content.*

TYPES AND CAUSES OF PRECIPITATION

Precipitation Processes and Forms

Condensation results in cloud formation, although not all clouds produce precipitation. The reason for this is that the water droplets or ice particles in most clouds are not large enough to fall freely through the atmosphere. Condensation takes place on *condensation nuclei*, very small particles in the atmosphere which are composed of dust or salt. When condensation initially occurs, the droplets or ice particles are very small and are kept aloft by the motion of the air molecules. For example, in a fog (a ground-level cloud) the water droplets remain suspended in the air. In order for the droplets to fall, they must grow by coalescing with one another. Once minute droplets have formed, the droplets themselves begin to act as condensation nuclei. Ordinarily, such droplets would tend to repel one another, but in the presence of an electric field in the atmosphere, they attract one another and grow larger, eventually falling through the atmosphere. Some of them may evaporate in the atmosphere between the clouds and the ground.

Most precipitation falls as *rain*, meaning that it arrives at the ground in liquid form. It may have actually condensed as water, or it may have condensed into a frozen state and melted while falling. If the dew point at the condensation altitude is below freezing, the water vapor in the air condenses into *snow* crystals and may fall to the ground in this state. Small amounts of liquid water may condense onto the snow crystal as it falls, and in this case the snow is said to be *rimed*. When a snow crystal collides with a rain drop, a small frozen ball, or *graupel*, is formed. *Sleet* is rain which freezes as it falls. *Hail* forms under conditions of strong atmospheric turbulence; a small snow crystal drops into the zone where temperatures are below freezing, collects additional water, and is then lifted into the frozen zone again and again. In this fashion hailstones can grow to large sizes.

Causes and Types of Precipitation

Since essentially all precipitation results when air cools by rising through the atmosphere, we can classify precipitation according to what causes the rise. Four principal causes are identifiable:

- *Orographic*—air flow over high terrain, usually a mountain range along which air is forced upward;

- *Cyclonic/frontal*—meeting of air masses of different densities, in which warm air is pushed upward by colder air;

- *Convectional*—instability of surface air, usually due to the intensive heating of air near the ground;

- *Convergent*—barometric or topographic sinks, where air is drawn into large low-pressure areas and rises.

Orographic precipitation. Orographic precipitation is the easiest type to describe. It results when moisture-laden air is forced to pass over a mountain range, inducing cooling and condensation. In areas where mountains lie in the paths of moist, prevailing winds, annual precipitation rates can be quite high. In fact, virtually all areas with rainfall above 500 cm per year are in such situations. Examples of areas of heavy orographic rainfall are the mountains of Hawaii, the Himalayan front, and the west coasts of North and South America. An extreme example often cited by climatologists is the town of Cherrapunji in the Assam Hills of northeastern India, where rainfall between October 1, 1860, and September 30, 1861, amounted to 26.5 meters (87 feet)! And 8.9 m (350 inches) fell in July alone!

Figure 8.9 illustrates the sequence of orographic precipitation. The leeward side of the mountain range is typically quite dry and is

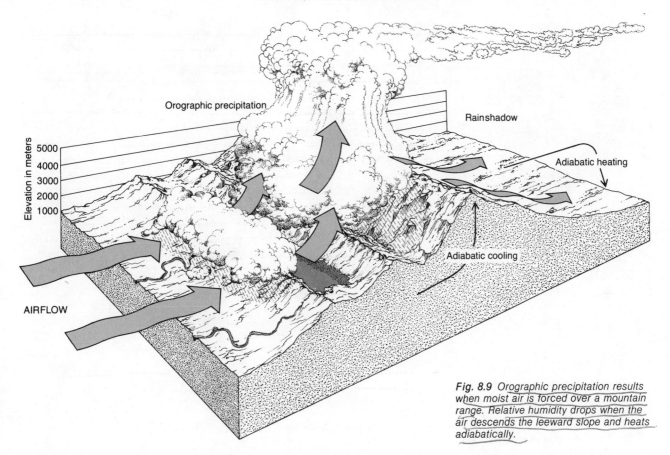

Orographic precipitation

Rainshadow

Adiabatic heating

Elevation in meters

5000
4000
3000
2000
1000

Adiabatic cooling

AIRFLOW

Fig. 8.9 *Orographic precipitation results when moist air is forced over a mountain range. Relative humidity drops when the air descends the leeward slope and heats adiabatically.*

termed the *rainshadow*. When the air descends the leeward slopes, it warms adiabatically and can become very hot and dry, resulting in rapid desication of the land. When such winds occur during the winter, they may produce a 20°–25°C rise in temperature in a matter of hours. The effects can be dramatic: rapid melting of snow, avalanches, and sprouting of spring flowers. In areas where these winds are common, they form significant events in the lives of people living in their path and apparently result in increased mental disorders and violent crimes. In Germany, Switzerland, and Austria such a wind is termed a *Föhn*, the German word for hairdryer; east of the Rocky Mountains it is the *chinook*.

Cyclonic/frontal precipitation. This type of precipitation results from the meeting of cold and warm air masses. The boundary between the two air masses is called a *front*, and it is designated a warm front or a cold front depending on which air mass is displacing the other. In a cold front the leading edge of the cold air mass drives under the warm

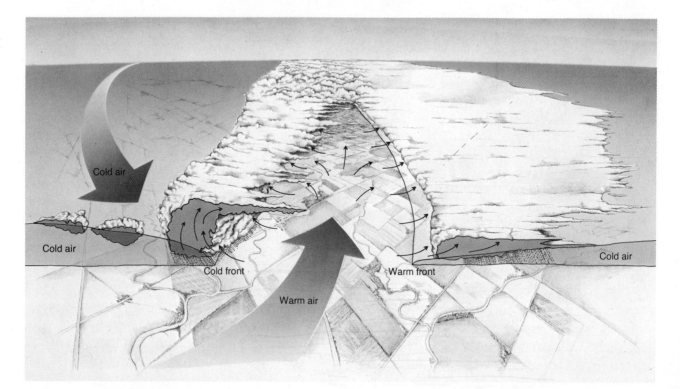

Fig. 8.10 *Cold front, warm front, and associated features of midlatitude cyclone. Turbulence is greatest along the cold front, but the more extensive cloud cover develops along the warm front. (Illustration by William M. Marsh)*

Fig. 8.11 *The stages of development of a midlatitude cyclone. The cyclone begins as a wave along the contact—polar front—between tropical and polar air. After the occluded stage, the cyclone dissipates. (Adapted in part from Godske et al., 1975)*

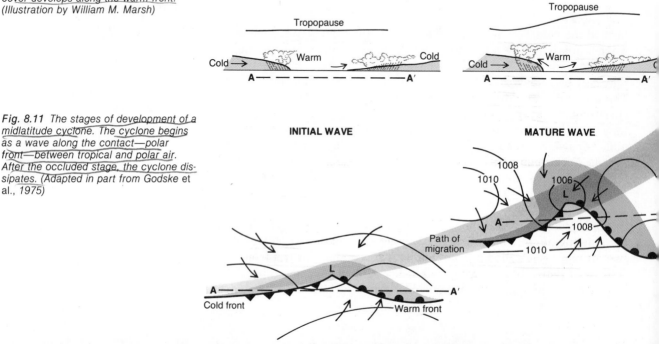

one, forming a steep contact slope over which the warm air is forced. In a warm front the warm air mass slides over the cold air mass along a gently sloping contact line (Fig. 8.10).

Frontal activity causes most precipitation in the midlatitudes, because it is here that warm maritime air masses meet cold arctic or polar air masses. The cold air serves as the catalyst for the discharge of the latent heat from the tropical air, which in turn gives rise to a cell of low pressure. In contrast to the low pressure in the ITC, which is relatively steady and continuous over a large belt, the low pressure in the midlatitudes is characterized by transitory systems which move from west to east. These are called cyclones, great storms ranging from 400 to 1500 km in diameter, into which the warm air and the cold air are drawn in the form of a huge swirl. Figure 8.11 shows the air move-

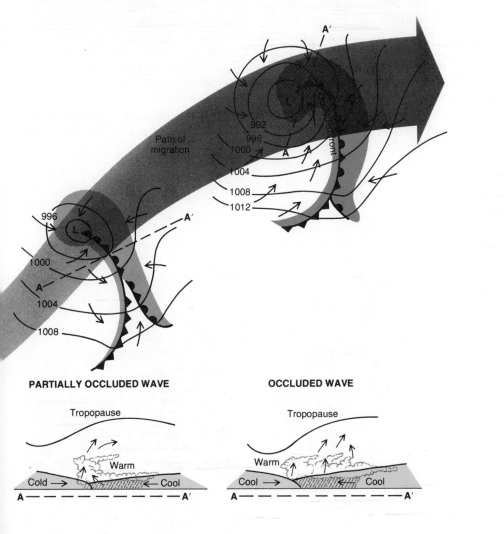

PARTIALLY OCCLUDED WAVE

OCCLUDED WAVE

ment around a Northern Hemisphere cyclone and the typical positions of the warm and cold fronts. As the cyclone moves, the relatively fast-moving cold front often "catches" the warm front, thus driving the warm air upward and forming what is called an *occluded front.*

Because cold fronts are steeper than warm fronts, the rate of uplift along them is greater. Consequently, precipitation and turbulence can be very intensive along cold fronts. In fact, strong cold fronts advancing on warm, moist air often produce a *squall line,* which is characterized by thunderstorms, gusty winds, hail, and hard rainfall (Fig. 8.10). Under extreme turbulence, especially when wind aloft is flowing cross-current to surface winds, *tornadoes* may form. These are the most intensive storms on earth, and they occur with greatest frequency in the plains and prairie states of the United States (Fig. 8.12).

Weather conditions along warm fronts contrast sharply with those along cold fronts. Cloud formation is mainly horizontal, leading to a broad zone of stratus clouds at the base of the warm air. Turbulence is negligible and precipitation prolonged. If the cold air under the front is below freezing, rain may freeze on contact with surface objects, resulting in the formation of glaze which damages vegetation and is hazardous to all modes of travel.

Temperature and moisture are the key properties used to classify air masses. Air masses that originate over water are wet and are denoted m (for maritime), whereas continental air masses are dry and

Fig. 8.12 Incidence of tornadoes per 10,000 square miles of area, 1953–1976. (From National Oceanic and Atmospheric Administration, Severe Local Storm Warning Service and Tornado Statistics, 1953–1976, Washington, D.C.: U.S. Department of Commerce, 1977)

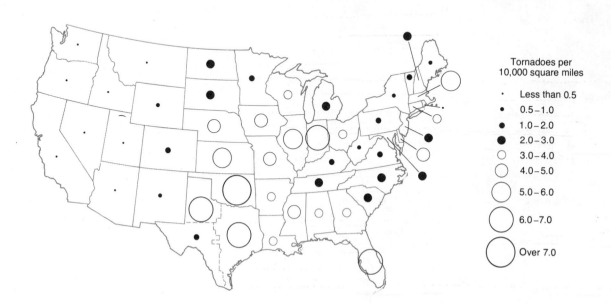

Tornadoes per 10,000 square miles	
·	Less than 0.5
•	0.5–1.0
●	1.0–2.0
⬤	2.0–3.0
○	3.0–4.0
○	4.0–5.0
○	5.0–6.0
○	6.0–7.0
○	Over 7.0

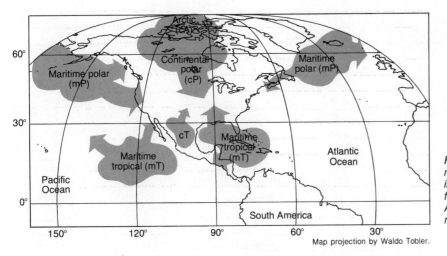

Fig. 8.13 Principal air-mass source regions and paths of air-mass movement in and around North America. (Adapted from maps of the National Oceanic and Atmospheric Administration, U.S. Department of Commerce)

are denoted c (for continental). The temperature characteristic is specified as arctic (A), polar (P), tropical (T), or equatorial (E) in order from coldest to warmest. The system is illogical, as polar really ought to be colder than arctic, but the convention has been established, so we shall have to use it.

The predominant air masses in North America are mP, cP, and mT. Most of the cold air masses are polar, but occasionally arctic air masses penetrate into the northeastern United States. The maritime polar air masses in North America originate mainly over the north Pacific, whereas the principal source of maritime tropic air masses is the Gulf of Mexico and the Caribbean Sea (Fig. 8.13). There are few cT air masses in North America, due to the small size of the source area in Mexico and Central America. In Eurasia, on the other hand, this source area is large enough to produce substantial cT air masses.

When air masses move from their source areas, both polar/arctic and tropical air masses migrate toward the midlatitudes, where they meet and are driven eastward by the westerly winds. The line of contact between tropical and polar/arctic air masses is called the *polar front*, and it often coincides with the pattern of geostrophic airflow aloft. Cyclones tend to move along the polar front.

The tracks of cyclones shift north and south with summer and winter. Weather forecasting in any season in the midlatitudes depends partly on predicting the paths of cyclones across the continent. Although the paths trend generally eastward, the particular direction taken depends on great waves of pressure in the upper atmosphere called Rossby waves. The polar front jet stream, a high-velocity geo-

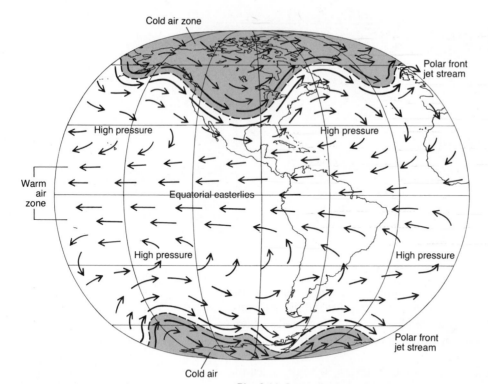

Map projection by Waldo Tobler.

Fig. 8.14 Generalized circulation in the upper troposphere, showing a typical configuration of Rossby waves along the edge of the cold air zones. The location of these waves is often marked by the polar front jet stream. The tracks of cyclones generally follow this airflow.

strophic wind so well known now because of its importance in air travel, follows these waves (Fig. 8.14). The jet stream provides for divergence of ascending air once it reaches high elevations, and this is essential for the maintenance of the surface-level cyclone. Otherwise, the cyclone would dissipate soon after formation.

Convectional precipitation. Because of the differences in the thermal properties of the materials that make up the landscape, pockets of hot air can develop in some areas. When such a pocket reaches a certain temperature and density, it becomes gravitationally unstable and floats into the atmosphere. (See the discussion in Chapter 3.) As it rises, it cools, and when the dew point is reached, condensation may set in. At this point instability is increased by the heat given off in condensation; in fact, the condensation and discharge of latent heat can be massive, leading to torrential rainfall and violent turbulence and the formation of a *thunderstorm* (Fig. 8.15). Although most thunderstorms are convectional in origin, in the midlatitudes they may also be initiated by cold fronts.

A thunderstorm functions as a huge chimney, with a strong upflow of air cells in its interior. Aloft, the air expands with condensation and cloud formation, and the top billows out to form a huge head. Precipitation droplets form quickly, but are usually delayed in falling because they are held in the cloud by the powerful updrafts. When they do grow heavy enough to fall, the weight and drag of the falling droplets create downdrafts in sections of the thundercloud. On the ground, one can often anticipate the onset of rain by sensing the cool downdrafts.

The distribution of convectional precipitation corresponds to the intensity of surface heating and the availability of moisture. Accordingly, the heaviest concentration of this precipitation is over land in the

Fig. 8.15 *Formation of a thunderstorm can be described in three stages: cumulus, mature, and dissipating. The first is characterized by updrafts, the second by updrafts, downdrafts, and heavy rain; and the third by downdrafts and light rain. (Illustration by William M. Marsh)*

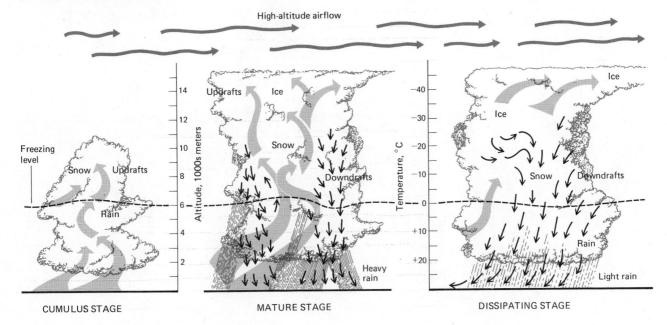

equatorial regions and in the ITC zone may produce as much as 300 cm (125 inches) of rainfall yearly. Poleward convectional precipitation declines in frequency and intensity, and in the midlatitudes it is limited to four or five months in the summer (Fig. 8.16).

Convergent precipitation. Convergent precipitation results when air moves into a low-pressure trough or enclosed topographic area from which it can escape only by rising upward. The ITC zone, which is fed by the trade winds, is such a trough of low pressure. Although individual storm cells may be convectional in origin, the net upward flow in the ITC zone is fundamentally convergent.

The most intensive convergent precipitation occurs in hurricanes, which are tropical cyclones that develop from disturbances in the trade-wind belt in both hemispheres. The storm is set into whirling motion by the Coriolis effect, drawing in moisture-laden tropical air as it travels eastward and poleward into the subtropical areas. When a hurricane enters a land area, it begins to subside, because continental air provides insufficient moisture, and thus latent heat, to power the system (Fig. 8.17).

Fig. 8.16 Mean annual number of days with thunderstorms. The incidence of thunderstorms in the Gulf Coast is two to three times higher than in the Northern states and as much as fifteen to twenty times higher than at comparable latitudes on the West Coast. *(From John L. Balwin,* Climates of the United States, *Washington, D.C.: U.S. Department of Commerce, 1974)*

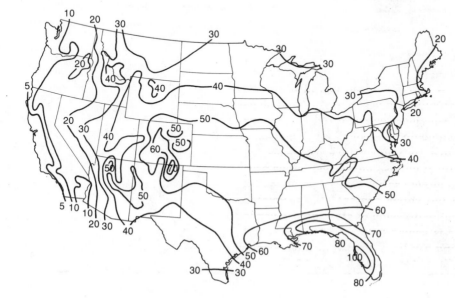

Florida

Fig. 8.17 Hurricanes are large tropical cyclones fed by massive amounts of latent heat from maritime tropical and equatorial air. Air pressure drops rapidly toward the center of the storm, where air ascends, inducing the development of a huge cloud bank. The very center, called the eye, is cloudless and relatively calm. Figure 32.9 shows hurricane tracks along the east coast of North America. (Illustration by William M. Marsh)

SUMMARY

Water vapor comprises less than 4 percent of the gas in the troposphere, is highly variable geographically, and is subject to rapid phase changes associated with energy exchanges between the earth's surface and the atmosphere. The total amount of vapor that can be held in air is dependent on air temperature; as temperature increases, the total vapor that can be held at saturation also increases. Vapor pressure, absolute humidity, saturation mixing ratio, and specific humidity are conventional expressions for the vapor content of air. Relative humidity is the most commonly used expression of atmospheric vapor.

Precipitation occurs when air is cooled below its dew point, and moisture falls to the surface. Cooling can be produced by advection or convection. Most precipitation is related to convection and falls in the form of rain. The types of precipitation are classified according to the cause of cooling. Orographic yields the heaviest annual averages of precipitation, exceeding 500 cm in many locations. Cyclonic/frontal precipitation is prevalent in the midlatitudes where warm air and cold air masses meet; convectional precipitation develops with greatest intensity and frequency in areas of intensive surface heating; and convergent precipitation results from the upflow of air when opposing winds meet as in the ITC zone.

CHAPTER 9 CLIMATES AND CLIMATIC CHANGE

INTRODUCTION

Climate is the word we use to describe the representative conditions of the atmosphere at a place on earth. Climate is more than an average of the weather over a period of time, because extreme and infrequent conditions, which are not evident in averages, are important traits of climate. In geography, climate refers to the conditions of the lower atmosphere that most directly affect the landscape and the organisms in it. To the Western world, the popular notion of climate has mainly to do with comfort and aesthetics. "It's warm and sunny there, but every now and again there's a hell of a thunderstorm." "Nice place for most of the year, but the winters can get mean; lots of snow in January and February some years." "Spring is lovely, but there's always the threat of tornadoes." "It's eternally hot, the humidity's high, and it rains nearly every day." These are descriptions of four different climates as they were perceived by persons who lived in them. They are informative remarks, but for two reasons are not very meaningful portrayals of the subject climates. First, they are inconsistent in terms of what aspects of climate are mentioned; second, they are subjective in that no numerical reference is given for hot, cold, rainy, and so on.

Climate Classification

This points up a basic problem in classifying the world's climates: What features, processes, and events of the atmosphere do we use to define climate, and what quantitative levels or values in terms of heat, precipitation, humidity, and so on are most appropriate? For example, heat is clearly an important characteristic of climate, but what temperatures are meaningful for the definition of a climate is not so clear. Only one temperature actually comes to mind immediately, namely, 0°C, to distinguish climates that experience frost from those that do not. But how should frost, or freezing temperature, be defined? Should it be defined on the basis of monthly mean temperature, daily minimum, or daily mean temperature? The answers are not easy to come by, and geographers have debated over the years about what categories are most meaningful. All we can conclude is that there is no single classification scheme that serves all purposes well; some are   **129** better for agriculture, others for human comfort, and others for the study of natural vegetation.

Our concern here is mainly with climatic processes and the controls on these processes. In particular, we want to examine these controls at a global perspective, in order to understand the variety of climates that exist over the earth. These climates result from land-water effects and locational and seasonal variations in the sun angle and atmospheric and oceanic circulation.

Classification of climate according to climatic processes and controls does not lend itself to the definition of very exact climatic zones. This tends to be true for any classification scheme, no matter what criteria are used, because the conditions of the atmosphere tend to change gradually over space, and zones or regions merge imperceptibly into one another. Thus we should bear in mind that climatic maps usually imply more precision than is warranted by the nature of atmospheric phenomena.

To reference the locations of the climates discussed in the following section, we will use a modified version of a standard climatic map. The original map was developed in the early part of this century by Wladimir Köppen, a German climatologist, and later was refined with the help of Rudolph Geiger, whose work we mentioned in Chapter 4. The Köppen-Geiger system classifies climates mainly according to temperature and moisture, with emphasis on the seasonal character of each. For example, among the warm or tropical climates (defined as climates with a mean monthly temperature of 18°C or greater in all months) are wet (rainfall in all months), winter dry, and winter short dry, and each is given a specific numerical definition. Although climatologists have identified several shortcomings in the Köppen-Geiger system, including the fact that the borders for some climates were drawn on the basis of vegetation patterns rather than climatic data, it does provide a useful framework for a description of global climate.

MAJOR CLIMATE TYPES

The major climates can be categorized as follows: tropical, midlatitude, and subarctic and arctic. Each of these three types can be further subdivided. Figure 9.1 presents annual temperature and precipitation data representative of each major climate type, and Fig. 9.2

Fig. 9.1 Average annual precipitation and temperature data for stations representative of the world's major climates. (Data from Environmental Data Service, Washington, D.C.: U.S. Department of Commerce)

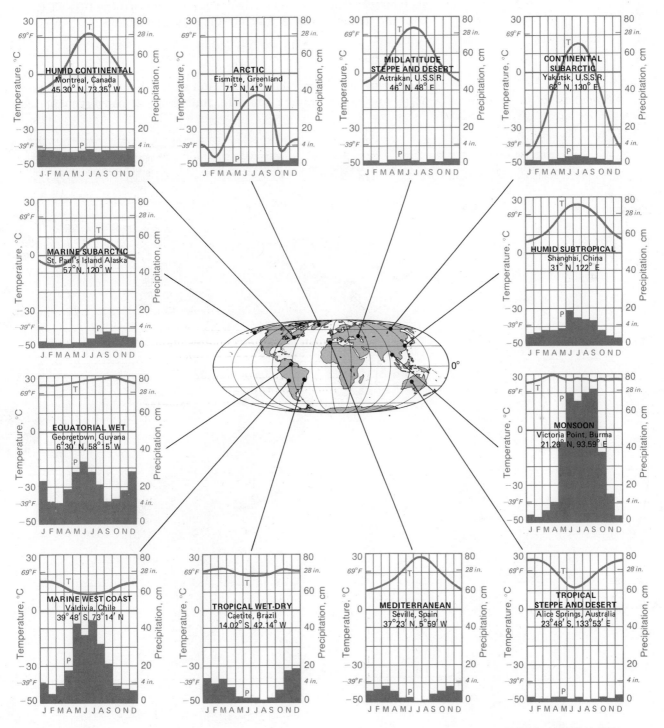

Map projection by Waldo Tobler.

shows a world map of climate distribution. Refer to these figures as you read the following descriptions of the world's climates.

Tropical Climates

Equatorial wet. Along the equator and as much as 10 to 15 degrees north and south from it, the ITC (intertropical convergence) zone

Fig. 9.2 World map of climate distribution.

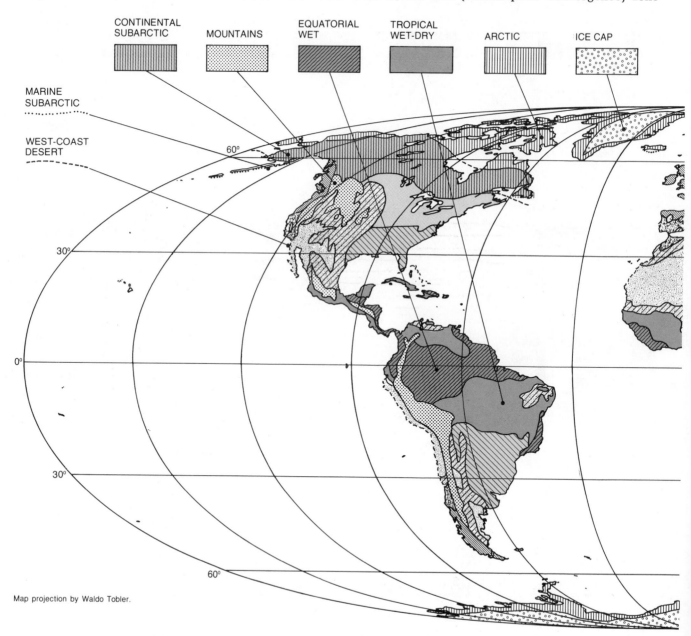

Map projection by Waldo Tobler.

continuously influences the weather because it is overhead or nearby in all seasons. Temperatures show remarkably little variation, with monthly means within a few degrees of 30°C. Every month has precipitation, and clouds are prominent for a part of most days in the year. Rainfall is generally due to convergent and convectional mechanisms. On the poleward margins of the equatorial-wet climatic zone, the easterly trade winds bring moisture to coastal areas, and during the

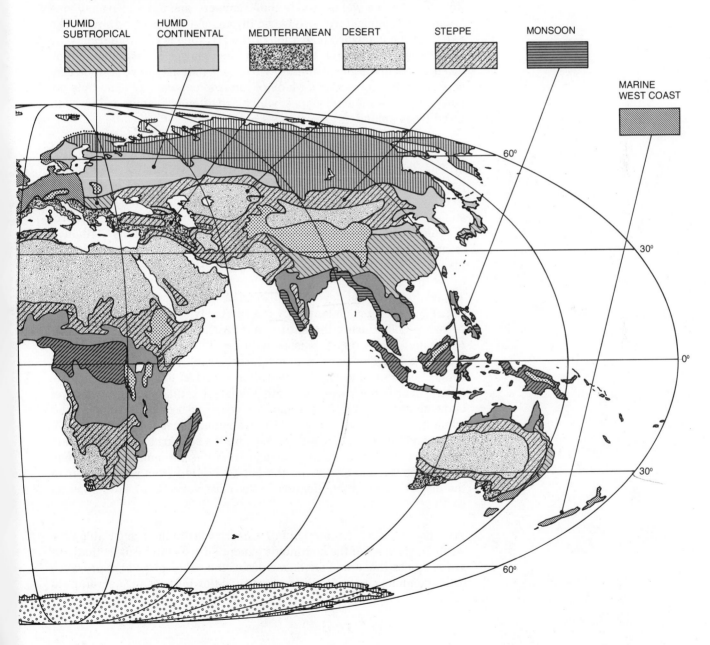

HUMID SUBTROPICAL

HUMID CONTINENTAL

MEDITERRANEAN

DESERT

STEPPE

MONSOON

MARINE WEST COAST

months around the equinoxes tropical cyclones may migrate onto these coasts (see Fig. 8.17). The landscape of the equatorial-wet climate is predominently heavy forest, called rainforest.

Tropical wet-dry. North and south of the equatorial-wet climatic regions are areas that are in the ITC zone during the high-sun season, but in the subtropical high-pressure cell during the low-sun season. Thus the areas have a wet season in the "summer" and a dry season during "winter." Temperatures are warm throughout the year, however. The extent of the poleward migration of the ITC zone is variable from year to year. Hence areas on the poleward boundaries of this climate type are likely to undergo substantial variations that may have serious consequences for agriculture. Much of the tropical wet-dry climate is occupied by savanna landscapes, which are often characterized by scattered trees among large expanses of grass.

Monsoon. The monsoon was described in Chapter 7. As a consequence of this winter-summer reversal of airflow, winters in South Asia particularly are dry, whereas summers consist of a series of storms (see Fig. 7.10). The warm, moisture-laden air from the Indian Ocean is subject to convective forces as it blows over the land, as well as orographic forces when it encounters the Himalayan foothills. Although monsoon-type circulation occurs in other places, nowhere does it produce as distinct a wet season and dry season as it does in Asia.

Tropical steppe and desert. Poleward of the tropical wet-dry climatic regions are areas dominated by the relatively stable subtropical high-pressure cells during the entire year. Airflow is downward and outward, and rarely do air masses with precipitable moisture penetrate these regions. The world's major deserts, such as the Sahara, Rub Al Khali, Sonoran, and Great Australian, are located in these regions. Cloud cover is the lowest on earth, and solar radiation at ground level is the highest (see Fig. 6.9). Temperatures are very high in summer and might be cool on winter nights. The transition zones between these climates and the tropical wet-dry areas are the semiarid steppes, characterized by a gradual spatial change from the perennially dry on one side to the seasonably dry on the other. Rainfall from year to year in the desert and steppe climates is the most variable of any place on earth.

West-coast desert. Located in fairly narrow latitudinal zones along the west coasts of all of the continents except Eurasia (and Antarctica) are areas where the stable eastern sides of the subtropical high-pressure cells combine with cold, equatorward-flowing ocean currents to produce cool, exceptionally dry climates. In the Atacama of southern Peru and the Namib of southwest Africa, the recurrence interval between rain events may be as much as one or two decades. Despite the paucity of rainfall, the combination of cool marine air and cool nights produces frequent fog, which may dampen surfaces (see the 20°C summer isotherms in Fig. 7.13).

Midlatitude Climates

Midlatitude climates lie in the zone of prevailing westerly winds. Climatic variation on a locational or seasonal basis depends on the relative positions of the subtropical high-pressure cells and the polar front, as superimposed on the general pattern of land/water controls. The polar front weakens and shifts poleward in the summer as the subtropical high-pressure cells intensify and expand poleward. Most of the precipitation is related to the cyclonic/frontal mechanism.

Mediterranean. On the west coasts of the continents, the stable eastern sides of the subtropical high-pressure cells effectively prevent frontal precipitation. Because these cells shift poleward during the summer, the summers are warm and dry. Temperatures are cool in the winter, and cyclonic storms, which pass farther equatorward with the shifts in the pressure zones, bring precipitation. This climatic type occurs in the Mediterranean (of course), California, Chile, South Africa, and western Australia.

Marine west coast. Poleward of the Mediterranean climatic regions are the marine west-coast areas. These areas lie in the path of cyclonic storms that originate over the ocean and drift onto the continents with the westerlies. Where the coasts are mountainous, as in North and South America, the orographic mechanism induces heavy precipitation, often exceeding 500 cm. The heavy rainfall and moderate coastal temperatures are conducive to the formation of heavily forested landscapes such as those of the American Northwest.

Humid subtropical. In the subtropics on the southeastern sides of large continents, cyclonic and convectional storms are the principal sources of precipitation. Midlatitude cyclones are predominant in winter, but in spring and fall tropical cyclones occasionally cross the coastal areas. In North America and Asia, incursions of polar air masses may take place many times each winter (see Fig. 8.14). As the polar front weakens in summer, the frontal precipitation is largely replaced by convectional precipitation, and the monthly averages of precipitation rise. In coastal areas, especially in China, a monsoon-type airflow also augments summer precipitation. In contrast to Mediterranean areas on the west coasts, these areas lack the thermal influence of the oceans; therefore, seasonal temperatures tend to be more extreme.

Humid continental. Poleward of the humid subtropical climate, winters are more severe, with temperatures averaging below freezing in several months. Midlatitude cyclones are the principal cause of precipitation, although convection contributes appreciable rainfall in summer. Temperatures are subject to sudden changes associated with the passage of polar/arctic or tropical air masses, especially in fall and spring. Because of the absence of sizable land masses in the midlatitude of the Southern Hemsiphere, humid continental conditions are not found there.

Midlatitude steppe and desert. In North America, Eurasia, and the southern part of South America, precipitation declines on the interior sides of the humid continental and humid subtropical climates, eventually giving way to steppe. Cut off from maritime air by mountain ranges, conditions grow even drier farther inland, and the steppe graduates into desert. Midlatitude steppe and desert differs from its tropical counterparts in that it tends to be less severely arid, especially in winter, and more seasonal in temperature, with winter temperatures often falling below freezing.

Subarctic and Arctic Climates

Continental subarctic. In the northern interiors of North America and Eurasia, winters are fairly dry and very cold. The dryness is due to the stability of the high-pressure cells that develop in response to the winter cold. These are the source areas of continental polar air masses, which move southward and eastward into the humid continental zones and beyond. At depths of a meter or more in the ground, permafrost can be found through much of the continental subarctic zone (see Fig. 3.7). Summers may be mild, with long hours of sunlight, but there are only two to three months without frost.

Marine subarctic. Along the shifting polar front, particularly on west coasts, the climate remains cold and wet throughout the year. Due to the sea, however, temperatures are not nearly as cold as in the interiors and hover around freezing throughout much of the winter. These coastal regions, however, have some of the cloudiest, windiest weather on earth. Indeed, in the language of the Aleuts, the native inhabitants of the Aleutian islands, there is no way to actually say "It's a nice day." The phrase used, when literally translated, corresponds more closely to "The wind isn't blowing so hard today." The continental sub-arctic climate includes both boreal forest and tundra landscapes, a distinction made on the basis of vegetation. Tundra is the colder of the two, but the climatic processes are the same in each.

Arctic and ice cap. In areas of permanent snowfields and ice caps, average air temperatures do not exceed 0°C in any month. Despite the abundance of snow and ice, little snow is precipitated from the dry arctic air; the annual snowfall in liquid water is less than 25 cm. The equatorward margin of this zone gives way to the tundra landscape, where most snow melts away in summer but permafrost underlies the entire landscape. Greenland, Antarctica, and North America and Eurasia above 70° N latitude are the principal areas of arctic climate. Outside these areas, arctic conditions are also found in isolated high-mountain areas, some of which lie in the tropics (see Chapter 29).

CLIMATIC CHANGE

Background

This topic has received increased attention in recent years, because it has become evident that climatic changes of a magnitude sufficient to affect large populations of plants and animals are taking place. We know that at times during the earth's history, the global distribution of climates was much different than it is today; further, we know that such changes will continue and that they will influence the capacity of the planet to support life.

How do present climatic conditions compare with those at various times in the past? At present the average earth temperature is warmer than it has been over most of the last million years, but colder than it has been over most of the last billion years. We are probably still in an Ice Age period, as current temperatures are below those of the inter-glacial stages by about 1°C.

During the period of about 70 to 200 million years ago, climate on the continents was wetter and warmer than at the present. Dinosaurs lived, and the lush plant growth that now is the source of our coal and oil was widespread. But how much of this can be attributed to continent drift and the location of the continents in the tropical latitudes at that time or to conditions peculiar to the atmosphere itself is difficult to say. (See the discussion in Chapter 25 for details on continental drift and plate tectonics.)

About 2 million years ago the climate cooled, and the present Ice Age began. This period has been marked by dramatic expansion and contraction of the world's ice volume. Over the past million years, global temperature fluctuations have amounted to about 4°C; even since the last major retreat of the ice, there have been major temperature fluctuations, about 2°C over the last 10,000 years (Fig. 9.3). After the last glacial peak, about 16,000 years ago, the climate became warmer until about 4000 B.C., then gradually cooler with intervening warming trends.

During the several centuries before the birth of Christ, relatively large populations of nomads lived in areas of Africa and Southwest Asia where today they could not. We know of the existence of caravan routes over areas that today are not passable. This does not mean that these areas were not deserts then. They were, but there are degrees of dryness, and oases were spaced more closely then. Portions of the Sahara still have beneath them large bodies of groundwater (in the Nubian sandstone, for example) inherited from this more favorable climate.

Several identifiable trends in climate temperature have occurred during the last 2000 years. A major warming trend culminated in the period A.D. 800–1200 (Fig. 9.3d). During this period the Vikings colonized Iceland, Greenland, and North America, and grapes grew in England. The vines are there still, but they do not bear fruit. Increased storminess and more severe winters began in the centuries following A.D. 1200.

In the thirteenth and fourteenth centuries the Viking settlements in North America and Greenland were driven out, and a severe outbreak of the Black Death (plague) in 1348–1350 reduced the world population by almost half. Glaciers advanced; famines were frequent; and the period from the fourteenth to the eighteenth centuries is sometimes called the "Little Ice Age." Around 1750 a warming trend began, apparently lasting until the 1940s, when temperatures began to get cooler, although perhaps not in the Southern Hemisphere.

In addition to the rather dramatic historical events mentioned above, many gradual changes in human populations have been associated with climatic fluctuations in the 3000 past years or so. Though some of these changes were related to climatic variations only coincidentally and were actually caused by political and cultural events, many were undoubtedly related to climate in one way or another. We can only guess about such relationships, however, because climatic records go back only a hundred years or so for much of the world.

Modern Implications of Climatic Change

With the burgeoning of human populations in the past several centuries, we have grown increasingly susceptible to fluctuations in climate. One reason for this is that large numbers of people have been forced to seek out a living in lands that are considered to be marginal for agricul-

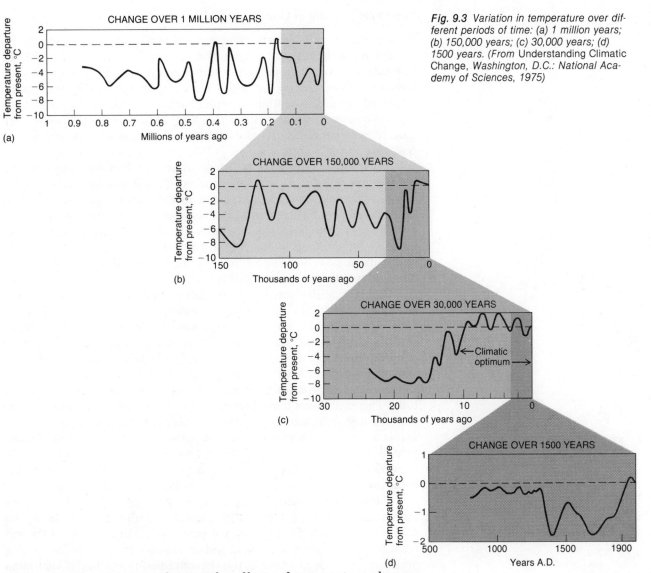

Fig. 9.3 *Variation in temperature over different periods of time: (a) 1 million years; (b) 150,000 years; (c) 30,000 years; (d) 1500 years. (From* Understanding Climatic Change, *Washington, D.C.: National Academy of Sciences, 1975)*

ture and settlement. In evaluating the effects of temperature changes on the order of 2°C or so, it is essential to realize that much of the major agricultural area of Canada and the Soviet Union is currently marginal. That is, in especially cold years, which may occur once or twice a decade, the growing season is just long enough to sustain a crop. If the mean temperature drops by 2°C, it might easily shorten the growing season thirty days or so, causing failure in certain crops. Since the farming economy is already marginal, a few crop failures can depress it below the survival threshold, and farmlands will fall out of production.

Search for a Cause of Climatic Change

Why is the earth's climate change a subject of intensive, ongoing research in the geophysical sciences? A major difficulty in answering the question is that we do not understand very much about the long-term feedback mechanisms between the atmosphere and the ocean. Although we might be able to explain what effects would result from certain causes, predicting the magnitude of these effects is very difficult; hence evaluating the net result of conflicting effects is often not possible.

In trying to present some possible mechanisms for climatic change, we will focus our attention on the last million years. During this period some significant changes have occurred, characterized by the advance and retreat of great glaciers, yet there has not been appreciable change in the locations of the continents. We know, of course, that shifting in the latitudinal position of a continent would cause its climate to change, but we also know that changes do not necessarily depend on such shifts.

The last million years are also interesting because the continents have been favorably positioned for widespread glaciation. The basic requirement is, of course, that more snow falls than melts in some lands areas. A suitable geographic arrangement of the continents is one that has a great deal of land at high and middle latitudes, preferably some of it at high altitude above the Arctic Circle. This by no means ensures glaciation, because if winters are too cold, little snow falls. The conditions most favorable to glaciation are actually warm winters and cool summers. This necessitates the presence of ocean water at high latitudes, because the water provides a source of moisture in winter and helps to depress temperatures in summer. In cold winters, sea ice builds up, causing even cooler summers, a fact that helps us account for the coincident glaciation in both hemispheres.

But what forces drive the changes in climate? A first place to look for mechanisms of change is in the amount of radiation received by the earth from the sun. There is no evidence that the total output of the sun itself has varied significantly; however, regular changes in the earth's orbit would cause changes in the seasonal and latitudinal distribution of solar radiation. There are three orbital variations to consider:

1. The eccentricity varies from a more elliptical orbit to a nearly circular one, with a period of 90,000–100,000 years. At present the orbit is nearly circular and becoming more so, making seasonal differences less pronounced.

2. The obliquity—the angle between the earth's axis and a line normal to the plane of the orbit—varies from 21.8° to 24.4°, with a period of about 40,000 years. At present the angle is

23°27″ and is decreasing. As the angle decreases, seasonal differences become smaller.

3. The precession of the equinox has a period of about 21,000–23,000 years. This means that the time of year when the earth is closest to the sun, currently January 5, varies with this period. At present the sun is closer to the earth during the Northern Hemisphere winter, hence making the winters relatively warmer and the summers relatively cooler there. In the Southern Hemisphere this position makes the seasons more different.

Figure 9.4 shows the variation in total ice volume over the last million years; evident is a periodicity between maxima of about 100,000 years, with a very short time required for a decay to a minimum, only about 10,000 years. In addition, analysis of the periodicities of the temperatures at which deep sea sediments were formed show strong peaks at 100,000, 41,000, and 23,000 years.

For variations on scales shorter than 21,000 years, the evidence is not so compelling for astronomical influences. Variations in solar output with the eleven-year sunspot cycle have not been convincingly related to any widespread changes in climate. In fact, there is no evidence of any *regular* variations in climate on a time scale of hundreds or tens of years. For these reasons we must seek the answer in the earth-ocean-atmosphere system. There are many plausible mechanisms, and we will mention only a few of them here.

1. The climatic system has natural variation, and the "feedback" between the effects is poorly understood. Here "feedback" means that the consequences of certain effects will either amplify or dampen a variation. For example, increased volcanic activity will lead to more dust in the atmosphere, which will reduce solar radiation, leading to lower temperatures and more ice, causing increased albedo and lowered net solar radiation. Spring comes later, cyclone

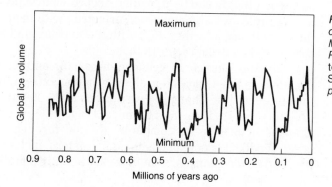

Fig. 9.4 Estimate of the global ice volume over the last million years. (From B. J. Mason, "Towards the Understanding and Prediction of Climatic Variations," Quarterly Journal of the Royal Meteorological Society **102**, 433 (July 1976): 473–498. Reprinted by permission.)

tracks are closer to the equator, and ice-covered seas and a lowered sea level reduce the poleward transfer of heat, causing still lower temperatures in the higher latitudes. These effects are what we often call "positive feedback," in that the original cooling trend caused effects which led to further cooling. However, there are checks and balances that would tend to prevent the development of another ice age. Cool temperatures and increased cloudiness would reduce the amount of longwave radiation to space, and the increased storminess in the midlatitudes would lead to increased heat exchange with the tropics.

2. Patterns of land use and changes in land cover cause climatic changes. (In Chapter 4 we examined the differences between urban and nonurban climates.) Over larger areas the alteration of vegetation, through clearing for agriculture or grazing by cattle, affects the heat balance of a region. (See Chapter 22 for additional discussion on this topic.)

3. Changes in atmospheric gases and particulates will have climatic effects. Increased particulates (called *aerosols*) from volcanic, industrial, or agricultural sources will reduce solar radiation, change atmospheric temperature profiles by raising the air temperature at altitudes where absortion takes place, and form condensation nuclei. Increased carbon dioxide would reduce outgoing longwave radiation in the atmosphere, thereby cooling the stratosphere and warming the surface. Chlorofluoromethanes added to the atmosphere in the past few decades from refrigerators and spray cans are notorious for their possible effects on the ozone layer, which is responsible for absorption of a large portion of the incoming ultraviolet radiation.

SUMMARY

Climate is the representative conditions of the atmosphere at a place on earth. Climate can be classified according to many different factors; in geography, we are concerned primarily with temperature and moisture conditions, how they vary with the seasons, and what climatic processes and controls regulate them.

Climatic change is common to earth, and in the past several decades we have come to realize its importance in human affairs and the quality of life. Changes with periods of 20,000 to 100,000 years are probably related to astronomical factors, whereas shorter-term changes appear to be complexly related through earth-atmosphere-ocean feedback mechansims to many factors, including volcanic activity, land-cover changes, urbanization, and air pollution.

UNIT II SUMMARY

- Earth is nearly a true sphere, and locations on it are referenced by a coordinate system of parallels and meridians.

- The receipt of solar radiation on the earth varies with latitude, season, and atmospheric conditions. Given a clear atmosphere, the equatorial zone would receive the most solar radiation, but because of heavy cloud cover near the equator, the largest annual receipts at the surface occur in the subtropical deserts, where clouds are scarce.

- The large annual variation in sun angle in the midlatitudes contributes to the seasonal character of the climates there, especially over large land masses, which heat and cool quickly in response to changes in incoming radiation.

- The atmosphere can be subdivided vertically according to temperature. The lowest zone is the troposphere, which ranges in thickness from 7–17 km; above the troposphere is the stratosphere, where most of the absorption of ultraviolet radiation by ozone takes place.

- The pressure exerted on the earth by the atmosphere is equal to 1013.25 mb at sea level. Because air is highly compressible, pressure falls rapidly with elevation, reaching about 500 mb at 5500 m elevation.

- Air is a mixture of many gases, over 99 percent of which are nitrogen and oxygen. Of the minor constituents, water vapor, carbon dioxide, and ozone are very important because of their capacity to absorb radiation.

- The atmosphere is set into motion in response to pressure differences, which are produced mainly by differential heating of air. The particular geographic patterns assumed by wind systems are strongly influenced by the Coriolis effect, which causes air to deviate from a pressure-gradient route and to flow along great S-shaped routes when moving from high-pressure cells to low-pressure cells.

- Circulation in the upper atmosphere tends to counterbalance surface circulation. According to the principle of angular momentum, circum-global airflow in the upper atmosphere becomes faster as it moves pole-ward from the ITC zone; however, the system breaks down over the sub-tropics, and the air descends. This results in the formation of Hadley cells in the tropics (or intertropics), with trade winds on the surface and anti-trade winds aloft.

- The global pattern of air pressure is a result of: (1) differences in atmospheric heating related to latitude and the distribution of land and water;

and (2) the dynamics of airflow in the upper atmosphere, as illustrated by the subtropical highs.

■ Oceanic circulation largely follows that of the prevailing wind systems, and a gyre forms in each major ocean basin north and south of the equator. As in the circulation of the atmosphere, ocean currents also effect a large latitudinal transfer of energy.

■ The amount of water vapor contained in air when it is saturated increases with air temperature. When air is cooled to its dew point, condensation will set in if condensation nuclei are available. Cooling in the atmosphere occurs primarily by means of convection and secondarily by means of advection.

■ The types of precipitation are classified according to the principal causes of cooling and condensation. The causes are airflow over mountainous terrain, meeting of cold and warm air masses, ascent of unstable air, and the upward flow of air from convergent wind systems.

■ Convectional precipitation occurs with greatest intensity and frequency over land masses in the low latitudes, where there is a supply of moist air. Orographic precipitation occurs at any latitude where mountains lie in the path of moisture-laden winds. Frontal/cyclonic precipitation occurs mainly in the midlatitudes, where cold and warm air masses meet, and convergent precipitation is most common in the equatorial and tropical zones, where wind systems meet and where air is drawn into tropical storms.

■ The geographic patterns of climate are the result of atmospheric processes controlled largely by land-water effects, locational and seasonal variations in sun angle, and atmospheric and oceanic circulation. In general, world climates can be divided into tropical, midlatitude, and subarctic and arctic types.

■ Changes in climate have been common in the earth's past and will continue to be so in the future. Humans are growing increasingly susceptible to climatic change because large numbers of people have been forced to live in marginal lands where small changes in temperature or moisture can have drastic effects on agriculture.

■ The causes of climatic change are a topic of much research and debate today. Changes with periods of 20,000 to 100,000 years appear to be caused by variations in the receipt of solar radiation related to astronomical factors such as the shape of the earth's orbit and the tilt of the earth's axis. For shorter-term changes, however, the principal causes appear to be associated with variations in the earth-ocean-atmosphere system, which stem from natural and human-made changes in air quality and land cover.

FURTHER READING

National Academy of Sciences, *Understanding Climatic Change*, Washington, D.C.: U.S. Government Printing Office, 1975. *A review of findings on regional and global climatic change as of 1975.*

Oliver, John E., *Climate and Man's Environment: An Introduction to Applied Climatology*, New York: Wiley, 1973. *An interesting examination of a wide variety of topics, including climate and landforms, climate and industry, and climatic change.*

Perry, A. H., and J. M. Walker. *The Ocean Atmosphere System*, London: Longman, 1977. *A review of the processes and energy exchanges between the oceans and atmosphere. Contains many useful maps. Requires some advanced work in physical science.*

Trewartha, G. T., *Introduction to Climate*, 4th ed., New York: McGraw-Hill, 1968. *A standard text on climatology with a good description of world climates.*

REFERENCES AND BIBLIOGRAPHY

Ackerman, A. E., "The Köppen Classification of Climates in North America," *Geographical Review* **31** (1941): 105–111.

Bowen, I. S., "The Ratio of Heat Losses by Conduction and by Evaporation from Any Water Surface," *Physical Review* **27** (1926): 779–787.

Budyko, M. I., *Atlas of the Heat Balance*, Leningrad: Gidrometeoizdat, 1955.

Budyko, M. I., *et al.*, "The Heat Balance of the Surface of the Earth," *Soviet Geograph. Rev. Transl.* **3**, 5 (1962): 3–16.

Byers, H. R., *General Meteorology*, 4th ed., New York: McGraw-Hill, 1974.

Climates of the World, Washington, D.C.: U.S. Department of Commerce, Environmental Data Service, 1969.

Climatic Atlas of the United States, Washington, D.C.: U.S. Department of Commerce, Environmental Data Service, 1968.

DeCandolle, A., *Géographie botanique raisonnée, ou exposition des faits principaux et des lois concernant la distribution géographique des plantes de l'époque actuelle*, vol. 2, Paris: Masson, 1855.

Defant, A., *Physical Oceanography*, 2 vols., New York: Pergamon Press, 1961.

Dietrich, G., *General Oceanography*, New York: Wiley/Interscience, 1963.

Donn, W. L., and M. Ewing, "A Theory of the Ice Ages, III," *Science* **152** (1966): 1706–1712.

Dregne, H. E., "Desertification: Man's Abuse of the Land," *Journal of Soil and Water Conservation* **33**,1 (1978): 11–14.

Fairbridge, R. W., ed., *Encyclopedia of Atmospheric Sciences and Astrogeology*, New York: Reinhold, 1967.

Godske, C. L., T. Bergeron, J. Bjerknes, and R. C. Bundgaard, *Dynamic Meteorology and Weather Forecasting*, Boston: American Meteorology Society, 1975.

Goody, R., *Atmospheric Radiation*, London: Clarendon Press, 1964.

Hare, F. K., "Future Climates and Future Environments," *Bulletin American Meteorological Society* **52** (1971): 451–456.

Huntington, E., *Mainspring of Civilization*, New York: Wiley, 1945.

Kendrew, W. G., *The Climates of the Continents*, London: Oxford University Press, 1922.

Köppen, W., ''Versuch einer Klassifikation der Klimate, vorzugsweise nach ihren Beziehungen zur Pflanzenwelt,'' *Geograph Z* **6** (1900): 593–611, 657–679.

Köppen, W., and R. Geiger, *Klima de Erde* (map), Darmstade, Germany: Justus Perthes (and Chicago: Nystrom), 1954.

Starr, V. P., and J. P. Peixoto, ''On the Global Balance of Water Vapor and the Hydrology of Deserts,'' *Tellus* **10**, 2 (1958): 188–194.

Shapley, Harlow, ed., *Climate Change*, Cambridge Mass.: Harvard University Press, 1953.

Thornthwaite, C. W., ''The Climates of North America According to a New Classification,'' *Geographical Review* **21** (1931): 633–655.

_____ ''Problems in the Classification of Climates,'' *Geographical Review* **33** 2, (1943): 233–255.

U.S. Nautical Almanac, Washington, D.C.: Government Printing Office, 1976.

WATER IN THE
LANDSCAPE

UNIT III

KEY CONCEPTS   OVERVIEW

hydrologic cycle
moisture variability
infiltration
capillary water
field capacity
permeability
gravity water
porosity
water table
hydraulic gradient
recharge
transmission
aquifer
artesian flow
cone of depression
cone of ascension
hydrologic equation
runoff
drainage basin
drainage network
overland flow
interflow
baseflow
pruning
concentration time
rational method
hydrograph method
outflooding
inflooding

Water is brought to the land by the atmosphere. The patterns in which water is precipitated on the continents, however, is irregular over both time and geographic space. As a result, the amount of water on the land, in the form of streams and lakes, and below the surface, in the form of soil water and groundwater, tends to be uneven from place to place and from time to time over most continents.

The disposition of precipitation once it has reached the land depends largely on the particular characteristics of the ground itself, including whether it is vegetated or barren, rock or soil, flat or sloping. These factors determine how much water infiltrates the soil and how much runs off the surface. Of the water that moves into the soil, part is taken up by the soil, eventually to be lost to evaporation and plants, and part moves to greater depths to become groundwater.

Groundwater resides in the empty spaces between soil particles and in the cracks in rock. The upper boundary of the groundwater mass is called the water table, and it usually slopes with the slope of the land. As a result, groundwater tends to flow downhill, where it may intercept a stream channel and discharge into the stream.

During dry periods groundwater usually constitutes the only source of water for streams. During and after rainstorms, however, streams may receive large amounts of water from surface runoff, called overland flow, and water that seeps out of the soil above the water table, called interflow. Streamflow, therefore, can vary radically, depending on the amount and intensity of rainfall and snowmelt. Changes in the landscape made by humans usually result in increased overland flow; urbanization is especially critical in this respect.

Chapter 10 briefly examines the hydrologic cycle, providing some historical background on this concept. Chapter 11 deals with infiltration, the movement of water in the soil, and soil moisture. Chapter 12 is concerned with groundwater and its dynamics. Runoff, streamflow, and flooding are described in Chapter 13.

THE HYDROLOGIC CYCLE

INTRODUCTION

Hydrology is the study of water. Since water in one form or another influences most processes on the earth's surface, it is studied in virtually all fields in the natural sciences, humanities, and social sciences. Physical geography is concerned with the distribution of water and its influence on the landscape. Water is an important agent in the energy balance, particularly in the transfer of latent heat. In the form of streams and glaciers, water erodes earth materials, thereby shaping the land surface; in the form of soil water, it is important in rock weathering as well as in plant growth.

In this chapter we begin our discussions of water by introducing the "hydrologic cycle." This is an idealized model of the land-ocean-atmosphere water exchange, or cycle, which includes evaporation from the sea, movement of water vapor over the land, condensation, precipitation, surface runoff, subsurface runoff, and so on. In reality, the flow of water from atmosphere to land, land to atmosphere, and land to oceans is complex and irregular over time and geographical space. This fact is not widely appreciated, however, because of the acceptance of a "standard" hydrologic cycle which has become more or less a norm of modern academic thought. Since this hydrologic cycle is a fundamental concept in natural science today, it is appropriate here to outline briefly its origins so that we may better understand its meaning and scientific utility.

DEVELOPMENT OF THE HYDROLOGIC-CYCLE CONCEPT

Humans have long puzzled over the origins of rainfall, rivers, streams, springs, and their interrelations, but our understanding of the true nature of the hydrologic cycle is comparatively recent. Prior to the sixteenth century, it was generally believed that water discharged by springs and streams could not be derived from the rain, for two reasons:

1. Rainfall was thought to be inadequate in quantity;

2. The earth was thought to be too impervious to permit penetration of water very far below the surface.

The ancients did, however, recognize that the oceans did not fill up and that rivers continued to flow. They recognized then that somehow the water got from the sea into the rivers and in the process lost its salt content. The Bible says:

All the rivers run into the sea, yet the sea is not full; unto the place from whence the rivers come thither they return again.
Old Testament—Ecclesiastes, 1:7

Generally the removal of salt was attributed to various processes of either filtration or distillation. The elevation of the water above sea level was ascribed to vaporization; subsequent underground condensation, to rock pressure, to suction of the wind, to a vacuum produced by the flow of springs, and other processes.

The recognition of the role of infiltration in supplying water to springs and rivers began in the sixteenth century. Leonardo da Vinci, an exceptional genius who was in charge of canals in the Milan area, is generally credited with one of the earliest accurate descriptions of the hydrologic cycle:

Whence we may conclude that the water goes from the rivers to the sea and from the sea to the rivers, thus constantly circulating and returning, and that all the sea and rivers have passed through the mouth of the Nile an infinite number of times. . . . The conclusion is that the saltness of the sea must proceed from the many springs of water which, as they penetrate the earth, find mines of salt, and these they dissolve in part and carry with them to the ocean and other seas, whence the clouds, the begetters of rivers, never carry it up."

John P. Richter, *Literary Works of Leonardo da Vinci,* 2 vols. London: Phidon Press, 1970.

In the seventeenth century the French scientist Pierre Perrault measured rainfall for three years in the drainage basin of the Seine River above a point in the province of Burgundy. He computed that the total volume of the rainfall was six times the river flow. Although his measurements were crude, he was able to disprove the fallacy that the rain was inadequate to supply the flow in rivers. Edmund Halley, the English scientist after whom Halley's comet was named, made estimates of evaporation from the Mediterranean Sea and demonstrated that it was as great as the flow of all rivers into the Mediterranean.

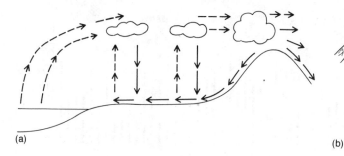

(a)

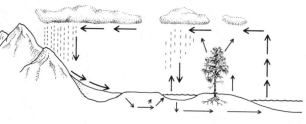

(b)

This was a period in history of energetic dialogue between Christian theologians and natural scientists, and new ideas about nature were carefully scrutinized by the Church. Although the theory of the hydrologic cycle was initially rejected by theologians, the modern geographer Yi-Fu Tuan tells us that the theory gained favor when it became clear that it could be used to support the doctrine of the Divine plan of nature. This doctrine held that the earth was created by God expressly as the home of humans and that all of its processes were parts of a great ordered scheme with humans at its center. In early versions of the hydrologic cycle, natural theologians saw verification by science of a portion of the Divine plan of nature. Eventually they adopted and idealized the model, and over the course of the past several centuries a more or less standard hydrologic cycle evolved. This model appears in academic texts today in a form little changed from that, for example, presented by theologians such as John Ray in the seventeenth century. Figure 10.1 shows presentations of the hydrologic cycle in seventeenth- and twentieth-century publications.

Fig. 10.1 Conceptions of the hydrologic cycle: (a) in the style of that presented in the seventeenth century by natural theologian John Ray (from Yi-Fu-Tuan, The Hydrologic Cycle and the Wisdom of God: A Theme in Geoteleogy, *Toronto: University of Toronto Press, 1968); (b) as fashioned in a modern earth science textbook in which the authors admit to oversimplification, but nonetheless present a cycle little changed from that of John Ray.*

WATER SOURCES AND DYNAMICS OF THE HYDROLOGIC CYCLE

Table 10.1 shows the proportions of water in various phases of the hydrologic cycle. The oceans, ice caps, and glaciers together are enormous reservoirs which contain 99.5 percent of the world's water. The amount in the atmosphere is surprisingly small. If *all* of the water held in the atmosphere at a given moment condensed out and rained evenly all over the earth, it would result in only 2.5 cm (1 inch) of precipitation. Contrast this figure with the worldwide average for precipitation of about 100 cm (39 inches) per year. This means that there is a complete exchange of atmospheric moisture about forty times every year.

Thus the atmospheric phase of the hydrologic cycle is dynamic, with atmospheric moisture being recycled an average of once every

Table 10.1 Proportions of water in the major phases of the hydrologic cycle.

PHASE	PROPORTION (%)
Oceans	97.6
Ice caps and glaciers	1.9
Ground- and soil water	0.5
Rivers, lakes	0.02
Atmosphere	0.0001

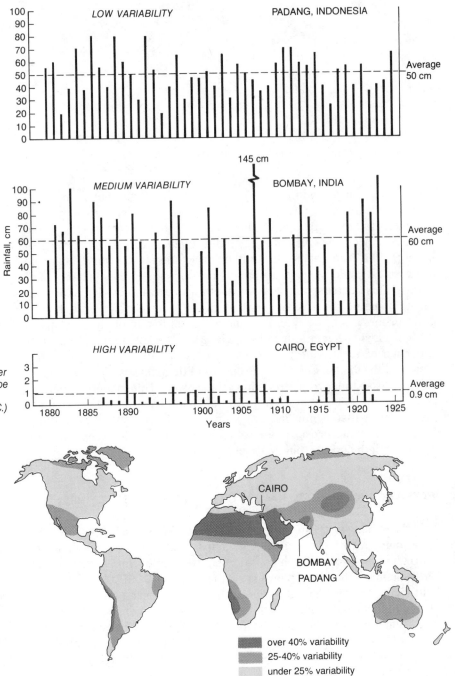

Fig. 10.2 *The graphs show the total annual rainfalls of Padang, Indonesia; Bombay, India; and Cairo, Egypt, for the period 1880–1925. Rainfall at Cairo is highly variable, which is the norm for deserts, as the world map suggests. If precipitation is highly variable, most other parts of the hydrologic cycle must also be highly variable. (Data from the Smithsonian Institution, Washington, D.C.)*

nine days. It is not surprising, then, that the cycle is variable in time and space over the planet. Careful examination of scientific data reveals, in fact, that the hydrologic cycle is not always a complete cycle. Under natural conditions water does not always pass through cycles; that is, it does not circulate through the system and return to the same source and form at regular time intervals. Evaporation from the oceans is highly uneven over the planet and absent altogether in some regions. Climatologists tell us that most of the water vapor blown onto some continents from the oceans is carried completely over the continents only to precipitate back on the ocean. Even for the water that does fall on the land, the cycle may be incomplete or highly variable. Small amounts of water locked in rock formations may not return to the oceans for millions of years. Similarly, large amounts of water are held in glaciers for hundreds and thousands of years before being released to the atmosphere or the oceans. During the Ice Age, continental glaciers formed a hydrologic bottleneck of sorts, holding enough water on land to lower sea level around the world by more than 75 meters. (See Chapter 30 for details about the Ice Age.)

Variability in the phases of the hydrologic cycle is also great over short segments of time. As Fig. 10.2 indicates, precipitation in many regions is highly changeable in quantity from year to year or from season to season. In deserts, for example, the annual rainfall varies more than 40 percent from the average in most years. Although the average may be 30 cm (12 inches), it may commonly total 15 cm or 50 cm in a given year. It follows that runoff is also highly variable, and in most years deserts generate insufficient runoff to reach the sea. Thus the land-ocean segment of the hydrologic cycle is often missing in arid lands. In short, many lines of evidence support the argument that the planet's water system is best characterized by variability rather than uniformity over time and space.

SUMMARY

Given this variability, what, then, is the value of the hydrologic-cycle concept? The best answer is that it establishes an orderly scheme from which we can begin to systematically examine and analyze the movement of water through the landscape. Moreover, it provides a construct against which we can compare scientific findings and ultimately sharpen our understanding about the nature of water on the land. Let us begin by examining the forms and processes of water on and within the soil (Chapter 11). In Chapter 12 we will consider groundwater, and in Chapter 13 we will conclude our discussion of the hydrologic cycle by examining runoff and streamflow.

INFILTRATION AND SOIL MOISTURE CHAPTER 11

INTRODUCTION

For our purposes, it is useful to begin with a brief look at a small segment of hydrologic cycle. Let us start with the flow of water in a humid setting from the lower atmosphere to the vegetative cover, soil surface, into the soil, and finally back to the atmosphere. Figure 11.1 shows generalized and detailed versions of this segment of the cycle.

Interception

With the exception of areas such as plowed fields and hard, artificial surfaces, such as streets and roofs, a measurable portion of rainfall is intercepted by plants before reaching the soil surface. As the plant surfaces become wetted, downflow (called stemflow) and drippage from the foliage are initiated (Fig. 11.2). A dense forest canopy may

Fig. 11.1 The hydrologic cycle: (a) the major inflows and outflows of water from a parcel of landscape; (b) a detailed version of these flows in a forested site within that parcel.

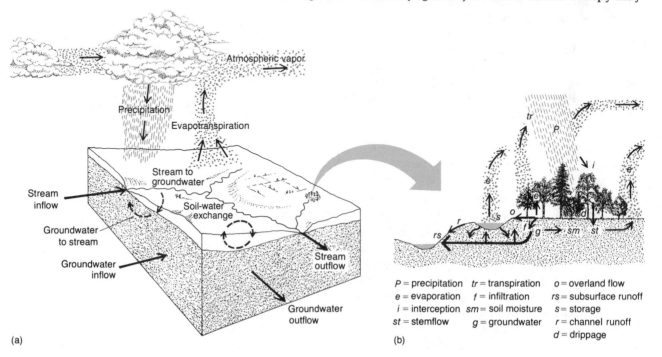

(a)

(b)

P = precipitation $\quad$ tr = transpiration $\quad$ o = overland flow
e = evaporation $\quad$ f = infiltration $\quad$ rs = subsurface runoff
i = interception $\quad$ sm = soil moisture $\quad$ s = storage
st = stemflow $\quad$ g = groundwater $\quad$ r = channel runoff
$\quad$ d = drippage

Stemflow

Fig. 11.2 Drippage and stemflow resulting from intercepted rainfall. Measurements show that conifers are somewhat more effective than broadleaf trees in intercepting rainfall.

intercept the total fall of a light rain, precluding receipt of moisture on the ground. Under these conditions, the moisture evaporates into the atmosphere, thereby completing the cycle short of the ground.

WATER ON THE GROUND

What happens to precipitation once it reaches the ground? There are three possibilities: It may (1) sit on the surface in puddles, ponds, or snow; (2) soak into the soil; or (3) run off over the surface into streams, lakes, and wetlands. Which takes place depends on both the infiltration capacity of the soil and the intensity and duration of rainfall.

In examining the energy balance, we frequently noted that the surface of the ground is a very important interface where energy is absorbed, reflected, and converted to other forms. The role of the ground surface is no less important in hydrology. Indeed, the work of the American hydrologist Robert E. Horton in the 1930s and 1940s showed that through its regulation of surface-water absorption, the soil surface has profound influence on runoff, streamflow, floods, and related processes.

The water-absorption function of the soil surface is termed *infiltration capacity* (*f*) and is defined as the rate at which water is received by the soil. Infiltration capacity is equal to the depth of water

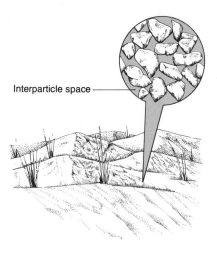

Interparticle space

Fig. 11.3 Large interparticle spaces commonly found among sand-size particles.

Table 11.1 Representative infiltration values.

PARTICLE SIZE (DIAMETER)	INFILTRATION RATE (MM/HOUR)
Sand (0.05–2 mm).	10–100
Silt (0.002 to 0.05).	2–15
Clay (smaller than 0.002 mm).	0.2–7

received divided by the time taken to receive it and can be expressed by the formula:

$$f = \frac{d}{t},$$

where d is depth of water received in inches or centimeters and t is time in minutes or hours.

Factors Influencing Infiltration Capacity

Interparticle spaces. One important control on infiltration capacity is the size and interconnectedness of interparticle spaces in the soil surface. Large spaces that are linked with extensive networks of intrasoil spaces tend to maximize infiltration. As any builder of sand castles knows, coarse-grained materials, such as beach sand, tend to have high infiltration capacities, especially if the sand is well sorted, i.e., of relatively uniform grain sizes (Fig. 11.3). Conversely, fine-grained soils, such as clay, have low infiltration capacities, sometimes less than 4 mm per hour. Table 11.1 gives representative infiltration values for sand, silt, and clay.

Vegetation. Another important set of controls on infiltration is vegetation and the associated organisms which reside in the soil. Burrowing creatures, such as worms and moles, build passageways which serve as water-entry routes into the soil. The roots and stems of plants loosen soil, further facilitating water penetration of the surface. This fact is strongly supported by the data in Table 11.2; for a given soil, the infiltration capacity of an herb-covered field (old permanent pasture) is nine to ten times greater than that of barren ground. In addition, plant stems and organic litter increase the roughness of the soil surface. This in turn decreases the speed of overland flow, thereby allowing more time for infiltration to take place. In addition, we find that on bare soil, the physical impact of the rain drops compacts the soil and reduces infiltration capacity. Vegetation virtually eliminates this effect. Figure 11.4 illustrates the influence of vegetation on the infiltration capacities of a sloping surface of three different soil textures.

Rainfall intensity. Rainfall intensity is a measure of the amount of rain that strikes the ground per minute or hour. If intensity is so great that the infiltration capacity on a sloping surface is exceeded for a short time, water is lost to runoff. Under the same conditions, but with a gentler, longer duration of rain, however, more water may be absorbed. But as rainfall continues, interparticle spaces become filled with water as well as with fine particles dislodged from the surface,

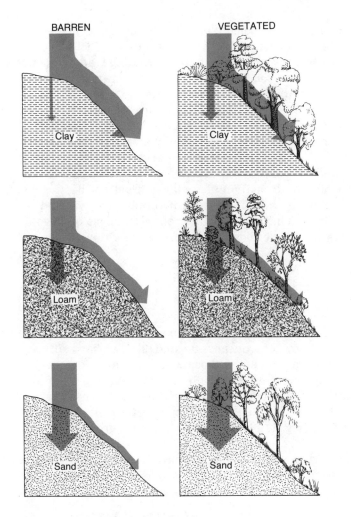

Fig. 11.4 *The differences in infiltration on vegetated and nonvegetated soils. The widths of the arrows indicate the relative amounts of water that infiltrate and run off.*

Table 11.2 Representative infiltration capacities of various ground covers.

GROUND COVER	INFILTRATION RATE, CM PER HOUR
Old permanent pasture	6.0
Permanent pasture, moderately grazed	2.0
Strip-planted crops	1.0
Weeds or grain	0.8–10
Clean, tilled soil	0.6–0.8
Bare ground, crusted	0.4–0.6

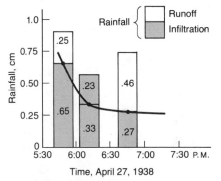

Fig. 11.5 *Infiltration rates for three short rainshowers that fell on a 2.7-acre plot of ground within about seventy minutes. Note that the amount of water lost to infiltration decreases sharply from the first to second shower, but decreases only slightly from the second to third shower. This trend, shown by the graph line, is typical of most soil surfaces. (Based on tests conducted by the U.S. Soil Conservation Service in 1938.)*

resulting in a decline in infiltration. Figure 11.5 depicts a sharp decrease in infiltration during the first fifteen minutes or so of rain. As infiltration decreases with each succeeding rainfall, the proportion of the rain converted to runoff increases correspondingly.

Seasonal factors. Several seasonal factors may have a strong influence on infiltration. If freezing takes place when the upper soil is at or near saturation, infiltration can be reduced to zero. Under such condi-

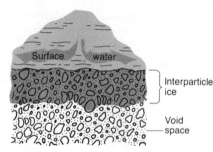

Fig. 11.6 *Only a thin layer of interparticle ice can reduce infiltration capacity to zero. In the northwestern United States such conditions have been known to contribute greatly to flooding by locking out water from snowmelt and rain, thereby forcing essentially all of it into streams and rivers.*

tions, interparticle spaces become blocked with ice and divert essentially all rainwater or snowmelt water into runoff (Fig. 11.6). In the Arctic, permafrost has a similar effect on percolating water. It is not surprising, then, that the soil layer above the permafrost is usually saturated or covered with ponded water in summer.

As the water content of the upper soil changes seasonally, so may infiltration capacity. In clayey soils, particularly those comprised of the clay mineral montmorillonite, addition of water may produce particle swelling and closing of interparticle spaces. Summer drying, in contrast, may produce contraction and cracking, although this may also be accompanied by formation of a clayey crust on the surface. These factors, combined with the seasonality of plant growth and the ground frost of winter, characteristically render changes of 25 to 50 percent in infiltration for periods ranging from a few days to several months.

SOIL WATER

Infiltration water has two directional components as it passes into the topsoil: (1) horizontal flow, or *transmission;* and (2) downward *diffusion* from pores of high water content to those with low water content (Fig. 11.7). As this water enters air-filled interparticle spaces, some of it clings to the walls of particles. Microscopic examination shows, in fact, that it attaches itself to an existing film of water molecules on the walls of the particles (Fig. 11.8). This film is termed *hygroscopic water*, and it is held to a particle surface by molecular forces so powerful that it cannot be removed from the surface by natural processes. In terms of pressure, these adhesive forces exceed 31 bars and may be as great as 10,000 bars (atmospheric pressure at sea level = 1013.2 millibars, i.e., about 1 bar).

Fig. 11.7 *Soil-water movement during infiltration. Diffusion constitutes downward movement; transmission, lateral movement. Since most terrain is sloping, both types usually occur simultaneously.*

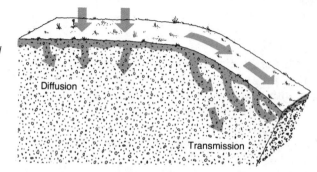

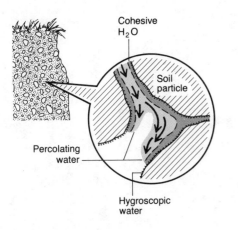

Fig. 11.8 *A film of cohesive water being deposited on the hygroscopic film as infiltration water percolates between soil particles. (Illustration by William M. Marsh)*

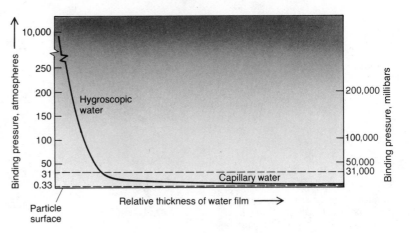

Fig. 11.9 *The relationship between the pressure under which molecular water is held to the particle surface and the thickness of the water film. Hygroscopic water is tightly bound at pressures exceeding 31 bars, whereas capillary water is loosely bound at pressures between 31 and 0.33 bars. (One bar of pressure exerts a force of 10^5 newtons per square meter, or 14.7 pounds per square inch.)*

Capillary Water

Cohesive forces between the ever-present hygroscopic water and the percolating water draw additional water molecules to the particles. As the film of cohesive water, or *capillary water* as it is more commonly called, grows in thickness, the forces that hold it to the particle weaken. At a binding pressure of less than 0.33 bars, the water molecules can no longer be held to the film and are removed by gravity. The relationship between binding pressure and thickness of the film of molecular water on a soil-particle surface is shown schematically in Fig. 11.9.

Field Capacity

Capillary water acts much the same as the meniscus, or "skin," on fluid in a tube in that the water surface clings together in a rounded shape. In small, confined soil-pore spaces, capillary water tends to coalesce and to concentrate along particle contacts (Fig. 11.10). The greater the number of particle contacts in a given volume of soil, the greater the capillary water holding capacity. It follows that fine-textured soils can hold more capillary water than can coarse-textured soils. The capillary water holding capacity of soil is termed *field capac-*

Soil particles

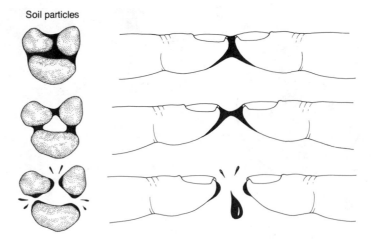

Fig. 11.10 *The tendency for capillary water to concentrate at the contact between particles is illustrated by this demonstration using wetted fingers.*

Table 11.3 Field capacities of five types of soil.

TEXTURE	CAPILLARY WATER CAPACITY*
Sand	7
Sandy loam	15
Silt loam	25
Clay	40
Peat	170

*Expressed as a percentage based on the weight added to dry soil by capillary water.

ity. Table 11.3 gives the field capacities for four soil textures and an organic soil, peat. Peat has an extraordinary holding capacity because the organic particles themselves absorb water in addition to holding capillary water around and between them.

If water is allowed to percolate into a soil until the capillary reservoir is full, the soil is said to be at *field capacity.* If even more water is added, it must pass through the soil, since the molecular cohesive tension in the pore spaces is less than 0.33 atmosphere. This water is called *gravity water* because it flows with the gravitational gradient. In contrast, capillary water moves in response to molecular gradients, which exert a stronger force than gravity. Since soil-water losses are usually greatest in the uppermost part of the soil—because evaporation and transpiration are concentrated there—capillary moisture gradients are typically upward, and hence the predominant directions of capillary moisture flux must also be upward.

To capsulize, three types of water are found in soil: hygroscopic, capillary, and gravity (Fig. 11.11). Hygroscopic is tightly bound, immobile molecular water and is of virtually no importance in hydrologic problems. Capillary water is loosely bound molecular water and *is* an important consideration in hydrologic problems. This water moves, evaporates, and can be utilized by plants and is thus a critical link in the subsurface water system.

Gravity Water and Soil Permeability

Gravity water occurs in the liquid state and flows in the part of interparticle space not occupied by capillary water. In coarse-textured

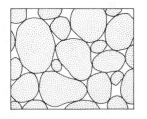 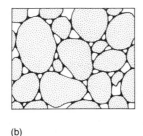 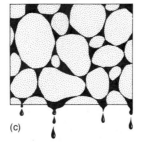

(a)   (b)   (c)

Fig. 11.11 The three types of water found in the soil: (a) hygroscopic water, which is totally imperceptible to any of the naked human senses; (b) capillary water, which in soil at field capacity is damp to the touch; (c) gravity water, which is readily apparent at a glance or a touch, for in most soils it drips from the sample when held in the hand.

soils, where these spaces are large and well connected, gravity water is transmitted rapidly. Such soils are said to have high *permeability*. It follows that soils with high permeability should also have low field capacities, since they lack the vast area of particle surfaces provided by fine particles.

Permeability is determined in the laboratory by measuring the time of passage of water through a specified volume of soil. A standard field measurement of permeability used in the environmental health sciences is called the *percolation test.* This involves first excavating a small pit and filling it with water. The water is allowed to drain, thereby equalizing the water content in the surrounding soil. The pit is then refilled with water, and the drop in water level is timed until the pit is empty, as is illustrated in Authors' Note 11.1. This test is usually used to determine the suitability of a soil for domestic septic-tank drainage in rural and suburban areas. The map shows the principal areas in the coterminous United States where soil drainage poses a limitation to septic-system disposal of domestic sewage.

SUMMARY

In a vegetated landscape precipitation is intercepted by plants before reaching the ground. On the ground, water may lie on the surface, run off, or infiltrate the soil. Infiltration capacity is the rate at which water

AUTHORS' NOTE 11.1
Soil Percolation and Sewage Disposal

The set of three diagrams illustrates the fall of water level in a percolation-test pit. This particular soil is well suited for septic drainage, as it received 60 cm (24 in.) of water in 38 minutes, a rate of 1.58 cm per minute. In 1970 approximately 50 million people in the United States relied on septic systems (soil-absorption systems) for sewage disposal. Failure of clayey soils, for example, to receive waste water rapidly has resulted in surface seepage of contaminated water, causing serious health problems as well as pollution of rivers and streams in many areas. The map shows the major areas in the coterminous United States with soil limitations for wastewater disposal by conventional septic systems.

3 minutes

18 minutes

38 minutes

Severe limitations   Moderate limitations

passes into the soil from the surface, and it is influenced by soil-particle size, vegetation, rainfall intensity, and other factors. Within the soil percolating water may be taken up as capillary water or trickle on through the soil as gravity water.

CHAPTER 12 GROUNDWATER

INTRODUCTION

If water percolates into a soil that is already at field capacity, it flows slowly through the soil into the subsoil, eventually reaching a zone where all of the interparticle spaces are filled with water. This zone is called the *zone of saturation*, or *groundwater*. As it moves toward this zone, gravity water tends to flow mainly downward, but it may also move laterally under certain conditions, especially where the terrain slopes steeply or impervious materials impede its downward flow. Where the soil is deep, several belts of this water, each representing a different rainfall, can often be found moving through the capillary zone, as shown in Fig. 12.1.

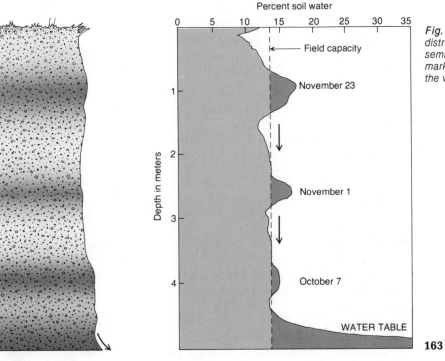

Fig. 12.1 Schematic portrayal of what the distribution of gravity water might resemble after three rainstorms. Each date marks a belt of gravity water in transit to the water table.

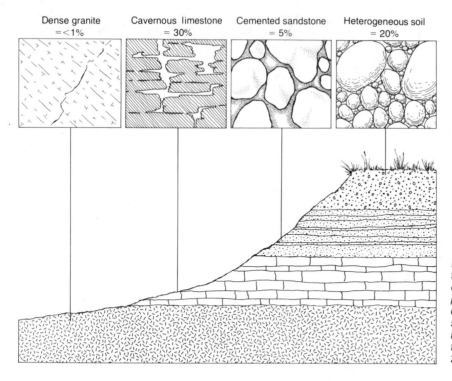

Fig. 12.2 Porosities of four materials near the surface. In all materials porosity decreases with depth, because the great pressure exerted by the soil and rock overburden tends to close the voids. Porosity can be as high as 45 percent in some near-surface bedrock, but is rarely higher than 2–3 percent at depths exceeding 2000 m.

FACTORS AFFECTING GROUNDWATER

Porosity

Groundwater is an accumulation of gravity water which fully saturates the interparticle voids. The total amount of groundwater that can be held by soil is controlled by soil porosity, which is a measure of the sum total of soil-void spaces. This can be determined by measurements of dry and saturated weights of a soil sample, as shown in the formula:

$$\text{Porosity} = \frac{(\text{Mass of saturated sample}) - (\text{Mass of dry sample})}{(\text{Density of water}) \times (\text{Volume of sample})} \times 100 = \%.$$

Density must be taken into account because soil particles typically have a density at least 2.5 times that of water. Porosity commonly varies from 5 to 20 percent in soils and near-surface bedrock. Figure 12.2 depicts the porosities of various materials.

The Water Table

The boundary between the soil and groundwater zones is the *water table*. In coarse-grained soils the water table approximates a boundary line, but in fine-grained soils it has more the aspect of a transition zone. This zone, called *capillary fringe*, represents a layer ranging from centimeters to several meters in thickness across which moisture is transformed from gravity to capillary water as it reenters the soil during dry periods. The movements of capillary and gravity water in relation to the water table are schematically illustrated in Fig. 12.3.

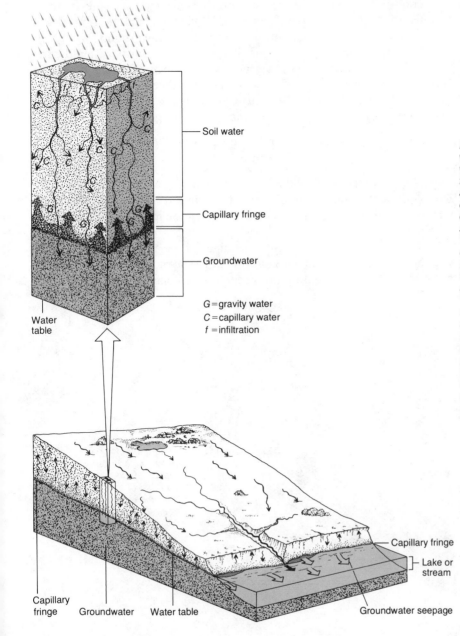

Soil water

Capillary fringe

Groundwater

Water table

G = gravity water
C = capillary water
f = infiltration

Capillary fringe

Lake or stream

Capillary fringe   Groundwater   Water table

Groundwater seepage

Fig. 12.3 As downward-percolating gravity water approaches the top of the groundwater, called the water table, it enters a transitional zone between capillary water and groundwater. Here the capillary force draws groundwater upward to form a capillary fringe. (These processes are shown in detail in the enlarged section.) With the addition of new gravity water, the capillary fringe and water table are raised. In some settings the inflow of gravity water is sufficient to raise the water table and the capillary fringe to the soil surface. This is often the case in areas of permafrost, as is evidenced by the saturated state of the soil surface in summer. In most areas, however, the water table slopes with the overlying terrain, forming a gradient that induces lateral groundwater flow. Ultimately, this water may outflow back to the surface in a river channel, swamp, or lake. (Illustration by William M. Marsh)

GROUNDWATER FLOW

The Hydraulic Gradient

If we trace the water table across the land, we will find that where the subsurface material is fairly homogeneous, the elevation of the water table changes with the elevation of the land. Barring special subsurface conditions where it may be elevated, or *perched*, on a soil strata of low permeability, the relief of the water table (i.e., variations in elevation) typically is much less than that of the overlying ground (Fig. 12.4). Because of the sloping nature of the water table, gravitational gradients are set up along which groundwater slowly flows. These are termed *hydraulic gradients* (*I*), and an individual gradient can be calculated by dividing the elevation difference between two points on the water table by the horizontal distance separating them:

$$I = \frac{e_1 - e_2}{d}, \text{ or } I = \frac{\text{Elevation of high point} - \text{Elevation of low point}}{\text{Distance between these two points}}$$

Figure 12.5 portrays this calculation graphically. Measurements of the water table elevations required for this computation, by the way, are usually attained by drilling test holes at two or more elevations on a hillslope.

Given a particular hydraulic gradient, the rate of flow of groundwater is regulated by the permeability of the subsoil, a fact discovered more than 150 years ago by the French hydrologist Henri Darcy. The

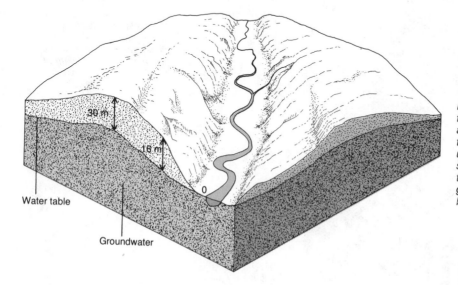

Fig. 12.4 The general relationship between the configuration of the water table and that of the overlying terrain. The variation in the elevation of the water table is usually less than that of the terrain. In low spots the water table is often higher than the land surface, resulting in an outflow of groundwater. (Illustration by William M. Marsh)

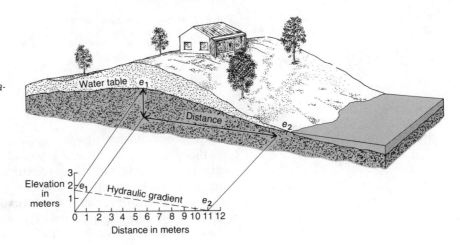

Fig. 12.5 The hydraulic gradient. (Illustration by William M. Marsh)

expression for the velocity (V) of groundwater flow has since come to be known as Darcy's law:

V = Permeability × Hydraulic gradient.

How fast does groundwater flow? Compared to surface water, flow rates are very slow in most materials, and hydrologists consider that 15 m per year is an average velocity for most groundwater, with wide variation from this value. In certain highly permeable materials, tests reveal that groundwater can move at velocities as great as 125 m per day, but this is uncommon.

Recharge and Transmission

In all materials the hydraulic gradient tends to vary with permeability and groundwater recharge. *Recharge* is the term given to gravity water supplied to the water table from surface sources, such as swamps and lakes. If material of low permeability receives rapid recharge, groundwater builds up because outflow is relatively restricted. As a result, the hydraulic gradient steepens, but as it does, so does flow velocity, according to Darcy's law. Under such conditions, the hydraulic gradient will continue to rise until the rate of flow, or *transmission*, is equal to the rate of recharge. Of course, the opposite is true where recharge is low and permeability is high.

Under these conditions the hydraulic gradient falls to a level at which the pressure exerted by the weight of the water higher on the gradient is just adequate to sustain a balance between recharge and groundwater flow. This is a good example of the dynamic-equilibrium principle, whereby a flow system is continuously trending toward a

UNCONFINED FLOW

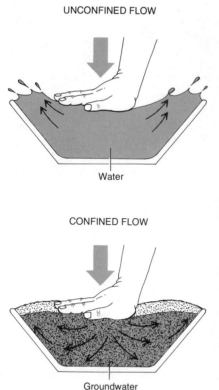

Water

CONFINED FLOW

Groundwater

Fig. 12.6 The response of unconfined water and confined water to downward pressure. Under the confined conditions of groundwater, flow resembles that of a plastic such as potter's clay. (Illustration by William M. Marsh and Peter Van Dusen)

state of equilibrium, or what in physics is called a *steady state.* Owing to this principle, it is normal in most areas to find steep hydraulic gradients in materials of low permeability and gentle hydraulic gradients in materials of high permeability.

In nonstratified materials groundwater tends to flow in broad arcs under the hydraulic gradient. The pattern is similar to one that might be expected if you were to apply pressure with your hand to the top of a tart, causing the filling to ooze out around the rim of the pan. The similarity can be traced to the fact that groundwater moves in highly confined interparticle spaces, and the pressure exerted by the mass of water is transferred both downward and outward, resulting in the semicircular flow lines similar, incidentally, to the flow of a glacier (see Fig. 12.6). In contrast, stress exerted on the surface of free water produces a direct lateral displacement, since unconfined water has no resistance to such stress.

Aquifers

In many areas the subsurface materials occur in the form of diverse layers or mixed aggregates of various water-holding capacities. Consequently, it is typical to find zones with especially heavy concentrations of groundwater. These zones are known as *aquifers.* Many different types of materials may form aquifers, but porous material with good permeability, such as beds of sand and conglomerate formations, are usually the best. In the central part of North America, for example, where glacial deposits are the predominant surface material, the composition of the subsoil typically is highly varied, ranging from dense clay to sand and gravel. The occurrence of aquifers is correspondingly varied. This is illustrated in Fig. 12.7, which shows cross-sectional diagrams of acquifers in glacial deposits.

Artesian Wells

Some aquifers produce a pressurized outflow of groundwater. This phenomenon is known as *artesian flow* and results from a geologic condition in which a tilted aquifer is sandwiched between two impermeable formations, called *aquicludes.* In Fig. 12.8 the groundwater in the

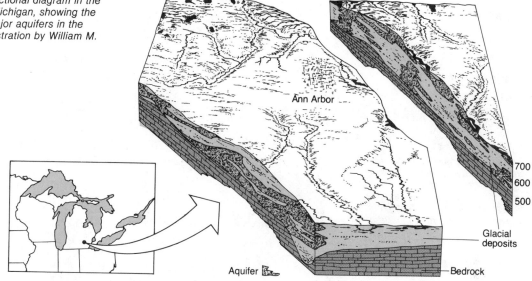

Fig. 12.7 A cross-sectional diagram in the area of Ann Arbor, Michigan, showing the distribution of the major aquifers in the glacial deposits. (Illustration by William M. Marsh)

lower part of the aquifer exists under the pressure of the weight of the water upslope. This pressure is called *hydrostatic pressure*. If the aquifer is tapped by a well, hydrostatic pressure will cause the groundwater to flow upward until it reaches a height approaching the elevation of the aquifer water table near the surface. It does not reach the same elevation as the top of the aquifer because of energy loss to friction as the water moves through the aquifer and up the pipe. The artesian-well condition illustrates the concept of the *piezometric surface.* This is the theoretical surface to which the water table would

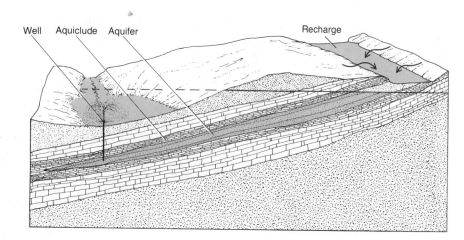

Fig. 12.8 A simplified illustration of an artesian well. (Illustration by William M. Marsh)

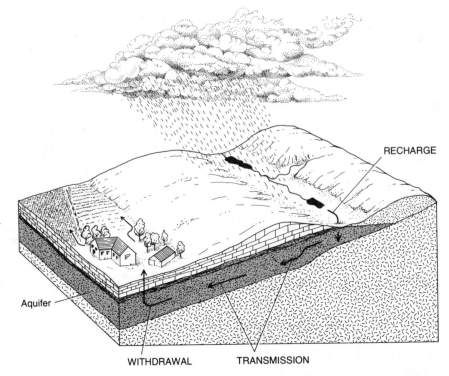

Fig. 12.9 Groundwater flow through an aquifer, beginning with recharge and ending in withdrawal some distance away. In order to maintain an equilibrium in this system, the rates of withdrawal and transmission cannot exceed the rate of recharge. (Illustration by William M. Marsh)

adjust should all groundwater be released from differential hydrostatic pressure related to variations in the structure and permeability of subsurface materials.

GROUNDWATER WITHDRAWAL

Maintenance of the Groundwater Supply

From the standpoint of water supply for human use, the best aquifers are those containing vast amounts of water that are easily withdrawn through wells. Two conditions must exist if an aquifer is to yield a dependable supply of water over many years. First, the rate of withdrawal cannot exceed the transmissibility of the aquifer. Under such circumstances, the water is pumped out faster than it is resupplied, and the reserve around the well dwindles (Fig. 12.9). Second, the transmissibility of the aquifer cannot exceed the rate of inflow of recharge water, which enters the groundwater zone from surface features such as lakes, ponds, swamps, sandy soils, and porous rock formations. If recharge is low, as it usually is in arid and semiarid regions, groundwater transmission may outrun recharge, resulting in the decline of aquifer reserves. This, in fact, has been the case in arid regions throughout the world. In the American West aquifer transmissibility is sufficient to allow high rates of pumping, but recharge is insufficient to maintain aquifer reserves. This results in a negative mass balance of the groundwater supply, leading to an eventual depletion of aquifers unless pumping rates are adjusted to recharge. In parts of Arizona the water table is declining as much as six meters a year because of heavy pumping for irrigation and domestic uses. Figure 12.10 shows the daily average use of groundwater in the United States. Those states with relatively small dots but large population, e.g., New York, Michigan, and Illinois, draw most of their water supply from rivers, lakes, and reservoirs (Authors' Note 12.1).

Formation of Cones of Depression

In urbanized areas the water table and the level of water in aquifers often have an especially irregular configuration due to uneven rates of groundwater withdrawal for municipal and industrial uses. This is caused by a flow imbalance in which pumping by individual wells exceeds aquifer transmissibility, thereby depressing the water table immediately around the well casing. The water table takes on a funnel

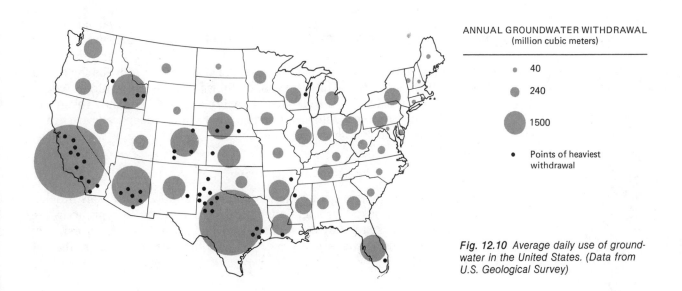

ANNUAL GROUNDWATER WITHDRAWAL
(million cubic meters)

40

240

1500

Points of heaviest withdrawal

Fig. 12.10 Average daily use of groundwater in the United States. (Data from U.S. Geological Survey)

AUTHORS' NOTE 12.1
Water Witching

There persists the belief in many parts of the world, including Europe and North America, that groundwater flows in underground streams and that in order to dig a successful well, one must intersect one of these streams. Many centuries ago this belief gave rise to the practice of "dowsing," or "water witching," and town fathers placed great confidence, and in some cases their political careers, on the ability of a dowser to locate a site for the town well. The dowser would walk about tightly grasping his special forked stick until the stick dipped violently toward the ground. He would proclaim that to be the spot of an underground stream, and digging would usually reveal a supply of groundwater there. Such discoveries, however, have neither a scientific nor a supernatural basis, owing to the fact that groundwater is so widespread that it is difficult to miss, regardless of where the stick may point.

   The diagram shows a water witcher, or dowser, stylized from sixteenth-century woodcuts.

(Illustration by James G. Marsh)

---

shape, called a *cone of depression* (Fig. 12.11a). As the cone of depression deepens with pumping, the hydraulic gradient increases, causing faster groundwater flow toward the well. If the rate of pumping is not highly variable, the cone usually stabilizes in time. However, if many wells are clustered together, as is often the case in urbanized areas, they may produce an overall lowering of the water table as the tops of neighboring cones intersect each other as they widen with drawdown (Fig. 12.11b). If the wells are of variable depths, this may result in a loss of water to shallow wells as the cones of big, deep wells are drawn beyond the shallower pumping depths.

Salt Water Intrusion into Groundwater

Salt water intrusion into the groundwater supply is a serious problem in some coastal areas. Under natural conditions, the groundwater under the sea is salt water. If there is no barrier along the coast, such as a ledge of impervious rock, to hold the salt water back, the salt

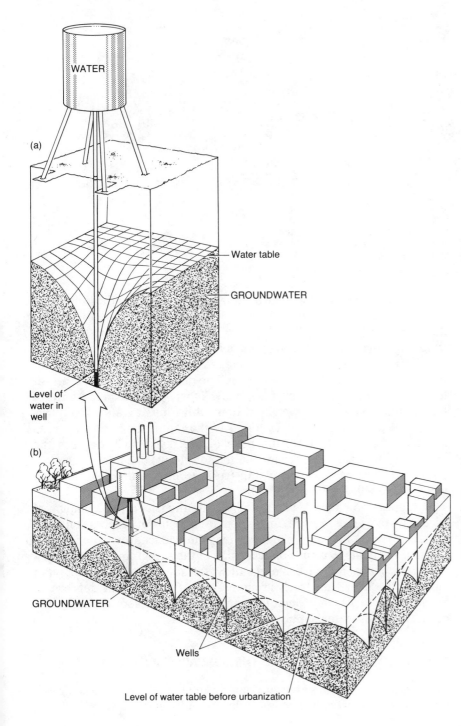

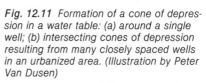

Fig. 12.11 Formation of a cone of depression in a water table: (a) around a single well; (b) intersecting cones of depression resulting from many closely spaced wells in an urbanized area. (Illustration by Peter Van Dusen)

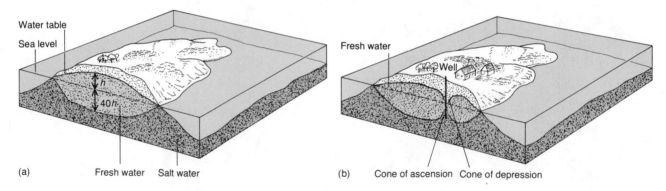

Fig. 12.12 The intrusion of salt water into groundwater: (a) The relative positions of fresh groundwater and salt groundwater under an island. The pocket of fresh groundwater floats on the denser salt groundwater. (b) If the weight of the fresh groundwater is reduced by pumping, the salt groundwater will adjust to a new equilibrium by rising under the well. (Illustration by Peter Van Dusen)

groundwater will form a wedge under the fresh groundwater of the land. This happens because the salt water is denser. This principle is illustrated in Fig. 12.12. The fresh water table is above sea level, and the salt water table is below sea level.

Because it is lighter, the lens-shaped pocket of fresh groundwater floats on top of the salt groundwater. The density ratio of fresh water to salt water is 1:1.025; therefore, it takes forty-one volumetric units of fresh water to equal the weight of forty units of salt water. It follows, then, that as gravity water is added to the top of the pocket of fresh groundwater (Fig. 12.12b), the base of the pocket will depress the underlying salt water in proportion to the relative densities of the two fluids. For 1 m of elevation of the fresh groundwater table, the salt water table will be depressed 40 m.

Suppose now that a well is drilled into the island; the pumping of groundwater leads to the formation of a cone of depression (Fig. 12.12b). Below the cone of depression, the reduced elevation of the water table leads to a *cone of ascension* of the salt groundwater as the boundary between the fresh and salt water adjusts to the reduction in mass of the overlying fresh groundwater. If the cone of depression becomes too large, the cone of ascension may intersect the well; if it does, the fresh water supply is lost to the well.

RETURN OF GROUNDWATER TO THE SURFACE

Although some groundwater becomes locked in rock formations for millions of years, most of it remains underground for periods ranging from

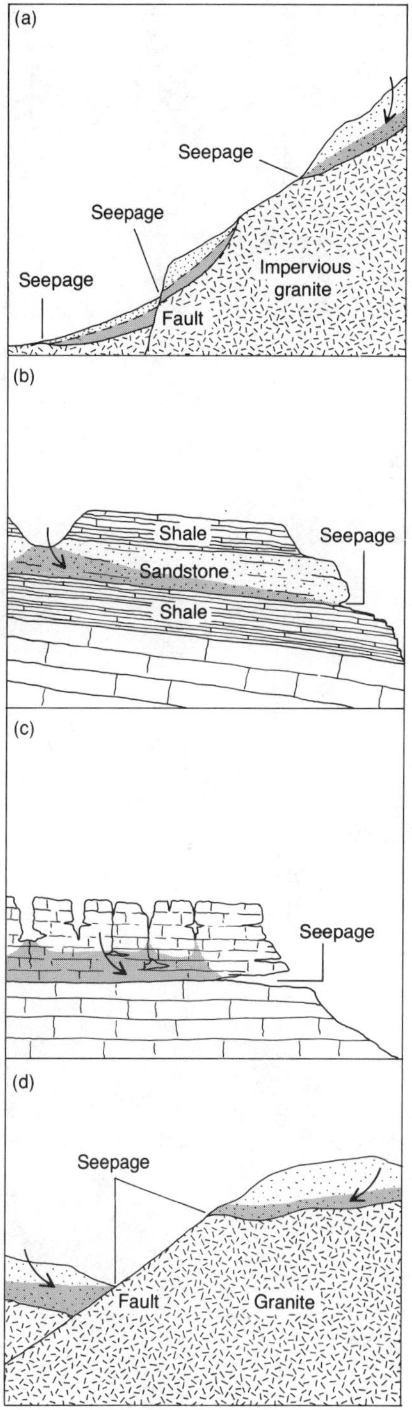

Fig. 12.13 Seepage zones associated with different rock formations: (a) unconsolidated sediments over impervious bedrock; (b) inclined sandstone over shale; (c) cavernous limestone over unweathered limestone; (d) unconsolidated materials broken by a fault. (Illustration by William M. Marsh)

several months to several years. Aside from pumping, groundwater is released to the surface mainly through: (1) capillary rise into the soil from which it is evaporated or taken up by plants; and (2) seepage to streams, lakes, and wetlands from which it runs off and/or evaporates.

Seepage

Many different geologic and topographic conditions can produce seepage. In mountainous areas, for example, it can be traced to fault lines and outcrops of tilted formations (Fig. 12.13). In areas where bedrock is buried under deep deposits of soil, groundwater seepage is usually found along the lower parts of slopes. Steep slopes represent "breaks" in elevation that may be too abrupt to produce a corresponding elevation change in the water table. Under such conditions, the water table may intercept the surface, especially in humid regions where the water table is high. If the resultant seepage is modest, a spring is formed, but if it is strong, a lake, wetland, or stream may be formed (Fig. 12.14). Which of these features develops depends on a host of conditions, including the rate of seepage, the size of the depression, and the rate of water loss to runoff and evaporation. Most of the thousands of inland lakes of Minnesota, Wisconsin, Michigan, and Ontario, for example, are "seepage"-type lakes whose water levels fluctuate with the seasonal changes in the elevation of the water table.

SUMMARY

Groundwater is an accumulation of gravity water in the subsoil and bedrock. The supply of groundwater varies geographically, owing to differences in climate, topography, and subsurface geology. The velocity of groundwater flow is controlled mainly by the hydraulic gradient and the permeability of the water-bearing material. Where ground withdrawal exceeds the transmission rate, cones of depression

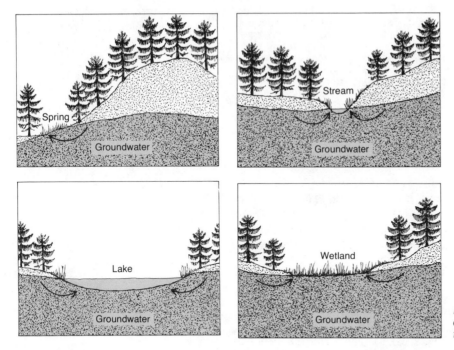

Fig. 12.14 Four types of groundwater outflow in areas of deep soils and hilly terrain.

form in the water table, and where both the rate of withdrawal and transmission exceed the rate of recharge, aquifers will be drawn down. This is common in urbanized areas and agriculture districts, particularly in dry areas. Most groundwater, however, returns to the surface through capillary rise in the soil or seepage into streams, wetlands, and lakes.

CHAPTER 13 RUNOFF AND STREAMFLOW

INTRODUCTION

The movement of water over the surface of the land is called *runoff*. This water may flow directly over the surface in small trickles, rivulets, or ditches or in large streams or major rivers. The rate of runoff is typically irregular, and one of the key questions in hydrology is how to forecast, or predict, changes in runoff. This is of great practical importance because of our dependence on streams and rivers for water supplies and our wariness of the floods they produce.

THE HYDROLOGIC EQUATION

Precipitation is the sole source of water for runoff. But not all precipitation ends up as runoff, because most of it is lost to the atmosphere in evaporation and plant transpiration. The amount of runoff from any parcel of ground (Fig. 13.1) is a residual, representing the water left

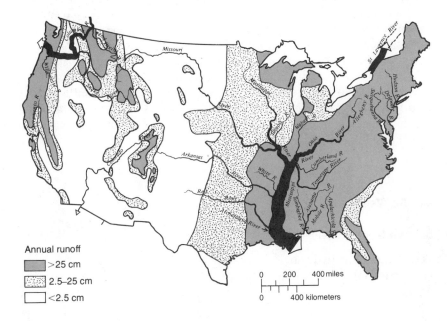

Annual runoff
- >25 cm
- 2.5–25 cm
- <2.5 cm

0 200 400 miles

0 400 kilometers

Fig. 13.1 Average annual runoff in the coterminous United States, including the annual average flows of major rivers. In all parts of the country, annual runoff is less than half of the annual precipitation. (Adapted from Kathaleen T. Iseri and W. B. Langbein, "Large Rivers of the United States," Geological Survey Circular, 686, Washington, D.C.: U.S. Geological Survey, 1974.)

EXPLANATION

600 cubic meters per second
1,400 cubic meters per second
3,000 cubic meters per second
7,000 cubic meters per second
14,000 cubic meters per second

Rivers shown are those whose average flow at the mouth is 500 cubic meters per second or more. Average flow of Yukon River, Alaska, is 6,800 cubic meters per second.

after losses to the atmosphere, with allowances made for additions or depletions from storage water (groundwater and soil water). We can express this concept in a formula called the *hydrologic equation:*

RUNOFF = Precipitation — Evaporation and ± Change in
transpiration   storage water

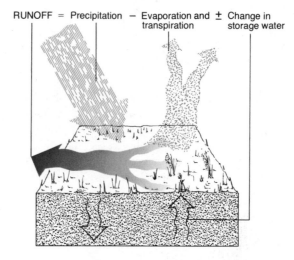

The hydrologic equation holds true over any period of time. Over long periods, say, several years, the net change in storage water should be zero, barring a sharp change in climate or large withdrawals of groundwater. During short periods of time, however, the "change in storage" portion of the equation can vary considerably, especially on a seasonal basis. This accounts for the fact that runoff is not as irregular in time and space as precipitation is. When rainfall occurs, much of it infiltrates into the soil and percolates into the groundwater. During periods of dry weather, groundwater gradually drains into springs and streams. The enormous storage capacity of the surface layers of the earth may keep large streams flowing during several months of dry weather. Recall that aside from the oceans and the glaciers, groundwater and soil water together represent the largest source of water on the planet (Table 10.1).

MEASURING RUNOFF

Units of Measurement

The terms in the hydrologic equation are all expressed as a *volume* per *time* or as a *depth* per *time*. The flow of water in a river expresses the rate at which water is leaving a drainage basin. Typical units in which this quantity is measured are cubic meters per second (m^3/sec, or "cumec") or, in the United States, cubic feet per second (ft^3/sec, or "cfs").

These are both volume-per-time units, and any volume-per-time unit can be converted to any other, if the appropriate conversion factors are known. Other examples of volume-per-time units are: liters per second, millions of gallons per day (mgd), acre-feet per day (or month or year). If a volume-per-time unit is divided by the *area* of a drainage basin, a depth-per-time unit results (because depth × area = volume).

Thus rainfall rates, usually expressed in depth-per-time units, can be compared to runoff rates. For example, a rainfall rate of 3 cm/hour over a basin of 240 square kilometers represents an input of water into the drainage basin at a rate of 2000 m^3/sec. Because much of the rainfall is absorbed in soil infiltration, the actual runoff will be spread out over a time period longer than the rainfall itself. Therefore, the *rate* of runoff will actually be somewhat lower than 2000 m^3/sec, but at least it gives us a "ballpark" figure of how much flow to expect from this rainstorm.

Rate of Runoff

Runoff that flows in streams is also called *discharge*. The discharge passing through a section of a stream is equal to the product of the cross-sectional area and the mean velocity of the stream (Fig. 13.2). Unfortunately, this equation cannot be used in a one-step computation of discharge, because mean velocity cannot be measured at a single point in the stream. Instead, discharge is determined from measurements of velocity taken at many points (20 to 100) in the stream, using an instrument called a current meter. It is a tedious task, requiring one to two hours for a stream 10 to 30 meters wide.

The Geological Survey, the government agency in charge of discharge measurements in the United States, does not have the financial resources to undertake velocity measurements every day on each of the country's major streams. To obtain the necessary flow data, the Geological Survey has established river-gaging stations where the elevation of the water surface, called the *stage*, is automatically recorded. With this system, current meter measurements need be made only occasionally and then correlated with the stage readings. The sample discharge values are plotted against their respective stage heights, and a generalized line, called a *rating curve*, is drawn through the points. The resultant graph, shown in Authors' Note 13.1, enables one to interpolate a discharge value for any stage.

Fig. 13.2 The discharge of a stream is equal to the cross-sectional area of water in the channel times the mean velocity.

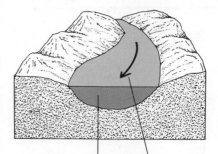

Discharge = Area × Mean velocity

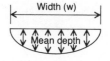

Width (w)

Mean depth

AUTHORS' NOTE 13.1
The U.S. Geological Survey Gaging Station

Monitoring the environment is an expensive and often thankless task, but it is absolutely essential if we are to achieve a scientific understanding of the landscape and hence the knowledge necessary to establish a workable balance between land use and environment. Streams and rivers are among the most prized and variable of our natural resources, and the federal government has charged the Department of the Interior with the responsibility for maintaining streamflow records for the nation. The Geological Survey carries out this charge with the aid of more than 6000 gaging stations on the major streams and rivers of the country. At each station, the stage of a river can be continuously recorded, using a stilling well, shown in the diagram. From the rating curve and a continuous record of the stage, hydrologists can estimate the discharge on a continuous basis at that section on a river. Because erosion and deposition may change the cross-sectional area of the channel, the rating curve must be checked by an occasional current-meter measurement and readjusted as necessary.

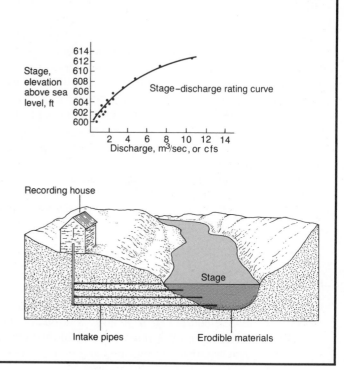

DRAINAGE BASINS AND STREAM NETWORKS

The Drainage Basin

A *drainage basin* is that area of land which contributes runoff into a specific river system. A *drainage divide* is the topographic boundary that partitions runoff into different drainage basins. Drainage basins are typically arranged in a hierarchy in which large basins are made up of smaller basins, each of which is made up of even smaller basins, and so on (Fig. 13.3).

Fig. 13.3 *The hierarchy of a drainage basin. This arrangement is often described as a "nested" hierarchy because each basin is situated within a series of progressively larger basins.*

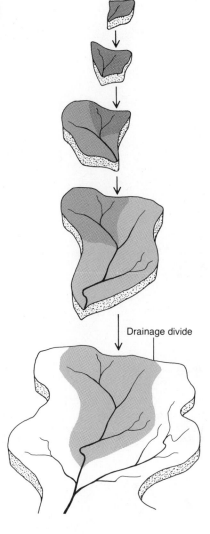

The runoff of small drainage basins drains into small stream channels, which in turn join together to form larger channels. As this occurs, the discharge of the channels progressively increases, because farther downstream in a given river system, the total area contributing runoff increases. In fact, it is often the case that average flood size increases at an increasing rate with drainage area (Fig. 13.4).

Stream Order and Drainage Density

A stream channel and the basin it drains may be classified according to their relative position within the drainage hierarchy. The classes of streams or basins are called *orders*. A first-order stream has no tributaries; a second-order stream is formed by the joining of two first-order streams; a third-order stream is formed when two second-order streams join; etc.

The system of channels formed by the various order streams in a basin is known as a *drainage network* (Fig. 13.5). It is interesting to note that many other natural systems have similar hierarchical networks to transport fluids. Examples include the veins and branches in

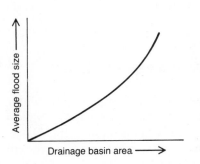

Fig. 13.4 *The relationship between average flood size and the area of the basins drained.*

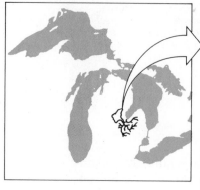

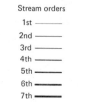

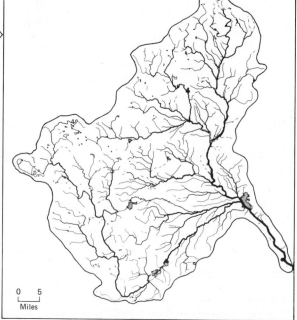

Stream orders
1st ———
2nd ———
3rd ———
4th ———
5th ———
6th ———
7th ———

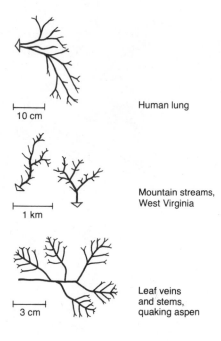

Fig. 13.5 A drainage network formed of a hierarchy of streams from first through seventh orders.

☒ Communities

0 5
Miles

Human lung

10 cm

Mountain streams, West Virginia

1 km

Leaf veins and stems, quaking aspen

3 cm

Fig. 13.6 Examples of natural drainage networks.

most trees and the blood vessels in the human body (Fig. 13.6). In any drainage network the number of streams in each order decreases with higher stream orders. The geometric pattern of the network, however, appears to have little to do with the stream-order principle.

Drainage density is defined as the total length of channels per unit area of drainage basin. Basins of high density consist of a large number of tightly interfingered channels, whereas basins of low density have relatively few channels, with more water transported beneath the surface. Densities range from 0.5 km to 300 km of channel per square kilometer of area. Where drainage density is high, stream discharge tends to show a more direct response to rainfall than where drainage density is low, because runoff gets into the channels more quickly.

THE RELATIONSHIP BETWEEN RUNOFF AND PRECIPITATION

By a variety of processes, collectively called precipitation, water arrives at the surface of the earth. The two major sources of precipitation are rainfall and snowfall. Rain is already in the liquid form, whereas snow must melt in order to become liquid. In this section we will examine the relationship between the rate at which liquid water is produced at the surface, through either rainfall or snowmelt, and the discharge from a drainage basin.

Sources of Discharge

There are four major sources of discharge: channel precipitation, overland flow, interflow, and groundwater flow. *Channel precipitation* is rain or snow which falls directly into streams, lakes, or other waterways of a drainage basin. It is rapidly converted into runoff, but typically the areas involved represent only a small portion (less than 5 percent) of a drainage basin, unless the basin includes large areas of lakes or swamps.

Overland flow occurs when the rainfall intensity exceeds the surface infiltration capacity. The excess water travels over the surface, collecting in small rivulets and trickles, and flows into stream channels relatively quickly. In vegetated areas infiltration capacity is normally high, and overland flow is therefore relatively rare. On bare or sparsely vegetated soils, on frozen ground, and in urban areas, it becomes more important, often representing more than 50 percent of the total rainfall. This form of runoff is also called *Horton overland flow*, after Robert E. Horton, the first to demonstrate the relationship among infiltration capacity, rainfall intensity, and surface runoff.

Interflow, which is also called subsurface soil-moisture flow, takes place when water infiltrates into the soil more rapidly than it can percolate through to the main groundwater level. Under these circumstances the percolating water will move laterally in the downhill direction (Fig. 13.7). In forested areas interflow often accounts for the major

Fig. 13.7 The four main sources of streamflow. The enlargement shows the transmission of interflow from the soil to the surface of a slope.

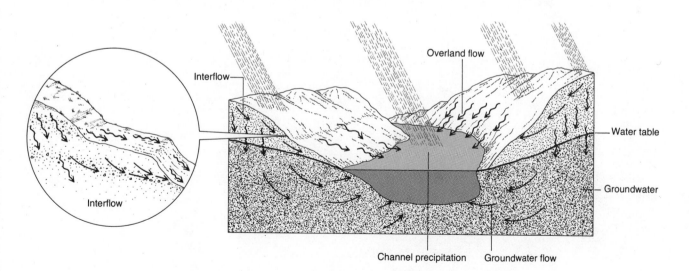

portion of the runoff. This fact was not really appreciated by most hydrologists until J. D. Hewlett's analysis of storm flows in the southern Appalachians demonstrated the importance of interflow. Interflow may flow to the stream channels as subsurface flow, or it may emerge onto the surface on the lower parts of slopes. These circumstances are quite common, because soil is usually loosely compacted near the surface and more compressed at depth.

Groundwater flow supplies water to streams where the stream channel intersects the water table. The water table follows the surface topography in a subdued form, and stream channels occupy valleys. Thus the hydraulic gradient generally slopes toward the stream channels. Since groundwater flow is usually very slow, the lag between precipitation and flow into the streams from the groundwater is considerable. In humid areas groundwater flow thus accounts for most of the steady flow of streams during dry periods (Fig. 13.7).

Channel precipitation and overland flow will deliver water to stream channels very rapidly after precipitation begins. Once in the channels, this runoff flows rapidly out of the drainage basin. Where such components form a major portion of the runoff, the response of the stream to a rainstorm will be rapid and peaked. Interflow, on the other hand, is considerably slower than overland flow and tends to smooth the stream's response to precipitation considerably. Groundwater flow responds to precipitation even more slowly than interflow does. Thus it is possible to divide streamflow into *stormflow*, or *quickflow*, comprising the runoff that occurs relatively soon after a storm,

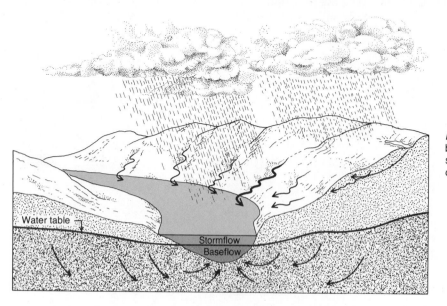

Fig. 13.8 *The two types of streamflow:* baseflow, *derived from groundwater, and* stormflow, *derived from overland flow and channel precipitation.*

and *baseflow,* which is slowly released over a longer time period. Channel precipitation and overland flow always contribute to stormflow; groundwater flow is always part of baseflow; and interflow can contribute to either type of flow (Fig. 13.8).

Variations in Discharge Related to Basin Conditions and Precipitation Events

The relative amount of water contributed to streamflow by each of the sources discussed above depends on two sets of factors. The first set is related to the characteristics of the drainage basin; the second, to the precipitation event. Some of the characteristics of the drainage basin are relatively permanent, whereas others are relatively transitory. Among the permanent characteristics, which change only slowly or sporadically, are such factors as soil type, proportion of area urbanized, vegetation, shape of the drainage basin, and the nature of stream channels. Transitory characteristics include the amount of soil moisture and the elevation of the water table. The important characteristics of a precipitation event are rainfall intensity, area of coverage, and duration.

Let us examine the discharge of a stream as the product of different precipitation, soil water, and groundwater conditions.

Streamflow without rain. If no rain or snowmelt has occurred for some time, any discharge that does exist will be baseflow, derived from groundwater flow. Under these circumstances, the slope of the water table and the level at which it intersects the stream channels determine the stream's discharge. In highly permeable materials, such as sand, the height of the water table and stream level correlate very closely, often within a few centimeters of each other. In less permeable channel material, the water table and the capillary fringe may extend more than a meter above the stream surface. In either situation, fluctuations in the water table produce corresponding fluctuations in stream level.

Measurements show that over a rainless period, stream baseflow will decline progressively. As the groundwater is depleted, the elevation and slope of the water table decrease, thereby reducing the rate of groundwater flow into the channel. The sketch and the hydrograph in Fig. 13.9 portray this situation.

If the dry period continues, the water table may drop below the bottom of the stream channel. We classify streams as *perennial* if the water table is above the stream bottom during the entire year and as *intermittent* if the water table drops below the stream bottom during the dry season. A third type of stream, *ephemeral,* receives no base-

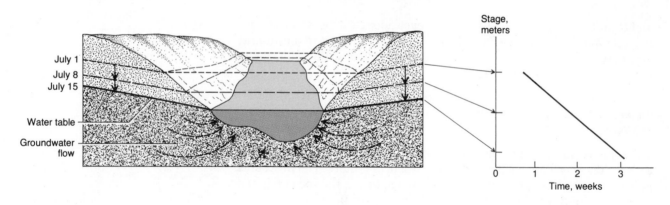

Fig. 13.9 *During rainless periods, stream-flow is derived from groundwater flow. Called baseflow, it declines with the fall of the water table.*

flow at all and therefore contains water only during and shortly after rainfall or snowmelt.

There are some situations in which the water table slopes away from the stream, and the stream actually contributes water to the groundwater (Fig. 13.10). This may be a temporary condition during periods of storm runoff, but it also happens in some areas where streams flow from humid, mountainous areas into and through dry areas. While flowing through the dry area, the stream is continuously

Fig. 13.10 *Conditions that exist below exotic streams, such as the Colorado River. Streamflow is lost to the ground-water, resulting in a water table that slopes away from the stream channel.*

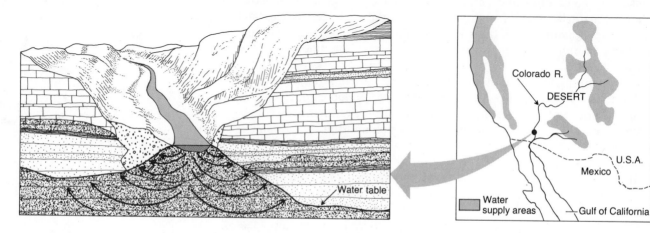

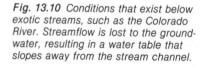

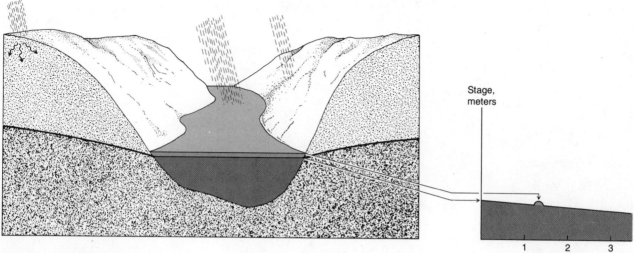

Fig. 13.11 *When the water from a light rainfall is taken up entirely in the soil, the only water that gets to the stream is channel precipitation. This makes a small bump in the hydrograph.*

losing water to the groundwater, but the flow from upstream may none-theless be sufficient to sustain flow. Such rivers are called *exotic*. Well-known examples are the Colorado and the Nile; the Nile has no tributaries over the last 2000 km to the Mediterranean.

Streamflow from a light rainfall. A light rainfall, whose intensity exceeds neither infiltration capacity nor percolation rate and whose total amount is less than the soil-moisture deficiency (the amount of water needed to raise soil moisture to field capacity), will cause only a small temporary rise in streamflow. The rise results from the rain that falls directly on the stream surface, and it is reflected in a tiny bump in the hydrograph (Fig. 13.11). This rain will not generate overland flow or interflow, and no water will percolate to the groundwater; virtually all rainwater enters and is taken up and held by the soil. Following the rain, the stream returns to the prerain trend of baseflow.

Streamflow from a long drizzle. If a light rainfall continues until the soil moisture reaches field capacity, gravity water will be produced. The gravity water percolates to the water table, causing it to rise. As

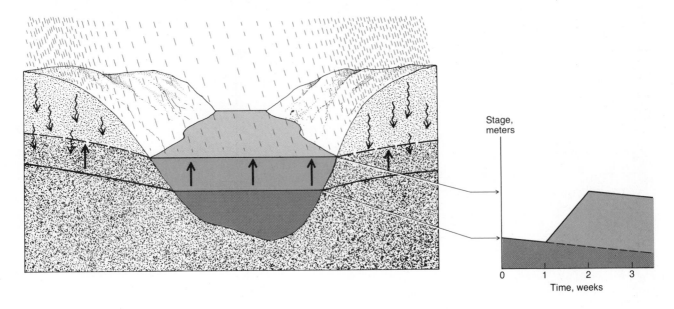

Fig. 13.12 An extended rainfall of low intensity can bring the soil to field capacity and then add enough water to the groundwater to raise the water table and the level of baseflow.

the water table rises, stream baseflow will increase. This is shown in the hydrograph portion of Fig. 13.12 as a step up to a new baseflow.

Streamflow from a cloudburst. Rainfall of high intensity and short duration often exceeds the infiltration capacity of the surface layers of the soil and the maximum percolation rate of the lower layers, resulting in strong overland flow and interflow. Typically, such rains occur during summer thunderstorms, when soil moisture is at its lowest value of the year. Even if a large fraction of the rainfall is lost to infiltration, it is usually insufficient to make up the large soil-moisture deficiency. The hydrograph rise, therefore, reflects only overland flow and interflow contributions to the stream. Once this source of water is dissipated—which may be only a matter of hours—the stream returns to its prestorm baseflow, as shown in Fig. 13.13. This hydrograph is somewhat similar to that in Fig. 13.11, except that the rise is much greater and not as immediate (Fig. 13.13).

Streamflow from long, hard rain. A high-intensity, longer-duration rainfall or an extended snowmelt period produces a combination of the effects described above. In urbanized areas or areas of bare soil, water reaches the surface at a rate that far exceeds the soil-infiltration capacity, thus producing strong overland flow. In forested areas, on the other hand, such a storm does not produce overland flow directly; however, because of the high infiltration rate in forested

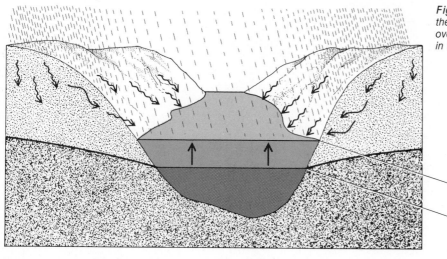

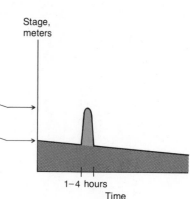

Fig. 13.13 Intensive rains often exceed the soil-infiltration capacity and produce overland flow, resulting in a sudden jump in the hydrograph.

areas interflow may be appreciable. Once the soil-moisture deficiency is exceeded, water percolates through the soil to the groundwater, raising the water table and steepening hydraulic gradients. Low on the slopes, small ephemeral stream channels capture a portion of the inter-flow, leading to a rapid rise in the hydrograph (Fig. 13.14). Under extreme conditions, the water table along the streams is raised to the soil surface, establishing baseflow at the bankfull level.

Fig. 13.14 A long, hard rain can produce an increase of all four sources of runoff to the stream. During the storm, a sizable peak flow is generated, and baseflow is raised to a new level.

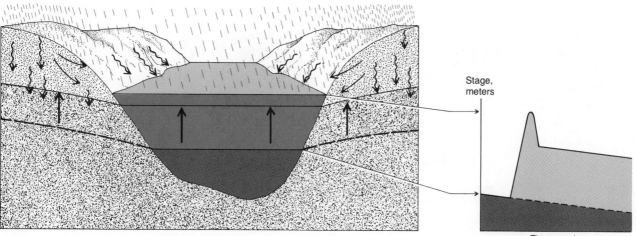

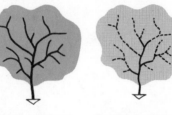

Fig. 13.15 *Drainage networks expand and contract under different precipitation and storage-water conditions: (a) during a storm; (b) during a drought. Therefore, the drainage patterns shown on maps actually represent one instant in a geographic system which is otherwise highly dynamic.*

DURING A STORM
(a)

DURING A DROUGHT
(b)

Spatial Fluctuations in a Drainage Network

So far, most of our discussion in this chapter has focused on the variations in streamflow at a single point in the drainage basin, usually the outlet. But let us also recognize that besides the ups and downs of flow in the channel, there are also corresponding expansions and contractions in the drainage net with different rainfalls, snowmelts, and storage-water conditions. As a result, the order of a stream in the network may not be constant over time, but may increase and decrease with the seasons or with precipitation events. For example, during an intensive rainstorm, the drainage net enlarges as the intermittent and ephemeral channels take on water (Fig. 13.15a). On the other hand, during a summer drought, baseflow may decline until only the trunk stream in a drainage net is left with flow, at which time this stream and its basin qualify as first-order (Fig. 13.15b).

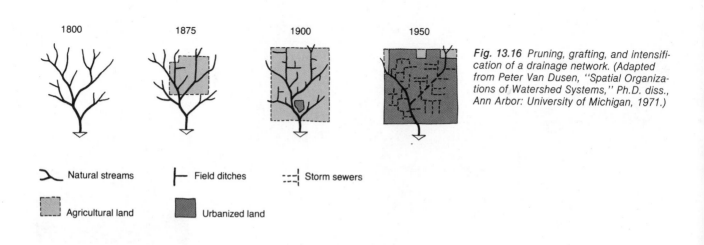

1800   1875   1900   1950

Fig. 13.16 *Pruning, grafting, and intensification of a drainage network. (Adapted from Peter Van Dusen, "Spatial Organizations of Watershed Systems," Ph.D. diss., Ann Arbor: University of Michigan, 1971.)*

⌇ Natural streams   ⊢ Field ditches   ⌐--⌐ Storm sewers

▨ Agricultural land   ▪ Urbanized land

Permanent changes can also take place in drainage networks. Most of these changes are a result of urbanization and agricultural development, and they tend to fall into three classes: (1) "pruning," in which part of a drainage net is severed from its natural basin; (2) "grafting," which involves the incorporation of new channels draining areas outside the natural basin; and (3) intensification, which involves adding new channels within the basin, resulting in a higher drainage density (Fig. 13.16).

In urbanized areas most of the changes in drainage networks are related to the construction of storm sewers. The purpose of storm sewers is to get rid of water that may flood basements, streets, and highways. To do this, engineers build ditches and lay drain pipe in areas where natural channels do not exist or where existing channels do not carry runoff away quickly enough. These artificial drains are incorporated into the natural drainage network (Fig. 3.16).

One of the chief hydrologic effects of storm sewers is to increase the responsiveness of the receiving streams to rainfall. Added to this is the increased overland flow produced from the impervious surfaces (roofs, asphalt, etc.) of urbanized areas. Hydrologist Luna B. Leopold notes that these factors combine to produce a higher and faster hydrograph response to a given storm; therefore, flooding tends to occur with greater frequency and magnitude after a basin has been urbanized (Fig. 13.17).

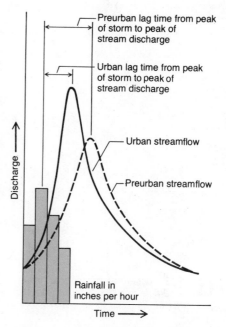

Fig. 13.17 The response of a drainage basin to a given storm before and after urbanization. (Adapted from Luna B. Leopold, "Hydrology for Urban Land Planning," U.S. Geological Survey Circular, 554, 1968.)

HYDROGRAPH ANALYSIS AND FLOW FORECASTING

For a variety of reasons it is necessary to analyze the response of a stream hydrograph to an input of water. Such analyses allow us to forecast floods, to estimate the flow magnitudes associated with different sized rainstorms or snowmelt periods, or to estimate what effect land-use changes in a basin will have on hydrograph response to a given storm.

Concentration Time

Our analysis must begin with an examination of the sizes of basins. Basin size can be expressed in different ways; for hydrograph analysis, the most meaningful description is in terms of *concentration time*, or *travel time*. This is defined as the time required for a raindrop falling on the perimeter of the basin to flow—as overland flow, interflow, and streamflow—entirely through the basin to the outlet (Fig. 13.18). Depending on the size of the basin, travel time can vary from a

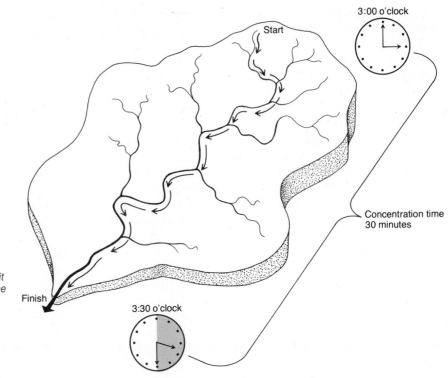

Fig. 13.18 Concentration time, the time it takes a particle of water to move from the perimeter of the basin to the outlet.

Fig. 13.19 In a small basin, both the entry and departure of storm water may occur simultaneously during a one-hour storm and at comparable rates of flow.

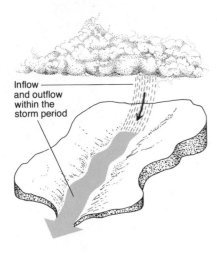

few minutes to several weeks. For our purposes we must distinguish among three different sizes of basins:

1. *Small basins,* which have such a short concentration time that the storms producing maximum flows actually last longer than concentration time. It is therefore possible that during a rainstorm the entire basin will produce runoff at the same time (see Fig. 13.19).

Let us examine the response of a small basin to a rainfall that exceeds the infiltration capacity on percolation rate. As the rain begins, only that portion nearest the outlet of the basin will contribute flow to it. As the rain continues, a greater portion of the basin will contribute to the flow, so the flow at the outlet will increase. At the end of one hour, essentially the entire basin is contributing to the flow, and at this time the flow is maximized. (Actually, only a part of the basin contributes overland flow because of high infiltration capacities or surface detention of runoff at certain sites.) If the rain continues, water will be leaving the basin as runoff at a rate comparable to the rate at which it is entering the basin as rainfall, so the streamflow rate will not increase further.

2. *Medium-size basins,* which have a long enough concentration time so that the storms producing maximum flows are generally of shorter duration than the concentration time. This means that the storm is usually over before the water from the perimeter of the basin has reached the outlet.

3. *Large basins,* which are so expansive that a single storm is unlikely to cover all of the basin at the same time. Thus one part of the basin may be producing water from a storm while the other parts are quiet.

Computing Flood Flows

In order to evaluate the magnitude of a large flood, we need to ask two questions: (1) What portion of the rainfall will contribute to runoff; and (2) What amount of rainfall is likely to fall on the basin within the concentration time?

We answer the first question by assigning a "coefficient of runoff" to the basin. This coefficient is based on a combination of surface characteristics (vegetative cover, soil, slope, and land use) and is expressed as a dimensionless number between 0 and 1, which represents a ratio between overland flow and rainfall. For surfaces of unfractured granite or highly urbanized areas which are mostly concrete, for instance, the coefficient may be as high as 0.95, indicating that 95 percent of rainfall is converted to runoff during most storms, whereas the remainder is lost to infiltration, surface wetting, and evaporation. Generally, the coefficient of runoff increases with slope angle and decreases with the density of vegetation (Fig. 13.20). Level areas with forested, sandy soils have coefficients of less than 0.10, indicating that they produce very little overland flow. In fact, recent studies have shown that such areas may produce no measurable overland flow, even during very intensive rainstorms. For most nonurban areas, the coefficient of runoff can be estimated on the basis of vegetative cover, soil texture, and slope. Some fairly standard coefficients for some important types of surfaces are shown in Fig. 13.20.

Rainfall intensity is the second key determinant of overland flow and is usually expressed in terms of the depth of water deposited on the surface per hour. Rainfall intensity varies markedly over North America, the greatest intensities occurring along the Gulf Coast, where thunderstorms are most abundant and powerful. Northward and west-

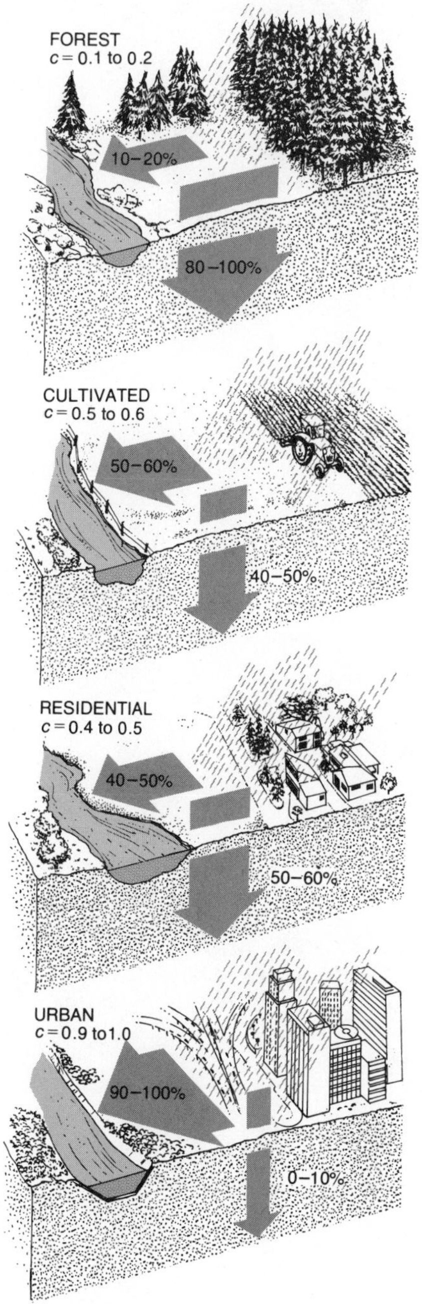

FOREST
c = 0.1 to 0.2
10–20%
80–100%

CULTIVATED
c = 0.5 to 0.6
50–60%
40–50%

RESIDENTIAL
c = 0.4 to 0.5
40–50%
50–60%

URBAN
c = 0.9 to 1.0
90–100%
0–10%

Fig. 13.20 Coefficients of runoff for four major types of surfaces.

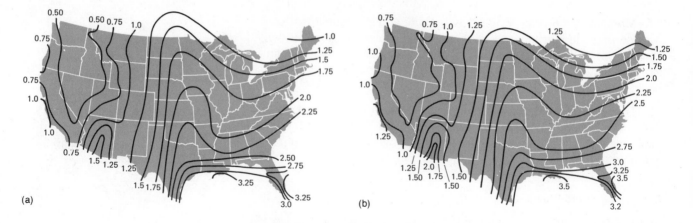

Fig. 13.21 *The amounts of rainfall produced by the (a) five- and (b) ten-year storms in the coterminous United States. The values represent one-hour rainfall intensities. (Adapted from U.S. Department of Agriculture data.)*

ward from this area, the maximum intensity of rainfall expected in any period declines considerably.

Based on climatic records from the past century or so, we are able to determine the intensity and duration of the strongest rainstorm that occurs at different time intervals, for example, two, five, ten, and twenty-five years. (For a discussion of how these frequencies are determined, see Chapter 30.) The two maps in Fig. 13.21 show the five- and ten-year storm intensities for the United States in inches of rain within one hour.

The rational method. Combining rainfall intensity, the coefficient of runoff, and the area of the surface drained, we can compute an estimate of the volume of water discharging from a small basin for any rainstorm as follows:

$$Q = c \times I \times A,$$

where Q is the discharge in cubic feet or cubic meters per second, c is the coefficient of runoff, I is the intensity of rainfall in feet or meters per second, and A is the drainage area in square feet or square meters. This method is called the "rational method," and it is commonly employed in civil engineering and environmental impact analysis for the computation of runoff from areas of less than 2 or 3 km².

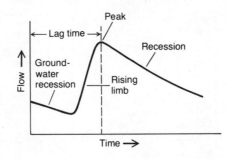

Fig. 13.22 The main parts of a stream hydrograph.

The hydrograph method. For medium-size basins, severe floods can and generally do result from storms of much shorter duration than the concentration time. In our analysis, in fact, we consider that the rainfall amount in excess of infiltration capacity can be assumed to be placed on the basin virtually instantaneously and then allowed to flow through the drainage network. The basic hydrograph resulting from a given storm for flows that are likely to cause problems looks like the one in Fig. 13.22. The main elements of the hydrograph are baseflow, rising limb, peak, and recession. The time and flow scales are not explicitly specified and would depend on the size and drainage density of the basin.

The shape of the hydrograph will vary for different basins. For medium-size basins, which are small enough so that most storms cover the entire basin, the hydrograph for a given basin will have a very similar shape for all short-duration storms. The height of the peak, of course, depends on the size of the storm, but the lag time is relatively independent of rainfall amounts, as long as the intensity is great enough to cause either overland flow or significant interflow.

This fact allows us to "normalize" the hydrograph, to express it in such a way that the size of the storm is filtered out. There are a number of ways to do this. One way is to express the hydrograph in terms of runoff per inch or centimeter of rainfall. Another way, the one we use in this text, is to draw instead a *distribution graph*, where time units are used for the horizontal axis, but the units for the vertical axis are percentage of total storm runoff for a short-duration storm. The example in Fig. 13.23 uses data from a storm in October 1937 on the Youghiogheny River above Connellsville, Pennsylvania, with a drainage area of 3434 km^2. Within twenty-four hours, a storm dropped slightly over 5 cm of rain on the basin. Of this, about 0.72 cm percolated into the groundwater or was held in the soil and 4.28 cm became runoff. The hydrograph and the percentages of the total runoff are

Flow period	Total flow	Storm-flow
0-12 hrs	63 m³/s	6 m³/s
12-24	192	127
24-36	1065	991
36-48	1101	1019
48-60	714	623
60-72	453	354
72-84	275	170
84-96	194	85
96-108	144	28

Peak=1253; 1175 at hour 36

Flow period	Percentage of total flow
0-12 hrs	0.2%
12-24	3.6
24-36	29.0
36-48	29.8
48-60	18.3
60-72	10.3
72-84	5.0
84-96	2.4
96-100	0.8

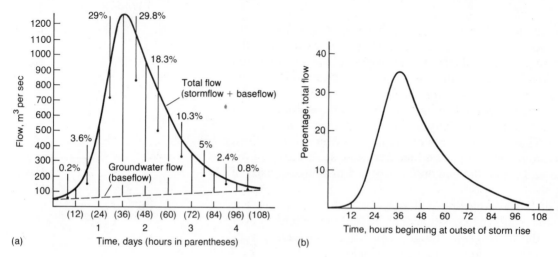

(a)

(b)

Fig. 13.23 *Analysis of a hydrograph: (a) Baseflow and stormflow are separated by the broken line. Stormflow is divided into twelve-hour units, and each unit is expressed as a percentage of total flow. (Data from C. O. Wisler and E. F. Brater, Hydrology, 2d ed., New York: Wiley, 1959.) (b) A distribution of the percentages computed for the flow periods in (a).*

shown in Fig. 13.23(a). Figure 13.23(b) indicates these percentages in a distribution graph.

FLOODS

Floods have long been of concern to humans, for they represent one of the most destructive natural processes to areas of human settlement in virtually every climatic region of the world. Despite the millenia of observations and folk records on floods, society as a whole seems to have gained comparatively little insight into the variable and hazardous nature of river flow. Indeed, floods render massive damage to America every year, and the annual averages have generally increased since 1905 (Fig. 13.24).

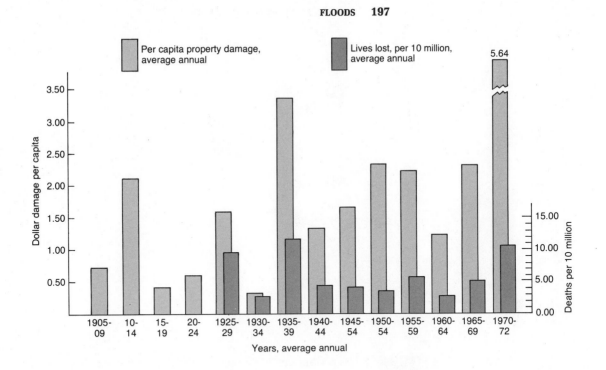

Fig. 13.24 Losses in United States floods, 1905–1972. (From Gilbert F. White and J. Eugene Haas, Assessment of Research on Natural Hazards, Cambridge, Mass.: M.I.T. Press, 1975. Reprinted by permission.)

Outflooding and Inflooding

Although not widely realized, there are actually two different types of flooding caused by runoff, and both types are shown in Fig. 13.25. The most spectacular type can be termed "outflooding," and it results when a river overflows its banks and spills into nearby lowlands, usually the river floodplain. "Floodplain" is the low, flat ground that borders the river channel. When such flooding occurs on large rivers, such as the Mississippi and the Ohio, thousands of square kilometers may be inundated, causing tremendous damage to farms, towns, and cities. Most outflooding occurs in the lower reaches of a drainage net where the floodplain is often tens of kilometers wide.

Flooding also occurs in upstream areas when surface water collects in low spots before reaching a stream channel. This is called *inflooding*, and it is especially prevalent in areas of flat ground with low infiltration capacities or where drainage is closed in hilly terrain. Inflooding usually begins with the formation of "mud puddles" and reaches its maximum during or shortly after the rainstorm with the formation of shallow, but extensive ponds.

In small drainage basins damage caused by inflooding is often more serious than that caused by outflooding. Crop damage, late planting, stunted crops, septic-field malfunction, cellar flooding, and

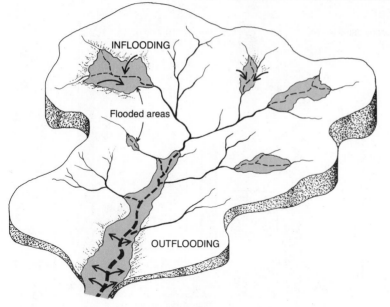

Fig. 13.25 *The concepts of inflooding and outflooding. (Illustration by Peter Van Dusen)*

lawn damage are some of the land-use problems resulting from this process.

The Timing of Floods

In large drainage basins river flow and flooding may be influenced by conditions and events that are particular to only part of the basin. This is so for the Nile River, for example, which rises under the tropical savanna climate and ends under a Mediterranean-type climate. The tropical savanna climate yields its maximum rainfall in the summer, producing a flood crest on the Nile in late summer when the Mediterranean climate, several thousand kilometers away, is at its driest. This peculiarity in the Nile's flow puzzled the ancient Greeks, who did not know about the climate at the river's headwaters.

In North America the Missouri River flows through several different climatic zones with different runoff characteristics. Although much of the Missouri's drainage area lies in the semiarid Great Plains, the headwaters of most large tributaries lie high in the Rocky Mountains and therefore are strongly influenced by late-spring mountain snowmelt. As a result, the seasonal frequency of floods on the Missouri River in the Great Plains is often more directly related to runoff conditions in the Rockies than in the Great Plains (Fig. 13.26a). On the other hand, smaller watersheds within a single climatic zone should show a

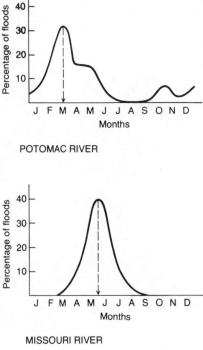

Fig. 13.26 The monthly distribution of peak annual flows on the Potomac and Missouri rivers.

seasonal distribution of floods more consistent with the seasonal character of the climate. The Potomac River is a case in point, with most of its floods corresponding to snowmelt, spring rains, and high soil water and groundwater conditions of spring (Fig. 13.26b).

SUMMARY

Precipitation is the source of all runoff, but not all runoff comes directly from rainfall or snowmelt. Groundwater is an important source of streamflow, providing baseflow for rivers and streams during rainless periods. Overland flow develops when the intensity of precipitation exceeds the infiltration capacity. This form of runoff enters streams quickly after a rainstorm or snowmelt, often adding a large surcharge of water to the baseflow. Overland flow usually increases substantially with both agricultural and urban development. In forested areas interflow appears to be far more important than overland flow as a source of stream discharge. Rivers flood when their stage exceeds bank elevation and water overtops the channel. In the upper parts of drainage basins a flood may also take the form of in-flooding. Annual flood damage has increased in the United States in this century.

UNIT III SUMMARY

- The movement of water among the atmosphere, land, and oceans can be described with an idealized model called the hydrologic cycle. In reality, the water system on earth is uneven in terms of both the geographical distribution of water and the rates at which it is delivered to the earth's surface.

- When water is precipitated on the land, some of it runs off, some goes into the ground, and some passes back into the atmosphere.

- Infiltration capacity is the ability of a soil to absorb surface water.

- The water that goes into the ground either is taken up in the soil, where it is held at the contacts between the soil particles as capillary water, or passes through the soil as gravity water and becomes groundwater at some depth.

- Groundwater constitutes the third largest supply of water on the planet (after the oceans and the glaciers), and it is important as a source of streamflow, soil water, and water for humans. In order for groundwater supplies to be maintained, pumping rates cannot exceed recharge rates; in many dry regions of the world, groundwater is being depleted at alarming rates.

- Streamflow is derived from four sources: groundwater, interflow, overland flow, and channel precipitation. Groundwater provides baseflow because it is released to streams slowly over long periods of time. Overland flow is released to streams quickly after a rainstorm, causing sudden increases in discharge called stormflow. Virtually all types of land uses increase the rate of overland flow.

- The area drained by a stream is the drainage basin, and the channels in a stream system form the drainage network. The number of streams in each order increases with stream order, a recurrent pattern in many networks in nature.

- Drainage networks change over space and time. During periods of high runoff, networks enlarge. Spatial changes are associated with agriculture and urbanization and may take the form of pruning, grafting, or intensification.

- Several methods are used to forecast river and stream discharges, including the rational method and the unit-hydrograph method.

- The magnitude and frequency of stream discharge are highly variable because of the variable nature of rainfall, snowmelt, and the soil-infiltration capacity; therefore, floods are difficult to forecast for most streams and rivers. In the United States the U.S. Geological Survey maintains flow records on thousands of streams, and these records are helpful in determining flood probabilities.

- Modern urbanization has effected an increase in both the magnitude and frequency of peak discharges in developed areas. Most of this is attributable to the increased coefficients of runoff and the introduction of storm sewers into the drainage net.

FURTHER READING

Bruce, J. P., and R. H. Clark, *Introduction to Hydrometeorology,* Oxford, England: Pergamon Press, 1966. *A useful handbook on techniques and methods in the hydrology of midlatitude lands.*

Dunne, Thomas, and Luna B. Leopold, *Water in Environmental Planning,* San Francisco: Freeman, 1978, 818 pages. *Exhaustive treatment of the role of water in planning and managing the environment, including water pollution, soil erosion, irrigation, and flooding.*

Satterland, Donald R., *Wildland Watershed Management,* New York: Ronald Press, 1972. *A synthesis of the diverse aspects of watershed management in nonurbanized areas.*

Wisler, C. O., and E. F. Brater, *Hydrology,* New York: Wiley, 1959. *A standard hydrology text with a balanced treatment of the various aspects of water science.*

REFERENCES AND BIBLIOGRAPHY

Benson, M. A., "Factors Influencing the Occurrence of Floods in the Southwest," *U.S. Geological Survey Water Supply Paper* 1580-D (1964).

Betson, R. P., "What is Watershed Runoff?" *Journal of Geophysical Research* 68 (1964): 1541–1552.

Dunne, T., and R. D. Black, "Partial-Area Contributions to Storm Runoff in a Small New England Watershed," *Water Resources Research* 6 (1970): 1296–1311.

EAAFRO, "Hydrological Effects of Changes in Land Use in Some East African Catchment Areas," *East African Agricultural and Forestry Journal* 27 (1962): 1–31.

Franke, O. L., and N. E. McClymonds, "Summary of the Hydrologic Situation on Long Island, New York, as a Guide to Water Management Alternatives," *U.S. Geological Survey Professional Paper 627–F* (1972).

Freeze, A. R., and J. A. Cherry, *Groundwater,* Englewood Cliffs, N.J.: Prentice-Hall, 1979.

Hewlett, J. D., "Soil Moisture as a Source of Baseflow from Steep Mountain Watersheds," *Station Paper 132,* Southeastern Forest Experiment Station, U.S. Forest Service, 1961.

Horton, R. E., "The Role of Infiltration in the Hydrologic Cycle," *Transactions American Geophysical Union* 14 (1933): 446–460.

_____, *Surface Runoff Phenomena,* Ann Arbor, Mich.: Edwards Brothers, 1935.

Kirby, M. J., *Hillslope Hydrology,* London: Wiley, 1978.

Leopold, L. B., and T. Maddock, Jr., *The Flood-Control Controversy,* New York: Ronald Press, 1954.

Rantz, S. E., "Suggested Criteria for Hydrologic Design of Storm-Drainage Facilities in the San Francisco Bay Region, California," *U.S. Geological Survey Open File Report,* Menlo Park Calif., 1971.

Strahler, A. N., "Quantitative Analysis of Watershed Geomorphology," *Transactions American Geophysical Union* 38, 6 (1957): 913–920.

U.S. Soil Conservation Service, "Soils Suitable for Septic Tank Filter Fields," *Agriculture Informative Bulletin No. 243,* 1967.

White, G. F., "Choice of Adjustment to Floods," University of Chicago: Department of Geography Research Paper No. 93, 1964.

SOIL:
THE
TRANSITIONAL
MEDIUM

UNIT IV

KEY CONCEPTS   OVERVIEW

solum

horizon

organic layer

weathering

parent material

residual soils

transported soils

soil texture

soil structure

**driving forces
in soil**

capillary water

field capacity

capillarity

**potential and actual
evapotranspiration**

moisture deficit

colloids

ions

leaching

eluviation

illuviation

pH

soil profile

podzolization

laterization

calcification

gleization

zonal and azonal soils

great soils

traditional soil classification

7th approximation

engineering classification

problem soils

Soil is the material that is transitional between the atmosphere and the bedrock of the continents. Soil is the part of the landscape that contains reserves of water, organic matter, heat, and mineral particles; together, these reserves are the principal terrestrial resources for plant growth. Therefore, soil is the foundation of the ecological food chains on the land.

Agronomists, engineers, and scientists have somewhat different perspectives on soil. The geosciences generally embrace all major perspectives; physical geography in particular is concerned with the interrelations among soil and water, vegetation, and atmosphere.

The formation of small layers, called horizons, in the upper several meters of soil results from the interplay of water with mineral particles, vegetation, and microorganisms. Water moves into and out of soil mainly in response to rainfall and surface heating and cooling, and as it does, it performs work by moving tiny clay particles and dissolved minerals. Where the movement of water in the soil is relatively unrestricted, the horizons that form reflect the bioclimatic conditions of the environment. Such soils are classified by agronomists as zonal soils. Under conditions of impeded drainage, as in a swamp, soil horizons are poorly developed, and the soils are called intrazonal; in highly active environments, such as sand dunes, no horizons develop, and the soils are called azonal.

Soils are problematic where they present serious limitations to land use. In urban areas organic, wet, and weakly consolidated soils can impede urban growth, but with investment of enough money, modern technology can overcome most problem soils. Soils underlain by permafrost and nutrient-poor soils in the rainy tropics are the most widespread problem soils in the world today. As for the least problematic soils in the world, modern agricultural practices have generally damaged them by facilitating erosion and depleting the nutrient content.

Chapter 14 opens with a comment on the principal concerns of different fields in the study of soil, followed with a description of the physical properties of soil and the energy that drives the soil processes. The movement of water in the soil and the soil-moisture balance are taken up in Chapter 15. The essential soil-forming processes are described in Chapter 16, and in Chapter 17 soil classification and problem soils are discussed.

CHAPTER 14 THE PROPERTIES OF SOIL

INTRODUCTION

Soil is the transitional material between the atmosphere at the earth's surface and the bedrock below the surface. As the material in which plant roots reside, and from which plants extract water and nutrients, soil is essential to the whole life-support system on the land. Moreover, soil is the material on which the landscape rests. Although this point seems so apparent that it hardly merits mention, it is nevertheless important to recall that buildings and other structures depend on soil for support and stability. Where supportive strength is weak, as in parts of Mexico City and Venice, Italy, this role of soil is deeply appreciated.

Although the formal study of soil is traditionally associated with agronomy, today it is also studied in many other fields, including civil engineering, geology, and geography. In every field the general study methods used to measure and to map soils are similar, but their specific objectives are quite different.

Agronomists, for example, would limit soil to the *solum*, the near-surface material capable of supporting plant life. The American soil scientist C. C. Nikiforoff suggests that "agronomy inherited this old concept of soil from the tillers of the land, for whom the soil is just the 'dirt' supporting their crops." This sort of soil contains gaseous, liquid, and solid portions in both inorganic and organic forms, and the agronomist is interested in what combinations of these materials are best suited for crops.

Civil engineers, by contrast, are concerned about the performance of soil when it is used as footing for buildings or highways. Therefore, civil engineers define soil on the basis of certain properties that control the soil's: (1) stability when molded into slopes; (2) compressibility under the weight of structures; and (3) capacity to limit the seepage of water in and around dams. Unlike agronomists, civil engineers do not limit their definition of soil to the root zone. Rather, all assemblages of loose materials that overlie the bedrock are recognized as soil.

In the geosciences there are several major objectives in the study of soil. Geologists have a traditional interest in the origins of soil materials as one way of deciphering the geologic history of the land. Another major objective of geologists is the identification of marketable soil materials such as gravel, sand, and groundwater. Geographers have long been concerned with the relationship of land use to soil and have, in the main, adopted the agronomists' definition of soil.

What is the physical geographer's objective in the study of soil? Actually, it incorporates parts of all those mentioned above, but three particular areas of interest can be singled out: (1) the role of soil in the landscape system as it interrelates with plants, water, and atmosphere; (2) the changes in soil due to erosion and deposition; and (3) the interplay of soil and land use, as in the case of urban development in areas of soft, wet soils.

In order to accommodate these objectives, our definition of soil must be broad enough to include most assemblages of ground materials above the bedrock, including part or all of what the geologist calls surface (or surficial) deposits. This soil may or may not contain air, water, and organic or inorganic materials and would, for instance, include rock rubble and ground ice in cold regions, muck and water in swamps, landfills in and around cities, as well as the soil that supports forest and crops. Although most earth scientists would disagree with this definition, most would, on the other hand, agree that the essential processes of the soil, such as moisture movement and chemical and heat flows, do not begin and end in the solum. The most pronounced effects of these processes are indeed found within a meter or so of the surface, but the soil-forming processes themselves extend tens of meters below ground level. Ultimately, though, it is usually the uppermost several meters of soil that are the locus of attention in physical geography.

PHYSICAL CHARACTERISTICS OF SOIL IN GENERAL

Before we examine the energy and processes involved in soil formation, let us briefly examine some of the physical characteristics of soil materials. The mass of loose material that rests on the bedrock, the stuff in which soil forms, varies greatly in thickness and composition from place to place. In the broad, flatish interiors of the continents, it is usually 25–50 m thick and very often several hundred meters thick. In contrast, in areas of rugged terrain, the soil materials may be absent altogether (Fig. 14.1). This is especially so in the high mountains, where erosional processes remove particles of rock from steep slopes as rapidly as they appear on the surface. These particles are moved downslope and accumulate in the valley bottoms, where they often create small areas of extraordinarily deep deposits. Higher on the mountain slopes, barren rock may give way to mixtures of snow, ice, and rock rubble, all legitimately soil materials in the broad sense of the term.

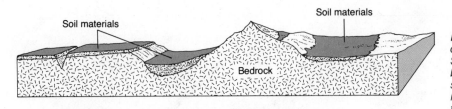

Soil materials

Soil materials

Bedrock

Fig. 14.1 The relationship between the distribution of soil materials and terrain. Steep slopes tend to lack these materials because powerful erosional processes, such as glaciers and runoff, remove the loose particles. The opposite tends to be true for adjacent valleys, where these particles accumulate to form deep deposits. Gently sloping terrain is often characterized by an intermediate thickness of soil materials which varies less in depth than in mountainous areas.

What does soil look like underground? Most of us have a faint notion about the features in the upper two or three meters, having seen soil exposed in road cuts, basement excavations, and gravel pits. But if we could descend in a glass elevator through the soil to the bedrock, we would probably be led to certain generalizations about the major features of soil (Fig. 14.2). First, in areas with plant cover, soils have a dark brown or black surface layer ranging in thickness from a few centimeters to several meters. This layer represents the transition between the zone of life on the surface and the deep zone without life (Fig. 14.2a). Here processes related to heat, organic material, chemicals, air, and water interact among the soil particles to form the lowermost layer, or zone, of life. In this zone live the roots of the plants that provide the supportive resources for the food chains on land. Generally, where the plant cover is dense and has fast growth rates, the organic content of soil is correspondingly high, especially if the rates of consumption of organic material by the small organisms that live in the upper soil are not intensive.

Below the organic layer many soils have a layered-cake appearance (Fig. 14.2b). The individual layers are distinguishable mainly on the basis of color, primarily white, brown, red, and yellow. The layers are called *horizons*, and they result from the segregation and dislocation of small particles and dissolved minerals by water moving up and down in the soil. This water moves primarily in response to heat, wind, plant growth, and rainfall on the surface. In some shafts, however, the horizons would be absent, especially in mountainous areas.

Under the horizons we would descend through a zone comprised almost exclusively of mixed inorganic particles (Fig. 14.2c). This material may be a stratified deposit, but often it is undifferentiated; that is, the particles are not arranged in layers or sorted into uniform sizes, but rather appear to be more or less randomly assembled. Although the spaces between particles are filled mainly with air, exceptionally damp spots are apparent here and there, and water may actually seep from some. This is the part of the soil that is most intensively mined for sand and gravel for road beds and the manufacture of concrete.

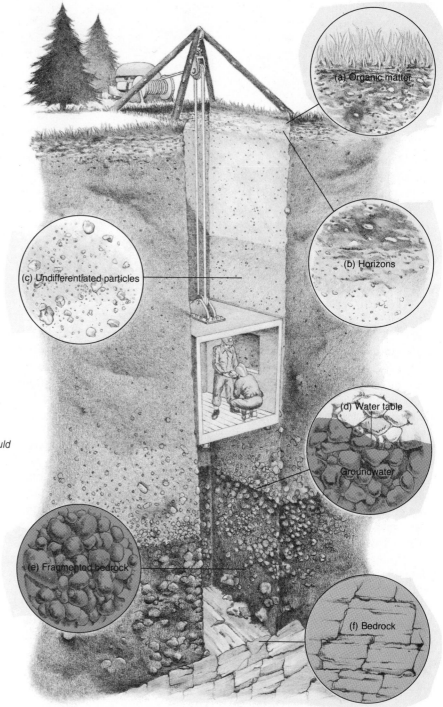

Fig. 14.2 *An elevator shaft extending from the soil surface to the bedrock would reveal many characteristics of soil that are difficult to appreciate from the surface. (Illustration by Peter Van Dusen)*

Within the undifferentiated zone or near the base of it, the moisture content of the soil increases perceptibly, and at a depth where the soil is fully saturated, we enter the zone of groundwater (Fig. 14.2d). Here water completely fills the interparticle spaces and drips or even pours into the shaft. In fact, if the shaft walls were not encased, this lower part of the shaft would fill with water in a few days. This, of course, is the zone from which water is withdrawn through wells for domestic, industrial, and agricultural purposes.

Finally, near the bottom of the shaft, large pieces of rock begin to appear, and at the very bottom, they give way to solid bedrock (Fig. 14.2e). If we could now step out of the elevator and inspect the bedrock itself, we would probably be disinclined to call the bedrock "solid," for its surface is often broken and riddled with pits and cracks (Fig. 14.2f).

We can summarize our observations in a series of graph lines showing the changes in soil composition with depth (Fig. 14.3). Graph lines A and B show the distribution of organic and inorganic particles from ground level to the bedrock, and graph lines C and D show the distribution of air and water over the same depth. From the graphs we can formulate a set of generalizations about soil:

1. Organic matter is limited to a thin surface layer.

2. Under the organic layer, particles and chemicals are often segregated into layers called horizons.

Fig. 14.3 Graph lines showing the relative volume of organic matter (A), inorganic particles (B), air (C), and water (D) from the surface to bedrock. The proportions are generally representative of soils in forested and grass-covered regions.

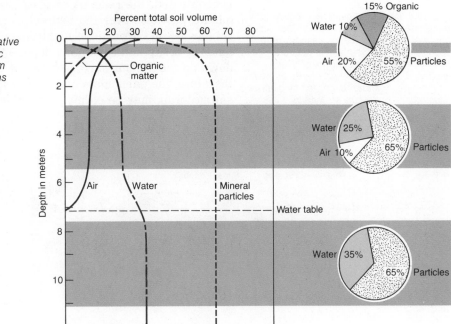

3. Water is most heavily concentrated at the base of the soil, where it fully saturates the interparticle spaces and forms groundwater.

4. Between the organic layer and the zone of groundwater is a zone of mixed particle sizes, where void spaces are occupied mainly by air and water.

THE PROPERTIES OF SOIL

Sources of Soil Particles

More than half of the total volume of most soils is made up of particles of different sizes and composition. The particles are lodged against one another such that they form a relatively solid skeleton capable of supporting not only the weight of the soil mass, but also material such as forests and buildings on the surface. Where do these particles come from? Originally, most were from the bedrock, but some came from organic matter. Through a complex set of decompositional processes, collectively known as *weathering,* rock and organic materials are broken down into smaller and smaller fragments (Fig. 14.4). Under moist climatic conditions, the more soluble minerals, such as salts, are dissolved and carried off in solution, leaving behind the less soluble minerals, such as quartz. Near the surface, where weathering is most concentrated, the remains of these stable minerals form residues of minute particles over the bedrock surface. Erosional processes then set to work on the particles and transport most of them away from their place of origin and eventually deposit them somewhere. Some are laid down on the land as surface deposits. Depending on the environment, therefore, the particulate material in which soils form, called the *parent material,* may be either surface deposits or the *in situ* residue of

Fig. 14.4 Bedrock weathering leading to the formation of loose particles. Chemical, physical, and biological processes are responsible for the weathering. Added to this material are the deposits of rivers, wind, and other erosion processes as well as the organic matter deposited by plants. (Illustration by William M. Marsh)

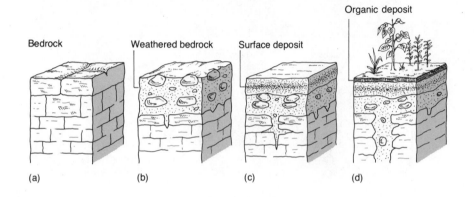

Bedrock

Weathered bedrock

Surface deposit

Organic deposit

(a)   (b)   (c)   (d)

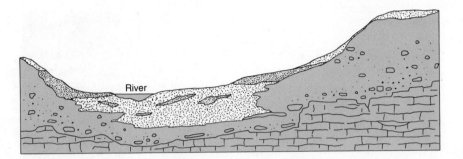

River

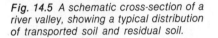
Transported soil material   Residual soil material   Bedrock

Fig. 14.5 A schematic cross-section of a river valley, showing a typical distribution of transported soil and residual soil.

weathered bedrock. (See Chapter 25 for a more detailed discussion of weathering.)

Soils composed of particles derived from the underlying rock are often called *residual soils*. Such soils generally exhibit characteristics that reflect the composition of the local bedrock itself. For example, sandstone weathers to yield sand and is therefore often associated with very sandy soils. In contrast, soils made up of particles from surface deposits may show little or no resemblance to the bedrock under them. These soils form in deposits made by rivers, glaciers, winds, and waves and can be called *transported soils* (Fig. 14.3c and d).

As you would imagine, it is not uncommon to find mixtures of residual and transported soil materials in most locales, especially at some depth below the surface. In river valleys, for instance, deposits of materials known as *alluvium* merge with residual materials on the sides of the valley as well as beneath the valley floor (Fig. 14.5). There are extensive areas of soils formed in surface deposits in the United States, and the map in Fig. 14.6 shows two examples. In fact, most of the world's soils have developed in surface deposits.

Texture

Soil particles vary greatly in size. Some are so small that they can be seen only with the aid of a microscope; others are so large that it is not apparent whether they are particles or part of the bedrock. Average diameter is the criterion used to classify individual soil particles, and three main size classes of soil particles are universally recognized: sand, silt, and clay. Agronomists, engineers, and earth scientists use

Soil of uniform particle sizes.

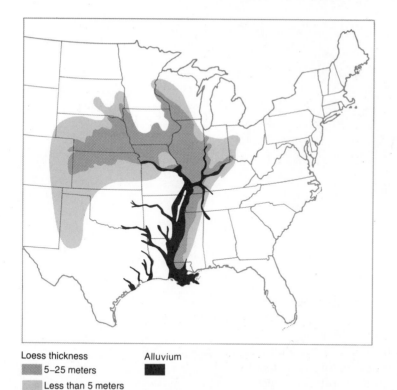

Loess thickness   Alluvium

▓ 5–25 meters   ■

░ Less than 5 meters

*Soil of mixed
particle sizes.*

*Fig. 14.6 Loess and major areas of
alluvium in the central United States.
Loess deposits are also found in central
Europe, Argentina, Russia, and China.
(Adapted from maps in Brady, 1974; Foth,
1978; and Hunt, 1974.)*

somewhat different size scales in classifying soil particles (Fig. 14.7).
Earth scientists often use the Wentworth scale, developed in geology.
Civil engineers use a variety of slightly different scales, including one
called the Massachusetts Institute of Technology scale, and agronomists use a standard scale developed by the U.S. Department of Agriculture.

Sand, silt, and clay each play an important role in soil formation.
For instance, sand enhances soil drainage, and silt and clay facilitate
the movement and retention of molecular water in the soil. Small clay
particles carry electrical charges that attract minute particles (called
ions) of dissolved minerals such as potassium and calcium. Attached to
the clay particles, the ions are not readily washed away; in this way,
clay helps to maintain soil fertility. Authors' Note 14.1 offers some additional information on clay.

The composite sizes of particles in a representative soil sample,
say, several handfuls, is termed *texture*. Soil texture is described as
the percentage of weight of particles in various size classes. Since it is
unlikely that all of the particles in a soil will fall within one textural

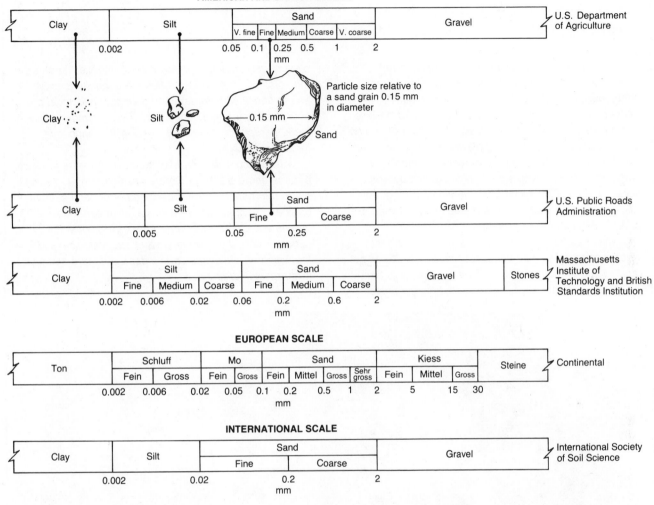

Fig. 14.7 Scales used for classifying soil particles: American and British, European, and international.

class, texture is usually expressed by a set of terms. At the center of this terminology is the natural class *loam*, which is a mixture of sand, silt, and clay. According to the U.S. Soil Conservation Service, the loam class is comprised of 40 percent sand, 40 percent silt, and 20 percent clay. Given a slightly heavier concentration of sand, say, 50 percent, with 10 percent clay and 40 percent silt, the soil is called sandy loam.

AUTHORS' NOTE 14.1
Clay

Clay is one of the most abundant materials on the earth's surface. It not only covers much of the continents, but also blankets most of the ocean basins. Although soil scientists define clay as a size of particle, it can also be defined on the basis of its mineral composition. Many different types of clays are recognized, and those in the silicate group are particularly important. These include kaolinite, montmorillonite, and hydrous micas. They are produced from the weathering of the silicate minerals such as mica and feldspars, which are primary constituents in granite.

Clay has three distinctive behavioral traits. First, when wetted, it has extreme plasticity; that is, it can be molded into virtually any shape without breaking. Second, it is highly cohesive, meaning that the particles cling to one another and to other substances, such as the soles of shoes. Third, it has the capacity to swell and to shrink with wetting and drying. This is most pronounced in montmorillonite, which can change volume by 50 percent or more with extreme fluctuations in moisture.

Throughout history, clay curiously appears as both a curse and a blessing. On the positive side, it is essential to the manufacture of earthen pottery and sun-dried brick and mud masonry buildings, all-important artifacts in both ancient and contemporary civilizations. On the negative side, it can muddy water supplies, sink under the weight of buildings, and hinder agriculture because it retards soil drainage.

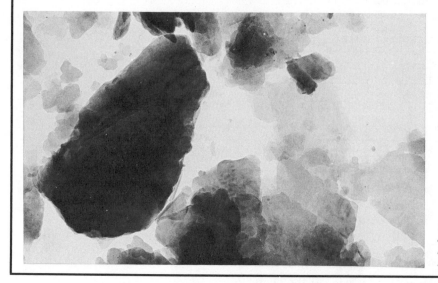

Kaolinite crystals (magnified 50,000 times) as they typically appear in soil. (Transmission electron micrographs by Kenneth M. Towe, Smithsonian Institution)

In agronomy, the textural names and related percentages are given in the form of a triangular graph (Fig. 14.8). If you know the percentage by weight of the particle sizes in a sample, you can use this graph to determine the appropriate soil name.

Soil Structure

Rarely do we find a soil that is comprised solely of free particles. Rather, particles are usually grouped into aggregates, especially if the soil contains clay and humus. Soil scientists refer to these aggregates as the soil *structure*. Individual structures may range from tiny granules about the size of your little fingernail to fist-size blocks. The four main soil structures usually recognized are given in Fig. 14.9.

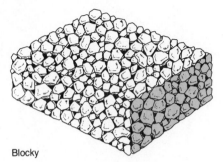

Blocky

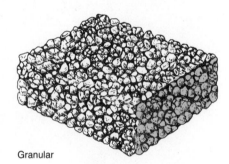

Granular

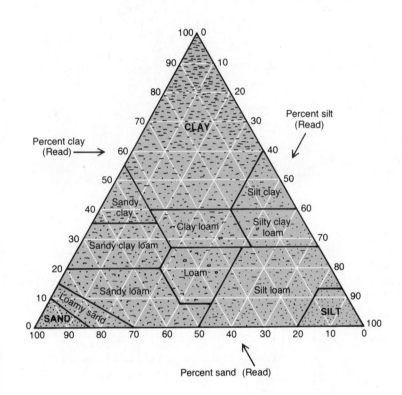

Fig. 14.8 Soil-texture triangle, showing the percentages of sand, silt, and clay comprising standard soil types as defined by the U.S. Department of Agriculture.

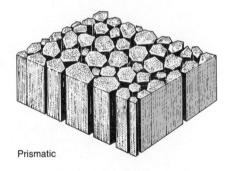

Prismatic

Fig. 14.9 The four main types of soil structure. Although the formation of specific structures cannot be easily explained, it appears that the process of particle aggregating is related to the clay, humus, and soluble salts.

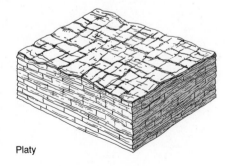

Platy

Fig. 14.10 (a) The primary, secondary, and tertiary consumers of organic matter in the soil constitute the primary animal food chain in the soil. (b) The progressive decomposition of organic matter with depth into the soil. (Adapted from Alan Burges, Micro-organisms in the Soil, London: Hutchinson/Hillary, 1958. Reprinted by permission.)

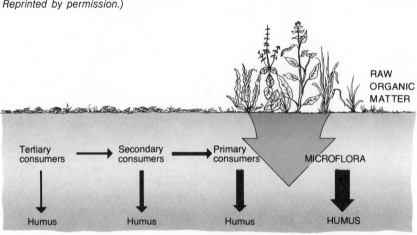

(a)

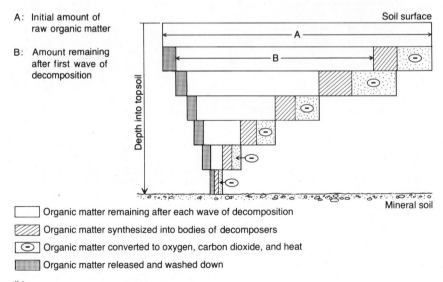

A: Initial amount of raw organic matter

B: Amount remaining after first wave of decomposition

Organic matter remaining after each wave of decomposition

Organic matter synthesized into bodies of decomposers

Organic matter converted to oxygen, carbon dioxide, and heat

Organic matter released and washed down

(b)

Although the processes that produce the different types of structures are complicated and not well understood, it appears that the presence of clay, humus, and soluble salts is important in controlling structure. Some of the influences of structure on soil processes are cited later in this chapter.

Organic Matter in the Soil

In addition to the mineral particles, most soils also contain particles of organic matter. These particles are composed of the partially decomposed remains of the plants that reside in and on the soil. Although certain soil environments, such as swamps and marshes, often contain massive accumulations of organic matter, the upper meter or so of most soils contains only about 3–5 percent organic matter by weight.

Although we tend to think of soil as a relatively lifeless part of the environment, careful inspection will usually show that it is rich in living organisms, especially in the upper 50 cm of soils that are well supplied with plant debris, heat, and moisture. This combination of ingredients fosters the growth of dense and diverse populations of small plants and animals that consume the organic matter. The chief consumers of organic matter are soil microflora, primarily microscopic-size bacteria, algae, and fungi, which are responsible for 60–80 percent of the soil's biological activity that results in the production of humus. Humus is a reduced form of organic matter that is somewhat more stable chemically than newer organic matter on the soil surface.

Helping the microflora to consume the organic matter are many small animals, principally insects and worms. These creatures are called primary consumers, and they eat not only leaves, twigs, and pieces of bark that fall to the ground, but the microflora as well. The droppings from primary consumers such as earthworms and snails constitute an important but minor source of humus compared to that produced by the microflora.

Ecologists also define several additional groups of animals that live in or on the soil and are dependent on the primary consumers. Secondary consumers are predators, such as spiders and centipedes, that feed on the primary consumers. They, in turn, are prey for another, smaller group of predators called tertiary consumers. Humus is produced by each consumer; however, the fraction of organic energy returned to the soil diminishes rapidly from consumer to consumer, because most of the energy consumed is dissipated by the animals in bodily heat and locomotion (Fig. 14.10).

Although primary, secondary, and tertiary consumers are less important than microflora as producers of humus, they perform several other important functions in the soil. Earthworms, for example,

AUTHORS' NOTE 14.2
Earthworms

The common earthworm is probably the most important of the large animals in the soil. Although more than 200 species are known, only 2 are found in the eastern and central United States. One of these species, Lumbricus terrestris *(a reddish worm), was transported to this country from Europe centuries ago. As farming spread westward, this worm replaced the native American worms of the forests and prairies, which could not withstand the changes in the soil environment brought on by cultivation.*

It is estimated that earthworms inhabiting only one acre of soil may pass as much as fifteen tons of soil through their bodies each year. During the soil's passage, both the organic and mineral particles are subjected to grinding action and to digestive enzymes. As a result, the droppings, or casts, left in the worm tunnel are more aerated and richer than the surrounding soil in bacteria, nitrogen, phosphorus, and potassium.

Earthworms prefer a moist habitat in which aerated organic material is plentiful. Under rich grasslands, such soil conditions may give rise to populations equivalent to more than 500 worms per square meter of surface area. In contrast, soils subject to drought or temperature fluc-

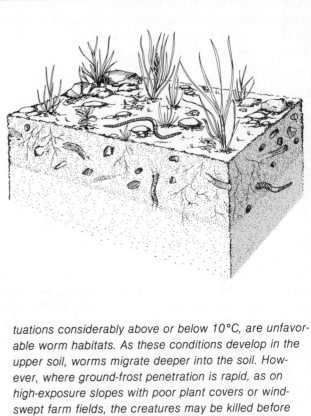

tuations considerably above or below 10°C, are unfavorable worm habitats. As these conditions develop in the upper soil, worms migrate deeper into the soil. However, where ground-frost penetration is rapid, as on high-exposure slopes with poor plant covers or windswept farm fields, the creatures may be killed before they can migrate below the frost line.

literally eat their way through the organic layer, leaving tunnels in their paths. These tunnels help to not only aerate the soil, but also provide routes for water penetration from the surface. Moreover, worm activity near the base of the organic layer mixes the humus with the underlying inorganic particles (Authors' Note 14.2).

Energy in the Soil

Several forms of energy are contained in the soil. Moving water represents kinetic energy capable of moving solid and dissolved particles as well as heat. The minerals that comprise both the organic and inorganic soil particles are a form of molecular chemical energy, which in the presence of water and heat may pass from the solid to dissolved state or vice versa. In the dissolved state chemical energy is represented by ions, each of which carries an electrical charge. Owing to this charge, ions are attracted to soil particles and to other ions, both of which can be transported by water moving through the soil.

Heat is one of the most important forms of energy in the soil because it has a strong influence on the movement, evaporation, freezing, and chemical activity of water. The bulk of soil heat is derived from solar radiation, but some is also derived from the earth's crust. Where the ground is not shaded by plants, solar radiation is absorbed directly by the soil surface. If the net radiation is strongly positive, very high daytime temperatures develop in the upper several centimeters of soil. In fact, the highest natural temperatures known on the planet, around 90°C, have been recorded directly on the soil surface. Despite such high surface temperatures, the downward transfer of heat is very slow compared with heat transfer in lake water, for example. This is due to the low thermal conductivity of soil in general and to the fact that apparently little heat is transferred into the soil by water percolating downward from the surface. (See Chapter 3 for details on heat flow in soil.)

The earth's interior also contributes heat to the soil, but its transfer upward through the bedrock can be even slower than the downward transfer of surface heat. This can be attributed to the fact that although the thermal conductivities of bedrock and soil are roughly comparable, the temperature gradient from bedrock to soil, at about 2.5°C per 100 m, is much lower than that between the soil surface and the soil just a few meters below. As a result, the inflow of geothermal heat to the soil, which averages about 0.06 joules/$m^2 \cdot$s, or 47 cal/$cm^2 \cdot$year, throughout much of the world is many hundreds of times less than typical heat fluxes from the surface. Nonetheless, geothermal heat is an important factor in the overall soil-heat balance. In the Arctic, for instance, geothermal heat helps to limit the penetration of permafrost to a relatively thin layer near the surface, usually less than 100 m thick. (See Chapter 23 for a detailed discussion of geothermal heat.)

The interplay of heat from the surface and heat from the earth's interior has given rise to several heat zones in most soils. As we discussed in Chapter 1, the upper 20 cm or so of soil is the zone of daily heat fluctuation, where temperature may vary as much as 25°C from day to night. Below this layer is a thin zone where heat varies from month to month, and under it is a zone about 2–3 m deep where temperature changes only on a seasonal basis (Fig. 14.11). In the lower part of the seasonal zone the change in heat is so gradual that temperature is usually a season out of sequence. Thus the highest temperature

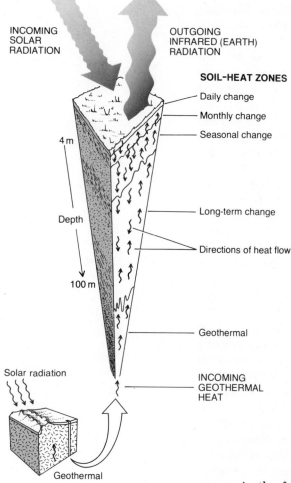

INCOMING
SOLAR
RADIATION

OUTGOING
INFRARED (EARTH)
RADIATION

SOIL-HEAT ZONES

Daily change

Monthly change

Seasonal change

4 m

Depth

Long-term change

Directions of heat flow

100 m

Geothermal

Solar radiation

INCOMING
GEOTHERMAL
HEAT

Geothermal

Fig. 14.11 Sources and directions of soil-heat flow. In the upper four meters or so, three heat zones can be defined on the basis of the period of temperature variation.

occurs in the fall rather than in the summer (Fig. 14.12). Below this zone, extending to a depth as great as 1 km in some places, is a zone transitional to the realm of geothermal heat. This is a very interesting zone because it responds to long-term changes in the earth's energy balance and is typically thousands of years out of sequence with the surface. For example, the Ice Age produced negative shifts in the soil-heat balance over broad sections of North America and Eurasia, and today, more than 10,000 years later, temperatures representative of those conditions still persist in the transition zone. The great lag in time necessary for the heat at this depth to equalize with the heat balance in the upper soil and the geothermal heat in the underlying bedrock is used to explain the extraordinary thickness of permafrost in parts of

Alaska, Canada, and Siberia. Here as much as 700 m of permafrost have been found in many locales.

Soil Water

Water is central to virtually every facet of soil. Erosion and deposition by runoff are important determinants of soil thickness and composition. Within the soil, water movement and associated chemical processes are responsible for the dissolution and precipitation of minerals, and at depth, the weathering of the bedrock. Soil moisture has a key control on plant growth and soil organisms and thus has an important influence on the humus content of the topsoil. Whether in arid or humid climates, these generalizations appear to be basically sound, and so it is important that we examine the origin, movement, and dissipation of water in the soil.

Water is deposited by precipitation on the soil surface, where it may run off or penetrate the soil mass. Recall from Chapter 11 that the rate of penetration is called the *infiltration capacity* and is expressed in millimeters or centimeters of surface water lost to the soil per hour. Suffice it to point out here that the infiltration capacity is controlled by many factors, including soil texture, plant cover, existing soil moisture, and ground frost.

As water moves into the soil, it joins existing water that is attached to the surfaces of soil particles in a thin, nearly imperceptible film. Recall that this water is called molecular water because unlike the infiltration water, which is controlled by gravitational force, it is controlled by molecular forces. There are two varieties of molecular water: hygroscopic and capillary. Hygroscopic water is represented by a thin film of water molecules attached directly to the surface of every soil particle. This film is bound to the particle by strong molecular forces, equivalent to 31–10,000 bars of pressure. Under this force, no natural processes are capable of moving or making use of this water. Thus it is of virtually no importance to soil formation, because it cannot perform physical and chemical work.

Capillary water, on the other hand, is fully capable of movement and is thus highly important in soil formation. Recall that this water resides on the hygroscopic film, where it is held under pressure ranging between 0.33 and 31 bars; hence it is often called "loosely bound" molecular water. Capillary water behaves in much the same way as does kerosene in a wick lantern. That is, it tends to move toward spots of relatively lower quantities of moisture, just as kerosene moves up the wick of the lantern. In soil of uniform-sized particles, the rate of movement is controlled primarily by the steepness of the moisture

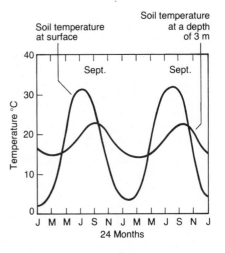

Fig. 14.12 *The difference in the monthly variation of soil temperature at the soil surface and at a depth of 3 m. The graph shows that: (1) the seasonal temperature variation, or range, decreases rapidly with depth; and (2) annual high and low temperatures at a depth of 3 m occur about a month later than the annual high and low monthly temperatures at the surface. If we projected this graph to greater depths, the seasonal temperature range would decrease, and at some depth it would disappear altogether.*

gradients. However, as we shall see in the next chapter, in heterogeneous soils particle size can be equally effective in controlling the rate of movement.

As water is added to a soil and the capillary film around the soil particles grows in thickness, the molecular force that binds it to the hygroscopic film decreases, and at a pressure less than 0.31 bar, the cohesive strength is insufficient to retain the water in the molecular form. This is extremely important to vegetation, because the availability of capillary water to plants depends in large part on cohesive pressure. As the capillary water is taken up by plant roots, the remaining fraction of the film becomes increasingly difficult to remove. When the residual capillary water is held too tightly for plants to draw off amounts sufficient for normal respiration, they may wilt and die.

For any soil, there is a maximum quantity of capillary water that it can hold. When this condition is reached, the soil is said to be at *field capacity*. Since capillary water is held on the surfaces of soil particles, the greater the number of particles (and thus surfaces) in a given volume of soil, the higher the field capacity must be. For example, clay can hold nearly six times more water at field capacity than sand can (Table 11.3). As a further aid in this discussion, you may find it helpful to review the illustrations concerning soil water given in Chapter 11.

SUMMARY

The solum is the part of the soil capable of supporting plant life, but a more general definition of soil would include most assemblages of ground materials above the bedrock. More than half of the volume of most soils is made up of inorganic and organic particles; the remainder is mostly air and water. The ratio of organic to inorganic content decreases sharply with depth, and organic material largely disappears beyond a depth of 1–2 m in most soils. Dead organic matter is consumed mainly by microflora near the soil surface. The amount of air in a soil varies with the water content, and below the water table virtually all air has been displaced by water. Texture is a critical soil property because of its influence on drainage, moisture-holding capacity, and bearing capacity, in particular. The energy that drives the internal soil processes is the heat from solar and geothermal sources, the kinetic energy of moving water, and the molecular energy of chemical processes. Water enters the soils by infiltration, and in the solum is taken up as capillary water.

CHAPTER 15 THE MOVEMENT AND BALANCE OF WATER IN THE SOIL

THE MOVEMENT OF SOIL MOISTURE

A remarkable quantity of water moves into and out of most soils in the course of a year. In the midwestern United States, for instance, it is not uncommon for the soil to receive 0.5 to 0.75 m of infiltration water annually and to release 0.25 to 0.5 m of water vapor to the atmosphere (Fig. 15.1). Inflows and outflows of soil water are variable over time; however, seasonal patterns are identifiable throughout most of the world where precipitation and/or heat tend to vary sharply from season to season. As the water moves up and down in the soil, it transports chemicals and particles, thereby altering the internal makeup of the soil. Let us trace the movement of water into and out of the soil over the year in the middle latitudes.

Outflow

As surface heat, soil heat, and plant transpiration increase with the onset of summer, soil moisture is rapidly lost from the uppermost several centimeters of soil. Should this trend continue without interruption by rainfall, the depth of this layer of depleted moisture would steadily increase, eventually extending throughout most of the zone occupied by the roots of small plants. Although most of the capillary water has been lost from this zone, careful inspection will usually reveal that not all has been lost. In fact, much of the more tightly bound molecules of the capillary film may remain in the soil. This fraction of water is held under molecular pressure greater than 15 bars, which is too strong to be broken by the water-drawing forces generated around the plant roots. No longer able to take up water from the soil, the plants may wilt and eventually die (Fig. 15.2a).

The description above is a very generalized statement on the capillary-water condition during drought. In reality, however, the upper soil is not totally devoid of the capillary water that can be used by plants. Rather, small pockets of it remain irregularly distributed in the lower part of the dry layer. Where the density of plant roots is high, moisture loss is greatest; the spots untouched by roots may be measurably wetter (Fig. 15.2b).

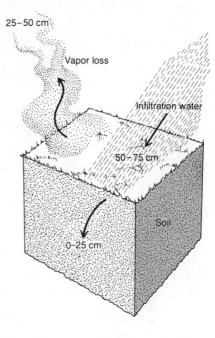

Fig. 15.1 The approximate amounts of water entering (infiltration) and leaving (vapor) the soil over a year in the midwestern United States.

223

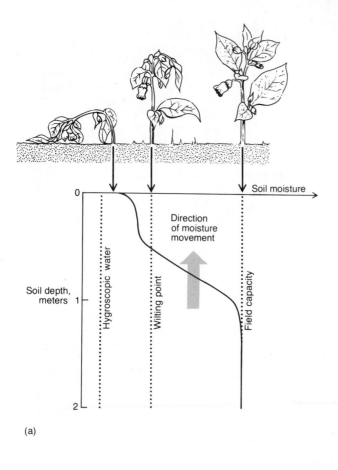

(a)

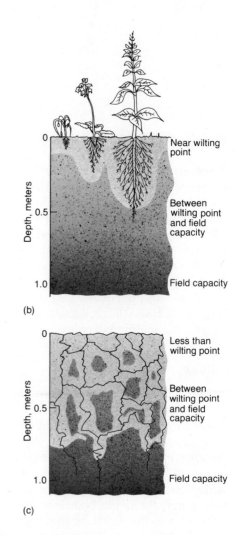

(b)

(c)

Fig. 15.2 Soil-moisture outflow: (a) In mid-summer soil moisture typically decreases toward the surface from a depth of about 1.5 m. Since evaporation and moisture uptake by plant roots are most concentrated in the upper 25 cm or so of the soil, capillary-water supply is typically lowest there. If this supply falls below wilting-point pressure (15 bars), the plants wilt and, if the condition is prolonged, eventually die. The wilting sequence of plants is often a crude indicator of root depth; grasses and other small herbs are the first to go during a drought; trees, usually the last.
(b) Variations in soil moisture as a response to root density. Moisture uptake by plants can be surprisingly high; large fir trees in the American West consume as much as 1500 gallons of water per day.
(c) The same cracks that enhance infiltration can also enhance moisture loss through evaporation. Here moisture gradients are shown between the interiors and perimeters of large soil blocks. In this context, soil structure can have an important influence on moisture loss.

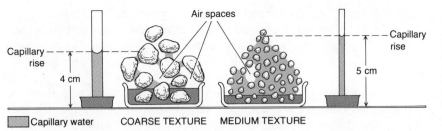

Fig. 15.3 *Differences in capillarity related to soil texture. The rise is much higher in medium textures than in coarse textures. This can also be illustrated with a simple experiment using large and small glass tubes standing in a pan of water.*

Further, the soil structure also affects differential rates of soil-moisture loss. Large aggregates, as in a coarse prismatic or blocky structure, tend to minimize the number of aerated passageways which serve as the release routes for vaporized water. Conversely, the small-particle aggregates of the granular or platy structure present many more fissures per column of soil and therefore tend to be more efficient in releasing moisture (Fig. 15.2c).

The loss of capillary water from the upper soil produces a moisture gradient toward the surface, which sets into motion the molecular transfer of water. Under a given drought condition, the rate of moisture transfer along the gradient is controlled principally by soil texture. Medium textures (silt, fine sand, and loam) have greatest capillary transfer capacity, or *capillarity*, because the particles are small enough to develop strong meniscuses, yet large enough to allow ready movement of the moisture film between particles. This is a critical determinant of a soil's propensity for drought. The capillarities of both the fine and the coarse textures are relatively low, though for different reasons. In coarse sand or pebbles, for example, the wide spacing between contact points places high stress on the capillary film, making it difficult for moisture to move (Fig. 15.3). In clays, on the other hand, capillarity is often limited by the smallness of the interparticle spaces, especially if the soil is tightly packed. In short, the low capillarities of the fine- and coarse-textured soils mean that moisture differentials are equalized less rapidly than they are in medium-textured soils. Therefore, under similar atmospheric and vegetative conditions, drought can be more severe in soils of the extreme textures.

Inflow

Returning to the drought condition represented by Fig. 15.2(a), we find that the drying trend has been broken by a rainfall sufficient to soak the soil surface. Immediately, of course, the uppermost segment of the moisture profile is reversed. And as gravity water is diffused down-

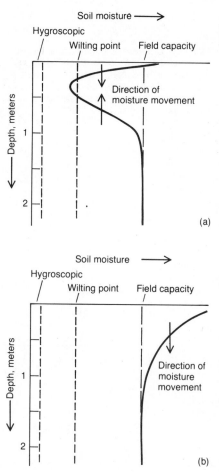

Fig. 15.4 *Soil-moisture profiles: (a) Immediately after a drought has been broken by a substantial rainfall. Infiltration water has recharged the upper 25 cm of soil, and with the upflow of moisture from the lower reservoir, the dessicated layer is being recharged from two directions. (b) After the soil has been recharged in excess of field capacity and is incapable of retaining additional water; the surplus is transmitted downward as gravity water.*

ward, it is quickly transformed into capillary water. Rarely, however, can a single rainstorm recharge the capillary reservoir, and thus the soil is left stratified, with an intermediate dry zone sandwiched between two moist zones. Double gradients are thereby formed (Fig. 15.4a).

Should rainy conditions, coupled with lowered rates of soil-moisture loss, pervade for an extended period of time, as is typically the case during the winter months in the northwestern United States and northwest Europe, the amount of water in the upper soil may greatly exceed field capacity. As a result, a great enough volume of gravity water can be produced to recharge the capillary reservoir as well as to generate through flow to the water table. Once the soil is at field capacity throughout, the only gradient that exists is the one formed by gravity water (Fig. 15.4b).

THE SOIL-MOISTURE BALANCE

As with radiation and heat, soil moisture can also be examined in the context of a balance problem. This involves accounting for the losses and gains of soil moisture. At the heart of the water balance are moisture flows between: (1) atmosphere and soil; (2) soil and plants; and (3) plants and atmosphere (Fig. 15.5). Precipitation constitutes the sole inflow of water to the soil. Evaporation from the soil and transpiration from plants constitute the sole outflows of capillary water from the soil. These are combined into the term evapotranspiration. The *soil-moisture balance* for any period of time is equal to *precipitation* minus *evapotranspiration*.

Potential and Actual Evapotranspiration

The moisture-balance problem is given additional meaning, particularly for application to agriculture, by the distinction between *potential* evapotranspiration and *actual* evapotranspiration. Actual evapotranspiration is the true quantity of moisture lost from the soil; potential evapotranspiration, the quantity that could be lost given an inexhaustible water supply. High levels of heat and wind can deplete an available supply of moisture early in the growing season, and although actual losses may be slight in the ensuing summer months, they are so only because there is no water left to lose. To create a moisture balance favorable to agriculture, for example, the gap between the actual and potential evapotranspiration, called the *moisture deficit*, must be made up through irrigation. The average annual potential evapotran-

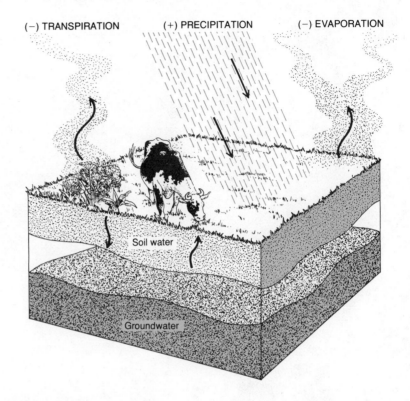

(−) TRANSPIRATION   (+) PRECIPITATION   (−) EVAPORATION

Soil water

Groundwater

Fig. 15.5 This basic model of the soil-water balance applies to the upper soil, i.e., the solum, and is designed to aid in computation of short-term (month-by-month) soil-moisture changes. The three fundamental processes comprising the model are precipitation, evapotranspiration, and capillary-moisture flow. For any time period, the net change in soil moisture is equal to precipitation minus evapotranspiration. (Illustration by Peter Van Dusen)

spiration values for the United States are given in Fig. 15.6. Comparing these values with the average annual precipitation highlights the problem areas for soil moisture in the United States.

By knowing the values of evapotranspiration, precipitation, and moisture content of the soil, we can compute fairly accurately the water balance of most soils in the midlatitudes. Figured on a monthly basis, the water balance often reveals seasonal soil-water surpluses, soil-water deficits, potential runoff rates, and changes in soil moisture available to plants, each directly or indirectly an important factor in soil formation.

More than any other "balance"-type problem that we have examined, the soil-moisture balance resembles an accounting problem. It begins with a reserve on account in the form of soil water. This account, however, is unique because it is limited to a maximum amount, fixed by field capacity. Into and from this account flows water. Summer is a time of big outflows, and the reserve is drawn down, perhaps even exhausted, but never overdrawn. During fall and winter the reserve is rebuilt, and in spring the fully recharged reserve rejects additional inflows, which are converted to runoff.

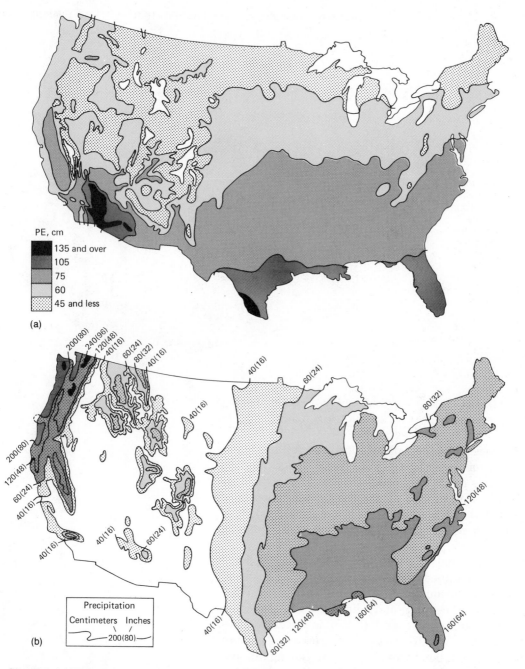

Fig. 15.6 (a) The average annual potential evapotranspiration (PE) for the continental United States. These values represent the amount of moisture which would be lost from the soil given an inexhaustible supply of soil water. (b) The average annual rainfall for the United States. By comparing the two maps, we can see that the Southwest would need more than 100 additional centimeters of water just to break even each year. (Reprinted from the Geographical Review **38** (1948), with the permission of the American Geographical Society)

Examples of the Soil-Moisture Balance

Figure 15.7 shows graphical examples of the annual water balance at four different locations in the United States. Note that the *actual evapotranspiration*, the true moisture-loss value when the capillary factor is taken into account, is fed by water brought up by capillary

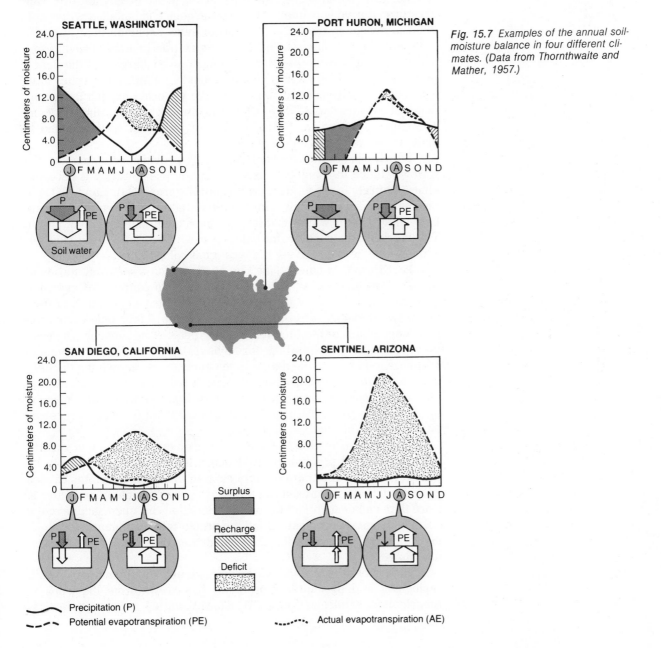

Fig. 15.7 Examples of the annual soil-moisture balance in four different climates. (Data from Thornthwaite and Mather, 1957.)

flow. But since this flow is usually too slow to meet the demands of plants and atmosphere, a moisture deficit develops, represented graphically in Fig. 15.7 by the area between the PE and AE curves. Note that AE can never exceed PE, as PE is the maximum possible evapotranspiration. *Recharge,* the replacement by precipitation of that soil moisture lost during a deficit period, is possible only until the soil-moisture reserve is refilled. After recharge is completed, a water surplus may occur in the form of runoff or snow accumulation.

Sentinel, Arizona (see Fig. 15.7), exemplifies an area of severe soil-moisture deficit. Note that there is almost no recharge and the deficit period is a year long. This means that existing plants must either have low water requirements or be artificially watered through irrigation. Sentinel has an annual PE of 119 cm and annual precipitation of 12 cm. This gives you an idea of the amount of irrigation water needed to maintain a lawn in this desert area.

San Diego, California (see Fig. 15.7), provides an example of a water balance in a highly seasonal climate. Precipitation is received almost entirely in the winter months, but the greatest demand (PE) is in the summer. This results in a large deficit that is only partially made up by recharge in the winter months, leaving a net annual deficit of 53 cm. This markedly seasonal water balance is characteristic of the Mediterranean climate found along the west coast of California.

Port Huron, Michigan (see Fig. 15.7), is also a seasonal station in that temperature varies widely from summer to winter. However, the precipitation is much more evenly distributed throughout the year than in the two previous cases. When the temperature falls below freezing, PE is negligible, contrasting with a sharp peak in the summer. The even distribution of precipitation throughout the year provides enough recharge to minimize the deficit in midsummer. By midwinter the deficit is made up; that is, recharge is completed, and the remaining precipitation is surplus water, much of which is stored in the form of snow until spring melt. Port Huron has an annual PE of 61 cm and receives 74 cm of precipitation.

Seattle, Washington (see Fig. 15.7), is an example of a site with an adequate annual precipitation (84 cm), which is quite seasonal in its distribution. The precipitation minimum in the summer months causes a strong deficit for that season, but heavy precipitation in the winter is more than enough to offset the deficit. Note that the steep decline in the AE in midsummer lags just behind the precipitation minimum, vividly illustrating their interrelation. This illustrates the time lag in soil-moisture flow.

The concept of the soil-water balance and its application to soil, ecological, and agricultural problems has grown from the work of American climatologist Charles W. Thornthwaite (1899–1963). Derivation of the method for computation of the soil-water balance involves the use of the energy-balance method described in Chapter 1. Appendix 2 provides a brief explanation for computation of the water balance on a monthly basis. Beginning with the monthly heat index, each step is explained; all calculations are in centimeters (cm) of water.

SUMMARY

Before going on to Chapter 16, it is necessary to underscore the importance of the soil-water balance to understanding the development of the solum. For through this concept we are able to account for three important controls on soil formation:

1. The directions and intensities of the soil-moisture movements, which relocate dissolved minerals and fine particles within the soil.

2. The rates of plant growth and organic-matter decomposition, which influence the organic content of the topsoil as well as the chemistry of infiltration water.

3. The availability of water for runoff, which, through erosion and deposition, influences soil thickness and organic cover.

SOIL-FORMING PROCESSES AND DEVELOPMENT OF THE SOLUM

CHAPTER 16

BIOCHEMICAL PROCESSES IN THE SOLUM

Chemical Activity

Both agronomists and geoscientists are interested in the development of the solum, the soil on which life most directly depends and that is most integrally connected to the rest of the landscape. The solum is, therefore, often a valuable indicator of past and present conditions of the landscape as a whole.

A moisture surplus in the upper soil results in a downward percolation of water; if clays are present, some of the particles may enter the water. The smallest of these particles, less than 0.001 mm in diameter, are called *clay colloids* and are so small that when dispersed in water, they will remain suspended indefinitely.

Important to soil formation is the fact that colloids are electrically charged and behave like tiny magnets, carrying both positive and negative charges. Because the negative charge is appreciably stronger than the positive one, colloids attract large numbers of positively charged ions, called *cations*. Cations of potassium, sodium, calcium, magnesium, and aluminum are called *bases* and are common minerals in soil. These bases are important nutrients for plants.

For each colloid there are a limited number of places, or *adsorption sites*, for cations to attach themselves. The process whereby cations are released and adsorbed at these sites is called *cation exchange*. The larger the particles, the fewer the adsorption sites per cubic centimeter of soil and the lower the cation-exchange capacity. Thus coarse-textured soils tend to be poor in bases, whereas clayey soils tend to be rich in bases.

In the absence of cations, soils tend to be relatively heavy in hydrogen ions. Hydrogen ion colloids are referred to as *acids*. The ratio of hydrogen ions in the soil solution to nonhydrogen ions, called hydroxl ions, determines the soil pH factor. The pH is a measure of the relative acidity or alkalinity of a soil. A pH of 7 is neutral; lesser values indicate that the soil is *acidic*, and values greater than 7 indicate that the soil is *basic*, or *alkaline* (Fig. 16.1). In agriculture the soil pH value determines whether or not a soil will need lime in order to be suitable for most crops. The lower the pH, the greater the amount of lime that must be

0 —

0.25 m —

0.5 m —

Caliche layer in a New Mexico soil

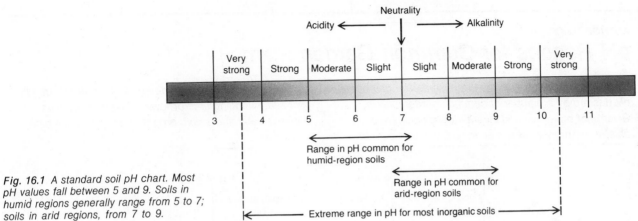

Fig. 16.1 A standard soil pH chart. Most pH values fall between 5 and 9. Soils in humid regions generally range from 5 to 7; soils in arid regions, from 7 to 9.

added to "sweeten" the soil sufficiently to return high yields on crops such as corn. Authors' Note 16.1 gives the recommended pH ranges for a number of common garden plants.

Eluviation

The removal of minerals in solution from a layer of soil is known as leaching. This involves the relocation, or "washing out," of ions to other levels in the soil. If the washing out involves colloids as well as ions, the term *eluviation* may be used to describe it. The layer, or zone, in the soil losing materials is called the *zone of eluviation* (Fig. 16.2a).

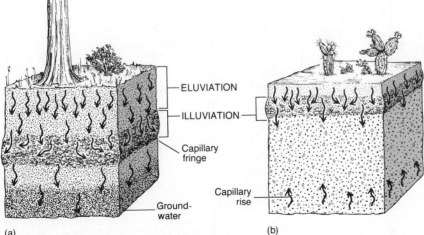

(a)

(b)

Fig. 16.2 Formation of zones of eluviation and illuviation under humid and arid moisture regimes: (a) Under a favorable moisture balance, eluviation is concentrated in the upper soil, and deposition of minerals and colloids is concentrated at a lower depth, called the zone of illuviation. (b) Under a negative moisture balance, the depth of penetration of surface water is slight, resulting in the formation of a zone of illuviation near the soil surface.

AUTHORS' NOTE 16.1
pH Ranges for Common Garden Plants

The following chart gives the approximate ranges of soil pH desired by a number of common garden plants. Given that other soil and climatic conditions are suitable, if the proper pH is maintained, you can expect normal plant growth and development. The pH can be raised by adding lime or lowered by adding clean quartz sand and acidic organic matter such as pine needles.

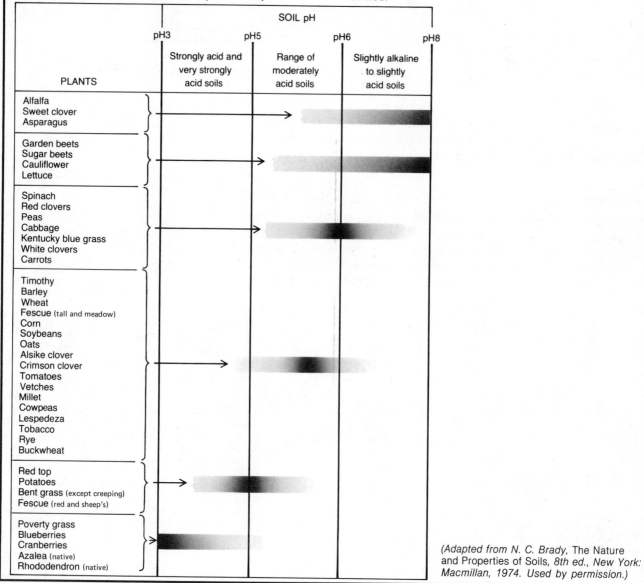

(Adapted from N. C. Brady, The Nature and Properties of Soils, 8th ed., New York: Macmillan, 1974. Used by permission.)

The downward transfer of ions and colloids often ceases at a level where the upward pressure of capillary water from the groundwater zone offsets that of the percolating gravity water. Here the materials may be deposited in a *zone of illuviation* (Fig. 16.2b). In time this zone can collect such a mass of colloids and minerals that the interparticle spaces become clogged, cementing the original particles together. It is not uncommon to find the zone of illuviation in heavily leached soils marked by a concretelike layer called *hardpan*.

In very wet climates, e.g., an equatorial rainforest, minerals may be rapidly leached from the upper soil and precipitated in the lower soil or carried by groundwater to streams and rivers. This process has gone on for so long and at such an intensive rate under the strongly positive moisture balance of the tropics that even particles of silica, the mineral that comprises most sand-size grains, have been largely leached from the soil. Elsewhere in the world, silica is considered a relatively insoluble mineral and is a common soil constituent. By contrast, under the negative moisture balances of desert climates, as in the southwestern United States, mineral-rich water does not pass through the soil, but penetrates only the first 10–30 cm below the surface. The water evaporates, leaving its ion load, commonly calcium carbonate, in the upper soil. In some places wind adds deposits of calcic dust to the surface. When rain falls on the dust, some of it is dissolved and washed into the soil, where it is deposited as the water evaporates. With each succeeding rainfall, more calcium is deposited, and eventually a hard crust, known by the Spanish word *caliche*, develops. Caliche is also known to form in other ways and is sometimes formed at depths of a meter or so in soil.

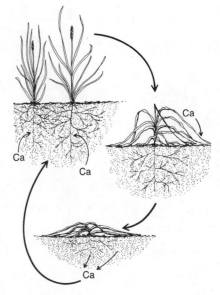

Fig. 16.3 The nutrient cycle for calcium (Ca) between plants and soil. A change to plants that are less efficient in holding nutrients can result in a change in soil chemistry.

Biological Activity

Soil is influenced by many types of biological activities, especially those related to the plant cover. Plant growth is affected by not only heat and moisture, but also the chemical makeup of the soil. For example, grasses need calcium and magnesium bases (alkalines) in order to grow well. If these minerals are present in the soil, they are taken up with the moisture absorbed by the plant roots and are stored in the plant tissue. When the plant dies, the minerals are returned to the soil, as Fig. 16.3 illustrates. In this way plants help to maintain the fertility of soil by bringing up the mineral bases and redepositing them at the soil surface in the form of dead leaves, stems, etc. Indeed, this cycle is critical to the maintenance of the lush rainforest vegetation of tropical and equatorial areas.

As we noted earlier, the biochemical decomposition of organic matter supplies *humus* to the soil, which produces not only the characteristic dark-colored topsoil, but also organic acids containing

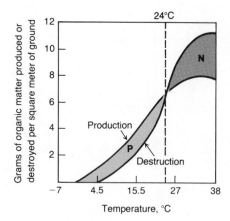

P = Positive balance; organic matter accumulates

N = Negative balance; potential exists for more
　　destruction than production

Fig. 16.4 Rates of production and destruction of organic matter as a function of temperature under humid climatic conditions. At a temperature of less than 24°C, the destruction of organic matter by microorganisms is not great enough to consume all of the organic material produced by plants, and so it accumulates as humus. But above 24°C, the rate of destruction by microorganisms increases much faster than the rate of plant production, and therefore all organic matter is consumed, leaving the soil with essentially no humus.

hydrogen ions. The organic acids are especially strong under certain plant covers, particularly the conifers. Organic acids aid in the weathering of soil particles and bedrock and tend to replace the leached calcium, potassium, magnesium, and sodium ions important to further plant growth. Thus in cold, damp climates, for instance, where conifers are abundant and the humus is inherently weak in bases, the soil tends to be strongly acidic. In contrast, deciduous trees growing under the same conditions may give rise to less acidic soils, with base-saturation levels two to three times higher in the upper soil.

The amount of organic matter in any soil is the product of the rate of plant productivity (total growth) less the rate of destruction of organic matter by organisms, leaching, and erosion. Recall that most of the organic matter is consumed by microflora, which are very sensitive to heat and moisture conditions. As a result, in a cold climate, such as the Arctic tundra, microfloral action is slow, and the rate of organic decomposition usually does not exceed the rate of plant growth; therefore, organic matter may build up over time. In arid regions the lack of moisture in the upper soil prohibits microfloral activity. But since plant productivity is exceedingly low in deserts, not much organic matter accumulates anyway. Under semiarid conditions, however, moisture is often adequate to foster substantial productivity by grasses and other herbs, but inadequate to foster comparable rates of destruction by microflora; hence considerable organic buildup may occur. Figure 16.4 shows the relationship between net organic production and temperature under humid conditions. In tropical and equatorial areas, where conditions are hot and moist, plant productivity is high, but microfloral activity is even higher; thus it is impossible for an organic layer to build up. Exceptions to this are found in swamps, where standing water limits microfloral activity, but not productivity, resulting in heavy accumulation of organic material.

Microflora are also known to increase soil fertility. In the presence of certain plants, some soil bacteria can convert gaseous nitrogen from the atmosphere into a form that can be utilized by plants. This process, called *nitrogen fixation*, is especially pronounced in plants such as the legumes (pea family) and is one of the most important means by which plants can increase the nutrient level of the soil well beyond its original level. As for larger organisms, such as insects and small animals, their role is largely one of reworking and mixing the soil through burrowing (Authors' Note 14.2). This activity tends to increase aeration of the upper soil and in turn may help to promote bacterial activity and moisture infiltration. Counteracting such activity, large herds of surface animals, such as cattle and reindeer, may pack the soil, decreasing both aeration and infiltration.

Soil Horizons

Working in combination, the chemical, biological, and physical processes discussed above produce vertical differentiation within the upper 5 m or so of soils in which there is comparatively free movement of gravity water and capillary moisture. This results in the formation of nearly horizontal layers, or *horizons*, as they are termed by the soil scientist, which are distinguishable on the basis of color, texture, and chemical composition. Generally speaking, four standard horizons are recognized, and they are designated top to bottom by the letters O, A, B, and C. The O horizon includes the humus-rich topsoil, where the organic content is greater than 20 percent. The A horizon corresponds to the zone of eluviation. The B horizon represents the zone of illuviation, and C is transitional to the parent material, usually some sort of deposit, where alteration by soil processes is markedly less advanced.

This sequence of horizons, called a *soil profile,* provides a suitable framework for describing most soils, particularly those formed in well-drained sites in regions of positive moisture balances. As with so many other natural phenomena, the boundaries between soil horizons are typically indistinct, as one horizon usually grades almost imperceptibly one into another. These transitional zones are classed as subhorizons, and several are usually identified in each of the upper two horizons (Fig. 16.5).

Factors of Soil Formation

Soil formation is influenced by a multitude of factors, some of which are apparent at a glance, whereas others are very subtle and apparent only through instrumental observation. In all cases, however, these factors can be related to driving forces, which in turn can be translated into basically four types of energy: potential, kinetic, chemical, and heat. The processes that operate within the soil, such as eluviation, illuviation, and microflora activity, are driven by this energy. Variations in the amounts and forms of energy are controlled by the larger environment within which soil formation occurs. This environ-

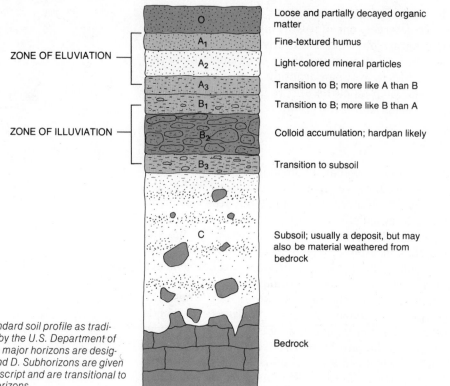

ZONE OF ELUVIATION

O — Loose and partially decayed organic matter

A_1 — Fine-textured humus

A_2 — Light-colored mineral particles

A_3 — Transition to B; more like A than B

ZONE OF ILLUVIATION

B_1 — Transition to B; more like B than A

B_2 — Colloid accumulation; hardpan likely

B_3 — Transition to subsoil

C — Subsoil; usually a deposit, but may also be material weathered from bedrock

Bedrock

Fig. 16.5 A standard soil profile as traditionally defined by the U.S. Department of Agriculture. The major horizons are designated A, B, C, and D. Subhorizons are given a numerical subscript and are transitional to and between horizons.

ment is classed into four major elements that soil scientists call *soil-forming factors*: (1) climate, (2) parent material, (3) vegetation and related biological phenomena, and (4) topography and drainage. Since these factors may vary considerably in terms of the rate at which they

Table 16.1 Major factors of soil formation.

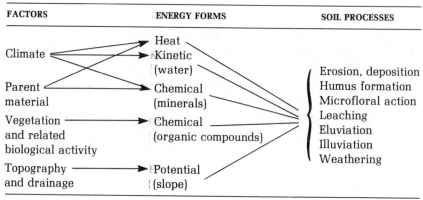

FACTORS	ENERGY FORMS	SOIL PROCESSES
Climate	Heat	Erosion, deposition
	Kinetic (water)	Humus formation
Parent material	Chemical (minerals)	Microfloral action
Vegetation and related biological activity	Chemical (organic compounds)	Leaching
		Eluviation
Topography and drainage	Potential (slope)	Illuviation
		Weathering

supply energy to the soil processes, time is usually also mentioned as a soil-forming factor. However, in reality time is only a measure of the rate of soil formation and not a factor in the same sense as climate, parent material, and vegetation. Table 16.1 provides a summary on the soil-forming factors and the energy and processes associated with each; Fig. 16.6 relates them to an environmental setting.

Each environment portrayed in Fig. 16.6 gives rise to one or more soil units, called a *pedon*. Pedons are the smallest geographic units of soil defined by U.S. soil scientists. Pedons of similar characteristics

Fig. 16.6 Soil-forming factors in an environmental setting. (Illustration by William M. Marsh)

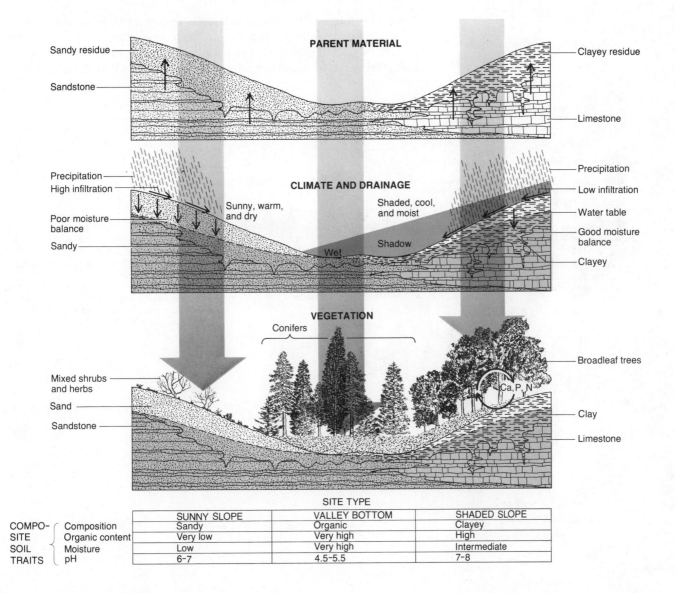

| | SITE TYPE | | |
	SUNNY SLOPE	VALLEY BOTTOM	SHADED SLOPE
COMPOSITE SOIL TRAITS — Composition	Sandy	Organic	Clayey
Organic content	Very low	Very high	High
Moisture	Low	Very high	Intermediate
pH	6-7	4.5-5.5	7-8

together form *soil bodies*, or *polypedons*. Polypedons vary in area, depending on the diversity of the environment, but usually cover several acres. Polypedons are in turn combined to form soil series; in the United States the U.S. Department of Agriculture has defined a total of 1050 soil series.

SOIL FORMATION AND THE BIOCLIMATIC ENVIRONMENT

General Soil-Climatic Relations

If we ignore the polar and mountainous regions, two general classes of soils can be identified in the world. These are called the *pedocals* and *pedalfers*. The *al* and the *fer* in pedalfer refer to the concentrations of iron (ferrous) and aluminium that are found in the B horizon; the *cal* in

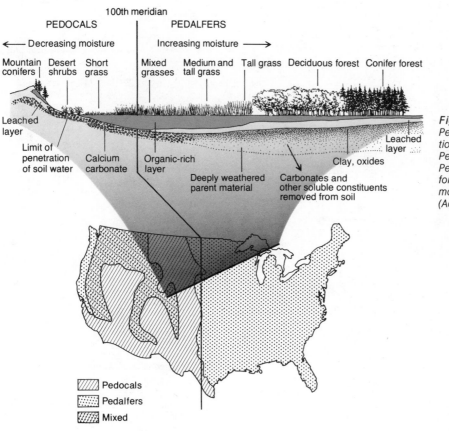

Fig. 16.7 The approximate distribution of Pedocal and Pedalfer soils. This classification is highly generalized, and within the Pedocal region are large areas of Pedalfer-type soils. This is particularly true for the Pacific coast, where the annual moisture balance is decidedly positive. (Adapted in part from Hunt, 1967)

pedocal, to the calcium retained in the soil as a result of weak eluviation due to the shallow penetration of water from the surface. The distributions of these two classes correspond to broad areas of negative and positive moisture balances. Pedalfers are generally found where precipitation exceeds 60 cm a year and the moisture balance is sufficient to produce pronounced leaching. Pedocals form under conditions of negative moisture balance, mainly areas that receive less than 60 cm of precipitation annually. In the United States a north-south line along the 98 degree meridian can be used as the approximate boundary between these soil types, with pedocals found in the drier west and pedalfers in the more humid east (Fig. 16.7).

Soil-Forming Regimes

Within this broad climatic framework, we can identify many combinations of different moisture, heat, and biological conditions. Each combination produces a characteristic set of soil-forming processes and is referred to as a *soil regime.* Although many regimes have been defined, four cover most major soil types. The *podzolization* regime is common to areas with a positive moisture balance in middle to high latitudes or at high elevations in mountainous areas. Conditions in these zones are cold enough to inhibit intensive microflora action, yet warm enough to support substantial forests. Coniferous forests, which require few bases for growth, are abundant and ineffective in sustaining a soil pH much above 5 or 6. Bases, colloids, and oxides are eluviated from the upper soil, leaving the A horizon with a sandy, gray character. This is the salient characteristic of a *Podzol* soil. Also characteristic of Podzols are the colloids and iron oxides that accumulate in the B horizon and form a dark, dense layer (Fig. 16.8).

Laterization (or ferratillization) occurs under conditions of heavy rainfall and warm temperatures such as those of the equatorial rainforest. Intensive microflora action in such regimes consumes most of the humus produced. Iron oxides form hard nodules, or rocklike layers called *laterites,* in the B horizon area. Some soil scientists argue that the iron oxides are relatively insoluble because strong organic acids are not abundant. Silica is leached out, and no distinct horizons form other than the B horizon laterites and the thin layer of humus in the O horizon (see Fig. 16.8).

Calcification occurs in arid and semiarid regions, where the moisture balance is decidedly negative. As there is little leaching by percolating precipitation, calcium and magnesium ions remain in the soil. Grasses use these bases and in turn restore them to the soil surface through their seasonal and annual life cycles. Colloids are not leached out, and they, too, remain in the soil and may weld it together into

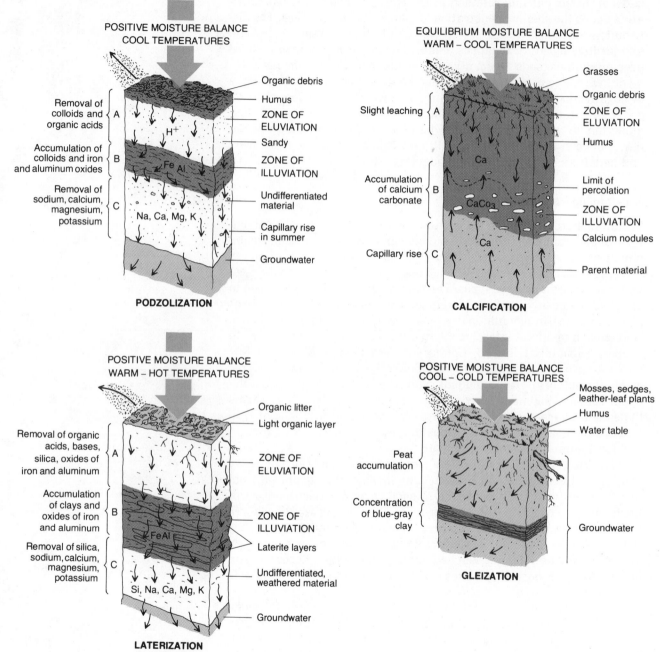

Fig. 16.8 The principal features and processes associated with podzolization, calcification, laterization (ferratillization), and gleization.

dense structures. This is accomplished mainly by calcium carbonate, which is brought up by capillary action during the prolonged periods of high evapotranspiration (Fig. 16.8). Microfloral activity is limited by low moisture, and where grasses are abundant, appreciable amounts of organic matter build up.

Gleization occurs in poorly drained areas, such as bogs and marshes, in cool to cold climates. A heavy organic layer, which is often acidic, forms at the surface and in time may metamorphose into peat. Decomposition is slow due to cold temperatures and water-saturated conditions. Beneath this a sticky blue-gray clay forms, called the *gley* horizon. It derives its color from partially reduced iron in this oxygen-poor environment (Fig. 16.8).

SUMMARY

Eluviation involves the relocation of ions and colloids in the soil. Where the soil-moisture balance yields a water surplus, this process results in the downward displacement of bases and colloids. In dry areas eluviation is ineffective, and bases are retained in the soil. Thus soil pH values are generally higher in dry areas than in humid areas.

Plants may recycle nutrients in the soil, contribute organic matter to the upper soil, and in case of nitrogen-fixing plants, actually improve the soil fertility. The combination of chemical, biological, and physical processes within the soil column produces differentiation of the soil into horizons. These processes are influenced by climate, parent material, vegetation and topography, and drainage. Soil-forming regimes are associated with broad bioclimatic zones.

SOIL CLASSIFICATION AND PROBLEM SOILS

CHAPTER 17

SOIL-CLASSIFICATION SCHEMES

The major soil-classification schemes in use today come from agronomy and civil engineering. For years, agronomists in the United States used a scheme developed by the United States Department of Agriculture, but that scheme has recently been replaced by a new classification scheme. Civil engineers use several schemes, all of which differ from the USDA schemes in that they emphasize the functional nature of soil, such as its bearing capacity, drainage, plasticity, and so on. Since the old USDA scheme is still widely used in the geosciences, we will begin with it. Later in the chapter we will introduce the new USDA scheme and examine some aspects of the engineering classifications.

The Traditional USDA Classification Scheme

The traditional USDA soil classification was developed at a time when it was popular in natural science to apply the concept of evolution to various parts of nature. This was reflected in the study and classification of soil through an emphasis on the origin and development of soils. Using concepts formulated by Russian soil scientists, soil was viewed as an organism that evolved into a maturity characterized by an equilibrium with the environment, primarily climate.

Three major levels of classification are used in the traditional USDA scheme: (1) orders, (2) great groups, and (3) individual soils that comprise each group. There are three orders of soil: zonal, intrazonal, and azonal. Zonal soils, the most common order, are distinguishable on the basis of their well-developed horizons. In any environment certain conditions are necessary for the formation of zonal soils:

1. Erosion and deposition on the surface *must be negligible.*

2. Moisture flux *cannot be impaired* by bedrock, groundwater, ice, or some other substance.

3. The conditions above must persist long enough to allow internal differentiation to take place. The actual amount of time necessary varies according to the *rates* at which the soil-forming pro-

ZONAL SOIL

AZONAL SOILS

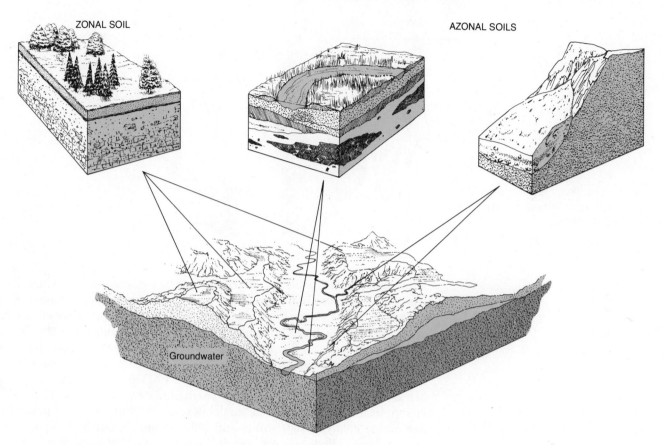

Groundwater

cesses operate. Some profiles appear to form within several centuries; others may take tens of millenia or more to form.

Because these conditions are prerequisite to the formation of zonal soils, we can fairly accurately identify those sites in the landscape where they should be found. Basically, the sites must be relatively flat, somewhat elevated, well drained, and geomorphically stable, such as those identified in Fig. 17.1.

The formation of zonal soils would be highly unlikely in the river floodplain, along the valley walls, in the tributary valleys, and on the steeper upland slopes. Rather, these sites give rise to *azonal soils*, which are characterized by the absence of horizons. This is because external processes rather than the internal processes that produce horizons dominate their formation. For instance, in the floodplain frequent episodes of erosion and deposition by floods and channel water mask the effects of the horizon-producing processes. Therefore, the

Fig. 17.1 *The formation of zonal soils is limited to the relatively flat upland sites. In highly diversified terrain, zonal soils may occupy only a small fraction of the total area. Azonal soils are associated with unstable sites, such as steep slopes and floodplains. (Illustration by William M. Marsh)*

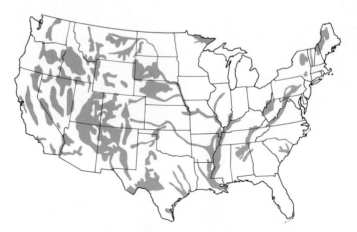

Fig. 17.2 Major areas of azonal soils in the United States.

alluvial (floodplain) soils are characterized by heterogeneous textures, which often appear in the stratified form in which they were originally deposited. Along steep slopes, such as the valley walls, mass movement and runoff bring large amounts of debris to the footslope, where it accumulates to form deep soils of mixed composition. Each of these powerful processes imparts a particular set of soil characteristics that is uniquely different from that of nearby zonal soils. The azonal soils are designated according to their particular characteristics. Highly mixed, unsorted deposits from landslides, debris flows, and soil creep are called *colluvial soils*; talus and other deposits that lack organic material are called *regosols*; and deposits composed of huge chunks of bedrock are called *lithosols*.

Because of the agronomist's emphasis on the zonal soils—due to the importance of this order to agriculture—there has been a tendency to attach the adjective "normal" to these soils. Although azonal soils are *not* called abnormal, there is a discernible trend in the annals of soil science to grant less than primary status to the external geomorphic processes as soil-forming agents. As a result, students of soils have often branded the azonal soils (or their equivalent soils in the new classification scheme) as "youthful" or "immature." However, it is well to remember that these regimes are dominated by highly active geophysical processes and that relative to *these processes*, they are no less "mature" than the neatly horizoned soils of nearby areas. Figure 17.2 shows the major areas of azonal soils in the coterminous United States.

Intrazonal soils form under conditions of impeded drainage. Because the ground in these areas is saturated, rates of organic weathering are low, and vegetal debris accumulates in thick deposits. The soil regime may be that of gleization or something similar to it. Although the up-and-down flux of moisture is limited or absent, crude horizons tend to form as a result of an internal differentiation of organic material and clay.

Descriptions of Some Great Soil Groups

Let us briefly describe the most widespread of the great soil groups. Within the zonal order are about a dozen great groups; the main characteristics of eight of the most common ones follow, and the locations of these eight are shown in Fig. 17.3.

Podzol soil from Massachusetts

Podzols. The predominant soil group in the northern continental regions is the Podzols. A cold winter and precipitation evenly distributed through the year appear to be necessary for the formation of podzols. The primary features of a Podzol are: (1) a dark colloid-rich O horizon; (2) an ash-gray leached A_2 horizon; and (3) a strong B horizon that is heavy in colloids and often forms a hardpan containing oxides of iron and aluminium. Coniferous forests and sandy parent material appear to be conducive to the formation of the distinctive Podzol horizons.

Gray-Brown Podzols. These soils differ from Podzols in that leaching is less intense, resulting in a gray-brown–colored B horizon. They are formed farther south than are Podzols, in areas of cool winters and evenly distributed annual precipitation. Deciduous forests comprised of trees such as maple, beech, and oak grow in this soil group, and the input of bases from these plants is greater than that of the conifers to the Podzols.

Red-Yellow Podzols. Located even farther south, in areas of warm temperatures and abundant precipitation, are the Red-Yellow Podzols. These soils are transitional between the Podzols and the Gray-Brown Podzols on one hand and the tropical soils on the other. Leaching is usually strong, and bacterial action reduces the humus content in the O horizon to a thin layer. The characteristic red and yellow colors of these soils, which are so distinctive throughout much of the American South, are due to the presence of hydroxides of iron produced by leaching. Oak and pine forests are the predominant vegetation of this soil group.

Latosols. Formed in the humid tropics, Latosols are typified by the following features: (1) thin topsoil, owing to the accelerated rate of destruction of organic matter; (2) great thickness as a result of an

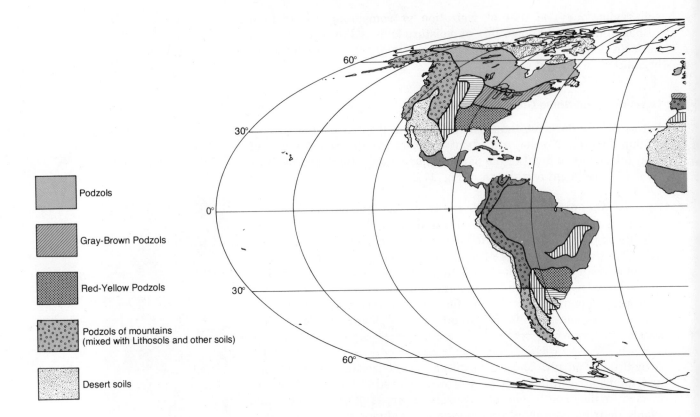

Podzols

Gray-Brown Podzols

Red-Yellow Podzols

Podzols of mountains
(mixed with Lithosols and other soils)

Desert soils

Fig. 17.3 Locations of eight common
great soil groups, according to the traditional USDA classification.

Map projection by Waldo Tobler.

advanced state of bedrock weathering; (3) absence of silica, due to
intensive and long-term leaching; (4) heavy accumulations of iron,
aluminum, and manganese oxides, which contribute to the formation of
laterite deposits that may form valuable ore reserves of bauxite (aluminum oxide), limonite (iron oxide), and manganite (manganese oxide);
and (5) a reddish color, owing to a concentration of hydroxides of iron.

Chernozem soils. These form under semiarid conditions in areas of
loess deposits. The main features of these soils are: (1) a humus-rich,
black O horizon up to 1 m thick; (2) a granular structure; and (3) a
brown or yellow B horizon containing colloids and bases. The leached
A_2 horizon found in Podzols is conspicuously absent in the Chernozems.
And in the B horizon calcium carbonate is abundant, often in the form
of pebblelike nodules.

Prairie soils. Usually formed in areas where annual precipitation is
60–120 cm per year, as in the center of the United States and in eastern
Europe, prairie soils are very similar in appearance to Chernozems,

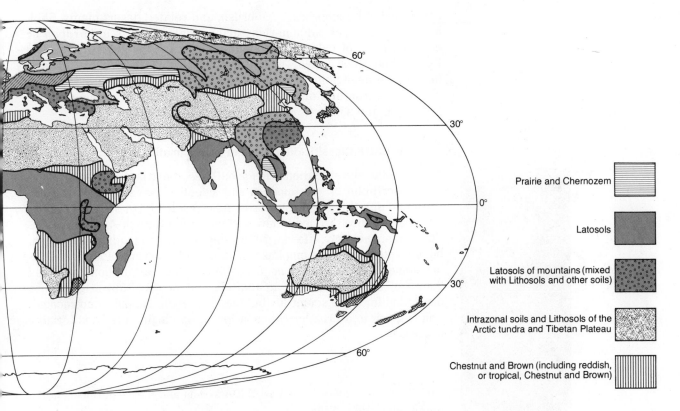

Prairie and Chernozem

Latosols

Latosols of mountains (mixed with Lithosols and other soils)

Intrazonal soils and Lithosols of the Arctic tundra and Tibetan Plateau

Chestnut and Brown (including reddish, or tropical, Chestnut and Brown)

except that calcium carbonate is less abundant, owing to a somewhat better moisture balance. Located on the humid side of the Chernozems, Prairie soils are a transition between pedocals and pedalfers. In the United States both Chernozems and Prairie soils have been extensively altered by intensive cultivation over the last century.

Chestnut and Brown soils. These soils are found in a climate more arid than that of the Chernozems, which in the United States is located west of the Chernozems in the wheat-growing areas. Their profile is similar to that of the Chernozems, but is less dark, as there is less plant life, owing to the lack of moisture. Soil structure is usually prismatic. Brown soils are lighter yet in color, due to even less humus. Grasslands on Brown soils are generally used for grazing.

Gray Desert and Desert Red soils. Soils in this group form in extremely dry, hot areas. Gray Desert soils occur in midlatitude deserts, as in Utah and Nevada. The color is light gray, and horizons are indistinct. The paucity of humus in these soils corresponds to a light plant cover comprised of bunchgrass, sage brush, and related desert plants. Shallow penetration of calcium-rich surface water results in *hardpan* or *caliche* crusts of calcium carbonate at or near the soil surface. The Desert Red soils appear in hotter tropical deserts. Vegetation and humus are negligible: The color—light to dark red—is produced by coatings of iron and other oxides on soil particles. Horizons are also indistinct, and soil textures are usually coarse.

The New USDA Classification: The 7th Approximation

Although the classification scheme outlined above is suitable for a general description of soil, modern soil scientists have found it lacking sufficient detail for their purposes. As a result, USDA scientists in the 1950s and 1960s devised a new system, called the *7th Approximation*. This system not only affords greater detail than the traditional system, but also places greater emphasis on diagnostic horizons, thereby avoiding some of the difficulties associated with designations such as "mature" and "immature." In addition, soil names of folk origins were discarded for more scientific (but less interesting) names, and recognition is given to a broader range of forces as "legitimate" soil-forming processes.

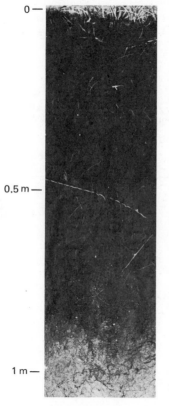

0 —

0.5 m —

1 m —

Mollisol from North Dakota

Table 17.1 Soil orders in the new USDA system and their equivalents in the traditional USDA system.

ORDER, NEW SYSTEM	MEANING OF NAME	TRADITIONAL-SYSTEM EQUIVALENT
Entisol	Recent soil	Azonal soils and some gley soils
Inceptisol	Young soil	Some Brown forest soils and gley soils
Aridisol	Arid soil	Mainly Red and Gray Desert soils
Mollisol	Soft soil	Mainly Chernozem, Prairie soils, and Chestnut and Brown soils
Spodosol	Ashy soil	Podzols
Alfisol	Pedalfer soil	Gray-Brown Podzolic, Prairie soils, weak Chernozems, and some intrazonal soils
Ultisol	Ultimate soils	Red-Yellow Podzolic, Reddish-Brown Lateritic, and some intrazonal soils
Oxisol	Oxide soil	Latosols
Vertisol	Inverted soil	No equivalent
Histosol	Organic (tissue) soil	Intrazonal Bog soils

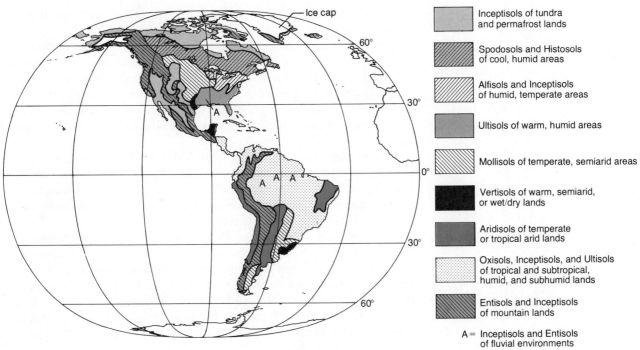

Map projection by Waldo Tobler.

The legend reads:

- Inceptisols of tundra and permafrost lands
- Spodosols and Histosols of cool, humid areas
- Alfisols and Inceptisols of humid, temperate areas
- Ultisols of warm, humid areas
- Mollisols of temperate, semiarid areas
- Vertisols of warm, semiarid, or wet/dry lands
- Aridisols of temperate or tropical arid lands
- Oxisols, Inceptisols, and Ultisols of tropical and subtropical, humid, and subhumid lands
- Entisols and Inceptisols of mountain lands

A = Inceptisols and Entisols of fluvial environments

Fig. 17.4 Soils of the Western Hemisphere, according to the new USDA classification. (From Soil Geography Unit, Soil Conservation Service, USDA, 1972)

The new USDA scheme provides six levels of classification, beginning at the broadest scale with *orders*, which are subdivided into suborders, great groups, subgroups, families, series, and on to pedons. In total, the system establishes about 7000 categories of soil for the United States. Table 17.1 lists the ten orders of the new scheme and identifies the equivalent class in the traditional scheme. The map in Fig. 17.4 shows their distributions in the Western Hemisphere.

The following is a brief description of the ten orders of the new classification. *Entisols* are soils of recent origin, principally the azonal soils of the traditional scheme. They are found in geomorphically active, or recently active, environments such as river floodplains, mountain slopes, and sand dunes. Horizons are very weakly developed or nonexistent, and these soils may be found in any bioclimatic region.

Inceptisols are soils in which horizons are just beginning to form; however, visible evidence of eluviation and illuviation is weak. They occur in all climatic zones and are especially widespread in the Arctic tundra. In the tropics many Inceptisols are found in volcanic ash.

Aridosols, as the name implies, are soils of arid lands. Eluviation is weak and shallow, and organic content is low. Calcium carbonate accumulates near the surface to form a calcic horizon (caliche in the traditional terminology). Erosion of the surface by wind and runoff is common in the Aridosols.

In the semiarid areas, bordering on the Aridosols, are the *Mollisols*. Grass cover is prevalent, and organic buildup is substantial. Although moisture is adequate to support the grasses, it is not adequate to promote strong leaching and therefore allows the buildup of bases, especially calcium. In the traditional scheme these soils are represented mainly by Chestnut and Brown, Chernozem, and Prairie soils.

Spodosols are soils with pronounced zones of illuviation comprised of colloids, iron, and aluminum. They form under humid climatic conditions in sandy parent material. Eluviation is strong, the A horizon may be ashy gray, and the solum is acidic throughout. The Spodosols are the Podzols of the traditional scheme.

Alfisols can be thought of as moist versions of the Mollisols. They develop under tree covers; bases are leached out, but are replaced from the organic layer at comparable rates. Illuviation is characterized by clay accumulations. In North America the Alfisols are found in the Midwest, where they correspond partially to the Gray-Brown Podzols and partially to the prairie soils of the traditional classification, and in southeastern Canada, where they correspond mainly to the Gray-Brown Podzols.

Ultisols are soils in an advanced state of development. Positive moisture balance provides a strong water surplus at least part of the year, and accordingly the effects of leaching are pronounced. Few bases are available for release into the soil, but aluminum is abundant. Trees recycle nutrients and are responsible for maintaining the fertility of the upper soil. Most Ultisols are Red-Yellow Podzols in the traditional scheme.

Also in an advanced state of development are the *Oxisols*. Owing to the long-term stability of the soil-forming environment and the prolonged leaching induced by heavy rainfall, essentially no reserve of bases is available. Nutrients in the upper soil are maintained by the vegetation, either rainforest or savanna grasses and woodland. Illuviation is characterized by concentrations of clay, iron, and aluminum, forming the oxic (laterite) horizon. In the traditional classification, Oxisols correspond to Latosols.

Vertisols contain the clay montmorillonite, which causes expansion and contraction with wetting and drying. Deep cracks form with drying, and pieces of soil slough off into the cracks. With rewetting, the cracks close. Many such cycles produce an overturning or gradual inverting of the soil. Vertisols are found in equatorial and tropic areas with pronounced wet and dry seasons.

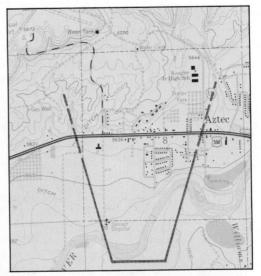

Fig. 17.5 Soil distribution at a local scale near Aztec, New Mexico. The letters on the photograph stand for the U.S. Soil Conservation Service names for soils; for example, RA is River Wash soil, WA is Walrees soil, and WR is Werlog soil. Note how the distribution of the soils corresponds to the terrain. The area shown in the photograph is outlined in the accompanying topographic map. (Photograph and map supplied by John B. Carey and Tommie L. Parham)

Histosols are organic soils that are saturated all or part of the year. The characteristics depend primarily on the nature of the parent vegetative materials. Histosols may be found wherever drainage and plant productivity are conducive to the buildup of organic matter. They are intrazonal soils in the traditional scheme.

Local Soil Mapping

Across the United States, the Soil Conservation Service, an agency of the Department of Agriculture, has carried out detailed soil surveys in agricultural areas. The results of each survey are published in a county soil report, which is comprised of large-scale soils maps accompanied by data and text. These reports represent the most comprehensive source of information on soil for any large region in the world today. Figure 17.5 shows two sources of information (aerial photographs and topographic maps) used in local soil-survey mapping. Each soil type is represented by a set of letters, as shown in the photograph.

Engineering Classification of Soil

Since one of the chief concerns of geographers regarding soil is how it is used, it is understandable that they have had an abiding interest in the agronomist's interpretation and classification of soil. In the past several decades, however, urban sprawl and increasingly elaborate construction schemes have necessitated that the geographer also appreciate the civil engineer's outlook on soil.

Engineers classify soil mainly according to how it performs when put to certain intensive uses, such as supporting buildings and highways or holding back reservoirs of water. As we indicated at the outset of this chapter, the engineer is not interested in the capability of a soil to support plants; in fact, the engineering part of the soil is usually considered to begin with the material beneath the solum. For this reason, agricultural soils maps, which disregard the soil below a depth of a meter or so, are of limited value to the civil engineer. Since the weight of most structures is supported by the soil mass several tens of meters below the surface, it is necessary for the civil engineer to be concerned mainly with what the agronomist would consider the subsoil.

Unified Soil Classification System

Many engineering soil-classification schemes were developed in the first half of this century. The most popular ones in the United States were those formulated by universities and agencies of the state and federal governments. After World War II, the respected civil engineer Arthur Casagrande initiated the development of an integrated system that has come to be widely used. It is called the Unified Soil Classification System and is based on two main criteria: (1) texture or composition, and (2) behavior when in a saturated state. The first criterion uses six categories (Table 17.2), as well as a qualifier related to grading. (Grading refers to the uniformity of the soil-particle makeup.) The behavior criterion applies only to the fines (silt and clay, both organic and inorganic) and is based on the relative plasticity and compressibility when placed under stress. (Plasticity is a measure of a soil's resistance to flow-type deformation relative to its water content, whereas compressibility refers to the loss of volume in a soil under pressure.)

Unlike agriculture, civil engineering projects are usually localized, and therefore soil mapping is undertaken only on a limited basis. As a result, soil maps based on engineering classifications are not available for broad, geographic areas. With proper qualification, however, the agricultural soil maps prepared by the Soil Conservation Service can be used to gain an indication of the engineering potential of soils. The soils that pose severe limitations to structures are generally the organic intrazonal soils, which have poor bearing capacity owing to a strong tendency to compress and deform under weight. Moreover, organic soils are often water-saturated and when drained, microfloral activity increases sharply, resulting in rapid volumetric loss.

The engineering potential of most zonal soils is generally better than that of the intrazonal soils, because they are well drained and have low organic contents. The azonal soils tend to be intermediate. Some azonal soils, such as sands and gravels in shore deposits or sand

Table 17.2 Unified soil-classification system.

LETTER	DESCRIPTION	CRITERION	FURTHER CRITERIA	
G	Gravel and gravelly soils (basically pebble size, larger than 2 mm diameter)	Texture	Based on uniformity of grain size and the presence of smaller materials such as clay and silt:	
S	Sand and sandy soils	Texture	W	Well graded (uniformly sized grains) and clean (absence of clays, silts, and organic debris)
			C	Well graded with clay fraction, which binds soil together
			P	Poorly graded, fairly clean
M	Very fine sand and salt (inorganic)	Texture; composition	Based on performance criteria of compressibility and plasticity:	
C	Clays (inorganic)	Texture; composition	L	Low to medium compressibility and low plasticity
O	Organic silts and clays	Texture; composition	H	High compressibility and high plasticity
P_t	Peat	Composition		

dunes, may be ideal for construction, whereas others, such as alluvial soils, may be very poor because they contain unstable materials such as loose clay and organic material.

PROBLEM SOILS

Soil and Urbanization

In his book *Design With Nature* (1969), planner Ian L. McHarg para-phases the following from a lecture by the anthropologist Loren Eisely:

Man in space is enabled to look upon the distant earth, a celestial orb, a revolving sphere. He sees it to be green, from the verdure on the land, algae greening the oceans, a green celestial fruit. Looking closely at the earth, he perceives blotches, black, brown, gray and from these extend dynamic tentacles upon the green epidermis.

These blemishes he recognizes as the cities and works of man and asks, "Is man but a planetary disease?"

A sobering thought, especially when we consider that since Eiseley made this statement in 1961, urbanization has advanced at rates unparalleled in the history of civilization. From our vantage point on the earth's surface, such descriptions tend to bring to mind an image of a great urban juggernaut indiscriminately devouring the landscape.

At a scale of observation distant from earth, this metaphor is appropriate; however, a closeup view of cities usually reveals that urbanization is not indiscriminate. Rather, urbanization occurs unevenly within most metropolitan regions, giving rise to a highly varied geographical pattern of development. Much of this is due to social, political, and economic factors, but much is also due to the physical conditions of the land.

Many physical factors influence urbanization, and although soil is usually not as important as major water features and rugged terrain, it is nonetheless important in some areas. Where it is inclined to give way under structures or where it is wet and requires special drainage facilities, soil has proved to be a serious limitation to urbanization. Nevertheless, when we compare old maps of soil conditions with modern maps of urban areas, we find that many locales with weak and poorly drained soils are now fully developed. What is more, these areas usually evidence no apparent problems stemming from the original soil condition. How is this so? The answer is found in economics, land use, and modern engineering technology. As cities grow and land values rise, investors are willing to spend money for site improvements because ultimately, the return on the land as a space for business will be high enough to amply offset the improvement costs. The problem soil may be excavated and replaced by a better one, or special building foundations may be designed to compensate for soil weakness. In any case, the influence of soil usually declines as the value of the land increases. Only in sites that are set aside for special uses, such as parks, are problem soils still apparent in the densely urbanized areas.

This is not to say that problem soils cease to be problematical once an urban area becomes built up. On the contrary, many cities are plagued by sinking ground because they unknowingly built over unstable soils centuries ago. Unfortunately, the problem became apparent only after the surface was heavily loaded with buildings, thus driving the cost or necessary technology of corrective measures out of reach. This is evident, for example, in parts of Mexico City, where centuries ago the city grew over deposits and fills underlain by compressible clays and ancient landfills. Many second stories are now at ground level, and buildings continue to sink.

Of the soils that pose severe limitations to development, none is as abundant as the tundra soil. This intrazonal soil is not only usually poorly drained and high in organic content, but more importantly, is also underlain by permafrost, which is highly sensitive to change in response to alterations in the soil-heat balance. Roads, buildings, and airfields have been shown to produce an increase in soil heat, resulting in melting of the upper permafrost. Since water occupies less space in the liquid state, the ground subsides as the permafrost melts. This may cause buildings to tilt, foundations to break, and facility lines to rupture. (See Chapter 26 for details on soil movement related to frozen ground.)

Soil and Agriculture

The food demands of a rapidly growing world population have placed unparalleled stress on soil as an agricultural resource. This has led to more intensive use of existing cropland as well as development of marginal farmlands. Coupled with chronically poor cultivation practices in many nations, this has resulted in serious deterioration of soils over much of the world.

Loss of fertility and erosion constitute the most serious damages to soil rendered by agriculture. Even in the United States, where agricultural practices are considered to be the most advanced in the world, both of these problems are widespread. The Department of Agriculture estimates, for example, that soil erosion is a serious problem on as much as 220 million acres of cropland, more than half of the total cropland acreage in the United States (Fig. 17.6).

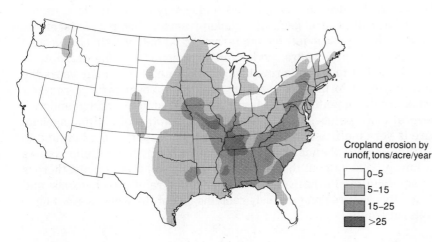

Fig. 17.6 Erosion of cropland by runoff in the United States, 1975. (From Cropland Erosion, *Soil Conservation Service, U.S. Department of Agriculture, 1977)*

Cropland erosion by runoff, tons/acre/year

☐ 0–5
▨ 5–15
▩ 15–25
■ >25

Erosion contributes to the loss of soil fertility, but the single greatest cause of soil impoverishment can be attributed to the fact that crops draw large amounts of nutrients from the soil without replacing them. When a natural plant cover is replaced by crops, two changes usually occur in the soil-nutrient system: (1) nutrients are drawn from the soil at an accelerated rate because most crops have higher nutrient requirements than do natural plants; and (2) little or no plant organic matter is resupplied to the soil surface, because the mass of the crops are removed in harvesting. In order to maintain productivity, fertilizers must be added to make up for this imbalance.

The amount of fertilizers required varies according to the crops planted and the soil conditions. Where soils are inherently weak in nutrients and the crops have high nutrient requirements, each year hundreds of pounds of fertilizer must be added to each acre in order to sustain production (Fig. 17.7). This is generally the case in the American South, where the Red-Yellow Podzols (Ultisols) are used for crops of corn, tobacco, and sorghum. In contrast, richer soils, such as the Chernozems of the Great Plains, that are planted to low nutrient-demand crops such as wheat, may require only 25–30 pounds (11.4–13.6 kg) of fertilizer per acre per year, even after decades of use.

The most severe soil-nutrient problems in the world occur in the Latosols (Oxisols) of the tropical rainforest. Despite the high soil fertility implied by the luxuriant rainforest vegetation, this group of soils is deceptively low in nutrients because of both the accelerated rates of organic matter decomposition and the rapid leaching under the heavy tropical rainfall. Unless massive amounts of fertilizer are added to these soils after clearing and planting, they are depleted of fertility after only several years of cropping. In addition, as the organic layer disintegrates and the crop cover diminishes, the clay beneath is exposed to intensive solar heating, which may render it into pavement hardness. Because the capital and technology necessary to maintain soil productivity are generally lacking among the peasants of the tropics, there is little left for the farmer to do but move to new ground every two or three years. Under this practice of *shifting agriculture,* the plot is left to second-growth vegetation, and after many years the nutrient input to the soil from the vegetative cover may be reestablished. In the meantime, the farmer has cleared, used, and abandoned several more patches of soil. Clearly, the practice of shifting agriculture is very inefficient, and it necessitates a larger amount of land to sustain a family than is needed to sustain a family in the United States or Europe. Yet without the means to maintain soil productivity, there is presently little alternative for the tiller of the rainforest Latosols, and hence this vast region curiously remains one of the world's great agricultural frontiers.

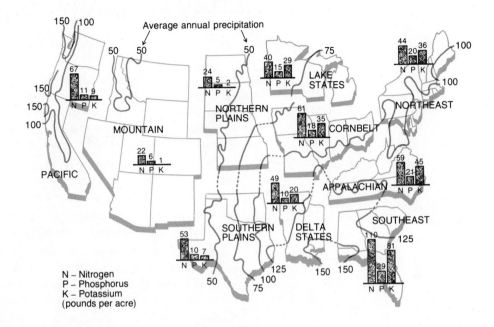

Finally, in arid lands there is the problem of salt saturation of irrigated soils. Where intensive irrigation is practiced, the water table under fields is often raised to the point where the capillary fringe reaches the surface. As the water evaporates, a salt residue builds up in the root zone, eventually rendering the fields useless for cropping. This process, called *salination*, has contributed significantly to the loss of highly productive agricultural land and is cited as one of the causes of desertification, the overall degradation of land due to problems associated with aridity.

Fig. 17.7 Average annual application of fertilizer per acre per year in the United States. Note the approximately sevenfold increase in quantities from the northern Plains to the Southeast. This corresponds to an increase in precipitation and in turn greater soil leaching and soil erosion.

SUMMARY

The major soil-classification schemes have come from agronomy and engineering. The traditional USDA scheme grew out of evolutionary concepts about soil formation. The new USDA scheme is based on diagnostic horizons and recognizes a broader range of "legitimate" soil-forming processes. Civil engineering identifies texture and behavior as the key properties in soil classification.

Soil is at the heart of human survival on earth, inasmuch as it is a key to agricultural productivity. Yet productive soil is being lost to urban development, and in most agricultural areas fertility is declining because of erosion, overuse, and salination of soil.

UNIT IV SUMMARY

- In agronomy, soil is defined as the solum; in civil engineering, as the bodies of inorganic particles over the bedrock. For the geographer, whose interests may range from soil productivity and land use to soil-moisture flux, a definition of soil should be broad enough to include essentially all assemblages of ground materials above the bedrock.

- The bulk of most soils is made up of inorganic and organic particles, but the organic material is limited to the upper 1–2 m for the most part. The remainder is made up of air and water.

- Soil texture is an important soil property because of its influence on field capacity, moisture movement, and nutrient retention.

- Heat is the driving force for many critical soil processes, including evaporation of water, microfloral activity, and chemical processes. In most areas soil-heat flux is greatest near the surface because of daily and seasonal variations in solar heating of the landscape.

- The movement of capillary water in the soil varies with the moisture gradient and soil capillarity. Moisture gradients change seasonally and over shorter periods in response to the receipt and loss of moisture in the upper soil.

- When evapotranspiration exceeds precipitation, moisture is drawn from storage water in the soil. As this reservoir is drawn down, available moisture falls behind evapotranspiration demands, and a soil-moisture deficit develops. In dry regions the moisture deficit is large and present in all seasons; elsewhere it tends to be seasonal or shorter-term.

- A moisture surplus results in downward percolation of water, which moves colloids and ions to a lower level in the soil. Where the moisture balance is negative, percolation is limited, and colloids and ions tend to remain in the upper soil.

- The organic content of soil varies mainly with plant productivity and microfloral activity. In the hot, wet climates plant productivity is high, but microfloral activity is even higher; therefore, the buildup of an organic layer is virtually impossible. In vegetated areas that are swampy, semi-arid, or cool and damp, microfloral activity may be retarded, allowing a buildup of organic matter.

- Working in combination, chemical, biological, and physical processes differentiate the soil into horizons. In a general way, the resultant profiles reflect the essential bioclimatic conditions, or regimes, under which the soil developed. Four soil-forming regimes are especially important: podzolization, calcification, laterization, and gleization.

■ Two major soil-classification schemes are recognized today: the traditional USDA scheme and the new USDA scheme. The new scheme emphasizes diagnostic horizons in the classification of soil, giving less importance to mature zonal soils and their relationship to bioclimatic regions.

■ Soil classification in engineering stresses texture and the soil's behavioral traits, such as compressibility and plasticity.

■ In most urban areas soils have only a modest influence on land-use patterns because economics allows for costly site improvements and special construction techniques to overcome soil constraints. In polar regions, however, soils underlain by permafrost have proved to be a serious problem to modern development.

■ Among the many land-use activities that influence soils, agriculture is the most pronounced. Most cultivation practices result in increased soil erosion and decreased soil fertility, and coupled with salination, the world's soil resources appear to be declining when the need for improved productivity is rising.

FURTHER READING

Birkeland, P. W., *Pedology, Weathering, and Geomorphological Research*, New York: Oxford University Press, 1974. *Soil formation as it relates to weathering processes, topography, vegetation, and climate.*

Eckholm, Eric D., *Losing Ground: Environment, Stress and World Food Prospects*, New York: Norton, 1976. *A strong statement on the declining condition of the world's soils and its implications in terms of food supply and the quality of life.*

Foth, Henry D., *Fundamentals of Soil Science*, New York: Wiley, 1978. *A basic text in modern soil science.*

Steila, Donald, *The Geography of Soils: Formation, Distribution and Management*, Englewood Cliffs, N.J.: Prentice-Hall, 1976. *Includes a fairly detailed description of the ten orders in the new USDA classification.*

REFERENCES AND BIBLIOGRAPHY

Baldwin, M., C. E. Kellogg, and J. Thorp, "Soil Classification," *The 1938 Yearbook of Agriculture*: *Soils and Man*, Washington, D.C.: Government Printing Office, 1938, pp. 979–1001.

Bidwell, O., and F. D. Hole, "Man as a Factor of Soil Formation," *Soil Science* **99** (1965): 65–72.

Brady, N. C., *The Nature and Properties of Soils,* 8th ed., New York: Macmillan, 1974.

Bridges, E. M., *World Soils,* 2d ed., Cambridge, England: Cambridge University Press, 1978.

Bunting, B. T., *The Geography of Soil,* Chicago: Aldine, 1967.

Cunningham, R. K., "The Effect of Clearing a Tropical Forest Soil," *Journal of Soil Science* **14** (1963): 334–335.

Foth, H. D., *Fundamentals of Soil Science,* New York: Wiley, 1978.

Hunt, C. B., *Physiography of the United States,* San Francisco: Freeman, 1967.

_____, *Geology of Soils,* San Francisco: Freeman, 1972.

Küchler, A. W., *Potential Natural Vegetation of Conterminous United States,* New York: American Geographical Society, 1964.

McHarg, I. L. *Design with Nature,* Garden City, N.Y.: Doubleday, 1969.

McNeil, M. "Lateritic Soils," *Scientific American* **207**, 11 (1964): 97–102.

Marsh, W. M., "Soils and Drainage," in *Environmental Analysis For Land Use and Site Planning,* New York: McGraw-Hill, 1978.

Mather, J. R., "The Moisture Balance in Grassland Climatology," in H. B. Sprague, ed., *Grassland,* Washington: American Association for the Advancement of Science, 1959, pp. 251–261.

Mather, J. R., and G. A. Yoshioka "The Role of Climate in the Distribution of Vegetation," *Annals of the Association of American Geographers* **58**, 1 (1968): 29–41.

National Cooperative Soil Survey, *Soil Taxonomy,* Washington, D.C.: Government Printing Office, 1970.

Nikioroff, C. C. "Reappraisal of Soil," *Science* **129**, 3343 (1959): 186–196.

Reeves, C. C., Jr., "Origin, Classification and Geologic History of Caliche on the Southern High Plains, Texas and Eastern New Mexico," *Journal of Geology* **78** (1970): 352–362.

Soil Survey Staff, *Supplement to Soil Classification, A Comprehensive System, 7th Approximation,* Washington, D.C.: Soil Conservation Service, U.S. Department of Agriculture, 1964.

Suomi, V. E., and C. B. Tanner, "Evapotranspiration Estimates From Heat Measurements over a Field Camp," *Transactions of the American Geophysical Union* **39** (1958): 298–304.

Thornthwaite, C. W., and J. R. Mather, "Instructions and Tables for Computing Potential Evapotranspiration and the Water Balance," *Publications in Climatology,* vol. X, no. 3, Centerton, N.J.: Drexel Institute of Technology and Laboratory of Climatology, 1957.

Young, A., *Tropical Soils and Soil Survey,* Cambridge, England: Cambridge University Press, 1976.

Zon, R., "Climate and the Nation's Forests," in *Climate and Man, Yearbook of Agriculture,* Washington, D. C.: U.S. Department of Agriculture, 1941, pp. 477–498.

PATTERNS AND PROCESSES OF VEGETATION IN THE LANDSCAPE

UNIT V

KEY CONCEPTS   OVERVIEW

plant geography
higher plants
floristic classification
life form
transpiration
photosynthesis
principle of limiting factors
net photosynthesis
production
energy pyramid
tolerance
life cycle
stress
disturbance
adaptation
succession
community
individualism
stress threshold
biochores
formations

Plant geography is concerned primarily with higher plants such as trees, grasses, and ferns. Important objectives in plant geography are to describe the earth's vegetation, to analyze the relations between plants and the physical environment, and to examine the nature of change in plant distributions.

Plant growth requires light, heat, water, and carbon dioxide. Where one or more of these resources is in limited supply, plant growth is restricted. In general, the wet tropics have the largest and most dependable resource supplies; therefore, the production of organic matter by the plant cover is greater there than anywhere else on the continents. In contrast, desert, alpine, and polar environments are deficient in one or more resources and produce very little organic matter.

The environment places stress on plants by not only limiting the resources for growth, but also producing disturbances such as wind storms, floods, and fire. Each plant has a certain tolerance to such stresses, and if that tolerance is exceeded, the plant is damaged or killed. However, over long periods of time the plants may improve their chances of survival through adaptation.

Changes in the distribution of plants are complex, and the causes are often not well understood. Plant scientists have developed several bodies of concepts to help them understand how change takes place. The most popular concept is the community-succession model, in which one group of plants succeeds another at a site, leading to a stable community of plants. The stress-threshold concept is based on the interplay between a fluctuating environment and plants.

This unit opens with a description of vascular plants and the classification schemes applied to them (Chapter 18). The processes of plant growth and the role of the environment in plant productivity are taken up in Chapter 19. Chapter 20 examines the disturbance and stress exerted on plants by the environment and how this relates to plant adaptation and distribution. Chapter 21 describes the global distribution of vegetation.

CHAPTER 18 CLASSIFICATION AND DESCRIPTION OF PLANTS

INTRODUCTION

The Plant-Soil-Atmosphere System

Our examination of the energy and processes of the atmosphere and the soil has at many points emphasized the necessity for considering a third major component of the landscape—the plant cover. Plants interact directly and indirectly with the soil and the atmosphere, each influencing the other. The atmosphere, the plant cover, and the soil together form a "system" linked by energy flows in the form of radiation, moisture, gases, and solid and dissolved minerals (Fig. 18.1). Because of these flows, the components of the system are highly interdependent, and a change in one can affect the other two. Once initiated, a sequence of changes may eventually affect the originating component, thereby "closing a loop" in the system.

This sort of feedback can be illustrated as follows. Suppose that incoming shortwave radiation at the soil surface is increased because of a reduction in forest cover from disease, lumbering, or wind damage. Certain ground plants with low tolerance to solar radiation, such as certain mosses and flowering herbs, may now be eliminated. This, in turn, may lead to the elimination of some other plants that are ecologically dependent on these ground plants. Several changes may thus be rendered in the soil. The input of plant matter to the soil may be altered, resulting in a decline of the soil organic mass as well as a loss of certain nutrients from the humus. With stronger ground-level sunlight, the soil surface is now more intensively heated, and coupled with higher wind velocities caused by a reduced roughness length, soil-moisture loss is accelerated. Acting together, these changes in soil and ground-level climate may lower the chances of regeneration of many remaining tree species. Over a period of several tree generations, the forest cover grows progressively thinner while ground-level radiation increases, ground plants perish, and soil dries out, placing greater and greater stress on remaining trees. Ultimately, the forest may be replaced by hardier shrubs and grasses.

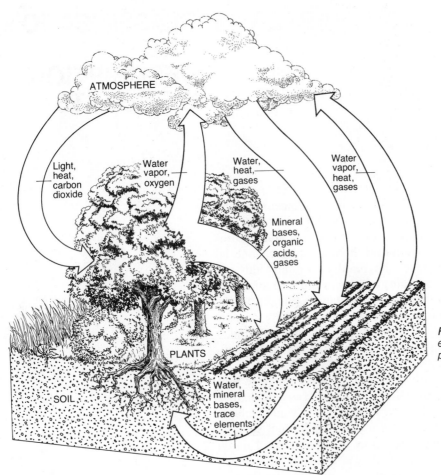

ATMOSPHERE

Light,
heat,
carbon
dioxide

Water
vapor,
oxygen

Water,
heat,
gases

Water
vapor,
heat,
gases

Mineral
bases,
organic
acids,
gases

PLANTS

SOIL

Water,
mineral
bases,
trace
elements

Fig. 18.1 The generalized pattern of the energy flow among the atmosphere, the plant cover, and the soil.

The Geographer's Perspective on Plants

This brief example touches on several of the physical geographer's concerns about vegetation. These include: (1) the physical character of the vegetative cover, including its *form* (e.g., trees or shrubs), *composition* (e.g., mosses or grasses), and *geographic distribution*; (2) the *relations* among the plant cover, climate, and soil; (3) the relationship of vegetation forms and *growth* rates on one hand to *energy* in the form of light, heat, water, and soil minerals on the other; and (4) the *nature of change* in the form, composition, and distribution of the vegetative cover. Our objective in this chapter is to describe the basic types of the largest and most abundant plants and the different schemes used to

AUTHORS' NOTE 18.1
Plant Geography

"Plant geography" is the term traditionally applied in physical geography to the study of the plant cover. Although the Greek Theophratus was probably the first plant geographer, plant geography did not emerge as a formal science until the 1800s. The main activity of early plant geographers was to identify, name, and map the distribution of plants in various regions of the world. The German Alexander Von Humboldt (1769–1859) and the Englishman Charles Darwin (1809–1882), for instance, studied the vegetation of South America; another Englishman, J. D. Hooker (1817–1911), explored and recorded the flora in the islands of the South Pacific; and the American Asa Gray (1810–1888) documented thousands of plants in North America. The knowledge contributed to science by these explorer-scientists was enormous and provided the foundations for early climatic and vegetational maps as well as many modern ecological principles.

Around 1900, plant geography gave rise to plant ecology, which eventually supplanted it, at least in name, as the recognized science of plant-environment relations. Today plant ecology is studied principally in the biological sciences. Here the emphasis is mainly on the interrelations of plants and the biological environ-

J. D. Hooker (The Bettmann Archive)

Asa Gray (Courtesy Harvard University Archives)

ment, that is, with animals and other plants. Plant ecology, or plant geography, in physical geography, on the other hand, emphasizes studies of plant–physical environment interrelations and plant distributions at global, continental, and regional scales, with special reference to geophysical processes and human activity as controlling agencies.

classify them. Authors' Note 18.1 presents a brief discussion of the history of plant geography.

THE HIGHER PLANTS

The plant kingdom is subdivided into three major groups: (1) vascular plants, (2) mosses and liverworts, and (3) algae and fungi. Plant geography is concerned mainly with the *vascular plants,* generally considered to be the "higher plants" because of their biological complexity, size, and evolutionary history. Plants in this group have bodies differentiated into stems, leaves, roots, and reproductive organs (Fig. 18.2a). The roots both anchor the plant in the soil and absorb water and nutrients. The stem holds the leaves and the reproductive organs aloft. Leaves are the plant's interface with the atmosphere and the medium through which it exchanges gases, releases water, and receives light for the manufacture of food substances. The reproductive organs provide the primary means by which the plant can regenerate its biological line.

All parts of plants are composed of minute, boxlike cells. The key feature of the vascular plants is that certain cells are arranged into vascular, or conducting, tissues, known as *xylem* and *phloem*. These tissues form a sort of pipe system in the plant through which water, nutrients, and manufactured foods are conducted. In the stems of woody plants, the xylem is surrounded by a cylinder of phloem and associated cells known as bark (Fig. 18.2b).

Botanists classify vascular plants into three main groups: (1) *angiosperms*, (2) *gymnosperms*, and (3) *pteridophytes* (Fig. 18.3). *Angiosperms*, the flowering plants, are divided into two major groups on the basis of whether the seed sprouts one leaf, called a cotyledon, or two. The largest group is the *dicotyledons*, comprised of the broadleaf trees, most shrubs, as well as familiar small plants such as peas, buttercups, and roses. The *monocotyledons* are mainly nonwoody plants with blade-shaped leaves; familiar examples include grasses, lilies, orchids, and palms.

Fig. 18.2 A vascular plant: (a) major anatomical features; (b) the major components in the stem.

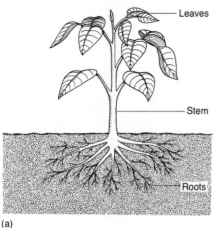

(a)

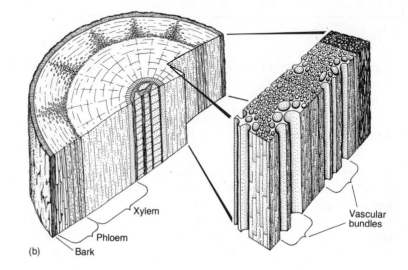

(b)

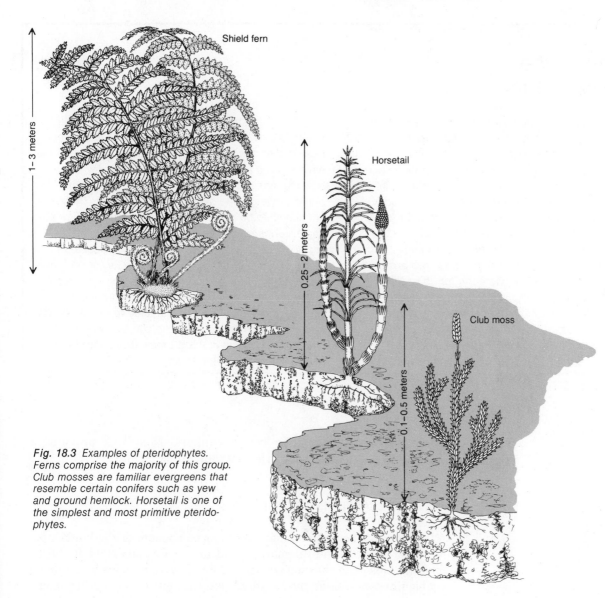

Shield fern

1 – 3 meters

Horsetail

0.25 – 2 meters

Club moss

0.1 – 0.5 meters

Fig. 18.3 *Examples of pteridophytes. Ferns comprise the majority of this group. Club mosses are familiar evergreens that resemble certain conifers such as yew and ground hemlock. Horsetail is one of the simplest and most primitive pteridophytes.*

Gymnosperms are plants that bear naked seeds, although we usually think of them as the cone-bearing plants. They are dominated by one group, the conifers, comprised of woody needleleaf plants such as pine, redwood, fir, and cypress. The *pteridophytes*, small, nonwoody plants, also are dominated by one group, the ferns. Unlike the angiosperms and gymnosperms, the pteridophytes regenerate by broad-

AUTHORS' NOTE 18.2

Origins of the Vascular Plants

The vascular plants are comparatively new to the earth, geologically speaking. The fossil record exposed in the strata of sedimentary rocks reveals that the pteridophytes evolved from aquatic plants around 350 million years ago. (The earth is considered to be about 4.5 billion years old; therefore, the vascular plants have been around for only about 8 percent of earth time.) Shortly thereafter, the gymnosperms appeared, but botanists are uncertain whether they evolved from the pteridophytes or along independent lines. These two groups of vascular plants dominated the earth for about 300 million years. Although the pteridophytes and gymnosperms are still abundant on the earth, they are most widely appreciated, at least in some quarters, for the small amounts of their remains that have been preserved in the form of coal, natural gas, and petroleum.

The angiosperms, considered the most advanced plants to ever inhabit the earth, appeared in the geologic record about 70 million years ago. Since then they have evolved into more than 250,000 species and are now the dominant plants over more land area than the pteridophytes or the gymnosperms together. The present seems to be part of the "evolutionary hour" of the angiosperms on earth, for their coverage on the planet is still expanding. Much of this expansion appears to be at the expense of the gymnosperms, as evidenced by the fact that in the last 50 million years or so, the gymnosperms declined to a total of less than a thousand species, many of which are restricted to refugelike environments, such as the subarctic and high mountains. The most extensive of these areas are the northern conifer forests of North America and Eurasia.

casting spores rather than seeds. Authors' Note 18.2 provides a glimpse at the evolutionary origin of the vascular plants.

WAYS OF CLASSIFYING PLANTS

Whenever we begin to examine a major component of the landscape, we must establish a systematic set of names for important features. This enables us to sharpen our treatment of certain topics by reducing ambiguities as well as the length of descriptions. Unfortunately, some scientific descriptive systems are so complex that they have themselves become points of confusion rather than aids to clarity, and botanical nomenclature provides good examples of this problem. There appear to be two reasons for this: (1) the botanical-names system uses Latin and latinized words; and (2) there are so many individual plants, each with its own name, that it is difficult for the nonspecialist in plant science to envision trends, generalizations, and examples.

Floristic Classification

The universal system of botanical names is based on a classification scheme called the *floristic* system. Under this system, the plant kingdom is made up of divisions, each of which is subdivided into classes, then into smaller and smaller classification units, or *taxons* (Table 18.1). Each taxon has a Latin name or a name with a Latin suffix (usually *-is, -us, -ae, -aceae, -i,* or *-a*). The more well-known plants also have common names, but they often vary from country to country or from region to region within a country. For instance, the plant popularly known in the United States as tamarack (*Larix laricina*) has at least two other names in English-speaking countries.

The *species* taxon represents the smallest classification unit in general use today. A species (pronounced speé-shees or speé-sees for both singular and plural) is comprised of individual plants that are able to freely interbreed among themselves, but are unable to breed with members of other groups. Although members of a species are not identical, they usually look so much alike that we would have trouble telling them apart. In fact, similarity of traits, or characters, especially in the reproductive organs and leaves, is the basis for classification at any level in the floristic system. This is based on the rationale that a high degree of physical similarity is indicative of a common heritage, that is, a genetic affiliation during the evolution of the plants. The importance of evolution studies and the need for an international plant-names system make the floristic scheme essential to the modern biological sciences. For the purposes of plant geography, however, other classification schemes are equally or more useful.

Life-Form Classifications

Life-form schemes are based on either the form of individual plants or the overall form, or structure, of the vegetative cover. The most familiar life-form scheme classes plants primarily according to individual

Table 18.1 Floristic classification, beginning at the class level.

TAXON	SIZE*	EXAMPLE
Class	Largest	*Gymnospermae* (cone-bearing plants)
Order	↓	*Coniferales* (evergreen, needle leaf)
Family		*Pinaceae* (pine family)
Genus		*Pinus* (true pines)
Species	Smallest	*Pinus strobus* (white pine)

*Based on number of members.

form and size. Trees are large woody plants with a main stem, or trunk, that supports branches from which the foliage grows. Most tree species are dicotyledonous angiosperms, although the very largest trees and some of the most extensive forests on earth are coniferous (Fig. 18.4).

Shrubs also are woody, branching plants, but are much shorter than trees and thus tend to be dominated more by branches than by stems. Most shrubs are dicotyledonous angiosperms, although a few popular conifers (e.g., yew and juniper) are also shrubs. Under stressful environmental conditions, certain trees can take on a shrub form. Such *dwarfism* is especially pronounced in mountainous, polar, and arid environments. Many ferns and monocotyledons (e.g., banana) have shrub forms and even tree heights, but because they are nonwoody, they may be considered to be herbs (Fig. 18.4).

Herbs are generally the smallest of the vascular plants. Having light superstructures, they are supported by stems comprised of nonwoody tissue. Herbs are limited exclusively to the pteridophytes and angiosperms and can roughly be arranged into three groups: (1) ferns, (2) blade-leaf monocotyledons, and (3) *forbs*, or broadleaf herbs (Fig. 18.4). Although most herbs are ground plants, some live as parasites on trees, and one group, called *epiphytes*, simply uses trees for support without rooting in soil. Epiphytes rest on tree branches and, as in the case of Spanish moss, dangle their roots into the air, where they are able to gain part of their moisture through condensation.

If we draw back from individual plants and allow the various forms in the plant cover to merge into a whole, we can identify differences in the general structure of the vegetative mass. These differences can be extraordinary, even over short distances, and we know the various formations as forests, grassland, tundra, or steppe, for example. In this respect, this scheme is little more than an extension of certain descriptive words from various folk cultures. A number of major vegetative formations are described in Chapter 21.

Classification According to Environment and Ecology

Since our concern is with the plant environment as well as with the plants themselves, we cannot overlook systems that classify plants according to the environments that sustain them. Such ecological classifications are based on the plant's environmental needs and limitations,

Fig. 18.4 Common tree, shrub, and herb forms. (Illustration by Peter Van Dusen and William M. Marsh)

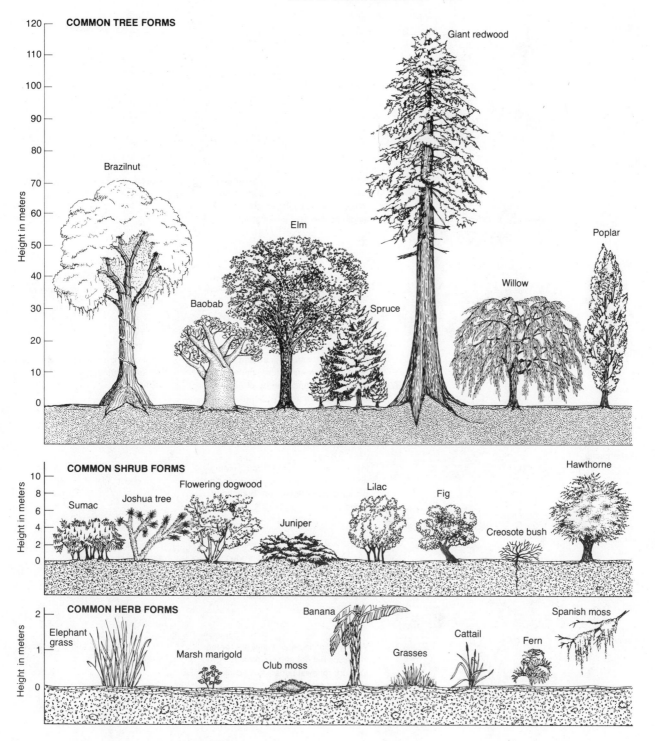

COMMON TREE FORMS

Height in meters

120
110
100
90
80
70
60
50
40
30
20
10
0

Brazilnut

Baobab

Elm

Giant redwood

Spruce

Willow

Poplar

COMMON SHRUB FORMS

Height in meters

10
8
6
4
2
0

Sumac

Joshua tree

Flowering dogwood

Juniper

Lilac

Fig

Creosote bush

Hawthorne

COMMON HERB FORMS

Height in meters

2

1

0

Elephant grass

Marsh marigold

Club moss

Banana

Grasses

Cattail

Fern

Spanish moss

CLASS	DEFINITION	ENVIRONMENT
Tall aerial plants Buds	Mainly trees and shrubs with renewal buds on shoots exposed to unfavorable weather	Abundant in tropics and subtropics; elsewhere species are few in number, but populations of species are great
Surface plants Buds	Herbs and small shrubs with renewal buds within 25 cm of the surface and thus protected by the roughness length as well as snow cover	Plentiful in boreal and alpine regions
Half-earth plants Buds	Mainly herbs with buds at ground level or within the upper several centimeters of soil; also protected by laminar sub-layer and snow cover	Abundant in arctic and high alpine areas; also plentiful in temperate regions
Earth plants Buds Buds	Mainly herbs with regenerating organs (such as bulbs and tubers) buried well within the soil and thereby protected from surface frost and other harsh atmospheric conditions	Most common in temperate regions but found farther north and south as well
Water plants Buds	Includes all water plants, both floating and rooted, but excludes microscopic plants known as plankton; buds on or in the water	Abundant in salt and fresh waters throughout most of the world
Annuals Buds	Herbs that complete their entire life cycle, germination to seed production, within a limited time period, say, days or weeks, and thereby minimize chances of exposure to harsh conditions	Especially common in arid and semiarid lands
Lianas Buds	Woody, climbing plants that are rooted in the soil but depend on trees for support; renewal buds elevated high above the ground	Limited to tropical and subtropical forests

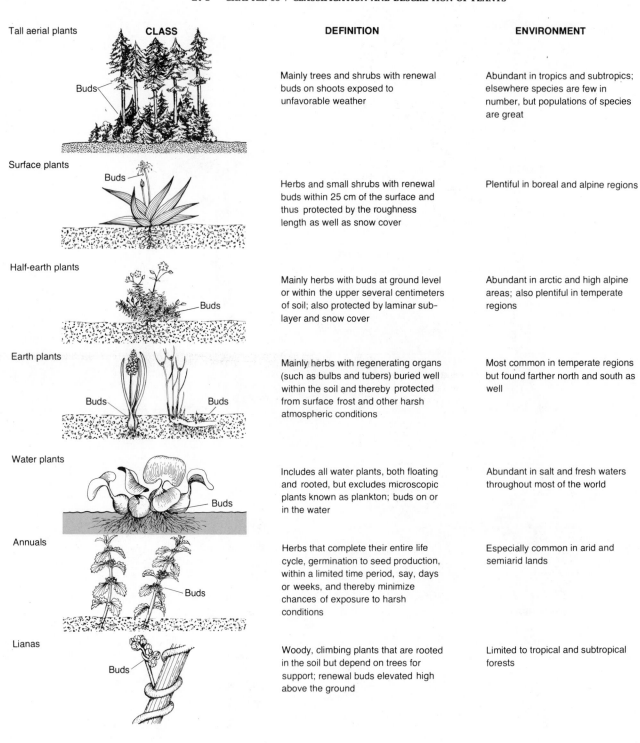

Fig. 18.5 Life-form classification of plants.

regardless of life-form or floristic affiliations. Although no comprehensive ecological system is presently complete, two simple systems based on temperature and moisture have been in use for many years. For example, the moisture-based system uses four main classes for virtually all plants. Those that live in water, such as water lily, are called *hydrophytes*; those that live in saturated soil, such as cattail, are called *hygrophytes*. In contrast are the drought-tolerant plants, called *xerophytes*, which are exemplified best by cactus. Plants that occupy intermediate moisture environments are termed *mesophytes*. In the midlatitudes mesophytes grow in well-drained sites that are neither saturated nor severely dry for more than several weeks a year. Maple, beech, elm, and hickory trees are good examples of mesophytes.

Finally, there is one system that combines ecological and life-form criteria. Developed by European botantists early in this century, this scheme is based on the distance above or below ground of the dormant-season buds on a plant. The rationale here is that bud position relative to the ground represents an adaptation to environmental conditions during the harsh season of the year. Figure 18.5 gives seven classes of this scheme. Studies reveal that the reliability of this scheme is questionable for some classes, e.g., earth plants. In contrast, it appears to apply well to plants in very harsh environments, particularly in arctic, high alpine, and desert settings.

SUMMARY

Plant geography is concerned with the composition, structure, and distribution of higher plants and with how these plants interrelate with other components of the landscape such as soil and climate. The angiosperms are the dominant plants in most landscapes, but the gymnosperms still cover extensive areas in subarctic and mountain environments. Ptedidophytes are mainly ground plants such as ferns. Plants may be classified according to their genetic affiliation, their life forms, or their ecological habitats. The floristic classification is the most widely used scheme in the plant sciences.

PHYSIOLOGICAL PROCESSES OF PLANT GROWTH

CHAPTER 19

INTRODUCTION

Now that we have looked at some basic types of plants and vegetation, we can begin to examine the processes of plant growth and the relationship of the plant to the physical environment. To do this we must take into consideration two fundamental sets of factors: (1) the physiological processes of the plant; and (2) the energy and processes of the physical environment which influence these plant processes, namely, light, heat, moisture, carbon dioxide, and certain soil minerals. *Photosynthesis*, *respiration*, and *transpiration* are the essential physiological processes of green plants. Photosynthesis is the process by which green plants convert energy in the form of light into chemical energy in the form of plant materials. Respiration is the set of internal biochemical processes that maintain the metabolism of the plant; transpiration is the process by which the plant releases water.

TRANSPIRATION AND MOISTURE

The Process

In transpiration water flows through the inner tissue of plant foliage, called *mesophyll*, and is released from the stomata as vapor (Fig. 19.1). This is the principal physiological mechanism for the internal water balance of the plant as well as an important regulator of leaf temperature. Water uptake by the plant begins as capillary water moves from the soil into the filamentous plant roots. Within the vascular tissues of the plant, the water moves upward to the foliage by a complex set of processes involving molecular transfer from cell to cell. These processes are exceedingly efficient, especially in large trees, where water is moved to heights of 75 meters above the ground. Once in the foliage, the flow of moisture from the mesophyll and stomata into the atmosphere is controlled by atmospheric humidity, the speed of wind moving over the leaf, and the temperature of the air and the leaf. As with evaporation of moisture from the soil, it is the combination of low humidity, high temperature, and fast wind that produces high transpiration rates in plants.

276

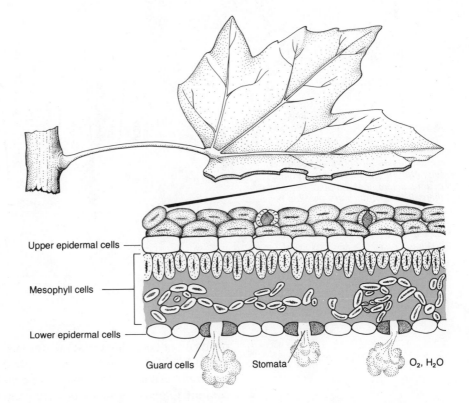

Upper epidermal cells

Mesophyll cells

Lower epidermal cells

Guard cells   Stomata O_2, H_2O

Fig. 19.1 An enlarged cross-section of a leaf, showing stomata, guard cells, and related features.

Influences of the Atmosphere

Incoming radiation influences transpiration in two ways. First, it initiates photosynthesis, which results in the opening of the stomata; second, it heats the leaf surface, inducing further stomatal enlargement and thereby greater moisture release. Conversely, as transpiration increases, latent-heat flow increases, lowering the temperature of the leaf and inducing some reduction in the size of the stomatal openings, followed, in turn, by a decline in transpiration. Sensible-heat loss due to wind generally produces the same effects. Figure 19.2 shows the changes in leaf temperature associated with sunny, cloudy, and windy conditions during a ten-minute midday period on a summer day.

Influences of the Soil

Although the atmospheric conditions immediately around the plant control the moisture loss through transpiration, it is the soil that controls the supply of water to the plant. The primary type of water used by most land plants is *capillary moisture* that is held in the upper half meter or so of soil. The amount of capillary moisture held in the soil de-

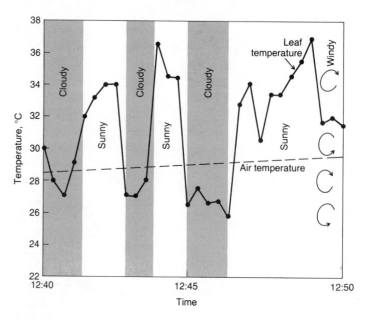

Fig. 19.2 Temperature variations of the leaf surface of a poplar tree relative to sunny, cloudy, and windy conditions in Colorado. Note that the temperature actually fluctuates far above and below air temperature. In the absence of direct-beam sunlight, leaf temperatures lower than air temperature are possible because heat is being released from the leaf through transpiration. (Adapted from D. M. Gates, "Leaf Temperature and Energy Exchange," Archiv für Metiorologie, Geophysik und Bioklimatologie, B12 (1963): 321–336. Reprinted by permission.)

pends on not only how much water is supplied by precipitation, but also the soil field capacity, which is controlled mainly by soil texture. Capillary moisture moves by molecular cohesion from spots of high moisture content to spots of low moisture content. During the growing season, the movement is usually upward toward the *soil root zone*. However, the rate of movement is often not fast enough to meet the water needs of the plants, especially in the middle and late summer, when plant demands are high. Therefore, a gap develops between the upflow of capillary water to the plant roots and the water demands of the plants. This differential is the *soil-moisture deficit*.

Two conditions can produce large soil-moisture deficits: (1) low soil capillarity, i.e., low capacity to move water via capillary processes, due to coarse soil texture; and/or (2) a small soil-moisture reserve or an especially deep moisture supply and thus a great distance of soil across which water must move to the root zone (Fig. 19.3). Most plants can tolerate weak moisture deficits without loss of vigor. However, strong deficits can produce *moisture tension* within the plant. As a result, respiration declines, and the leaf mesophyll loses water and begins to contract, or lose *turgor*. This is brought on by lowered water pressure in the mesophyll, which gives leaves a puckered appearance. For most plants, if soil moisture falls below a level at which only the more tightly bound capillary water (which is held under pressure equivalent to more than 15 bars) is available, wilting will result. If this condition is prolonged, the plant may ultimately die, but the particular

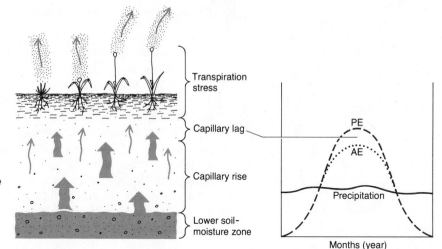

Fig. 19.3 Moisture flow from the soil through plants. Under conditions of low moisture, a gap (capillary lag) forms between the upflow of moisture and the plant roots. This is shown in the moisture-balance graph as the difference between the actual evapotranspiration curve (AE) and the potential evapotranspiration curve (PE).

soil-moisture tension, called the *wilting point*, at which this occurs usually varies from species to species.

RESPIRATION AND PHOTOSYNTHESIS

The Role of Heat

We have already mentioned one important aspect of heat in plant physiology: its relationship to transpiration. Even more fundamental, however, is the internal temperature of the plant organs, because heat is a primary control on the essential biochemical processes of photosynthesis and respiration.

Although freezing is the minimum survival temperature for many plants, it is noteworthy that a temperature of 10°C is necessary to initiate appreciable photosynthesis and respiration. At temperatures between freezing and 10°C, biochemical activity is negligible regardless of the availability of light, moisture, and carbon dioxide. Above 10°C, it doubles with each ten-degree increment until some optimum temperature is reached, around 30°C. At this temperature the rate of biochemical activity in the plant is at a maximum, given, of course, that all other necessary forms and amounts of energy are available. Beyond 30°C photosynthesis and respiration decline, and at some high temperature, around 40°C for many plants, the organism enters a stage of severe heat stress and suffers physiological damage (Fig. 19.4).

Fig. 19.4 Severe heat loss can also cause damage to plants, such as the frost splitting of this tree trunk. (Photograph by Michael Treshow)

Photosynthesis and the Principle of Limiting Factors

Photosynthesis is one of the truly astonishing natural processes, for it not only converts radiant energy into chemical energy, but also fixes that energy in the form of organic compounds that do not spontaneously break down and can thus be stored until needed by the plant. This is the essence of photosynthesis, which enables plants to survive and even grow during periods of little or no light.

Photosynthesis takes place only in the green parts of plants in the presence of light, water, carbon dioxide, and heat. Basically, the elements of carbon dioxide and water (two atmospheric gases) are combined with light energy, which is absorbed by the chlorophyll in the photo cells (called chloroplasts) of plant leaves. Molecular processes act on this combination to produce oxygen and plant materials (organic compounds) in the form of glucose (sugar) and carbohydrates (starch). The oxygen and water vapor are released into the atmosphere through the stomata (Fig. 19.1). Photosynthesis can be simplified into the following formula:

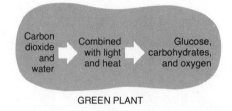

GREEN PLANT

Although light is the preeminent form of energy necessary in photosynthesis, scientific analysis shows that it is heat, carbon dioxide, and water acting together that regulate this process. This fact was discovered by plant physiologists in the early part of this century, and it formed the basis for the *principle of limiting factors*: When a process is influenced by several factors, the highest intensity or rate it can attain is controlled by the factor that is in shortest supply. Thus, as we mentioned earlier, without enough heat to maintain a temperature above 10°C, photosynthesis will be minimal no matter how much light, water, and carbon dioxide are available. The same holds true for carbon dioxide. Where it is reduced to levels below the atmospheric average of 0.03 percent, photosynthesis declines markedly (Fig. 19.5). With this principle in mind, let us examine photosynthesis further.

Photosynthesis and Radiation

Photosynthesis begins with the absorption of atomic particles of light, called photons, by the chloroplasts, which initiates photochemical ac-

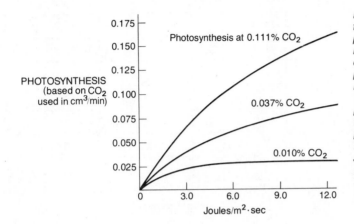

Fig. 19.5 *The rate of photosynthesis at three different levels of CO_2. The critical point to note is that at the 0.010 percent level, no matter how much light is increased above 300 or 400 footcandles, photosynthesis does not increase, thus illustrating the principle of limiting factors. This is in marked contrast to the 0.111 percent level, which is more than three times the atmospheric average, where photosynthesis increases steadily as light increases. (From W. H. Hoover, E. S. Johnson, and F. S. Bracket, "Carbon Dioxide Assimilation in Higher Plants," Smithsonian Institute Miscellaneous Collection 87, 16 (1933).)*

tivity, the first phase of photosynthesis. The rate of photochemical activity depends on, among other things, the wavelength of the radiation, the intensity of radiation, and the duration of the daily light period, called the *photoperiod*.

1 Longwave incoming (from sky)
2 Shortwave incoming (from sky)
3 Shortwave outgoing (reflected)
4 Longwave outgoing (reradiated)
5 Shortwave incoming (reflected from ground)
6 Longwave outgoing (from leaf to ground)
7 Longwave incoming (to leaf from ground)

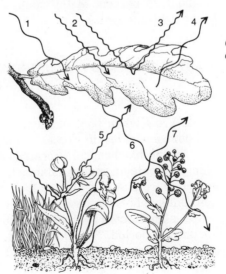

Components in the radiation balance of a leaf.

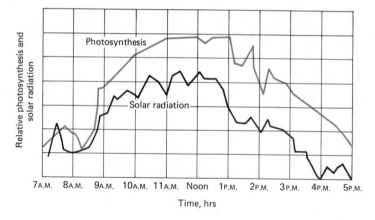

Fig. 19.6 The typical relationship between incoming shortwave radiation (S_i) and the rate of photosynthesis (measured by the rate of CO_2 taken in by the plant per hour). The S_i variations between 9 A.M. and noon are due to variations in cloud cover. (Adapted from Jen-Hu Chang, Climate and Agriculture, Chicago: Aldine, 1968.)

The rate of photosynthesis usually increases each day with the intensity of incoming shortwave radiation and vice versa. As radiation varies with cloud cover, for example, so does photosynthesis (Fig. 19.6). However, in order for any plant to achieve a peak rate of photosynthesis under conditions of enough H_2O, CO_2, and heat, an optimum intensity of light is necessary. Although we normally associate a reduction in the intensity of light or the length of the photoperiod with a decline in photosynthesis, light intensities surpassing the optimum range of a plant can also inhibit photosynthesis. For example, Fig. 19.7 shows the relationship between light intensity and relative photosynthesis in marine phytoplankton. The horizontal axis represents the full range of sunlight intensity. Note that photosynthesis is greatest at an intensity of about 25 percent of full sunlight and decreases with higher and lower light intensities.

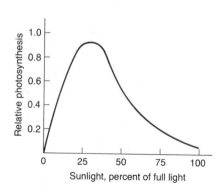

Fig. 19.7 The relationship between light intensity and relative photosynthesis in marine phytoplankton. Note that the intensity of photosynthesis is greatest at 25 percent of full light. (Adapted from J. H. Ryther, "Photosynthesis in the Ocean as a Function of Light Intensity," Limnology and Oceanography 1 (1956): 61–70. Reprinted by permission.)

PLANT GROWTH AND PRODUCTION OF ORGANIC MATTER

Net Photosynthesis

Plants grow when photosynthesis produces plant materials (glucose and carbohydrates) in excess of the rate of utilization of these materials by the biochemical processes of respiration. In other words, in order to create an energy balance favorable for growth, the plant must manufacture materials faster than they can be consumed in respiration. Unlike photosynthesis, which is limited by the photoperiod, respiration is unaffected by light conditions and thus continues throughout the diurnal period. Therefore, the *rate* of photosynthesis must greatly exceed the *rate* of respiration because of the shorter period of photosynthetic activity in each day. Combining the rate of *total photosynthesis* and the rate of respiration, we derive *net photosynthesis*, which is simply an expression of a plant's energy balance:

Net photosynthesis = Total photosynthesis − Respiration.

If *net* photosynthesis is positive, total photosynthesis exceeds respiration, and plant material is available for plant growth. If *net* photosynthesis is zero, respiration is consuming all plant materials, but still maintaining the plant's metabolism. If *net* photosynthesis is negative, an insufficient amount of plant material is available for respiration; respiration declines, and the plant loses weight. For many plants, these three states of net photosynthesis usually occur in early summer, mid to late summer, and fall, respectively.

Plant Production

Production refers to the rate of output of organic material by the plant cover. Production is equal to net photosynthesis and is measured in terms of the grams or kilograms of organic matter added to a ground area of one square meter per day or year. In general, where light, heat, moisture, and carbon dioxide are present in large and dependable quantities, plant life is most abundant and productive. Where one or more of these resources is in limited supply, we can expect, based on the principle of limiting factors, that production will be markedly restricted. Of the four essential resources, carbon dioxide is the least variable, at least at global and continental scales. The other three are highly variable over the planet and in many regions are also variable from season to season (Fig. 19.8).

From the map in Fig. 19.8, we can see that of all of the major climatic regions, the equatorial zone and the tropics receive the greatest combined total of heat, moisture, and light. In addition, the supply of each in this vast region is characterized by low variability; for example, the annual variability in rainfall is generally less than 15 percent from the average as compared with more than 40 percent for major deserts. With such ample and dependable supplies of energy, it is understandable that plant productivity should be higher in the tropics than in any other large region on earth. And measurements bear this out; the equatorial and tropical rainforests produce, on the average, 8.6 to 10.1 grams of organic matter per square meter per day, a rate forty to fifty times greater than that for the deserts.

Poleward from the tropics, one or more of the four resources declines in total availability and/or dependability. The deserts, for example, are severely deficient in moisture, but often amply supplied

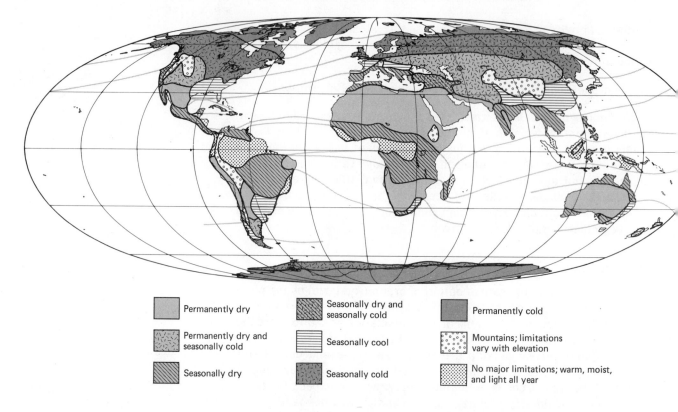

Permanently dry

Permanently dry and seasonally cold

Seasonally dry

Seasonally dry and seasonally cold

Seasonally cool

Seasonally cold

Permanently cold

Mountains; limitations vary with elevation

No major limitations; warm, moist, and light all year

Fig. 19.8 Generalized map showing the principal limiting factors to plant productivity worldwide.

Map projection by Waldo Tobler.

with light and heat, especially light. Despite the intensive light, the severe moisture deficiency limits photosynthesis, resulting in minimal growth rates. In fact, survival itself is possible only for plants such as xerophytes, which have special adaptive mechanisms for aridity.

Another severely limiting climate is the polar, where heat is insufficient to induce appreciable levels of biochemical activity for plant growth. In addition, water, though abundant in this environment, is in the frozen state most of the time and not available to plants. Light is plentiful during the summer, but severely limited during the winter. Except for a few lower plants, such as mosses, lichens, and liverworts, this climate is effectively devoid of a plant cover.

In the intermediate climates moisture is adequate, and temperature and light range seasonally from high to low. Although the plant cover in these midlatitude regions may be dense, growth rates are less than half those in the wet tropics, but much greater than those of the desert and polar climates. Table 19.1 gives the production rates of the plant covers of various major climatic zones. Note the magnitude of change from the tropics to the desert and Arctic tundra.

If we extend our examination of productivity to include coastal and marine environments, the contrasts are even more striking. The

Table 19.1 Net annual productivity of selected plant covers.

PLANT COVER	PRODUCTION, G/M²·DAY
Tropical forests	8.6–10.1
Subtropical broadleaf forests	6.7
Midlatitude beech, pine, and alder forests	2.9–4.4
Subarctic taiga spruce forests	2.6
Tall-grass prairie	1.2–3.3
Arctic tundra	0.3–0.7
Desert (harsh)	0.2–0.3

(Data compiled from a variety of sources)

most productive earth environments are actually the tropical coastlines—specifically, tidal marshes, estuaries, and coral reefs (Fig. 19.9). Here average daily productivity is as high as 25 grams per square meter. Only a few kilometers off shore, however, the rate drops to 0.5 to 3.0 g/m²·day, and beyond the continental shelf, over the deep

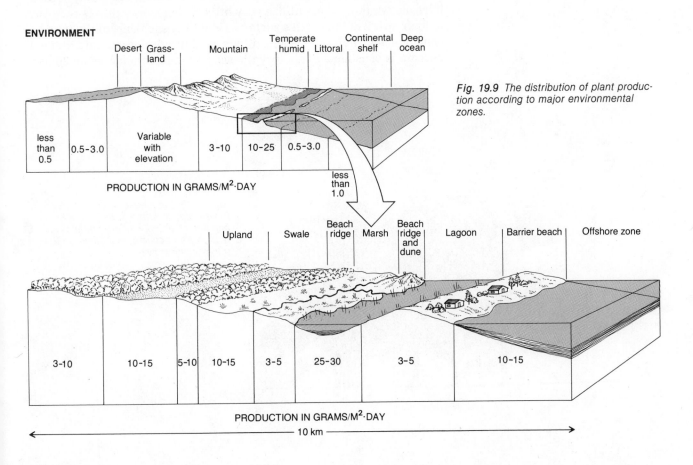

Fig. 19.9 The distribution of plant production according to major environmental zones.

ocean, the rate is less than 1.0 g/m² • day. An interesting aside relates to the productivity of polluted waters. Measurements from a South Dakota pond that received untreated sewage showed that productivity during some summer days reached 27 grams per square meter.

ENERGY FLOW IN THE PLANT COVER

Plants and the Atmosphere

As we noted in our examination of the energy balance, the vegetative cover has an important influence on heat flow and radiation near the ground. This effect is so profound in areas of dense forests that the effective boundary between atmosphere and land rests not on the soil surface, but on the forest canopy. Here most of the incoming radiation and rainfall are intercepted and the brunt of the wind received. As a result, a thin microclimate is created between the forest floor and the canopy.

But how effective an energy consumer is vegetation? Many studies have shown that plants utilize very little of the energy that is received in their space of the earth's surface. Generally speaking, of the total incoming solar radiation received by a midlatitude forest, only about 1 percent is consumed in photosynthesis. Of the remainder, some 20 per-

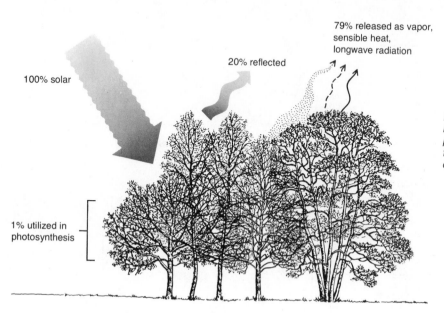

100% solar

20% reflected

79% released as vapor, sensible heat, longwave radiation

1% utilized in photosynthesis

Fig. 19.10 The breakdown of solar energy in a temperate forest. Note that only 1 percent is utilized in photosynthesis and that only a fraction of this actually produces plant growth.

cent is reflected off the tree crowns; 79 percent, absorbed by the crowns. The latter is dissipated as longwave radiation, sensible heat, and latent heat (Fig. 19.10).

The Energy Balance of the Plant Cover

The single unit of energy consumed in photosynthesis is converted into plant materials, which are used to produce growth and to maintain respiration of the plant. The fraction utilized in respiration is dissipated by internal biochemical processes and ultimately released in the form of heat. The fraction utilized in plant growth is transformed into living plant matter, namely, leaves, stems, roots, and reproductive organs. As the plants complete their growth cycles, most of this matter is deposited on the soil or consumed by animals. Most of the small fraction that is consumed by animals is released from them as heat; however, a small proportion of it may be passed along the biological food chains through a series of prey-predator relations in the *ecological pyramid*. The bulk of the organic matter that falls on the soil is decomposed by microorganisms and is also released as heat. In environments such as bogs, swamps, and lakes, the decomposition is incomplete, and a diminutive fraction of organic energy is retained or, more appropriately stated, detained on the earth's surface. Ironically, we have come to depend on this fragile reserve, in the form of peat, coal, and petroleum, as our chief energy resource. It is estimated that this reserve of fossil energy is equivalent to a total of only 3600 years of photosynthesis.

The Energy Pyramid

Because it is very difficult to isolate and measure the actual flow of organic energy from one group of organisms to another, we have relatively few complete sets of data on the energy balance of ecological systems, or *ecosystems*. To make such a determination, the total organic mass, called the *biomass*, must be measured for each level in the ecological pyramid, and an energy equivalent must be assigned to each unit of organic matter. The energy equivalent turns out to be 4 kilocalories (17,000 joules) per dry gram of organic matter for plants and 5 kilocalories (21,000 joules) for animals. Ecologist Howard T. Odum provided a good example of the energy pyramid of a fresh water ecosystem at Silver Springs, Florida (Fig. 19.11). His data demonstrate how rapidly energy attenuates from one level to the next. The respiration of soil microorganisms and herbivores alone accounts for the dissipation of almost 90 percent of the original mass of energy represented by vegetation.

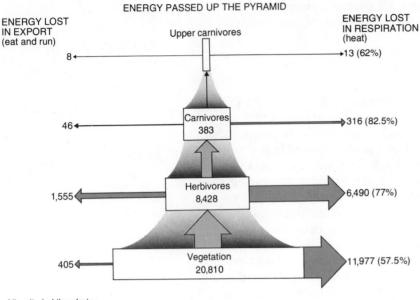

ENERGY PASSED UP THE PYRAMID

ENERGY LOST
IN EXPORT
(eat and run)

ENERGY LOST
IN RESPIRATION
(heat)

Upper carnivores

8 ← → 13 (62%)

Carnivores
383

46 ← → 316 (82.5%)

Herbivores
8,428

1,555 ← → 6,490 (77%)

Vegetation
20,810

405 ← → 11,977 (57.5%)

All units in kilocalories
per square meter per year

*Fig. 19.11   The energy pyramid for a
fresh-water ecosystem at Silver Springs,
Florida. Note that at each level, most of
the energy is lost in respiration as heat.
(Data from H. T. Odum, 1957)*

SUMMARY

Photosynthesis, respiration, and transpiration are the essential physiological processes of green plants. Certain amounts of soil water, light, carbon dioxide, and heat are needed for these processes to function. According to the principle of limiting factors, the one in least supply controls the rate of photosynthesis. Plant growth can take place only when total photosynthesis exceeds respiration. Production by land plants is greatest in the wet tropics and least in dry and cold regions. In all environments plants form the foundation for complex chains of organisms, called ecosystems, linked together by the flow of organic energy from one group to the next.

CHAPTER 20 THE INTERPLAY OF PLANTS AND THEIR ENVIRONMENT

TOLERANCE, STRESS, DISTURBANCE, AND ADAPTATION

Tolerance

The survival of a plant depends on not only proper supplies of the basic requirements for photosynthesis, respiration, and growth, but also its tolerance to a wide variety of influences from the environment. *Tolerance* refers to the range of stress or disturbance that the plant is able to withstand without damaging effects. *Stress* refers to the factors that affect photosynthesis—light, heat, water, carbon dioxide, and certain minerals. *Disturbance* refers to processes such as soil erosion, extreme wind, massive snow and ice accumulations, flooding, disease, and human activities that affect vegetation. Acting individually and collectively, forces representing stress and disturbance act on plants, causing damage, loss of reproductive capacity, or death itself. To appreciate fully the principle of tolerance, it is necessary to introduce two important concepts: (1) tolerance is variable from phase to phase in the life cycle of plants, as well as from group to group within a single species; and (2) over time the level of disturbance or stress produced by the environment, measured, for instance, by changes in the amount of soil moisture or the intensity of flooding, is also variable.

The life cycle. The life cycles of plants involve many phases, and for each phase the combination of growth requirements and tolerances may be different—often greatly different. In some phases, for example, a plant can withstand severe cold or drought that would kill it in every other phase of its life. Figure 20.1 shows the general phases in the life cycle of flowering plants. It begins with a mature plant and the induction of leaves, followed by the development of sexual organs, which in turn produce pollen that is dispersed by wind and insects. The flowers impregnated by the pollen then form embryos, which mature into seeds. Upon release from the plant, the seeds are dispersed and eventually deposited somewhere. Here they usually lie dormant for some time, usually over the winter season or dry season. When conditions are favorable, some of the seeds germinate and grow to seedlings. This is accompanied by photosynthesis, additional growth, and the development of various life forms.

289

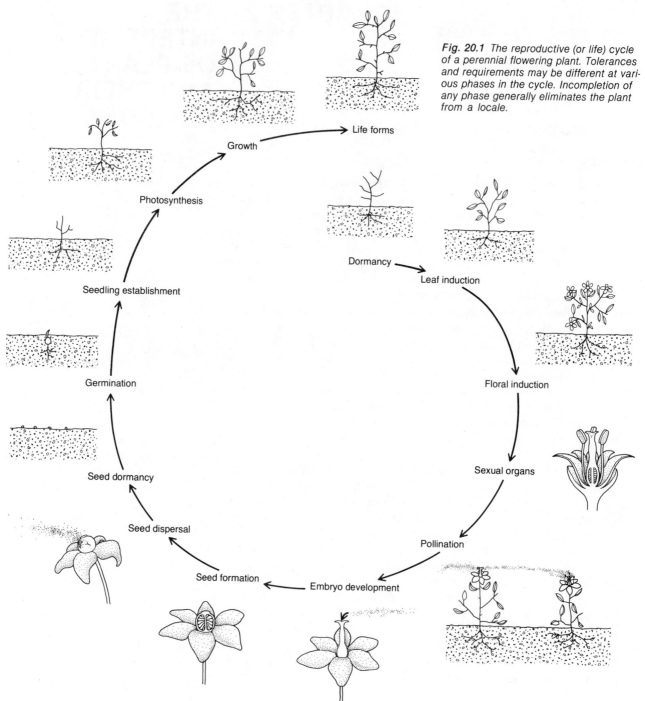

Fig. 20.1 The reproductive (or life) cycle of a perennial flowering plant. Tolerances and requirements may be different at various phases in the cycle. Incompletion of any phase generally eliminates the plant from a locale.

Growth

Life forms

Photosynthesis

Dormancy

Leaf induction

Seedling establishment

Floral induction

Germination

Seed dormancy

Sexual organs

Seed dispersal

Pollination

Seed formation

Embryo development

Interruption of any phase in the life cycle stops the reproduction process and hence can eliminate the plant from the locale. Of all the phases, seed germination is often identified by plant physiologists as one of the main "bottlenecks" in the cycle of many plants. To induce germination in most plants, specific combinations and intensities of light, heat, moisture, soil chemicals, and aeration as well as a certain level of ground stability are often required. Since these factors can be highly variable over small areas, it is easy to see that germination can be very irregular. Slight variations in topography, for example, can greatly influence germination potentials because sun angle varies with slope, and surface heating, snow melt, and evaporation in turn vary with sun angle.

The story becomes all the more complex when we recognize that among the members of a particular species, there is usually appreciable variance in tolerance and requirements. This can be referred to as *ecological amplitude* or *habitat versatility,* and has been demonstrated many times in tree farming, for example, when seeds are transported to a plantation several hundred kilometers from their birthplace. Although the new location is well within the species' range, the trees often do not do as well as those of the same generation that were left at home. The United States Forest Service recognizes the importance of habitat versatility in tree species and as as a rule will not move seed to plantations more than 100 miles or so from its place of origin.

The magnitude and frequency of disturbance and stress in the plant environment. The forces exerted on plants by physical processes show remarkable variation over time. Changes in the intensity of a natural process tend to follow a distinctive pattern in which relatively low intensities occur with highest frequencies and higher intensities occur with lowest frequencies. In fact, the magnitude of a process often increases at an increasing rate as frequency decreases. For example, the magnitude of a wind, measured in terms of the force exerted on vegetation, increases with the cube of velocity; therefore, a wind that occurs at an average frequency of once every ten weeks may not be just ten times stronger than the wind that occurs on an average of once per week, but perhaps twenty to fifty times stronger. Floods, precipitation intensities, and erosion tend to follow the same pattern. In short, the plant's physical environment is not more or less static, varying only a little from the average, but rather is highly dynamic, varying greatly from the average on a seasonal, annual, and longer-term basis. The probability of great variations in process intensities increases with time; thus chances of a truly powerful episode of a process are greater over a century than over a year, for example.

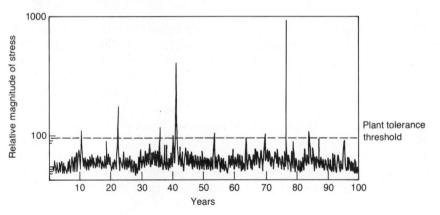

Fig. 20.2 Hypothetical frequency of various intensity levels of physical processes such as rain storms, river flow, drought, and wind. An intensity that exceeds the plant's tolerance limit eliminates the plant.

The point of the magnitude-and-frequency principle is that the level of disturbance or stress produced by the environment, as manifested through various climatic or hydrologic processes, is continually undergoing fluxes of intensity. Most of these fluxes fall below the tolerance thresholds of the various phases in the plant's life cycle, but every so often some exceed a tolerance threshold (Fig. 20.2). When this occurs, the population may be reduced in some way, e.g., by a lowering of population density or a dieback and loss of range. Although the magnitude-and-frequency principle is applicable to any environment, it is most apparent in certain high-energy environments, especially shorelands, sand dunes, and mountain slopes. (The magnitude-and-frequency principle is discussed in detail in Chapters 31 and 32.)

Increasing Tolerance Through Adaptation

Where plants are faced with a high intensity of stress or disturbance, they may improve their chances of survival through adaptation. Two types of adaptation are possible: *acquired* and *genetic*. Acquired adaptation occurs when an individual plant undergoes morphological or physiological change during its own lifetime in response to some stress or disturbance. For example, to offset a reduction in light caused by the growth of a forest canopy, a generation of ground plants may increase leaf size. Although this trait may be strongly developed, it is not passed on to succeeding generations.

This is in contrast to genetic adaptation, whereby new traits that appear in a population are part of the plant's genetic code and *are* passed on to succeeding generations. Biologists identify three major ways in which genetic adaptation, or *evolutionary change*, takes place: (1) *mutation*, whereby new traits are produced by a change in gene frequency; (2) *natural selection*, whereby environmental factors favor certain traits over others; and (3) *genetic drift*, whereby one group in a population develops traits that differentiate it from the rest of the population. Some combination of these three appears to be responsible for most evolutionary change in plant populations (see Authors' Note 20.1).

Common forms of adaptation. There are countless ways in which plants have adapted to stresses and disturbances posed by the environment, and we are limited to but a few general examples here. One of the most prevalent adaptations is the *annual habit* that has evolved in many herbs in the midlatitudes. These plants, called *annuals*, have developed a life cycle in which the seed-dormancy phase coincides with the high-stress season, normally winter. All other phases of the life cycle are completed during the spring and summer, leaving the dormant seed, which studies show to be the toughest phase in the life cycle, to carry the plant through the winter. The annual habit is a *phenological* adaptation; that is, it involves an adjustment of the timing of the life cycle to the rhythm of the seasons.

A similar type of adaptation involves the adjustment of the plant life cycle to a particular event in the environment that is favorable to the plant. This sort of adaptation is characteristic of desert herbs called *ephemerals*. These plants have evolved the capacity to maintain seed dormancy for extended dry periods, easily several years in duration, and then to suddenly germinate, mature, and flower in a matter of days or weeks when soil-moisture conditions are favorable. In the desert such brief episodes produce brilliant floral displays which are the subject of many popular nature films.

For the *perennial* desert plants, collectively known as xerophytes, two other forms of adaptation to drought are especially well developed. The first is the *succulent* habit, well typified by the cacti. Three physiological traits are pronounced in the succulents. (1) In order to reduce vapor loss, leaves are small and few in number, with small and widely spaced stomata. (2) Bodies are thick and fleshy for storage of water. (3) Root systems are shallow and fairly extensive in order to maximize water uptake when rainfall wets the upper soil. With these features, the succulent is able to withstand prolonged periods of drought that would kill most other plants.

The second type of adaptation in desert perennials is characterized by the formation of deep roots that enable plants to draw on water at depths far beyond the reach of most plants. Mesquite, a shrub found in the American Southwest, is a good example: Its roots have been traced to depths of 25 m or more. This is truly exceptional because even the roots of the largest trees, such as firs, redwoods, and pines,

Custion cactus, a succulent. (Photograph by W. T. Lee, U.S. Geological Survey)

Desert holly, a shrub, the most drought-resistant plant in Death Valley. (Photograph by John R. Stacy and Charles B. Hunt, U.S. Geological Survey)

AUTHORS' NOTE 20.1

Speciation and Geographic Change

The origin of species, or speciation as it is termed, is one of the most actively researched and controversial topics in natural science. The idea that one species of organisms originates from other organisms has not been with us for very long, and it is still rejected by many people who feel that plant and animal species were each uniquely created by God.

Our knowledge of processes of speciation has grown tremendously since Charles Darwin brought the subject to world attention in the nineteenth century. We know that it is a complex process that probably involves a combination of natural selection, genetic drift, and mutation. At the base of it all, however, is the relationship of an organism to its environment. In order to live and reproduce, an organism must possess the capacity to adjust, or adapt, to environmental change. This is necessary because over long periods of time, it appears that no environment is without major change. This we know from analysis of all kinds of data about the earth, especially geologic data on past environments. As the environment changes, the successful organism must

also change; otherwise, it will trend toward extinction. If part of the range of a species grows drier, for example, the individuals in that area may adapt by developing special characteristics that make them more drought-resistant.

One type of natural selection involves the geographical separation of a population into two or more isolated groups. This can happen where a barrier such as a mountain range or a desert forms across the geographic range of a species, thereby breaking the population into two separate groups. In time, the groups may evolve along such different lines that eventually they are no longer genetically compatible and are unable to interbreed if brought together. At this point they are two different species. The diagram shows the distribution of two varieties of a California shrub called Potentilla gladulosa. This mountain plant may be in the process of evolving into different species, as its environment is subdivided with the formation of the High Sierras and the arid San Joaquin Valley.

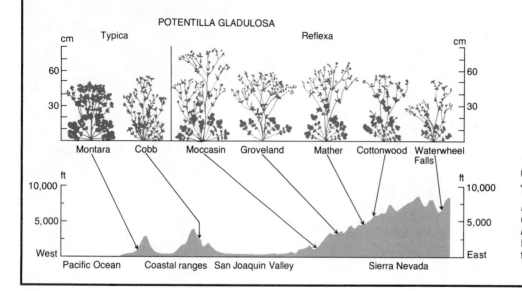

POTENTILLA GLADULOSA

(Adapted by permission from J. Clausen and W. M. Hiesey, "Experimental Studies on the Nature of Species IV, Genetic Structure of Ecological Races," Carnegie Institute Washington Publication No. 615, 1958.)

(a)   SUMAC CLONE

Parent plant

2 m

(b)   DUNE GRASS

Parent plant

1 m

are limited to depths of less than 10 m, with the bulk of the root mass within 1–2 m of the surface.

In certain environments it is difficult for plants to complete the reproductive phases of the life cycle. This is especially so in highly active environments, such as sand dunes, shorelines, and mountain slopes, where the soil surface is so changeable that seed germination is impossible for most plants. In order to survive in these environments, some plants have evolved the means to regenerate by nonsexual processes. This *vegetative regeneration* involves the production of new plants from a living part of the parent plant such as a runner or lateral root. Many such organs may radiate from a parent plant, and from them hundreds of stems can be produced. Together, the parent and its stems may form a single plant, called a *clone*, which has the aspect of many individual plants. Among the most successful vegetative reproducers are dune grass, willow, and sumac (Fig. 20.3). Sumac is particularly common along fence lines and in abandoned farm fields in eastern North America.

Fig. 20.3 Vegetative reproduction: (a) Sumac clone, the stems of which generally decrease in age outward from the parent stem. (b) Dune grass (marram), which spreads not only laterally, but also vertically in response to burial by wind-deposited sand. The photograph shows willow and sand cherry clones interspersed with dune grass. (Photograph by Charles Schlinger)

ENVIRONMENTAL INFLUENCE ON PLANT DISTRIBUTIONS

The landscape is a patchwork of vegetation types. The patches are all different sizes, but each represents an individual group of plants. Along their perimeters, the patches tend to merge into one another; ecologists use the term *ecotone* to describe the transition between two groups, or zones, of plants. Along lawns, plowed fields, or shorelines, ecotones are relatively sharp, more like borders, marking the edge of an area dominated by powerful forces which limit the area to only cer-

(a)

(b)

(c)

Fig. 20.4 *Three examples of ecotones: (a) wetland-forest; (b) Arctic fell-tundra; (c) alpine forest–alpine meadow. (Photographs by Deborah Coffey, Patrick W. Hassett, and William M. Marsh, respectively)*

tain plants. In most places, however, ecotones are gradual and tend to be rather complex in terms of actual changeover from one group to the other (Fig. 20.4). Our objective in this section is to examine some of the factors that control the location of a group of plants. Some of these factors are easy to ascertain, especially where human activity is involved, e.g., lumbering, agriculture, urbanization, and war. Other factors tend to be less so—in particular, climate, erosional processes, topography, and soils.

Climatic Controls

Centuries of experience by European vineyardists and farmers provided early plant scientists with ample evidence that plant growth and survival are related to climate. Such knowledge led nineteenth-century plant geographers to search for the critical relationships between broad patterns of vegetation and the major climatic zones. Some of the scientists, believing that they were able to identify certain correlations, went so far as to recommend that the vegetation in some regions be used to delimit climatic borders. Although later studies cast doubt on the reliability of some of these correlations, certain borders that were identified still appear on modern climatic maps (see Chapter 9).

The relationship of vegetation to climate is exceedingly complex, and so many variables are involved that attempts at broad correlations of the two have generally proved inconclusive. Three factors are particularly important in this regard. (1) Humans have drastically altered the distribution of vegetation in the past 10,000 years; thus in some regions vegetation borders have little to do with climate directly. (2) The earth's climatic patterns have undergone major shifts in the past

10,000 years, and some zones of vegetation are still in the process of adjusting their distributions to these shifts; therefore, their borders may not correspond well with their potential climatic ranges. (3) Plant migrations have been highly uneven in the past 100,000 years or so, with rapid diffusion in certain directions and none where barriers such as oceans or mountain ranges stand in the way; therefore, all vegetation has not had equal opportunity to respond to climatic patterns.

So what can we say about climate and the distribution of vegetation? Short of arm waving at a global scale, two statements are possible. First, the strongest relationships appear between individual plants and a key feature of climate such as the moisture balance. An example of the latter is provided in Fig. 20.5, which shows the east-west ranges of six plants as a function of soil-moisture conditions. Second, strong relationships also appear where climate is sharply differentiated geographically, as in mountainous regions. From the windward to leeward side of a coastal mountain range, for instance, vegetation may change from heavy forest to grassland or desert in response

Fig. 20.5 Tolerance ranges of various trees, shrubs, and grasses to water-balance deficit and surplus along an east-west United States transect at lat. 41°N. (From C. W. Thornthwaite and J. P. Mather, "The Water Balance," Publications in Climatology **8,** 1, Drexel Institute of Technology, 1955. Used by permission of C. W. Thornthwaite Associates, Elmer, New Jersey)

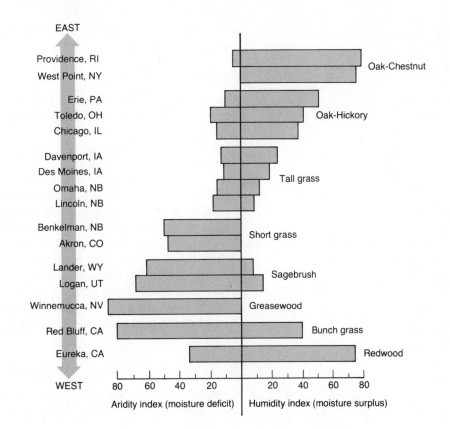

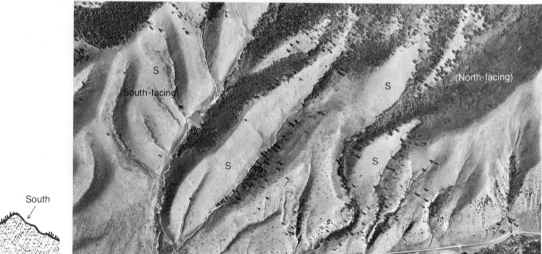

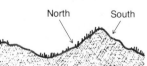

Fig. 20.6 Aerial view of the distribution of forest and grassland in the Sangre De Cristo Mountains of southern Colorado. Because the sun angles are more direct on the south-facing slopes, heating is more intensive, and the moisture balance is much poorer than on north-facing slopes. As a result, south-facing slopes are limited to grass covers, whereas north-facing slopes can support forests, as shown on the diagram.

to a sudden decline in total precipitation and an increase in evaporation. A similar change is also apparent from the sunny to the shaded slopes of mountains in semiarid regions, where the difference is attributed to change in soil moisture (Fig. 20.6).

Topography

Probably the most celebrated relationship of vegetation to mountain climates is found in the vertical zonation of vegetation on high mountains. Temperature declines with altitude, resulting in not only less heat higher up, but lower evapotranspiration rates as well. Coupled with stresses such as heavy winds, avalanches, and landslides, these factors produce different belts of vegetation at various elevations (Fig. 20.7). In the American West, the lower slopes of the Rocky Mountains are often dry and covered with grasses and shrubs, whereas the cooler middle slopes, having a better moisture balance, are usually inhabited by conifers and hardy broadleaf trees. At the upper limit of the forest, called the *tree line*, the trees are smaller and give way to alpine meadows of herbs. Here the limiting factor is heat. If the mountain is high enough, the alpine meadows will grade into a colder zone of snow and ice essentially devoid of plants. For another example of environmental control on plant distribution in mountainous terrain, see Fig. 22.16.

The Influences of Geomorphic Processes

As the geomorphic processes reshape the land by erosion and deposition, they may also influence the plant cover. Such influences are

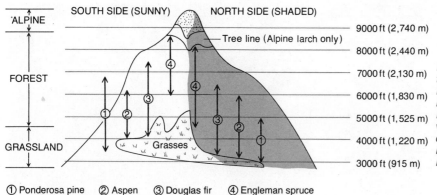

① Ponderosa pine   ② Aspen   ③ Douglas fir   ④ Engleman spruce

Fig. 20.7 *The vertical zonation of vegetation from the Bitterroot Mountains near Missoula, Montana. Three major zones are identifiable: alpine, forest, and grassland. Note that the altitude zone inhabited by various species is higher on the south side of the mountain, where there is more solar radiation. (Adapted in part from Arno and Habeck, 1972.)*

usually most pronounced at the local scale in places such as shorelines, sand dunes, mountain slopes, and river valleys, where geomorphic processes are very powerful. At the regional scale, the most evident influence of a geomorphic agent on vegetation is found in areas of continental glaciation. In North America the great sheets of Pleistocene ice destroyed vast areas of flora as they spread over the land. When the ice retreated, a new flora, the *boreal forest,* became established in the glaciated region, and the present borders of this forest cover are thought by some ecologists to coincide with the area occupied by the ice 6000 to 8000 years ago. It is suggested that the newness of the boreal forest is reflected in its low diversity of species. Beyond this example, however, we know very little about the relations of vegetation to geomorphic factors at a regional scale. Virtually nothing is known about vegetation and geomorphic systems at a global scale, although perhaps the recent discoveries related to plate tectonics will provide some important information about changes in the locations and climates of the earth's land masses, which will help to explain some puzzling similarities and differences in widely separated flora. Since the relationship of plants and geomorphic phenomena is best illustrated at the local scale, let us look at examples of two local settings: river valleys and sand dunes.

Rivers. River channels tend to shift back and forth on the floors of their valleys. As this takes place, vegetation is destroyed or damaged by bank erosion, powerful flood flows, and heavy sediment deposits. On the other hand, where fresh deposits, such as sand bars, form in the river channel, new plant habitats are created. Because of their nearness to the river, however, such sites are subject to intensive disturbance from flooding, erosion, and further deposition. As a result, it is difficult for plants to establish themselves on these sites and to main-

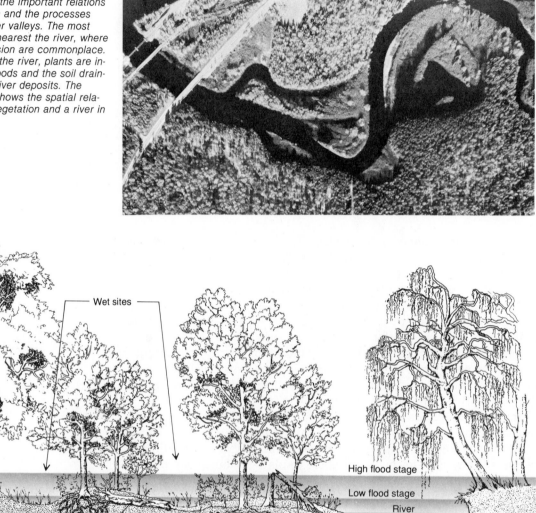

Fig. 20.8 Some of the important relations between vegetation and the processes and features of river valleys. The most stressful sites are nearest the river, where deposition and erosion are commonplace. Farther away from the river, plants are influenced by high floods and the soil drainage related to old river deposits. The aerial photograph shows the spatial relationship between vegetation and a river in flood stage.

Wet sites

High flood stage
Low flood stage
River

Tree downed by flood, then buried

Former soil surface now buried by deposits

Tree damage from high flood flows

Burial of plants by new deposits

Undercut bank

tain themselves once they are established (Fig. 20.8). Therefore, only those plants with high tolerance to these particular processes can grow on these sites. Willow (*Salix* spp.) and cottonwood (*Populus* spp.) are among the most successful trees. In fact, the pattern of these plants on the valley floor often reflects the distribution of channel deposits and/or the impacts of strong flood flows (see photograph in Fig. 20.8).

Sand dunes. Sand dunes are highly dynamic environments, character-ized by rapid sand erosion and deposition by wind. One of the chief fac-tors governing the survival of plants in the dune environment is their tolerance to burial by sand and root exposure by erosion. Figure 20.9 gives the approximate ranges of tolerance to erosion and deposition for six plants found in and around coastal dune fields in the midlatitudes. Dune grass (*Ammophila breviligulata*), willow (*Salix* spp.), and sand cherry (*Prunus pumila*) have the highest tolerance to burial. Studies show that dune grass actually prefers burial on the order of 10–20 cm per year for proper growth and development. At the other extreme is white birch, which seems able to tolerate only slight burial. This is one of the reasons that this lovely tree has limited success as a yard plant.

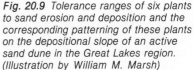

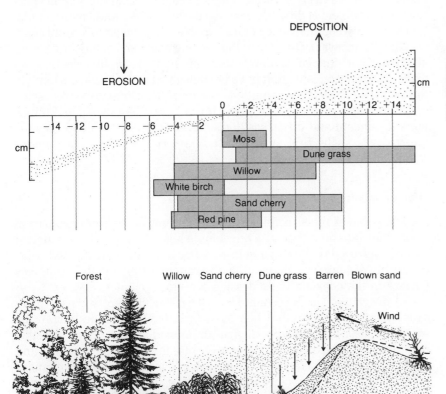

Fig. 20.9 Tolerance ranges of six plants to sand erosion and deposition and the corresponding patterning of these plants on the depositional slope of an active sand dune in the Great Lakes region. (Illustration by William M. Marsh)

Because dune plants have different tolerances to burial and erosion, they do not all compete for the same space in the dune environment. Where deposition rates vary down the lee slope of a dune, for example, the plants tend to pattern themselves correspondingly. Dune grass usually grows highest on the slope, followed by sand cherry, willow, mosses, and various trees (Fig. 20.9). As the pattern of deposition changes from year to year, the distribution of these plants tends to change as well.

CONCEPTS OF CHANGE IN THE PLANT COVER

The question of how change in the distribution of plants actually takes place has been a lively topic among plant scientists for most of this century. The problem is difficult because of the diverse and complex relationships between the various plant species and the environment. As a result, it is necessary to realize that no one scientific model or concept of vegetation change is universally applicable. In this section we examine three concepts that have been devised to describe and explain the nature of change in vegetation. The first, the community-succession concept, is the most popular of the three. The other two—the individualistic concept and the stress-threshold concept—provide interesting alternatives to the first, although they are less widely known among scientists.

The Community-Succession Concept

We know that plants are able to change various aspects of the environments they inhabit, especially the climate near the ground, the nutrient and moisture content of the soil, and the erodibility of the soil. This capability of plants to change the environment they inhabit is the key feature in the community-succession concept.

A description of the community-succession concept begins with an area of land or water that has been denuded, i.e., made barren, by some process, say, glaciation, fire, or human intervention. Into this environment migrate certain hardy plants, such as mosses and grasses, which are referred to as *pioneers*. These plants effect changes in the environment such as holding the soil against erosion and adding nutrients to the soil, thereby making it suitable for other plants. Once this has taken place, the pioneers are replaced, or *succeeded*, by a second group of plants, which in turn renders further change to the environment. Each group to inhabit this environment is called a *community* because it is comprised of plants that live together in an ecologically interdependent fashion. Wave by wave, more or less, one community of plants succeeds another until eventually, one of them achieves stability and is able to inhabit the environment on a relatively permanent basis. This group of plants is called the *climax community*, and its presence indicates that a state of equilibrium has been reached in the plant-soil-atmosphere system. In other words, the climax community represents a steady state between the plant cover and the physical environment. Should some outside force change the climate or soil, this state is interrupted, and a different climax community will probably evolve. Should the climax community be destroyed altogether, succession will begin anew and continue until a climax community is reestablished.

Many examples of plant succession have been described by modern scientists. One of the most frequently cited examples involves the filling of ponds, lakes, bays, and estuaries by plant succession. Initially, only aquatic plants, such as water lilies and algae, and semi-aquatic plants, such as reeds and rushes, inhabit the water body. As these plants grow and die, their remains are deposited on the bottom, and after many years a thick, organic layer may develop. In the shallow water near the shore, where plant productivity is greatest, the organic layer eventually builds up to the water surface, where it provides new habitats for certain mosses, grasses, and shrubs. These plants both stabilize the new soil and add organic matter to it. In addition, streams contribute sediment to the organic mass. Little by little, the community of plants encroaches on the open water, adding much of each year's production to the growing organic mass (Fig. 20.10a).

(a)

Fig. 20.10 The community-succession concept: (a) Seventy years of succession. This pond formed on the mass of debris produced by the Gros Ventre landslide near Jackson Hole, Wyoming, in 1906. Early accounts indicate that the original plant cover was completely obliterated by the slide. (Photograph by William M. Marsh) (b) Schematic example of a bog that has been completely filled and overgrown with a climax forest community.

(b)

Climax community

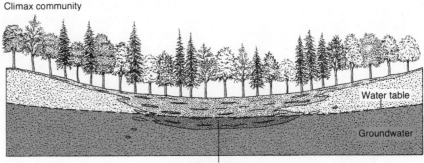

Water table

Groundwater

Former bog

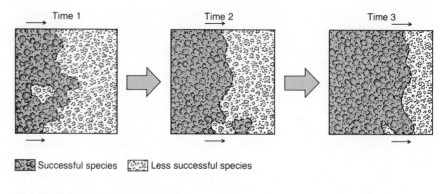

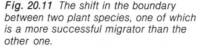 Successful species Less successful species

*Fig. 20.11 The shift in the boundary
between two plant species, one of which
is a more successful migrator than the
other one.*

When the water body is completely filled, the aquatic and semiaquatic
plants no longer have a place to live, and eventually the types of trees
that grew only near the original shore overgrow the entire environment, forming the climax (Fig. 20.10b).

The Individualistic Concept

In the 1930s the American plant taxonomist and ecologist Henry A.
Gleason developed a model counter to the community-succession concept. The main theme of his model, called the *individualistic concept,* is
that patterning and change in the plant cover can be explained on the
basis of the probability of recurrence of individual plants rather than
on the basis of the relationship of an entire community to the environment. Gleason argued that when a plant dies and space is made available for a new plant, it is a matter of probability as to which species
will fill the space. Those species that are (1) nearest to the site, (2) present in greatest numbers, and (3) most efficient in dispersal have the
greatest probability of inhabiting the vacant space.

Two points about the individualistic concept are pertinent: (1)
Changes in the plant cover that are commonly called succession are
attributable to the advantage certain species have over others in
establishing themselves at new sites. This advantage is often due to the
fact that some plants are more efficient dispersers than are others.
This is clearly the case for many sand dune plants, for example,
because they have the advantage of vegetative reproduction (Fig.
20.11). (2) Climax communities maintain stability because probabilities
favor recurrence of existing species, not because they are better suited

to the particular environment than are other groups of species. In other words, when a pine tree dies in the middle of a pine grove, chances are very high that it will be replaced by another pine.

Stress-Threshold Concept

According to the stress-threshold concept, the distribution of a plant population fluctuates in response to the magnitude and frequency of environmental forces around it. Every plant has a relatively fixed tolerance to stresses and disturbances in the form of geophysical processes, heat, light, soil moisture, disease, fire, and human actions. When the level of stress or disturbance exceeds the tolerance threshold of a plant species, the members living in the affected area are either eradicated or so severely damaged that they can no longer reproduce. But such powerful events usually occur so infrequently that in the intervening years, the plant may begin to reinhabit the area that was lost. While this is taking place, many smaller, but yet substantial, events are likely to occur which may delay or set back the plant's recovery of the area. In short, if we could compile a time-compressed film of the plant's distribution, it would appear to be in a state of continuous flux, expanding and contracting in response to the rises and falls in the magnitude of certain forces in its environment.

Let us look at lake or bog vegetation from the stress-threshold perspective. The water in bogs is derived mainly from groundwater, and the water level corresponds to the level of the water table in the surrounding land. We know that the height of the water table fluctuates with changes in the amount of precipitation; it follows, therefore, that the water level in the bog must also fluctuate. A rise of, say, a half meter in water level floods bog plants that are not hygrophytic; dieback occurs, and the bog and its water area may enlarge. Conversely, a comparable drop in water level eliminates many plants whose roots are adjusted to a limited range of water levels. More important, a drop in water level can expose the mass of dead organic material to the air, thereby accelerating its decomposition and in turn reducing the total organic mass of the bog (Fig. 20.12). We can thus envision a bog fluctuating in area with major water-level changes. During periods when water level is near average, the bog may fill with organic debris and shrink in area; during highs in the water level, the bog may lose certain plants and open up; during lows, it may lose organic debris and decline in total mass. In some cases two of these trends may occur in the same time period (Fig. 20.13).

This interpretation of bog formation is markedly different from the one customarily used in the community-succession concept. The standard succession interpretation is based on a plant-controlled change;

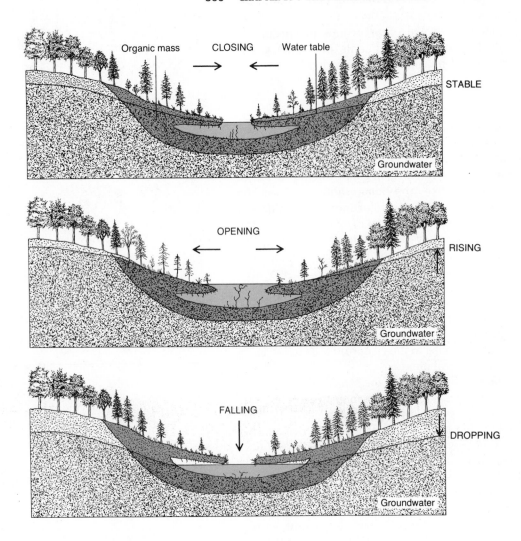

Fig. 20.12 Changes in bog vegetation with variations in water level. If the water table holds steady, the bog will grow inward and close out the pond. However, the opposite may occur when the water level rises or falls one or more meters from the average. Examples of such changes can often be seen where beavers have dammed the outlets to ponds and where established beaver dams have burst.

the stress-threshold interpretation, on environmentally controlled change. In areas where the environment remains fairly stable for centuries and millenia, succession is undoubtedly the predominant mode of change. The thousands of filled bogs across the Midwest, for example, attest to this. But the data gathered by scientists show that in many environments, there is a great variation in the energy flow that drives the essential processes in and around the plant community. Groundwater, stream flow, and soil moisture, for example, show such ranges of fluctuation that they are capable of drastically altering the

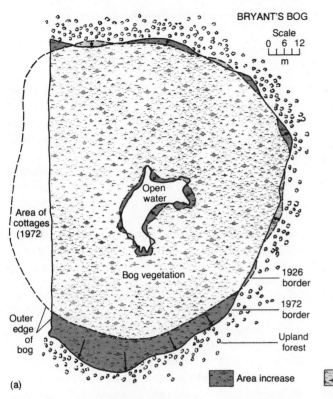

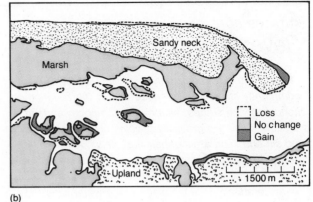

Fig. 20.13 *Simultaneous trends in bog fluctuation: (a) A bog near the University of Michigan Biological Station, just south of the Straits of Mackinaw, that has been studied by scientists for five decades. Recent mapping reveals that over the past forty-six years, the bog's outer perimeter has enlarged significantly (save for an*

area in which cottages have been erected), whereas the inner perimeter, around the open water, has decreased in area by 24 percent (excluding the fill area where the buildings are located. (Originally mapped by Frank C. Gates; remapped by Christa Schwintzer and Gary Williams) (b) Changes in a coastal marsh near Barn-

*stable, Massachusetts, between 1859 and 1957. Note that losses and gains are about equal. (From Alfred C. Redfield, "Development of a New England Salt Marsh," Ecological Monographs **42**, 2, 1972. Copyright 1972 by the Ecological Society of America. Used by permission.)*

plant cover. Therefore, it can be argued that the notion of plant succession, and its assumption that a stable physical environment is the norm rather than the exception, seems to be too simplistic a way to interpret vegetation change in environments as dynamic as some lakes, ponds, coastlines, mountain slopes, and river valleys. Some scientists feel quite strongly about the inadequacies of the community-succession concept. In Authors' Note 20.2 plant ecologist Hugh M. Raup of Harvard University remarks on this point, based on his own experience years ago in northern Canada.

AUTHORS' NOTE 20.2
Reverse Succession

"One of the most commonly used illustrations of plant community structure and succession is the zonation on pond and lake shores. Water lilies and pondweeds in shallow water are gradually building up peat and catching silt and sand around their root stalks. As the muck gradually nears the surface, other plants that like this kind of place, with its peaty substratum and shallower water, gradually take the place of the water lilies. Usually these are cattails or some kind of bulrushes. A little nearer the shoreline the sedges and cattails have so changed the substratum that they cannot live there any more, and perhaps some grasses can take their place. I described these things in great detail at Lake Athabaska, showing their neat zonal arrangement and the orderly change, or succession, that was presumed to be going on. My wife was along, collecting lichens and mosses, and she quickly found that there was a fine zonation of these plants also. It was particularly evident on vertical cliffs, where different-colored lichens made horizontal bands on the rock faces above the water level. Having been taught that wherever possible, one should be accurate and have actual figures and measurements, we carefully measured the heights of all the zones above the level of the lake.

"We went back to Lake Athabaska six years later, seeing again the things we had described and describing many more. Then we went back a third time, nine years after our first trip. In that field season we discovered that nearly everything we had described previously had vanished completely. Sometime during the three years that had elapsed since we had last seen it, the water level in the lake had risen no less than 6.5 feet, as shown at the places where we had made our careful measurements. A large part of the watershed that drains into the lake is in the northern Rocky Mountains, where snow and weather conditions had combined to send an extraordinary amount of water into the lake in a very short time. How often this sort of thing happens we do not know with any accuracy. From talking with the oldest inhabitants, the nearest we could come to it was at intervals of forty or fifty years.

"At Lake Athabaska, every so often, the whole system of shore vegetation is simply drowned and eliminated, and the plants have to start over again. If there is any succession at all, it is in little fragments that never make any real progress, at least in the way it is assumed that they do. The water lilies had been growing in shallow water offshore, the bulrushes and cattails in shallower water, the sedges and grasses on the wet shore itself, each of them for the simple reason that it had found for itself the place best suited to it. These simple observations could be refined to great detail; but to take the next step and say that because the different communities were growing next to each other, they were developing from one into the next would be going beyond the facts and into pure speculation.

"It can properly be asked, 'There must have been evidence of such extraordinary high water on Lake Athabaska. Why didn't you see it?' The answer is, I did see it. I described it and took pictures of it—large driftwood high on the old beaches and undercut banks high above the existing water level. But I was so entranced with the theory of succession that I bent all these facts around so that they would fit the theory—I thought they were evidence of unusually big storms or a gradual, though permanent, lowering of the lake level."

H. M. Raup, "Vegetational Adjustment to the Instability of Site," *Proceedings and Papers,* 6th Technical Meeting, International Union, Conservation of Natural Resources, Edinburgh, 1957, pp. 36–48. (Used by permission of the author.)

SUMMARY

Tolerance refers to the range of stress or disturbance a plant is able to withstand without damage and loss of reproductive capacity. Tolerance varies with the phases of the life cycle. In addition, the magnitude and frequency of stress and disturbance produced by the environment are highly variable, especially in dynamic environments such as coasts, river valleys, and sand dunes. Plants can, however, improve their chances of survival through adaptation. Notable examples of adaptation include the succulent habit, the annual habit, and vegetative regeneration.

The influence of the environment on plant distribution is complex. The climatic influences are manifested in the regional and global patterns of vegetation. The effects of other aspects of the environment, such as geomorphic processes, are most evident at the local scale. The community-succession concept, the most popular concept of change in the plant cover, is based on the notion that plant communities are capable of changing the environment, thereby giving rise to other plant communities until eventually a stable state is attained between one community and the environment. The individualistic concept argues that change is dependent on the probability of recurrence of individual plants. The stress-threshold concept recognizes the variable nature of environment, arguing that plant distributions fluctuate in response to the magnitude and frequency of environmental forces.

AN OVERVIEW OF THE EARTH'S VEGETATIVE COVER

CHAPTER 21

INTRODUCTION

Photographs of the earth taken from space can lead one to wonder seriously about the prominence of living matter on the planet. Animals are not evident at all; plants, which together constitute the vast majority of living matter on the land, are evident, but their distributions appear interrupted by large patches of barren soil, rock, ice, and snow. Standing on the earth's surface, however, we are often impressed by the sheer volume and diversity of the plant cover. And it is understandable that we should be, for despite its uneven distribution over the planet as a whole, the vegetative cover is truly massive, comprising hundreds of thousands of plant species. As we demonstrated earlier, we higher life forms cannot survive without this mass of greenery; therefore, as dependent organisms, it behooves us to know where the plants grow on the planet, and as students of the land, it behooves us to know what forms and roles they take on in the different landscapes.

Biochores: The Major Structural Classes

The description of vegetation can begin with the identification of three major biological systems, or ecosystems: (1) salt water; (2) fresh water; and (3) terrestrial. Our concern is mainly with the terrestrial, and it can be subdivided into biomes or *biochores*. Biomes are large regions, which correspond to climatic zones, characterized by a distinctive combination of vegetation and animals. Biochores are more or less the vegetative versions of biomes and are defined on the basis of the overall structure and composition of the vegetative cover over broad sections of continents. Four biochores are generally recognized: *forest, savanna, grassland,* and *desert* (Fig. 21.1). The forest biochore is characterized by a cover of trees that forms a continuous canopy over the ground. Although herbs and other plant forms are found in every forest, trees are the dominant vegetation. In the savanna biochore, a mixture of forest and grassland, patches of trees and shrubs are scattered among tall grasses. A blanket of grasses dominates the grassland biochore, and only infrequently is it interrupted, usually by trees and shrubs along stream courses. The desert biochore is characterized by

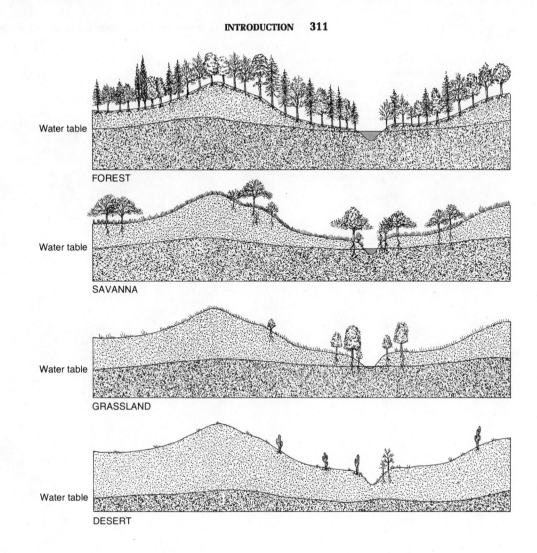

FOREST

SAVANNA

GRASSLAND

DESERT

a highly discontinuous plant cover comprised mainly of low herbs and shrubs. Within each of these biochores, a number of *formations*, or formational classes, can be defined (see Fig. 21.2 and Table 21.1).

Fig. 21.1 The four biochores.

Changing Geography of Biochores

The distribution of biochores appears to be controlled by many factors, chief of which are climate and people. Under natural conditions, the availability of heat and moisture probably has the greatest influence on the global patterns of biochores. Changes in climate as well as other factors such as fire and disease, however, help to produce complex border configurations between biochores that make scientific correla-

Fig. 21.2 Locations of some formational classes within biochores: forest—tropical rainforest, temperate forest, sclerophyll forest, boreal forest; savanna—thornbush savanna, tropical savanna; desert—shrub desert, dry desert. (Adapted from Küchler, 1967, and Polunin, 1967)

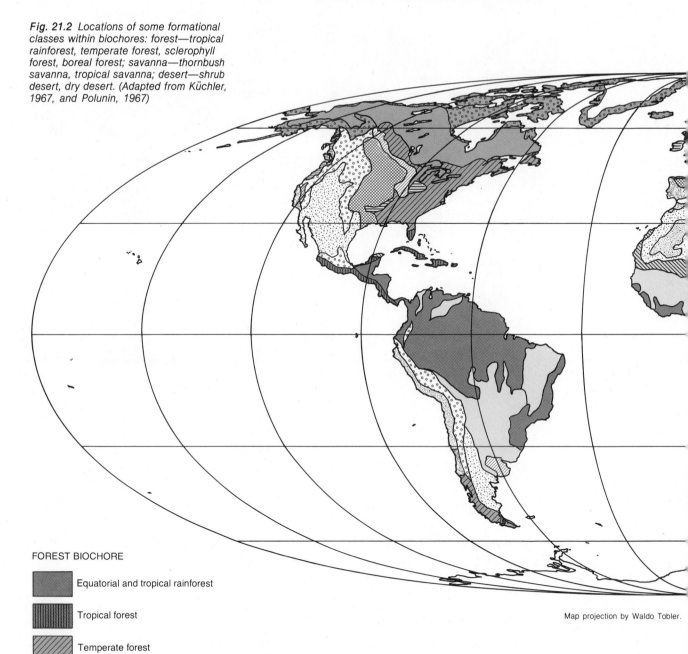

Map projection by Waldo Tobler.

FOREST BIOCHORE

 Equatorial and tropical rainforest

Tropical forest

Temperate forest

Boreal forest

West-Coast forest

 Sclerophyll forest

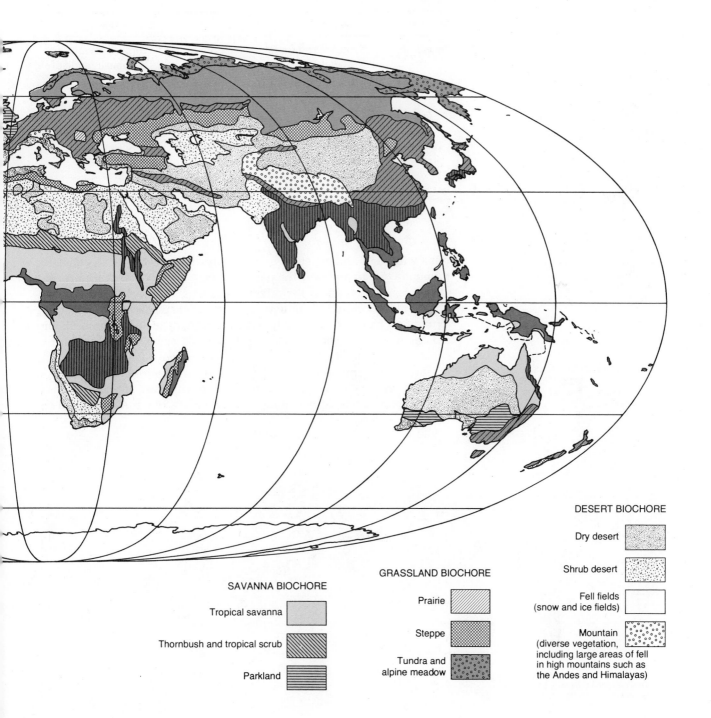

SAVANNA BIOCHORE

Tropical savanna

Thornbush and tropical scrub

Parkland

GRASSLAND BIOCHORE

Prairie

Steppe

Tundra and
alpine meadow

DESERT BIOCHORE

Dry desert

Shrub desert

Fell fields
(snow and ice fields)

Mountain
(diverse vegetation,
including large areas of fell
in high mountains such as
the Andes and Himalayas)

Table 21.1 The major formations of each biochore.

BIOCHORE	FORMATION	BIOCHORE	FORMATION
Forest	Tropical rainforest	Savanna	Tropical savanna
	Tropical forest		Thornbush (and scrub)
	Temperate forest		Parkland
	Boreal forest		
	West Coast forests		
	Sclerophyll forest		
Grassland	Prairie	Desert	Dry desert
	Steppe		Shrub desert
	Tundra		Fell fields
	Alpine meadow		

tions with heat and moisture exceedingly difficult. Added to this is the influence of humans over the past 10,000 to 20,000 years or more. We know a good deal about the effect of humans' actions in changing biochores in the twentieth century; however, we know very little about the extent and nature of such actions in the past, especially where historical records and archeological evidence are scarce. Some scientists reason, for example, that the savanna biochore of Africa is due largely to forest clearing and burning by earlier hunters, herders, and farmers. In the case of burning, how much was natural and how much was caused by people is mainly speculation. In North America, on the

Fig. 21.3 The distribution of cultivated land in the world today. Most is concentrated in areas of natural forest or grassland vegetation.

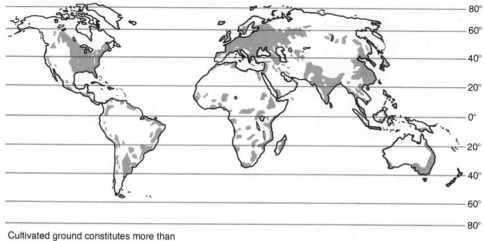

Cultivated ground constitutes more than 50 percent of the land cover

other hand, forest clearing has taken place mostly within the past 300 years. Therefore, it is possible to measure the gross changes in this biochore and to identify the causal factors. The extent of cultivated land in the world today gives a good indication of how much humans have changed the vegetative cover (Fig. 21.3).

THE MAJOR FORMATIONS OF VEGETATION

Forest Biochore

Equatorial and tropical rainforest. The wet tropics are characterized by a very dense, continuous plant cover dominated by an *open forest* formation, which reaches heights of 30 to 50 meters. The massive crowns of the trees, such as the Brazilnut (*Bertholletia excelsea*) of South America, intertwine to form an almost continuous roof of foliage, called a canopy. Although the canopy may appear to be comprised of equally tall trees, several different levels (or tree stories) can actually be discerned in most rainforests. The trees provide habitats for a rich variety of smaller plants, including epiphytes, thick woody lianas vines, and strangler vines (see Fig. 21.4).

In general, ground plants and shrubs are relatively scarce in the rainforest; the forest canopy is so dense that sunlight cannot penetrate in sufficient quantities to nurture a ground cover of shrubs and herbs. Most of the trees and associated plants are *evergreen,* that is, without a dormant season during which foliage is lost. Rather, foliage is shed and replaced during all months in a continuous cycle, so there is usually no period in the year when ground shade is interrupted.

The floristic composition of tropical vegetation is the most diversified on earth. Many species, commonly on the order of thousands per square kilometer of land, but each with small populations of individual plants, distinguish this plant cover. The northern forests are in sharp contrast, with relatively few species occurring in great populations that cover thousands of square kilometers.

Expansive areas of equatorial and tropical rainforest are found in the Amazon basin of South America, west-central Africa, and Southeast Asia. The climate in these areas is characterized by high temperatures, which from day to day vary little from the 20–25°C annual range, and ample precipitation throughout the year. Light is abundant, although the wet tropics as a whole receive somewhat less light than do the subtropical deserts because of a substantial cloud cover during most days of the year. All in all, the wet tropics are ideally suited to plant growth, and this is manifested in three principal ways: (1) high biomasses, (2) high rates of productivity, and (3) great variety of plant species.

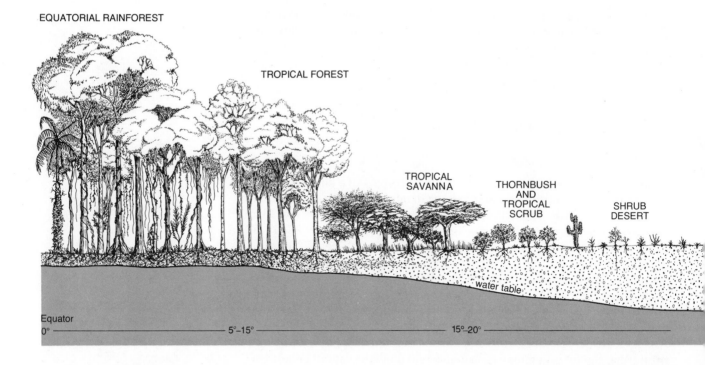

EQUATORIAL RAINFOREST

TROPICAL FOREST

TROPICAL SAVANNA

THORNBUSH AND TROPICAL SCRUB

SHRUB DESERT

water table

Equator
0°

5°–15°

15°–20°

Fig. 21.4 A schematic portrayal of the major formational classes of world vegetation. (Illustration by William M. Marsh)

Tropical forest. Where the rainfall regime of the wet tropics is broken by a dry period or a season of low rainfall, a less luxuriant forest may be found. Such *tropical forests* are typified by the *monsoon forests* of India and Southeast Asia. Trees tend to be smaller than those of the rainforest, and many exhibit a marked seasonal rhythm by dropping their leaves in the dry season, although they are actually evergreen given adequate moisture. Others, however, are deciduous. In addition, auxiliary vegetation in the form of epiphytes and vines is less abundant than in the tropical rainforest.

Where the tropical forests or rainforest have been destroyed and along edges of natural clearings, such as river channels, the open-forest structure is usually replaced by a lower and denser *closed forest.* The absence of a continuous canopy allows light to penetrate to the ground, where it fosters thick growths of herbs, shrubs, and small trees. This formation is the tropical jungle, and although Edgar Rice Burroughs may lead us to believe otherwise, the jungle is certainly not conducive to vine swinging.

Temperate forest. The original plant cover of the humid midlatitudes was usually comprised of continuous forest, but with a significantly

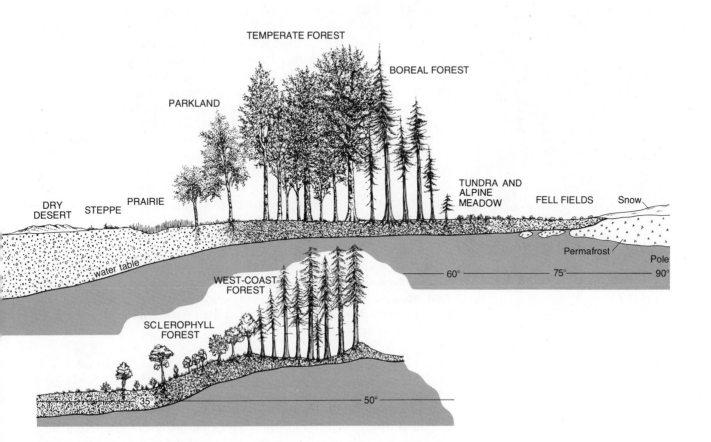

lower mass than that of the wet tropics. Three large regions of temperate forest were once found in the Northern Hemisphere: the eastern United States, Europe, and eastern China (Fig. 21.2). In the forest tracts that remain today, the trees are generally 15–25 meters high and are associated with less abundant life in the form of vines and epiphytes than is the case in the tropics. Mature temperate forests are usually open, but may support a light cover of ground plants: ferns, club mosses, and lovely flowering plants such as violets, geranium, and spring beauty.

The trees of temperate forests may be evergreen or deciduous, needleleaf or broadleaf, and all undergo marked winter dormancy when insufficient light and heat are available for growth. The most extensive tracts of the temperate forests are located in the Northern Hemisphere, where representative tree types include various kinds of

Temperate, deciduous forest.

oaks (*Quercus*), maples (*Acer*), and pines (*Pinus*) as well as ash (*Fraxinus*), beech (*Fagus*), walnut (*Juglans*), elm (*Ulmus*), and hickory (*Carya*). Although the needleleaf trees tend to be more abundant on the northern side of this forest zone, where they are often mixed with stands of broadleaf trees, they may also occur in appreciable numbers on the subtropical side as well. In addition, ground plants, vines, and epiphytes are more abundant in the subtropics, and in general the biomass is significantly greater than it is farther north.

For the most part, the temperate forests of Asia, Europe, and North America have been destroyed or drastically altered over the past several millenia by agricultural and industrial development. For example, China's forests were all but totally obliterated by A.D. 1000, and records show that England was practically treeless by 1600 or 1700. Although conditions in these countries have improved as a result of reforestation programs, reflecting what may be a new trend in some parts of Europe and Asia, very little of the original forest can be found today. In its place are farms, cities, and tracts of managed and planted forests. Managed forests are distinguishable from unmanaged ones because the former are often closed, a result of selective cutting practices which encourage multiple-storied woods, and may consist of a different mix of species.

Boreal forest. Poleward of about 45 to 50 degrees north latitude, the mixed forests of the midlatitudes give way to more homogeneous subarctic forests, where only three or four tree species may dominate extensive forest stands. Conifers such as spruce and tamarack are the principal trees, but a few hardy broadleafs, such as birch (*Betula*) and tag alder (*Ulnus*), are also common in these the great boreal forests of Canada and the Soviet Union (see Fig. 21.2).

On its northern boundary, the boreal forests grade into the treeless tundra. As is the case with essentially all borders between major vegetational formations, this border is characterized as a broad transition zone, where the lower tundra cover is interspersed with patches of trees, many of which exhibit dwarfism owing to the stressful Arctic environment. The same type of transition also appears on midlatitude mountains around an elevation of 3000 meters, where the forest grades into the grassy Alpine meadows.

West Coast forest. One notable exception to the humid midlatitude forests described above is the West Coast forests, which are often composed of very prominent trees such as the famous redwood (Sequoia), fir (Pseudotsuga), and pine (*Pinus*) of the American Northwest. Here high humidity, relatively abundant precipitation, and a moderate temperature regime that eliminates a long dormant season combine to create an environment that has nourished the largest trees on earth, some of which exceed a height of 100 meters. In locations where the West Coast climate yields especially heavy rainfall, say, more than 200 cm per year, the forests may be extraordinarily dense and harbor dense growths of ground mosses, epiphytes, and related plants. Such forests are often called midlatitude rainforests.

West Coast forest. (Photograph by I. C. Russell, U.S. Geological Survey)

Sclerophyll forest. Equatorward from the West Coast forests the climate grows drier, and the trees become much shorter and more widely spaced. This is the sclerophyll forest, or hardwood evergreen forest, found in areas of Mediterranean climate in both hemispheres. With a canopy cover of 25–60 percent, the sclerophyll forest is the weakest of the midlatitude forests; in fact, in many places it is more like a tropical savanna than it is a forest (see Fig. 21.2).

The sparseness of the sclerophyll forest is attributable to two factors. The first is the severe moisture stress of the warm, rainless Mediterranean summer, which has limited the sclerophyll forest to a light cover of drought-resistant trees. These include many types of oaks, notably live oaks (*Quercus spp.*), cork oak (*Quercus suber*), and white oak (*Quercus lobata*), several species of pines, and numerous shrub-sized trees, such as wild lilac (*Ceanothus sp.*) and olive (*Olea spp.*). The second factor is the long tenure of human settlement in Mediterranean regions, especially in the Old World. Where more luxuriant forests may once have stood, fires, agriculture, grazing, and wood harvesting long ago destroyed them, and the cover we see today is probably a very poor reflection of the past. This appears to be especially so in the case of the scrub vegetation of Mediterranean Europe and southern California. In Europe it is called *maquis* (or *macchia* or *garique*) and consists of dense shrubby thickets; in California, it consists of dwarf forests of oaks mixed with shrubs and is called *chaparral.*

Savanna Biochore

Between the forests and the open grasslands is a biochore comprised of grass and forbs, mixed with a light cover of trees and shrubs. This is the *savanna,* and although strict interpretation of the word "savanna" refers to a tropical vegetation, this biochore also includes a similar formation in the midlatitudes, called *parkland.* In general, the abundance of savanna grasses results from the strong drought conditions of the winter season, which limit the establishment of a forest cover, but as we suggested earlier, there are likely several other contributing factors as well, in particular, fires set by humans (see Fig. 21.4).

Tropical savanna. The savanna of the tropics is one of the celebrated vegetative formations of the world. Due in no small part to movies, travelogs, and magazines, we have come to stereotype this formation by the tall-grass savanna of Africa (see Fig. 21.2). Here extensive areas of grass such as elephant grass, which may reach heights of 3 m, are interrupted by scattered individual trees or groves of trees. The trees are umbrella-shaped and, like the grasses, tend to flourish in the wet season. During the dry season, by contrast, the grasses are brown, and the trees, which are drought-tolerant, may lose part or all of their foliage. This is the time when herders may burn large tracts of the savanna in an effort to improve grazing quality and to enlarge the grasslands at the expense of nearby forests.

Thornbush and tropical scrub. In many regions with a wet-dry tropical climate, short, thorny trees and shrubs are found in place of the classical savanna trees and grasses. This vegetation may form a nearly continuous cover, thereby eliminating most grasses, or it may be broken, allowing grasses and other herbs to fill the intervening space (see Fig. 21.2). The most impoverished thornbush formations are those where only barren soil is present between the woody plants. Generally, the thornbush and scrub savannas are thought to be responses to longer and more intensive dry seasons, but there are undoubtedly many other contributing factors in various regions, including fire and cultivation. Many different regional names are given to this formation; for example, in northeastern Brazil it is called the *caatinqa,* and in South Africa it is called the *dornveld.*

Parkland. Parkland can be described as prairie that is broken by patches and ribbons of broadleaf trees. It was apparently fairly common in Kentucky and Illinois when these areas were settled in the early 1800s. In both states most of the forest groves were subsequently destroyed as the land was converted to agriculture. A parkland formation, which the English call *deerpark,* is found in southern England, northwest France, and other regions of western Europe where pasture land is partitioned by hedgerows and interspersed with small groves of deciduous trees and shrubs. In a general sense, the abandoned farm fields in the eastern half of the United States also qualify as parkland, since they too are a mixture of grasses, forbs, trees, and shrubs.

Grassland Biochore

The great regions of grasslands in the world are located in the midlatitudes between the forests and the deserts. Summer drought is strong enough to prohibit the growth of trees and shrubs, but not severe enough to prohibit the growth of abundant grasses and forbs.

Prairie. Prairie is found on the humid side of the grassland biochore, where the annual moisture balance is just about even. Prairie consists mainly of tall grasses which grow over a lighter cover of smaller forbs. Trees and shrubs are not absent altogether, but are limited to depressions such as stream valleys and floodplains, where water is more plentiful. In North America grasses such as big blue stem (*Andropogon gerardi*) and little blue stem (*Andropogon scoparius*) were the predominant plants in the famous prairie of Illinois, Iowa, and adjacent states. Virtually all of this original cover, however, has been replaced by agriculture. This is also the case with the other prairies of the world, for example, the *Pampa* of Argentina and the *puszta* of Hungary.

Steppe. Farther toward the desert, the moisture balance grows poorer and the grass cover shorter and thinner. This formation is sometimes termed short-grass prairie, but it is more widely known by the Russian word *steppe*. In North America it consists of grasses such as buffalo grass (*Buchloe dactyloides*) and black gramma grass (*Bouteloua eriopoda*), which tend to grow in bunches that are often separated by barren soil. The most extensive areas of short-grass prairie are found in the Soviet Union and the Great Plains of North America.

Tundra and alpine meadow. Beyond the boreal forests of the subarctic and above the tree line in mountainous areas there may occur a grassy, treeless prairielike formation called, respectively, tundra and alpine meadow. Instead of a moisture limitation on tree growth, it is a lack of heat that prohibits establishment of a forest cover. In fact, most of the tundra is underlain by permafrost (see Fig. 3.7). Short grasses, such as cottongrass (*Eriophorum*) and arctic meadow grass (*Poa arctica*), and forbs are abundant. In protected sites, shrubs and dwarf trees can sometimes be found. Because of the geographic diversity and the dynamic character of arctic and alpine environments, grassy tundra and alpine meadow have localized distributions in many areas, particularly in mountain regions.

Tundra formation.

Desert Biochore

Dry desert. Desert formations are the lightest vegetative covers on earth and can generally be divided into those having virtually no plant cover and those having a conspicuous cover of shrubs and herbs (see

Fig. 21.2). In the harshest deserts, such as the Atacama of northern Chile, where in places measurable precipitation has never been recorded, the landscape is virtually barren of plants. Only in select microenvironments are a few plants able to survive, but they are very small and isolated. Moreover, survival is dependent on either special physiological traits, such as the succulent habit, or on special life cycle habits, such as ephemeralism. But even with the special adaptations, plant growth and reproduction in the desert are exceptionally slow, despite the implications conveyed by films such as Disney's *The Living Desert.*

In addition to severe drought, geomorphic factors can also restrict plant growth in dry deserts. Typically, desert surfaces are very active, owing, not surprisingly, to the absence of a strong plant cover to hold the soil in place. Running water is the most powerful agent, and during those infrequent periods of runoff, it is capable of eroding and depositing massive amounts of material, including plants. Wind is also an effective erosional agent in deserts. Active sand dunes, such as those of the Empty Quarter of Saudia Arabia, shift so rapidly that they preclude the establishment of plants altogether. In other areas severe wind stress and the absence of soil over the bedrock may be important limiting factors (see Fig. 21.4).

Shrub desert.

Shrub desert. At the other extreme are deserts such as those in the American West, which the respected plant geographer Nicholas Polunin calls *near-deserts*. These are characterized by diverse plant forms ranging in some places from sahuaro cactus (*Carnegia gigantea*), which reaches heights of 10–15 m, and various shrubs at heights of 1–2 m, to tiny forbs barely 3 or 4 cm above the ground. Together they may cover 10–20 percent of the ground, and in special locales, e.g., along dry river beds, where moisture is more plentiful, the coverage may be substantially greater. Although xerophytic vegetation represents one of the most interesting collections of plants in the world, it is not highly diversified floristically. Compared with forest formations at the same latitude, there are fewer types of plants in most deserts. Of these, cacti are probably the most famous, but they are limited to the New World. Other plant groups prevalent in deserts are the *Euphorbiales*, which often resemble cacti in their thorny, fleshy bodies, and the lily family (*Liliaceae*), which includes many desert palms and small succulents.

Fell fields. Some polar and high-mountain regions contain areas with a light cover of lichens, mosses, and flowering plants that resemble desert formations. Fell fields are geomorphically active sites where there is a heavy concentration of rock fragments. They may be places where frost action or deposition of rock debris is especially pronounced. Coupled with the harsh polar or high mountain climate, these sites limit plants to scattered patches which together constitute less than 50 percent coverage. Antarctica and Greenland, plus portions of insular Canada, Alaska, and Siberia, are occupied by fell-type landscapes.

SUMMARY

The major zones of terrestrial vegetation fall into four main biochores: forest, savanna, grassland, and desert. The global patterns of these biochores are mainly a response to climate, but human activity has undoubtedly been important in shaping the tropical savanna and the forests and grasslands of the midlatitudes. Forest cover the world over has been greatly reduced by humans, but extensive tracts still remain in the equatorial and tropical zones, the humid midlatitudes, and the subarctic. Savanna formations are most extensive in the wet/dry tropics, and the largest tracts of grassland are located in the semiarid zones of the midlatitudes. Tundra is limited to the Northern Hemisphere poleward of the boreal forests. The lightest plant covers are found in the deserts and polar regions, which together occupy about 25 percent of the world's land area.

UNIT V SUMMARY

- Plant geography, a branch of physical geography, is concerned with: (1) the distribution, form, and composition of the plant cover; and (2) the interrelationships among the plant cover and soils, climate, water, land use, and geophysical forces.

- There are three main groups of vascular plants: angiosperms, gymnosperms, and pheridophytes. The angiosperms are flowering plants and form the dominant cover in most major landscapes.

- The floristic classification scheme organizes the plant kingdom according to genetic affiliations. Other classification schemes are based on life forms and ecological relations; however, only the floristic scheme is universally recognized.

- Photosynthesis is the process by which plants convert energy in the form of light into chemical energy in the form of plant materials. Photosynthesis is dependent on heat, carbon dioxide, and water in addition to light and certain minerals. The maximum rate of photosynthesis obtainable is fixed by whichever of these resources is in least supply.

- Production is the rate of output of organic material by the plant cover. Terrestrial production is highest in the equatorial and tropical latitudes, where the supplies of the basic requirements for photosynthesis are generally greatest and least variable over time.

- Overall, vegetation utilizes relatively little of the solar energy received by earth; nonetheless, this energy represents the sole source of food for all other organisms on the planet.

- For any plant to survive, it must be able to withstand the various stresses and disturbances produced by the environment in which it is located. Tolerance varies with the phases in the plant's life cycle, and at the same time, the magnitude and frequency of stress and disturbance also vary.

- Plants can improve their chances of survival through adaptation. Genetic adaptation takes place by mutation, natural selection, and genetic drift and is exemplified by habits such as ephemeralism, annualism, and succulence.

- Global and regional variations in the makeup of the plant cover are related to many factors, including current climate, past events such as continental glaciation, human activities past and present, and the availability of different plants. The influences of geomorphic process, topography, and land use become apparent in plant distributions over much smaller areas.

■ The community-succession concept is based on the observation that plant communities have the capacity to alter the environment, making it suitable for other communities. The individualistic concept explains change in the plant cover on the basis of probability of recurrence of individual plants rather than of entire communities. According to the stress-threshold concept, the distribution of a plant population fluctuates in response to the magnitude and frequency of the environmental forces in its part of the landscape.

■ Forest, savanna, grassland, and desert are the principal biochores of the world. Extensive tracts of forest are located in the wet regions of the equatorial and tropical zones, the midlatitudes, and the subarctic. Savanna vegetation occupies much of the wet/dry tropics, and the large areas of grassland are located in the semiarid zones. Tundra, a grassland formation, is found on the poleward margin of the boreal forest. Desert formations include sparse shrub covers, fell, and dry deserts which are essentially devoid of plants.

FURTHER READING

Billings, W. D., *Plants, Man and the Ecosystem,* 2d ed., Belmont, Calif.: Wadsworth 1970. *A short introduction to plant ecology which highlights the biological, geophysical, and geographical aspects of the subject.*

Polunin, N., *Introduction to Plant Geography,* London: Longmans, 1967. *A comprehensive book on plant geography which includes adaptation, dispersal, evolution, economic plants, and the global distribution of vegetation.*

Raup, Hugh M, "Trends in the Development of Geographic Botany," *Annals Association American Geographers* **32,** 4 (1942): 319–354. (Reprinted by Bobbs Merrill.) *A good review of the development of plant geography and plant ecology through 1930.*

Watts, D., *Principles of Biogeography,* New York: McGraw-Hill, 1971. *A detailed survey of biogeography using an ecosystem approach.*

REFERENCES AND BIBLIOGRAPHY

Anderson, E., "Hybridization of Habitat," *Evolution* **2** (1948): 1–9.

Arno, S. F., and J. R. Habeck, "Ecology of Alpine Larch in the Pacific Northwest," *Ecology,* 1972.

Carpenter, P. L., T. D. Walker, and F. O. Lanphear, *Plants in the Landscape,* San Francisco: Freeman, 1975.

Clements, F. E. "Nature and Structure of the Climax," *Journal of Ecology* **24,** 1 (1936): 253–284.

Dansereau, P., *Biogeography: An Ecological Perspective,* New York: Ronald Press, 1957.

Eiseley, L., *Darwin's Century*, Garden City, N.Y.: Anchor Books, 1961.

Gleason, H. A., "The Individualistic Concept of Plant Association," *American Midland Naturalist* **21,** 1 (1939): 92–108.

Grime, J. P., *Plant Strategies and Vegetation Processes*, Chichester, England: Wiley, 1979.

Küchler, A. W., *Vegetation Mapping*, New York: Ronald Press, 1967.

MacArthur, R. H., and E. O. Wilson, *Theory of Island Biogeography*, Princeton, N.J.: Princeton University Press, 1967.

Odum, H. T., "Trophic Structure and Productivity of Silver Springs, Florida," *Ecological Monographs* **27** (1955): 55–112.

Polunin, N., *Introduction to Plant Geography*, London: Longmans, 1967.

Raunkiaer, C., *"The Life Forms of Plants and Statistical Plant Geography,"* Oxford, England: Clarendon, 1934.

Raup, H. M., "Species Versatility," *Journal of the Arnold Arboretum* **56** (1975): 126–163.

Ridley, H. N., *The Dispersal of Plants Throughout the World*, Ashford, England: L. Reeve, 1930.

Ritchie, D. D., and R. Carola, *Biology*, Reading, Mass: Addison-Wesley, 1979.

Santamour, F. S., H. D. Gerhold, and S. Little, eds., *Better Trees for Metropolitan Landscapes*, Washington, D.C.: U.S. Forest Service, 1976.

Treshow, M., *Environment and Plant Response*, New York: McGraw-Hill, 1970.

Wilsie, C. P., *Crop Adaptation and Distribution*, San Francisco: Freeman, 1962.

THE
STRUCTURE,
COMPOSITION,
AND
DEFORMATION
OF THE
EARTH'S CRUST

UNIT VI

KEY CONCEPTS   OVERVIEW

The rock platform on which the landscape and the oceans rest is the earth's crust. The rock composition and the form of the crust are highly variable from place to place on the continents. Since the crust influences the formation of soil, drainage features, climate, and vegetation, it is a very important component of the landscape, even though it is not always apparent in the landscape.

The crust is the upper section of the lithosphere, a zone of solid rock about 100 kilometers thick that envelops the entire planet. Immediately under the lithosphere lies a zone of partially melted rock having a somewhat plastic consistency. Driven by heat flowing from the earth, this rock flows very slowly; as it does, it exerts stress on the overlying lithosphere. The result is deformation of the lithosphere in the form of faulting (breaking and slipping), folding (bending), and volcanism.

Deformation of the lithosphere is not indiscriminately distributed over the planet. Rather, it is concentrated along the mountainous margins of the continents, along the major island arcs, and along the mountain chains on the ocean floors. These corridors of intensive geologic activity mark the edges of huge sections of lithosphere, called tectonic plates. The crust is partitioned into seven major plates, and each, with the exception of the Pacific plate, includes a continent.

The results of the first waves of scientific tests suggest that tectonic plates behave as independent units of the lithosphere, moving in different directions and at different rates. Neighboring plates may collide, draw apart, or slip alongside one another, producing earthquakes, faulting, folding, and volcanism.

Chapter 22 opens with a description of the gross topography and composition of the ocean basins and continents and follows with an examination of the nature of earth's interior and the rocks and minerals of the crust. Plate tectonics is described in Chapter 23, including the motions of the plates, the dynamics of the plate borders, and the apparent source of the energy which drives the plate movement. Chapter 24 explores the rock structures that comprise the mountain chains, namely, various folds, faults, and volcanic features, and some of their manifestations in the landscape.

CHAPTER 22

THE STRUCTURE AND MATERIALS OF THE EARTH

INTRODUCTION

The landscape rests on a platform of solid rock called the *crust*. Nowhere on the planet is the crust truly stable, and in many areas, especially along the margins of some of the continents and in the interiors of most oceans, it is particularly unstable. By unstable we mean that it shifts about frequently, say, often enough that one might expect to feel or to see shifts several times in a lifetime.

Crustal instability is displayed in four main ways: (1) sudden drops, rises, or lateral slips of huge sections of rock; (2) shaking of the ground, which usually accompanies the sudden movements; (3) outpourings of melted rock; and (4) slow bending of massive slabs of rock. We know about these movements because they have been observed and measured innumerable times throughout human history. We also know on the basis of recent scientific tests that the earth is very old, at least 4.5 billion years. Scientists also think that in the past, the rate of movement in the crust was probably comparable to the rate of movement today. Thus if we multiply the current rate of crustal movement by the age of the planet, we can gain a crude indication of the tremendous diversity that should exist in the crust beneath the landscape. Examinations of rock exposed on mountain sides, in mine shafts, and in the cores extracted from oil and gas wells support this contention and indeed often suggest that the crust is more diversified than we can imagine.

It is the diversity of the crust, as well as its instability, that make it so important in landscape formation and hence of great concern to physical geography. Physical geographers are interested in three major questions about the crust:

1. What is the configuration and composition of the crust where it meets the landscape?

2. What are the passive roles played by crustal form and composition in the development of soil, water features, vegetation, and climate (Fig. 22.1)?

3. What are the active roles played by the manifestations of crustal instability, such as volcanism and earthquaking, in shaping the landscape (Fig. 22.1)?

329

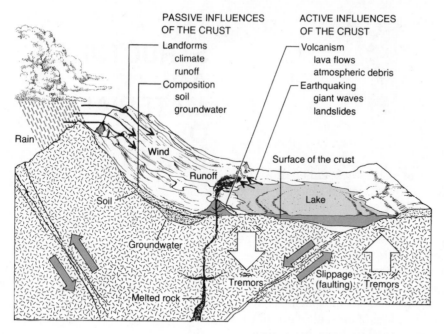

PASSIVE INFLUENCES
OF THE CRUST

Landforms
climate
runoff

Composition
soil
groundwater

ACTIVE INFLUENCES
OF THE CRUST

Volcanism
lava flows
atmospheric debris

Earthquaking
giant waves
landslides

Rain

Wind

Surface of the crust

Runoff

Lake

Soil

Groundwater

Tremors

Slippage
(faulting)

Tremors

Melted rock

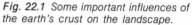

Fig. 22.1 Some important influences of the earth's crust on the landscape.

This unit outlines some of the elementary geophysical knowledge that can help us to understand not only the basic structure and composition of the crust, but also why the crust is unstable and how it acquired its present features. Part of our discussion will encompass elementary geology and geophysics. Although geophysics is traditionally not the territory of physical geography, this field in the past several decades has so significantly contributed to our understanding of the earth that its findings now form the backbone of modern earth science. The keystone of these findings is the theory of *plate tectonics*, or *continental drift*, as it is more popularly known.

THE STAGE: THE SHAPE OF THE EARTH'S SURFACE

The Ocean Basins

In order to establish a framework for study, we may begin at the broadest scale with a brief description of the major topographic, structural, and compositional units of the earth's surface and subsurface. If we eliminate ocean water for the moment, two major earth surface structures are immediately apparent: the continents and the ocean basins. The ocean basins occupy nearly 65 percent of the planet's surface and

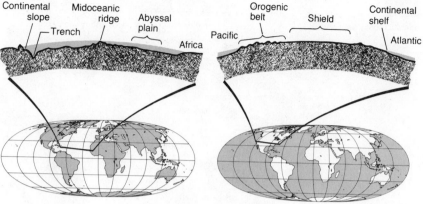

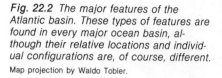

Fig. 22.2 The major features of the Atlantic basin. These types of features are found in every major ocean basin, although their relative locations and individual configurations are, of course, different.
Map projection by Waldo Tobler.

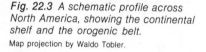

Fig. 22.3 A schematic profile across North America, showing the continental shelf and the orogenic belt.
Map projection by Waldo Tobler.

are comprised of several subdivisions. The largest of these in terms of area is the *abyssal plain*, or deep ocean floor; two smaller units are the *midoceanic ridge*, or mountain range, and the *ocean trench*, or deep, which does not appear in all oceans. On the margins of the ocean basins are the *continental slopes*, which lead from the abyssal plains up to the continental shelf, the outer, submarine margin of the continent (Fig. 22.2).

The Continents

The continents are comprised mainly of large, generally low-relief interiors called *shields*. These features are often characterized by broad dome shapes—hence the term "shield"—but on some continents large sections have a somewhat inverted shape. In North America, for instance, the central part of the Canadian Shield in the area of Hudson Bay forms a huge depression, the center of which is several thousand feet lower than the periphery (Fig. 22.3). On the margins of the shields are *orogenic belts* and *continental shelves*.

Orogenic belts are extensive mountain chains, such as the Rockies of North America and the Andes of South America. The highest orogenic belt is the Himalaya, where literally hundreds of peaks exceed an elevation of 6000 meters above sea level. Figure 22.4 shows the distribution of shields and orogenic belts for the large land masses of the world. Note that each shield has more or less the geographic aspect of a nucleus of a continent. The shields, in turn, are often fringed by orogenic belts.

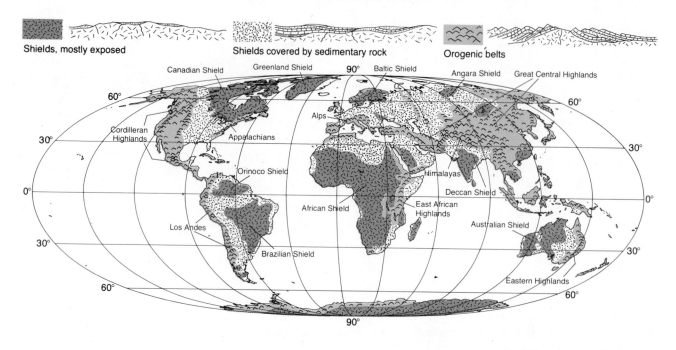

Shields, mostly exposed   Shields covered by sedimentary rock   Orogenic belts

Fig. 22.4 The major shields and orogenic belts of the great land masses of the world. (After maps by Murphy, 1968; Holmes, 1965; and Verhoogen et al., 1970)

Map projection by Waldo Tobler.

The continental shelves slope seaward from the continents under shallow water and together comprise about 5 percent of the earth's surface area. They range in width from several hundred kilometers, as in the Grand Banks near Newfoundland, to only several kilometers, along mountainous coastlines. At the outer edge of the continental shelf, the continental slope gives way to the deep ocean floor and in some locations to trenches situated near the foot of the slope. The continental shelves are becoming increasingly important as sources of food and minerals; Fig. 22.5 shows their distribution in the world.

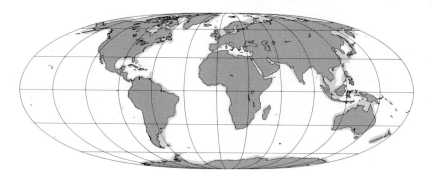

Fig. 22.5 The global distribution of continental shelves. The width of the shelf varies from several kilometers to several hundred kilometers and thus may have great geopolitical implications as nations increase their reliance on the continental shelves as sources of food and minerals.

Map projection by Waldo Tobler.

The Gross Topography of the Earth

Because we live within ten feet or so of the ground, the configuration of the planet's surface often seems to us extremely varied. Recently, however, with the perspective afforded us by spacecraft and high-altitude aircraft, we have often been impressed by the apparent smoothness of the earth's skin. And indeed, most of it is very smooth, more than 70 percent of it being water.

Actually, many of the largest topographic features of the earth are obscured by the oceans and have never been seen. The ocean trenches typically reach depths of 7500 meters or more below sea level. The deepest known is the Mariana trench, near Guam in the Pacific Ocean, which reaches about 11,000 meters below sea level and about 5000 m below the surrounding ocean floor. The most obtrusive topographic features of the ocean basins, however, are the great mountains that rise from the sea floor and the oceanic ridges. Many mountain islands exceed a height of 6000 m above the ocean floor, and some, such as Mauna Loa of Hawaii, exceed a height of 10,000 m. Geologists identify three kinds of oceanic mountains: (1) those that form island arcs, e.g., Japan; (2) those that rise as midoceanic ridges, e.g., Iceland; and (3) those that rise from the abyssal plains, e.g., Hawaii. The latter are called *seamounts*, and although those that form islands are very large, most seamounts do not reach more than 3000 m or so above the ocean floor.

In terms of topographic extremes the continents are generally less impressive than the ocean basins. The greatest difference is 9243 m (between Mt. Everest at +8848 m and the Dead Sea at −395 m), but these points are widely separated geographically. More typically, the extremes in and around the world's major mountain ranges are on the order of 3000–4000 m, roughly comparable to the differences between trenches and neighboring abyssal plains.

The prize for extremes in topography for points less than 500 km apart, however, must go to the combined areas of the ocean rims and continental coasts. Within this narrow ribbon of terrain lie both ocean trenches and high continental mountain ranges. Together these features commonly span a vertical distance of 12,000 m or more. Figure 22.6 presents some examples from around the world. Although these elevations look impressive, note that the average rate of slope is only about 2–3 m of rise per 100 m of horizontal distance. This is comparable to steep inclines on United States interstate expressways.

From the elevation maps of the continents and ocean basins, we can develop a more generalized topographic picture for the earth as a whole. How much surface area lies at different elevations is an interesting and meaningful piece of information, for it tells us something

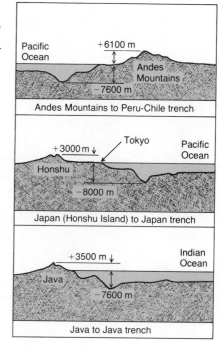

Fig. 22.6 Representative elevation differences between trenches and adjacent mountain ranges.

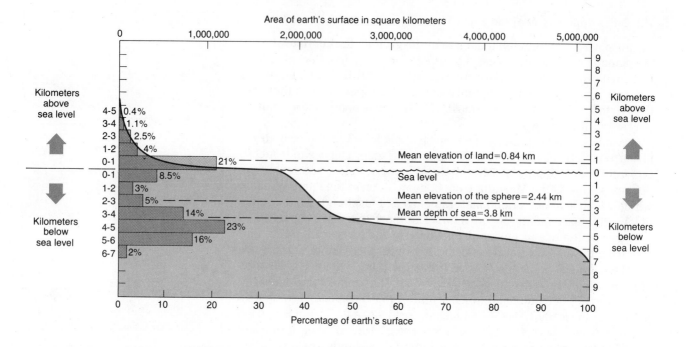

Fig. 22.7 A summary of the topography of the earth, including the areas above and below sea level. Note that most of the ocean basins lie between −3 km and −6 km, whereas most of the continental areas lie between sea level and 1 km. (Based on H. Sverdrup et al., The Oceans: Their Physics, Chemistry, and General Biology, Englewood Cliffs, N.J.: Prentice-Hall, 1942.)

about not only the distribution of the rock mass in the crust, but also the habitable area on the planet for humans and other life forms. The graph in Fig. 22.7 is a statistical summary of the distribution of the crustal surface according to elevation. The median elevation of the crust, corresponding to 50 percent on the graph, is well below sea level. Thus by virtue of their higher position, the materials of the continents tend to move toward the ocean basins. If, for example, the continents were made of a substance with the consistency of mud, they would creep into the ocean basins and stabilize at an elevation corresponding to the present depth of about 3300 meters. In that event, the oceans would rise with the volume of the mass displaced and would cover the entire planet.

The distribution of land elevations reveals that most of the land is near sea level. About 71 percent of the earth's land area lies between sea level and a height of 1000 m, and about 13 percent is higher than 2000 m. The latter figure is noteworthy because the conditions necessary to support life decrease rather sharply above 2000 m. If a larger percentage of land were at high altitude, the life-support capacity of the planet would definitely be lower than it is presently. See Authors' Note 22.1.

AUTHOR'S NOTE 22.1
Humans and Altitude

Human beings occupy a wider range of environments than do any of the other land-bound creatures. This is particularly so with respect to altitude. Currently, humans inhabit on a permanent basis a zone ranging from the lowest point on earth, the Dead Sea basin, at 395 m below sea level, to a point 5304 m (17,400 feet) above sea level, in the Andes Mountains of southern Peru. Although modern mountaineers with special survival apparatus have survived at altitudes above 7000 m for as long as two months, there is no evidence that people have ever lived permanently above this altitude.

What limitations do high elevations pose to humans? The most severe are related to climate. Ultraviolet radiation, which in large doses is toxic to the human skin, is very intensive at high altitudes. Temperature declines with altitude at an average rate of .65°C per 100 m; therefore, at 6000 m, air temperature averages around 40°C colder than at sea level. Air pressure and oxygen also decline with altitude, and at 6000 m, both are about half the sea-level quantity.

Although the effects of ultraviolet radiation and low heat can seriously limit human survival at high altitudes, it is low oxygen that places the greatest stress on the human physiology. In order to meet its oxygen demands, the body accelerates respiration. Breathing is deep and rapid, and above 6000 m, the human body is barely able to maintain a steady condition when at rest. The body adapts to this low-oxygen environment by increasing the number of the oxygen-carrying red blood cells and opening new capillaries. This generally requires about a week, and new arrivals at high altitude often experience severe discomfort, including severe headaches, sleeplessness, giddiness, and vomiting. More severe problems sometimes occur as well. High altitude causes an increase in the pressure of the cerebrospinal fluid, which bathes the brain. Cell walls of the capillaries in the brain and in the retina of the eye swell, often resulting in tiny hemorrhages. The result is called cerebral edema and can lead to rapid death if the afflicted person is not brought to lower altitude. High altitude can also cause fluid to seep into the alveolie of the lungs, resulting in pulmonary edema, a disorder characterized by coughing, breathlessness, gurgling in the chest, bloody sputum, and often death, unless the victim is brought to lower altitude.

It is also apparent that people cannot adapt permanently to altitudes above about 5500 m. Without an artificial supply of oxygen, the maximum period that can be spent above 6000 m is about 100 days; above 7000 m, 10 days; above 8000 m, 2 days. In South America attempts to induce miners to work and live at 5800 m failed. The miners preferred to live lower, at about 5000 m, and to travel to the mines when work required it. Among these very high-altitude dwellers in South America, full-term pregnancies are less frequent, and rates of mental retardation among children born at high altitudes are much higher than among those from mothers who travel to lower elevations for their pregnancies.

THE CRUST

The continents and the ocean basins together extend well below the surface and make up the earth's crust. Relative to the earth's 12,742 km (7918 miles) diameter, the crust is very thin, ranging in thickness from 8 to 65 km (5 to 40 miles). The crust is often described as the "skin" of the planet, which is especially appropriate when we realize that its thickness represents only about 0.005 times the radius of the earth. The geologic processes that change the earth's surface, such as volcanism and faulting, originate largely in the crust and in the upper part of the thick subcrust, called the *mantle*.

The Sima Layer

The crust can be segregated into two zones on the basis of rock types. The basal zone, called the *sima layer* (for the elements Silicon and Magnesium), is composed of a heavy, dark group of rocks called the *basaltic* rocks. The large amounts of iron and magnesium contained in these rocks give them fairly high densities, i.e., high masses per unit volume of rock. The densities of rock in the sima generally range from 2800 to 3300 kilograms per cubic meter. (Water has a density of 1000 kg/m^3; therefore, for a given volume, these rocks are 2.8 to 3.3 times heavier than water.) The sima layer circumscribes the entire planet, but is exposed only in the ocean basins. Elsewhere it is covered by the upper zone of the crust, called the *sial* layer (for the elements Silicon and Aluminum).

The Sial Layer

The sial layer occupies about 35 percent of the earth's surface, in places reaching a thickness of 40 km, and forms the continents. Measurements by geophysicists indicate that the rock of sial is lighter in weight than that of the sima, with densities generally in the range of 2700–2800 kg/m^3. This is due to the fact that the heavier elements of iron and magnesium are less abundant in the rock of the sial, which is made up mostly of silicon, aluminum, and other light elements. Granite is the representative rock type of the sial layer, and it has been traditional to refer to sialic rock in general as "granitic rock."

The sial should not be thought of as a flat, homogeneous slab resting on the sima. Actually, both its composition and structure are highly

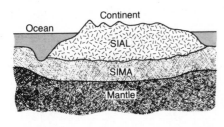

Fig. 22.8 Summary of the major features of the sial and sima.

	POSITION	SURFACE EXPRESSION	THICKNESS	COMPOSITION	DENSITY ks/m^3	LOWER BOUNDARY	GEOGRAPHIC COVERAGE
SIAL	Upper	Continents	0–50 km (0–30 miles)	Granitic	2700–2800	Upper sima (transitional)	Approximately 35% of earth's surface area
SIMA	Lower	Ocean basins	8–16 km (5–10 miles)	Basaltic	2800–3300	(Upper mantle)	Approximately 65% of earth's surface area exposed

diversified. The lower boundary appears to be highly irregular and is characterized by a transitional rather than an abrupt change to the upper sima. In composition, it includes large amounts of basaltic rocks, similar to those of the sima, that reside amongst the granitic rock. In sharp contrast, the sima, judging from the records of rock samples taken from the ocean basins, appears to be almost exclusively basaltic. Recently, a small exception was reported, however, when scientists discovered an area of granitic rock in the South Atlantic at a depth of about 3000 meters below sea level. Located about 2550 km (1600 miles) east of the South American mainland, this unique find is evidence of continental drift, a topic examined later. Figure 22.8 summarizes the major characteristics of the earth's crust as they are known to geologists today. Bear in mind that the continental crust consists of both sial and sima, whereas the oceanic crust consists of sima alone.

BELOW THE CRUST

Exploring at Depth

Geophysicists have learned about the composition and structure of the crust and subcrust by analyzing energy waves transmitted through the earth. The concept of energy waves in the earth can be envisioned by using the analogy of a struck bell. When a bell is struck, sound waves radiate from it. At the same time, waves travel through the solid part of the bell, and you feel these waves as vibrations. The waves are transmitted within the bell through the bumping together of myriads of atomic particles. As a wave travels along, successive particles bump into their neighbors and then return to their original positions. Because the particles always return to their original patterns after displacement, these waves are called *elastic waves*. The elastic waves studied in the earth are produced by sudden movement, or faulting, of massive quantities of rock or by artificial means such as an explosion from dynamite or a nuclear device. The branch of geophysics that studies earth waves, or tremors, is called *seismology*.

Seismic energy produces several types of waves, but two are particularly important in seismic analysis: *compressional waves* and *shear waves*. As compressional waves travel through rock, they generate a back-and-forth motion parallel to the direction in which energy is broadcast. Shear waves cause neighboring particles to move in directions perpendicular, or transverse, to the direction of energy transmission (Fig. 22.9). These motions can be detected with devices called seismometers and recorded with an associated instrument called a seismograph.

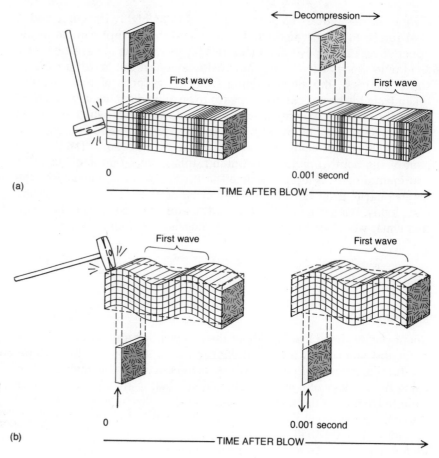

Fig. 22.9 Seismic waves: (a) Compressional P waves produce a back-and-forth motion along the line of energy propagation; as the wave passes through a material, this motion compresses and decompresses particles; (b) Shear (S) waves move in a direction that is transverse to the energy propagation and characterized by a rolling action. (Adapted from Owen M. Phillips, The Heart of the Earth, San Francisco: Freeman, Cooper, 1968. Used by permission.)

On the basis of the behavior of seismic waves as they pass through a mass of rock, geophysicists are able to learn some things about the composition and structure of the mass. For example, wave velocity generally increases with rock density; shear waves cannot penetrate molten masses; and when waves strike a boundary between two rock layers of different densities, part of the energy travels along the boundary and eventually back to the surface. The discovery of the base of the crust itself was accomplished through careful analysis of seismic waves obeying the latter principle. A Yugoslavian geophysicist, Mohorovičić (pronounced Mo-ho-ro-vee-chich), found that wave velocities increased rather sharply between what we now call the crust and the underlying zone, termed the mantle. The discontinuity has come to be

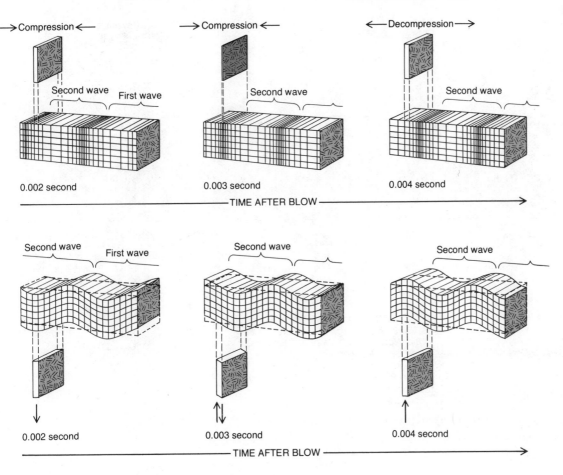

known as the M discontinuity, or more commonly, the "Moho" discontinuity.

The Moho Discontinuity

Using seismic techniques, geophysicists have determined the depth to the base of the crust in many regions of the world. The exact physical nature of this abrupt seismic discontinuity has been the subject of keen interest since its discovery. Some scientists hold that it is the point where sima-type minerals undergo an atomic reorganization, which at greater depths produces a new and much denser assemblage of minerals. Atomic reorganizations of this kind are called *phase*

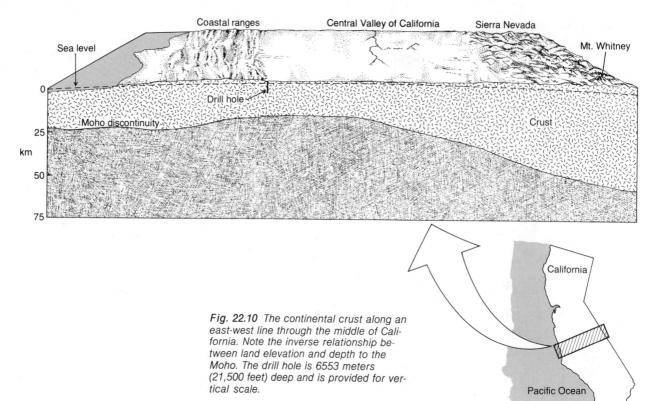

Fig. 22.10 *The continental crust along an east-west line through the middle of California. Note the inverse relationship between land elevation and depth to the Moho. The drill hole is 6553 meters (21,500 feet) deep and is provided for vertical scale.*

transitions. The change of water from ice to liquid and of carbon from graphite to diamond are examples of phase transitions. A phase transition does not entail a change in the chemical composition of the material. Other scientists, however, believe that the Moho reflects a drastic change in the chemical composition of the rock. These two explanations of the Moho are completely different, yet each accounts for the nature of the seismic observations. With increased research, it is becoming more apparent, however, that the Moho probably represents an abrupt change in chemical composition of the rocks.

Seismic data from all parts of the world indicate that the configuration of the Moho varies with the thickness of the crust. Where the crust is thick, as under the continents, the Moho shows a broad downward flexure; the opposite appears to be true under the thin oceanic crust. Directly under belts of large mountains, the bottom of the crust actually forms subtle rootlike protrusions. These, of course, are the points of greatest crustal thickness and weight (Fig. 22.10).

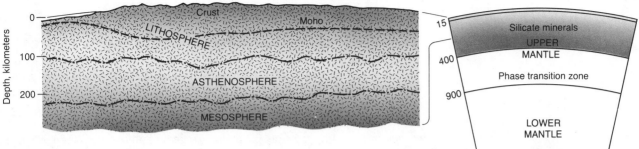

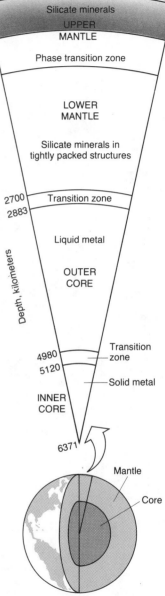

The Rest of the Earth: Mantle and Core

Just as the crust is defined by a seismic discontinuity, the remainder of the earth is subdivided on the basis of the changes in the velocity of seismic waves. The crust rests immediately on a massive zone called the *mantle*, which extends about 3000 km into the earth. The mantle contains about two-thirds of the total mass of the earth. For the most part, seismic-wave velocities increase steadily throughout this zone (Fig. 22.11b).

At the base of the mantle, a sudden change in wave velocities occurs. This marks the beginning of the *core*, which makes up the rest of the earth. Interestingly enough, the compressional-wave velocities *decrease* at the base of the mantle, and the shear-wave velocities drop to zero. Since shear waves cannot be transmitted through liquids, the outer portion of the core, which is about 2000 km thick, must exist in the liquid state. Seismologists have also determined that the innermost portion of the core is solid. Hence the core is divided into two units, the *inner core* and the *outer core* (Fig. 22.11b). The core contains one-third of the total mass of the earth and is composed of a material similar to iron. Due to the movement of the liquid iron in the outer core, the magnetic field of the earth is generated in this portion of the planet.

Returning to the mantle: It is further subdivided into the *upper mantle* and the *lower mantle*. The upper mantle, in turn, is subdivided into three smaller units. The uppermost 100 km, including the crust, is the *lithosphere*; the next 100 km or so is the *asthenosphere*; the remainder is the *mesosphere* (Fig. 22.11a). Both the lithosphere and mesosphere are solid, but the asthenosphere appears to consist of partially melted rock. Although it may actually be liquid at some points, most of the asthenosphere appears to have a consistency that approximates a weak solid.

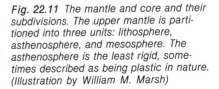

Fig. 22.11 The mantle and core and their subdivisions. The upper mantle is partitioned into three units: lithosphere, asthenosphere, and mesosphere. The asthenosphere is the least rigid, sometimes described as being plastic in nature. (Illustration by William M. Marsh)

ROCKS AND MINERALS: THE STUFF OF THE LITHOSPHERE

Now that we have examined the general structure of the earth, let us consider its composition. Our primary concern is with the rocks that are found at or near the surface of the continents, because these are the rocks that influence the formation of soils, the erosion of mountains, the availability of mineral resources, and so on. Subsurface rocks, especially those that occur at depths of a kilometer or more, are of secondary importance in physical geography, since they do not directly affect the landscape.

Although nearly 2000 minerals and dozens of rock varieties have been identified, relatively few of these are found in appreciable quantities at the earth's surface. Only one major rock type, basalt, comprises virtually all of the ocean basins. Five primary rock types occupy more than 90 percent of the continental area, in the following proportions:

Shale	52%
Sandstone	15%
Granite (and granodiorite)	15%
Limestone (and dolomite)	7%
Basalt	3%
Others	8%

If these rocks are lumped together according to mineral composition, the relative abundance of minerals in the continents is as follows:

Feldspars	30%
Quartz	28%
Clay minerals and micas	18%
Calcite and dolomite	9%
Iron oxide minerals	4%
Others	11%

Minerals

A *mineral* may be defined as a naturally occurring inorganic substance with a characteristic internal crystal (molecular) structure which is basically consistent in its physical and chemical properties from sample to sample. Not all minerals combine to form rocks. In fact, relatively few do.

The silicates are the most important group of minerals in the formation of rock. The silicate minerals are composed of a basic ion of silicon and oxygen combined with one or more elements. Depending on the particular elements that combine with the silicon-oxygen ion, the min-

Table 22.1 Silicate minerals (basic compound: element(s) + silicon-oxygen ion).

TYPE	IMPORTANT MEMBERS
Ferromagnesium silicates (silicate ion + iron and magnesium ions)	Olivine
	Hornblende
	Biotite mica
Nonferromagnesium silicates (also called aluminosilicates) (lack iron and magnesium ions)	Pyroxene
	Muscovite mica
	Feldspars
	orthoclase (potassic) plagioclase (sodic, calcic)
	Quartz

eral thus formed may range considerably in density and color and other features. Two subgroups of silicates are recognized by geologists: *ferromagnesian* and *nonferromagnesian* (or aluminosilicates). As the names indicate, these minerals are distinguished on the basis of the presence or absence of iron, magnesium, and aluminum (Table 22.1). Table 22.2 gives three examples of some other mineral groups and representative members that are fairly common in the upper crust.

Rocks

Rock is an assemblage of minerals. Rocks can be classified according to several different schemes. A traditional scheme used in geology is based on the manner and environment of rock formation. Under this scheme, three major divisions may be used for virtually all rocks: *igneous*, *metamorphic*, and *sedimentary*. Each of these has subdivi-

Table 22.2 Some nonsilicate minerals.

TYPE	BASIC COMPOUND	DESCRIPTION	COMMON EXAMPLES
Oxides	Element(s) + oxygen	Hardest minerals next to silicates	Limonite Hematite Magnetite Ice
Sulfides	Element(s) + sulfur	Commonly occurs with iron, silver, lead, zinc, copper, and mercury	Galena Pyrite Chalcopyrite
Carbonates	Element(s) + carbon-oxygen ion	Like silicates, basic ion is complex	Calcite Dolomite

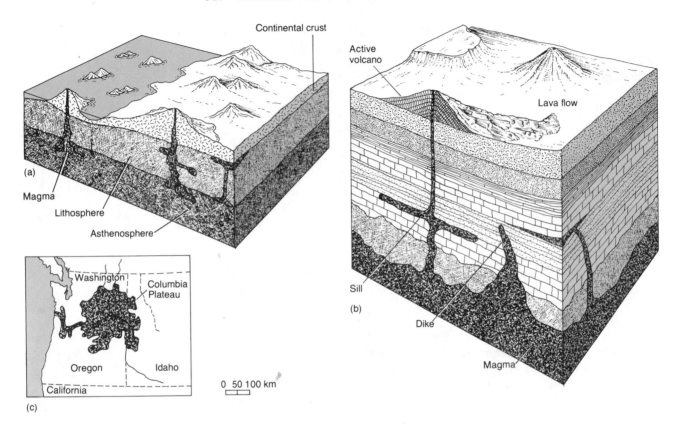

Fig. 22.12 *The magma of igneous rocks: (a) sources of magma in the lithosphere and asthenosphere; (b) from deep chambers, magma can make its way to the surface to form volcanoes and lava flows; (c) the Columbia Plateau, one of the largest areas covered by lava in North America.*

sions, which in turn may themselves have subdivisions—all an important part of the geologist's taxonomic system.

Igneous. Igneous rocks are composed of mineral crystals that form during the cooling of molten material. This material is called *magma*, and it originates in places of high temperature and pressure in the lithosphere and asthenosphere (Fig. 22.12a). Solidification of molten rock can occur at any level; in fact, most magma never reaches the surface.

In the lithosphere, magma is held in vast chambers from which channels, called *veins* or *dikes*, may radiate toward the surface. Where the crust is fractured, the hot, pressurized magma may extend dikes all the way through the crust and eventually flow forth on the surface as *lava* (Fig. 22.12b). If the lava builds up in a cone around the mouth of the dike, a volcano is formed; if it spills over the land in broad sheets, it is called *lava flow*. One of the largest areas of lava flows in North America is the Columbia Plateau of the American Northwest (Fig. 22.12c).

Geologists segregate igneous rocks into two main classes according to the depth of formation: (1) *extrusive* and (2) *intrusive* or *plutonic*. Extrusive rocks form at the surface, where the quick release of heat into the overlying air or water makes solidification rapid. Since the size of the crystals that form in the rock is inversely related to the rate of cooling, extrusive rocks are usually composed of microscopic-sized crystals and are thus known for smoothness of texture. By contrast, the magma that solidifies in dikes, veins, and related features forms *intrusive* igneous rocks. Because the surrounding rock insulates the magma, cooling is much slower than that of the extrusives, thereby allowing crystals to grow much larger. *Plutonic* rocks develop from the aggregation of large masses of magma called *batholiths*; thus cooling is even slower, resulting in the formation of huge crystals that may reach several centimeters in diameter.

The mineral composition of the newly formed igneous rock depends on the chemical composition of the magma, the geochemical environment through which it moved, and the heat and pressure present during solidification. If the magma is about 60 percent plagioclase feldspar and 35 percent ferromagnesian minerals, diorite or andesite may be formed. If the proportion of ferromagnesian minerals is 60–70 percent, gabbro or basalt may be formed. Together basalt and andesite comprise about 98 percent of all extrusive rocks. If the magma is composed primarily of quartz and 15–20 percent feldspar with an admixture of the heavy minerals mica and hornblende, the rock is likely to be granite or granodiorite. These rocks are intrusive and together comprise approximately 95 percent of all intrusive rocks (Fig. 22.13).

In Fig. 22.13 the important igneous rocks of the crust are classified according to mineral content and texture, i.e., crystal size. The principal members of the silicate group are given on the left; two textural classes representing intrusive and extrusive rocks are on the right. By combining these two features, we can identify two important trends in the igneous rocks. Note that granite, diorite, and gabbro each have a fine-grained extrusive counterpart. Although the mineral content of each pair is similar, the two rocks look and feel remarkably different. Furthermore, as we go from top to bottom of Fig. 22.13, we are actually reading the general change in rock types from the sial layer (granite and rhyolite) to the lower sial (diorite and andesite) to the sima (gabbro and basalt) to the lower lithosphere, represented by the very heavy rock peridotite.

Metamorphic. Since the temperature of the magma that invades the crust usually exceeds that of the crustal rock by hundreds of degrees centigrade, a great amount of heat is transferred into the resident

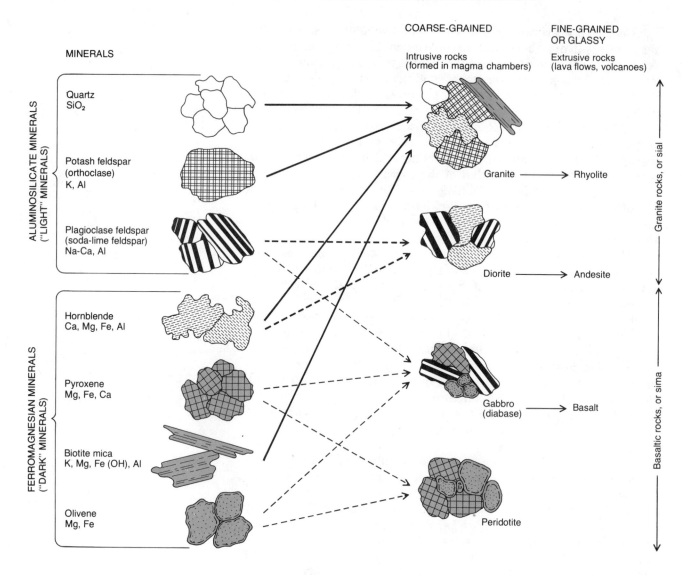

COARSE-GRAINED

FINE-GRAINED
OR GLASSY

MINERALS

Intrusive rocks
(formed in magma chambers)

Extrusive rocks
(lava flows, volcanoes)

ALUMINOSILICATE MINERALS
("LIGHT" MINERALS)

Quartz
SiO_2

Potash feldspar
(orthoclase)
K, Al

Plagioclase feldspar
(soda-lime feldspar)
Na-Ca, Al

FERROMAGNESIAN MINERALS
("DARK" MINERALS)

Hornblende
Ca, Mg, Fe, Al

Pyroxene
Mg, Fe, Ca

Biotite mica
K, Mg, Fe (OH), Al

Olivene
Mg, Fe

Granite → Rhyolite

Diorite → Andesite

Gabbro
(diabase) → Basalt

Peridotite

Granite rocks, or sial

Basaltic rocks, or sima

Fig. 22.13 Common intrusive and extrusive rocks produced by different combinations of light and dark silicate minerals. The drawings resemble the appearance of the minerals as you would see them under the high magnification of a polarizing microscope. (Adapted by permission from Arthur N. Strahler, Physical Geography, *2d ed., New York: Wiley, 1960.)*

rock, or *country rock,* as the old-time geologists called it. This heat may produce partial melting, increased pressure, and superheating of water in the surrounding rock. As a result, the crystal structure as well as the mineral composition of the country rock may be changed or metamorphosed. Such change may result from factors other than the heat of magma, including the heat and pressure produced in rock fracture zones. *Metamorphic* rocks, then, are those that have been changed due to heat, pressure, and related factors. In granite, for example, metamorphism produces a realignment of the minerals into bands, and the resultant rock is called *gneiss* (pronounced "nice"). Although metamorphosed igneous rocks are common, most of the metamorphic rocks found on the surface of the continents are derived from sedimentary rock, described below.

Sedimentary. Water, ice, waves, and wind as well as biological and chemical processes weaken and disintegrate the rocks on or near the surface of the crust. From this action, sediment residues such as clay, sand, and biological debris are produced and accumulate in lowlands, basins, and on the continental shelves. As layers of sediment grow to thicknesses of hundreds of meters, the lower layers are compressed and in time may consolidate into sedimentary rock.

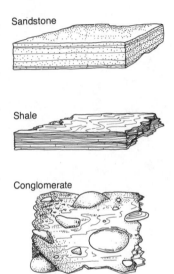

Sandstone

Shale

Conglomerate

Sedimentary rocks are divided into two main groups: *detrital* and *chemical*. Detrital rocks are formed of particles that have been transported to their resting place by erosional processes. Clay, silt, sand, and pebbles are the most common constituents of detrital rocks. These particles provide the key identifying feature for these abundant rocks, e.g., sand in sandstone, clay in shale, and particles of all sizes in a rock called conglomerate.

The chemical group of sedimentary rocks is produced mainly by the precipitation of minerals out of water. Limestone and dolomite are the most abundant chemical rocks. Limestone is formed mainly from either: (1) animal bodies and shells composed of calcium the creatures extracted from seawater; or (2) chemical precipitation of calcium directly from seawater. These two modes of limestone formation often occur together, making it difficult to assess the relative importance of each.

The origin of dolomite is disputed, but it is probably an altered form of limestone in which some of the calcium has been chemically replaced by magnesium. Also included in this group are rocks called evaporites. As the name implies, these are the residues left after the evaporation of mineral-rich water. Various kinds of salts, including rock salt, gypsum, and borax, are the most abundant evaporites.

Most sedimentary rocks form under the shallow waters on the continental shelves, where huge quantities of material eroded from the land are deposited by rivers. This material ranges in size from large particles, 25 cm or more in diameter, to fine clays, less than 0.002 mm in diameter, and the ions of dissolved minerals. When a river enters the ocean, it slows down and drops its sediment load, generally according to particle size. Large particles such as pebbles and sand are deposited nearest shore, whereas the clay particles, which can remain

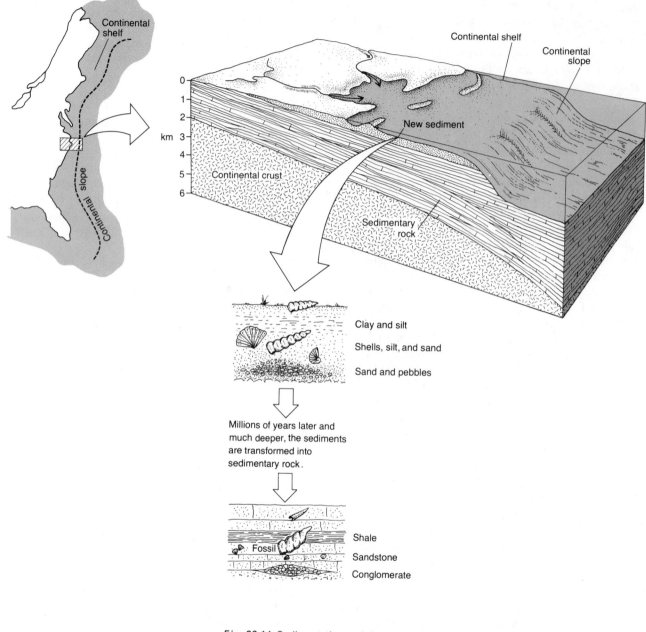

Fig. 22.14 Sedimentation and the formation of sedimentary rocks on the continental shelf. Both detrital and chemical go into this sediment mass.

suspended in sea water for many days or weeks, are often deposited far out on the shelf, on the continental slope, or beyond (Fig. 22.14).

Throughout much of the world, the continental shelves are also areas of abundant marine life, because light, heat, and various forms of plant and animal nourishment are concentrated there. As a result, the remains of shell creatures and other organisms may constitute an important source of sediment. Coupled with the precipitation of chemicals from sea water and the inflow of detrital sediments from the land, the sediment makeup on the continental shelves is very complex, with sand, pebbles, and shells interbedded near shore and mud, calcium ooze, and small shells interbedded some distance off shore.

The rate of sedimentation leading to the formation of chemical and detrital rock can be very rapid on the continental shelf. Sedimentologists' studies reveal that in bays along the coast, the rate can be as high as two meters a year. Simple multiplication tells us that even at one-tenth this rate, it would take only 5000 years for sediment to reach a thickness of one kilometer. But as the sediment builds up, it is compressed under its own weight, and much of it is washed farther out on the shelf and beyond. As a result, after millions of years, the sedimentary rocks on large continental shelves, such as that off the Atlantic coast of the United States, grow to a total thickness of five to six kilometers and may extend over a zone of several hundred kilometers in width (Fig. 22.14).

Sedimentary rocks are also subject to metamorphism. This usually results in an increase in density and hardness as well as a change in texture. For instance, the metamorphism of sandstone into quartzite results in a welding together of the sand grains to the extent that individual grains are difficult to distinguish. Table 22.3 lists the metamorphic counterparts of several common sedimentary rocks.

Table 22.3 Metamorphic counterparts of four sedimentary rocks.

SEDIMENTARY	METAMORPHIC
Shale	Slate
Sandstone	Quartzite
Conglomerate	Metaconglomerate
Limestone	Marble

Rock Type and the Landscape

We opened this section by stating that surface rocks are important to physical geography. Let us close by citing a few examples of the role of rocks in shaping the landscape. Bedrock is the major source of soil particles, and variations from place to place in rock composition have in some areas produced corresponding variations in soil type. Figure 22.15 shows the relationship between sandstone and sandy soils and chalk or limestone and clay soils. Similarly, the chemical properties imparted to soil by bedrock can have a strong influence on the distributions of plants. For example, in forested mountain regions, outcrops of serpentine rock are usually marked by treeless patches, or "holes," in

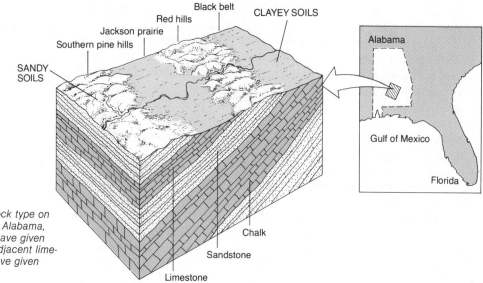

Fig. 22.15 The influence of rock type on soil is evident in south central Alabama, where sandstone formations have given rise to sandy soils, whereas adjacent limestone and chalk formations have given rise to clayey soils.

Fig. 22.16 The relationship between the distributions of limestone and bristlecone cone above 3000 m elevation in the White Mountains of California. Note the exact boundary in the central part of the photograph, where the bristlecones grow immediately adjacent to the rock boundary. (Photograph by William M. Marsh and Jeff Dozier)

the forest. Bristlecone pine in the upper parts of the White Mountains of California appears to prefer one rock type over others. Above an elevation of 3000 m, on the other hand, it grows well in soils derived from limestone, whereas adjacent areas underlain by other rock types are treeless (Fig. 22.16).

Rock type also has a major influence on erosion rates. Depending on mineral composition and the internal structure of crystals or sediments, rocks vary considerably in their resistance to weathering and

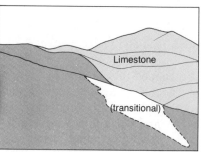

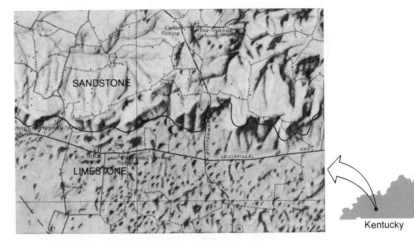

Fig. 22.17 In this area in central Kentucky, near Mammoth Cave, sandstone and limestone have given rise to distinctly different landforms. The limestone has not only been lowered more by weathering and erosion, but also developed a fine, pitted texture as a result of dissolution of the rock by underground drainage. The sandstone, on the other hand, has developed into large hills and valleys as a result of surface erosion by streams. (See Appendix 3 for a description of contour maps.)

erosion. Therefore, the particular forms into which the land is sculptured, a favorite topic of geographers, is often closely tied to rock type. For example, in humid areas some limestones are highly susceptible to chemical breakdown. Groundwater dissolves and washes away the calcium carbonate, leaving the rock riddled with pits, caverns, and underground drainage ways. By contrast, adjacent areas of sandstone, for example, show none of these effects, because the silicon dioxide that makes up the sand particles is relatively resistant to dissolution by water. Because of these differences, limestone and sandstone will often yield markedly different landforms when subjected to the same environmental conditions (Fig. 22.17).

SUMMARY

The ocean basins and the continents are the major units of the crust. The ocean basins are made up of basaltic rocks which are both darker and heavier than the granitic rocks of the continents. The lithosphere rests on a somewhat plastic layer called the asthenosphere; both layers are a part of the upper mantle. Seismological data reveal that the mantle extends 2700 km into the earth; near a depth of 2900 km, it gives way to the outer core, which is made up of liquid metal. The inner core, a mass of solid metal, is 2500 km in diameter.

The crust is composed of thousands of minerals, some of which combine to form rocks. Sedimentary rocks occupy nearly 75 percent of the surface of the continents, whereas virtually all of the oceanic basins are occupied by volcanic rocks, principally basalt and its relatives. Among the influences of rock types on the landscape are soil composition and vegetation distribution in certain areas.

PLATE TECTONICS: A UNIFYING MODEL FOR THE PLANET

CHAPTER 23

INTRODUCTION

Continental Drift

Almost two decades ago the Canadian earth scientist J. Tuzo Wilson wrote:

Geology has reconstructed with great success the events that lie behind the present appearance of much of the earth's landscape. It has explained many of the observed features such as folded mountains, fractures in the crust and marine deposits high on the surfaces of continents. Unfortunately, when it comes to fundamental processes—those that formed continents and ocean basins, that set the major periods of mountain-building in motion, that began and ended the ice ages—geology has been less successful. On these questions, there is no agreement, in spite of much speculation.

"Continental Drift," *Scientific American*, April 1963.

Intensive geophysical research in the 1960s and 1970s, however, has transformed the speculation about the fundamental processes responsible for the formation of continents, ocean basins, and major mountain ranges into a substantial body of theory supported by scientific data. These data reveal that the continents and the ocean basins are parts of great mobile sections of crust which move not only vertically but also laterally and that over geologic time (hundreds of millions of years), the continents have moved thousands of kilometers across the earth's surface. The idea of moving continents has come to be widely known as *continental drift*. The development of this concept has been one of the truly exciting stories in earth science. In Authors' Note 23.1, geophysicist Bruce D. Marsh of the Johns Hopkins University presents a brief outline of the early history of the drift concept.

Actually, the term "continental drift" is misleading in light of the body of information that has emerged on the subject. We are aware that the lithosphere consists essentially of two major units: the lighter continental blocks and the heavier basal layer that extends from the ocean floor past the Moho to the top of the asthenosphere. It is important to remember that this basal layer underlies both the continents

352

and the ocean basins, because it is this layer, not the continental blocks, that is involved in the drift movement. Here the edges of the drifting sections, or plates, of the lithosphere are defined. Since the continental blocks rest on this layer, the continents simply go along for the ride, so to speak, when it moves. Thus the displacement and deformation of the crust caused by movement of lithospheric plates have come to be known as "plate tectonics" instead of continental drift.

"Why Is It So Skinny Between North and South America?"

This was a question asked of a graduate student in physical geography during an oral examination at the University of Michigan in the 1960s. At the time, it was intended to be amusing, and as expected, it earned a chuckle but no answer. Today, however, such a question would be seriously entertained, and an earth scientist would be expected to shape a response based on the growing body of knowledge about the plates and their movements in and around Central America. We know that a small plate, called the Cocos, is moving eastward and is sliding under another small plate, the Caribbean. This has given rise to a string of volcanoes along the western edge of the Caribbean Plate that have coalesced to form the Central American land bridge between North and South America. The geographic implications of this link are extraordinary, not the least of which are the human and animal migrations between the two continents.

Important Questions for Physical Geography

For the scientist interested in the earth's surface, plate tectonics provides the key to the origins of most major crustal features. For the geographer, it offers not only a unifying model for the major systems of landforms on the continents, but also a means of understanding inter-relations between the continents themselves. Most of the traditional theories advanced by geologists concerning the development of orogenic belts and the growth of continents assumed mechanisms of crustal deformation that were far less extensive than we now know to be associated with plate tectonics. In general, these theories provided a poor framework for understanding the functional implications in the broad geographic patterns of crustal features such as island chains and active volcanoes. The case is quite the opposite with plate tectonics, and the way now appears open to further our understanding of geographic phenomena the world over.

What are some of the important questions that plate tectonics raises for physical geography? One relates to the past changes in proximity of land masses, including the formation and destruction of land

AUTHORS' NOTE 23.1
The Origin of the Continental-Drift Idea

"The main motivation to propose large continental displacement across the earth's surface seems to have come from the ability to fit the continents back together as pieces of a puzzle. The remarkable coincidence between the coastlines of Africa and South America, by itself, is convincing. It is said that Francis Bacon (1561–1626) commented on this correlation in his treatise Novum Organum. *As the subject became a matter of debate in the early twentieth century, many names came to light. Today it seems a matter of an author's nationality as to who was essential in the "drift" conception. However, it is to the German Alfred Wegener (1880–1930), who worked so prolifically and debated so cunningly, that the credit must go for the first pronouncement of the hypothesis of continental drift. When Wegener's ideas were published in his book* The Origin of the Continents and Oceans *(1915), they were, like all radical propositions, accepted by some scientists and vigorously rejected by others. Almost needless to say, the following debates were heated.*

"Alfred Wegener produced compelling evidence to show that the continents were once united into a single supercontinent that he called Pangaea (pronounced "pan-gee-a"). Pangaea existed 600–225 million years ago and gradually drifted apart in the ensuing several hundred million years to form the present continents and oceans.

"The maps reproduce Wegener's conception of the stages of the breakup of Pangaea. Though he relied fundamentally on the puzzle-like geometry of the continents, he carried the case of drifting continents further by drawing on diverse bodies of evidence from the fields of geodesy, geography, geophysics, geology, paleontology, biology, paleoclimatology, and physics. Among the strongest evidence he produced was that of striking similarities in rocks, fossils, and ancient geologic events in continents now widely separated geographically.

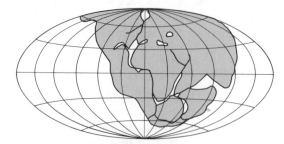

About 300 million years ago

About 50 million years ago

About 1 million years ago

"For example, working with the well-known climatologist Wladimir Koppen, who, by the way, was his father-in-law, Wegener deduced a number of climatic indicators of former geographic linkages between Africa and Antarctica, India, South America, and Australia.

Alfred Wegener (1880–1930). (Photograph courtesy of Dover Publications, New York)

These indicators showed that some continents not only migrated poleward, but have split apart as well. Coal and salt deposits that formed under tropical or subtropical conditions were found in climatic regions that are presently much too cold for such formations. Good evidence for Pangaea was also provided by signs of an ancient glaciation that crossed parts of South America, Africa, and India, now separated by distances of more than 5000 km. The deposits left in Africa by this glaciation are presently situated in a subtropical environment where there is no evidence of the existence of past mountain ranges of sufficient height to foster glaciation.

"In the decades following 1915, advocates of the hypothesis refined and extended Wegener's ideas, but without means to generate data on the subcrust, the hypothesis could not be tested and so to most geoscientists remained largely an interesting body of speculation. Following World War II, however, the intensity and sophistication of geophysical research increased greatly, and in the 1960s data were assembled that generally confirmed Wegener's hypothesis. As for Alfred Wegener, he perished on an expedition to Greenland in 1930. Sadly, he never really gained the satisfaction of seeing his ideas win the applause of scientists the world over."

(Bruce D. Marsh, personal communication.)

bridges and how this has influenced the dispersal of the plants and animals from which modern populations are derived. Another relates to climatic changes. Oceanographers speculate that the oceanic circulation patterns probably changed drastically with the breakup of Pangaea, the opening of the Atlantic, and the positioning of Antarctica. Changes in the flow of warm and cold currents northward and southward should have affected the global transfer of energy, which in turn would produce climatic change. How severe were such changes? Could they have been the source of major biological changes on the planet, including extinctions such as that of the dinosaurs? And could the oceanic circulation and climatic changes have contributed to the origin of the Ice Age? We may learn the answers to some of these questions in the next decade or two as the theory of the plates is further unfolded.

THE FUNDAMENTALS OF PLATE MOVEMENT

The Ice Analogy

The basic processes involved in plate tectonics are really quite easy to comprehend; however, the geographic scale at which they operate is so vast that to many people, the whole idea seems incredible. Let us, therefore, start at a much smaller scale with the assistance of an analogy drawn from another part of nature. Imagine a large ice-covered lake that is sectioned into several large sheets (Fig. 23.1). Moving along the bottom of each sheet is a current of water strong enough to set the sheet into motion. Each current flows in a different direction, and as a result, the various sheets move into, away from, or along one another.

For sheets moving against one another, compression is produced along the contact. If this force exceeds the strength of the ice to hold itself intact, the edges of both sheets will crumble and buckle, forming what is called a pressure ridge (Fig. 23.1a). The ridge, which is a crude facsimile of some types of mountain ranges, protrudes not only above the surface of the sheets, but into the water below them as well.

Now let us give the sheets different densities so that one rides much higher in the water than the other. As the two sheets collide, the higher sheet will tend to override the lower one, forcing the denser ice downward (Fig. 23.1b), perhaps tens of meters into the water under the pressure ridge. Surrounded by water at a temperature slightly above freezing, the slabs will begin to melt; if the condition is prolonged, they may melt away altogether. This example appears to be a reasonably accurate representation of the movement of an oceanic plate of the lithosphere against a plate capped by a continent. The heavier, oceanic plate plunges under the leading edge of the continental plate, sending a huge slab downward into the asthenosphere, where it is eventually reabsorbed. The entire process is referred to as *subduction*, about which we will say more later.

Back on the ice, we can find places where there is movement opposite the compression in pressure ridges. As sheets slide apart, water fills the vacated space and freezes into new ice (Fig. 23.1c). As the sheets are driven farther apart, more new ice is formed. Eventually, a sequence of progressively older and thicker ice can be seen extending outward from the zone of separation (Fig. 23.1d). In the realm of plate tectonics, this example resembles certain conditions that develop along

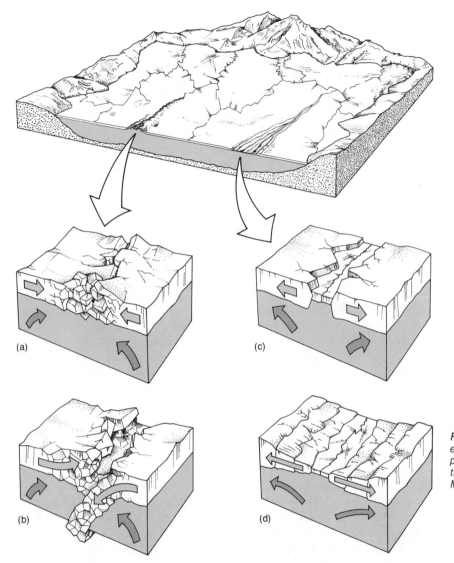

Fig. 23.1 Changes that can occur at the edges of moving ice sheets. Lithospheric plates appear to behave similarly. (Illustration by Peter Van Dusen and William M. Marsh)

the midoceanic ridges. Here the crust separation is probably due to divergent currents in the asthenosphere, and magma is filling the vacated space. Moreover, as the crust is drawn away from the fracture zone, it grows thicker. For both crust and ice, the immediate zone of separation and upflow has a relatively thin cover and thus tends to be warmer than average.

Continuity of the Crust

Although the ice model has a number of shortcomings, especially in relation to volcanism and earthquakes, it does help to convey some key concepts about plate tectonics. Foremost is the principle of crustal continuity, according to which the production and consumption of lithosphere counterbalance each other. At any moment the lithosphere must cover the entire planet; therefore, if it is consumed at some rate in subduction zones, it must also be produced at a similar rate along the oceanic divergence zones. Understandably, these two zones are the foci of stress in the lithosphere, where earthquaking, volcanism, and crustal deformation are most concentrated.

How quickly is the crust being consumed and produced? This can be estimated on the basis of the rates of plate movement that have been deciphered in various parts of the world. The fastest movement appears to be about 10 cm per year, but for most plates the rate ranges from 1.5 to 6 cm per year. If we assume that an average rate of movement for all plates is proportional to the rates of the global subduction and crustal emergence, it would take the subductable lithosphere about 160 million years to replace itself. Therefore, it seems safe to say that the lithosphere under the oceans is probably nowhere more than 200 million years old. On the other hand, the continental lithosphere is much older, apparently as much as four billion years, according to the results of measurements of rock ages. Owing to their greater thickness, lower density, and higher position in the lithosphere, the continental blocks are not consumed by subduction.

THE WORLD GEOGRAPHY OF TECTONIC PLATES

The data recorded at seismic stations around the world enable us to pinpoint the locations of virtually all earthquakes that have occurred in recent years. Plotted on a world map, the epicenters tend to arrange themselves into linear patterns that mark the borders of the tectonic plates (Fig. 23.2). From such maps, it is possible to identify as many as twenty plates in the lithosphere.

Plate Size and Location

Tectonic plates tend to fall into three classes according to size. Seven major plates and four or five minor plates are widely recognized. The remaining eight or nine plates are small, on the order of 40,000 square miles in area, and are often referred to as "platelets" by geophysicists. Platelets are usually found in the destructive zones between major

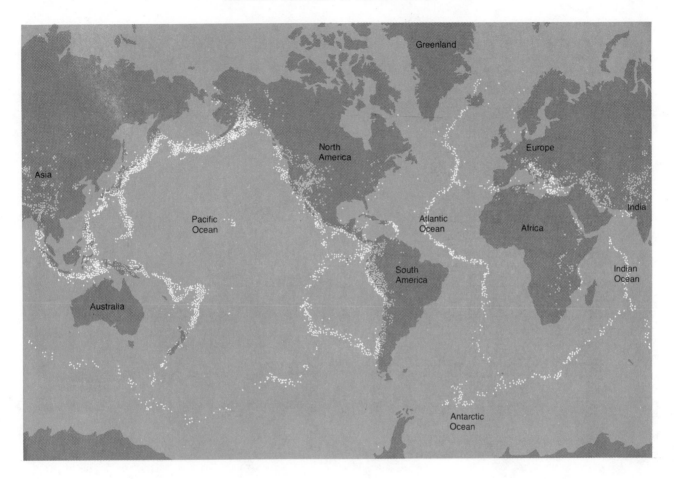

plates, and several are concentrated between the Eurasian and African plates (Fig. 23.3).

Unfortunately, we have very few geographic data on tectonic plates. As far as we know, the areas of the plates have not been accurately measured, but a rough estimate places the size of major plates around 25,000,000 square miles (65×10^6 km²) each. In addition, the true locations of former plate positions on the planet are not known and perhaps never will be. This statement is based on the fact that cartographers reference geographic locations according to the international geographic grid, the global system of coordinates (intersecting lines) known as parallels and meridians. This system is fixed on certain landmarks, one of the most important being the Royal Observatory at Greenwich, England, which is designated 0 degrees longitude. Al-

Fig. 23.2 The distribution of earthquakes with foci at depths of 0–700 km during a recent seven-year period. The linear patterns define the boundaries of tectonic plates. (From U.S. Geodynamics Committee, National Academy of Sciences)

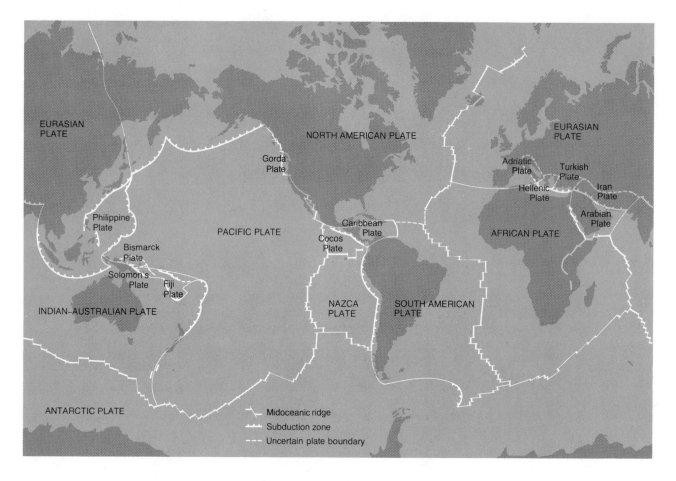

Fig. 23.3 Tectonic plates of the world. Platelets are concentrated in the destructive zones between major plates; for example, between the African and Eurasian plates. (Adapted from U.S. Geodynamics Committee, National Academy of Sciences)

though the founders of the global coordinate system had no way of knowing this, these landmarks are moving with the plates on which they are situated, and since the plates are moving at different rates and in different directions, the geographic grid of the world is being stretched out of shape like a huge rubber map. To plot accurately the future locations of plates on the globe, we will have to turn to an extraterrestrial reference system, probably the stars.

Projecting into the geologic past to, say, the time of Pangaea, as you can see, we really have no way of deciphering precisely where that land mass was situated on the globe. However, we can decipher the *relative* locations of the various sections of Pangaea by matching rocks, magnetic patterns, fossils, shapes, and so on. And then by differentiating the distances between the present locations of continents and their former positions in Pangaea, we can determine *net movement* and the *relative* directions of movement.

The distribution of seamounts on the ocean floor, according to one view, also seems to provide a fairly accurate reference to the former locations of certain plates. These volcanic rises appear to form over hot spots in the asthenosphere, and as a plate passes over them, a string of volcanoes is formed. Assuming that the locations of hot spots remain fixed for long periods of geologic time, the past positions of these plates can be reconstructed from the distribution of seamounts. This is possible with the Pacific plate, using the Hawaiian seamount chain (Fig. 23.4). However, there are problems with this method also. Not all plates have hot spots that have been continuously active long enough to provide such long-term data, and recent drilling in the ocean floor has failed to show a connection between some seamounts and hot spots.

The Principle of Plate Movement

The movement of plates across the globe is governed by a basic geometric principle described in a theorem developed by the Swiss mathema-

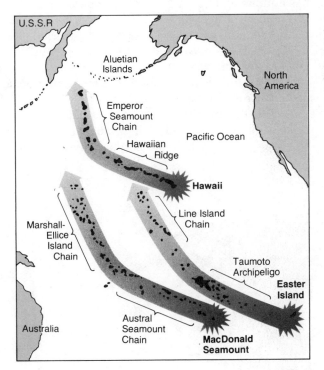

Fig. 23.4 *The trend of three strings of seamounts in the Pacific Ocean suggests that the Pacific plate has changed direction by 40–60 degrees from an initial direction of movement. Not all of the seamounts shown form islands. (After maps by Morgan, 1972; and Dalrymple et al., 1973.)*

tician Leonard Euler (pronounced "Oiler") (1707–1783). According to Euler's theorem, a plate moving on the surface of a sphere rotates about its own pole along *small-circle* routes. (Recall that a small circle is any circle route around the earth whose circumference is less than the earth's maximum circumference.) A small circle can be of any size, but like a hat that is too small, no matter how it is arranged, it never fits precisely around the sphere.

When a tectonic plate is set into motion, its vast area moves as a single unit along many small-circle routes, all of which center on one Euler pole. Those parts of the plate that move along the larger of the small circles move faster than those parts moving along the smaller of the small circles. Thus over a *specified* amount of time, the distance that the plate travels will not be geographically uniform throughout, but will vary according to the size of the circle route that is followed. For example, suppose that a plate extending from the Arctic Circle to near the equator is moving westward. Since the southern end of the plate is moving on a *large* small circle, it must travel a greater distance to cover, say, a 15-degree sector of that circle than the northern end does, because a 15-degree sector of the Arctic Circle is a much smaller distance than a 15-degree sector of the equator.

By tracing a plates's movement, its Euler pole, which, incidentally has nothing to do with the earth's geographical or magnetic poles, can be found by projecting the differential movement of the plate to a point of intersection, as is shown for South America in Fig. 23.5. According to Fig. 23.5, the movement of the plate can be described by the principle of angular velocity about the Euler pole. The angular velocity is constant throughout the plate, but the actual velocity increases with distance from the pole.

With this explanation, it is possible to account for differential rates of plate movement along an oceanic ridge. And this, in turn, is necessary in order to explain in part how the general form of a single

Fig. 23.5 The differential movement of South America about an Euler pole at 45° N and 30° W. The arrows mark small-circle routes.

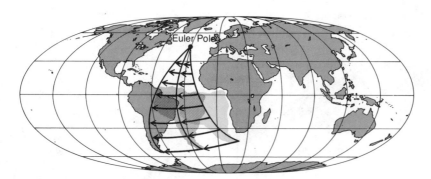

Map projection by Waldo Tobler.

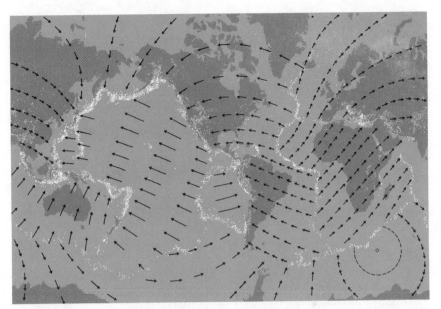

Fig. 23.6 The worldwide pattern of tectonic plate movement. Each arrow represents 20 million years of movement; the longer the arrow, the greater the rate of movement. (After Judson, Deffeyes, and Hargraves, 1976; map from U.S. Geodynamics Committee, National Academy of Sciences)

ridge can range from high and narrow at one end to low and wide at the other end, a topic addressed in the next section.

Although the worldwide data on the contemporary rates and directions of plate movements are not abundant, geophysicists have learned enough to map the general patterns of movement in the major plates. One such map is presented in Fig. 23.6. Each arrow on this map represents 20 million years of movement, and it is immediately apparent that the Pacific, Australian-Indian, and Nazca have been the most active plates in the past 20 million years.

EVIDENCE FOR PLATE MOVEMENT

What actual evidence is there for plate movement? Scientists have been able to generate several bodies of evidence that reveal both the amount and relative directions of the geographic displacement of plates. One of the strongest bodies of evidence was born with the discovery of parallel bands of magnetic alignments in the rocks on the floors of the Pacific and Atlantic oceans.

The Case of the Atlantic

Let us examine the case of the Atlantic. The mid-Atlantic ridge is a volcanically active rise that runs south from the Arctic through Iceland to near Antarctica. For some time fresh flows of basaltic rock have

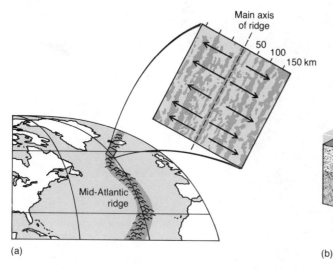

(a)

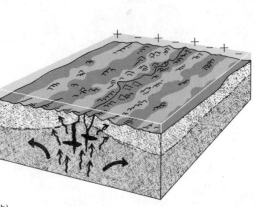

(b)

Fig. 23.7 The pattern of magnetic anomalies in a section of the mid-Atlantic ridge south of Iceland (a) and a schematic interpretation of this pattern (b). The dark areas represent the rock with positive magnetization; the light areas, the rock with negative magnetization. (After Cox et al., 1967)

Map projection by Waldo Tobler.

appeared from the various fissures in the ridge. Such flows have been frequently observed on Iceland and other islands along the ridge. When the new igneous rock solidifies, the polarity of its magnetic minerals is permanently frozen in place and should be consistent with that of the earth's magnetic field at that time. Over periods of time ranging from tens of thousands to millions of years, the polarity of the earth's magnetic field undergoes reversals, that is, changes from negative to positive and vice versa. Analysis of volcanic rocks formed in various parts of the world indicates that nine major reversals have occurred in the past 3.6 million years.

Geophysicists generally agree that the bands of magnetic deviations in the Atlantic floor correspond with the reversals in the earth's magnetic field which have been imprinted in volcanic rocks as they formed along the midoceanic ridge (Fig. 23.7). As the crust spread from the ridge, the rock, with its magnetic polarity fixed, was carried outward across the ocean floor. The reasoning follows that the older rock farther from the ridge shows alignments with former positions of the earth's magnetic field, whereas the newer rock nearer the ridge shows bands of alignment with relatively recent magnetic fields. If the crust were not spreading from the ridge, the newer volcanic rock would cover the older rock as it emerged, and the magnetic reversals would be piled on top of one another. Therefore, the striking geographical pattern of magnetic polarities that has been documented in the ocean floor would not be discernible. However, this is not the case, and we believe that new rock is appearing while the ocean floor is spreading from the ridges.

Along each ridge system, the rate of spreading and the emergence of volcanic rock vary markedly over time and from place to place.

Along some midoceanic ridges there is no evidence of any present activity, but much evidence of former activity. Along the mid-Atlantic ridge, the segment around Iceland may be spreading at a fairly slow rate. This can be deduced from the fact that here the ridge tends to be very narrow and high in elevation, suggesting that sea-floor spreading cannot draw new rock away from the zone of emergence before it has time to pile up. The opposite interpretation can be applied to the wide, low sections of the ridge, and this is often reflected in broad geomagnetic banding in the ocean floor.

ACTIVITY ON THE EDGES OF PLATES

Three major types of contacts can be identified along the edges of tectonic plates, and all are associated with earthquakes and volcanism. Two of these, *constructive* and *destructive*, were mentioned in the context of the ice sheet analogy. Whereas these contacts tend to be located at the opposite ends of a plate that is moving essentially in one direction, the third type of contact is found along the sides of the plate. Here movement is mainly lateral as the plate slips past its neighbor. Crust is neither destroyed nor created, but conserved; hence, such contacts can be termed *conservative*. Let us briefly describe how earthquakes and volcanism are related to plate movement in these three contacts.

Destructive Zones

Where plates converge to form a subduction zone, enormous amounts of energy are available to heat and to deform the lithosphere. As an oceanic plate subducts under its neighbor, the surface of the crust is drawn down to form a trench (Fig. 23.8). Beneath the trench, the subduction process is manifested in a jerking downward motion as the plates slip along the angular contact between them. This motion is the result of frequent small displacements of massive blocks of rock as the plate bends into the earth. Each displacement, or *fault*, releases seismic energy into the adjacent plates, part of that energy traveling to the surface to produce an earthquake of some strength.

From our vantage point on the surface, the size of an earthquake depends on many factors in addition to the amount of energy released from a displacement. Among these other factors are the depth at which the fault takes place and the physical nature of the rock around the zone of displacement. If the fault is shallow and the rock is rigid, the surface tremors are likely to be very strong. This is clearly the case in the upper 50 km of the lithosphere, where compressional stress in the brittle rock of the crust generates not only the strongest tremors, but

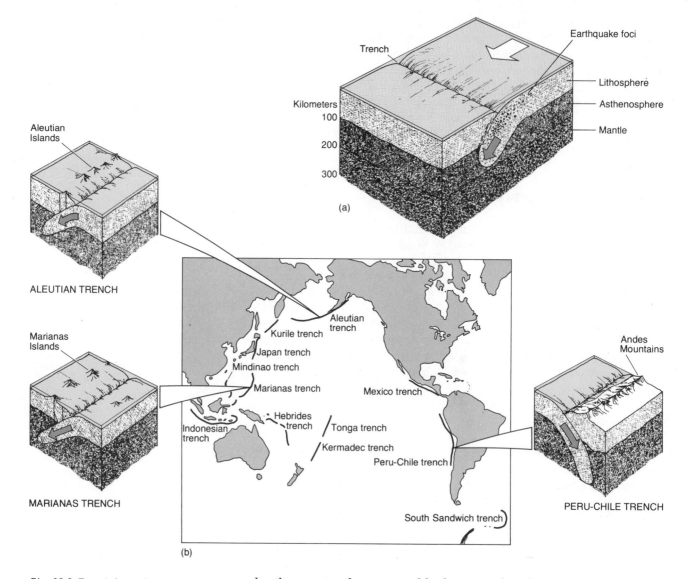

Fig. 23.8 *Trench formation: (a) the early phase in the formation of a subduction zone, showing the trench and the concentration of shallow earthquake foci; (b) the locations of trenches in and around the Pacific basin. (Illustration by William M. Marsh)*

also the greatest frequency of faulting. Farther down, the rock appears to grow less brittle, and the magnitude and frequency of faulting decline markedly.

Not all of the energy in subduction zones is spent in displacement of rock. Part of it also goes into the production of heat, which gives rise to an elongated body of magma along the contact zone during the early stages of subduction. Some of this magma works its way through the overlying plate, forming conduits called *diapirs*, and then issues onto the surface to form volcanoes (Fig. 23.9). According to one interpreta-

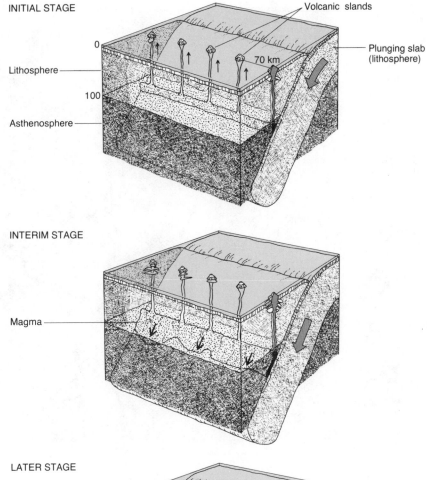

INITIAL STAGE

Volcanic slands

Lithosphere

0

70 km

100

Asthenosphere

Plunging slab
(lithosphere)

INTERIM STAGE

Magma

LATER STAGE

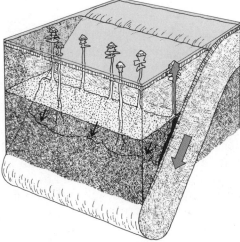

Fig. 23.9 An interpretation of the volcanic
activity in a subduction leading to the formation of an island arc. The magma tube
that develops in the early phase gives rise
to diapirs which produce volcanoes. The
spacing of the active volcanoes is remarkably uniform along the zone, around 70
km between neighbors. As subduction advances and the magma is dragged down,
more diapirs develop, and the island arc
grows. (Illustration by Peter Van Dusen;
after Marsh and Carmichael, 1974; Marsh,
1979)

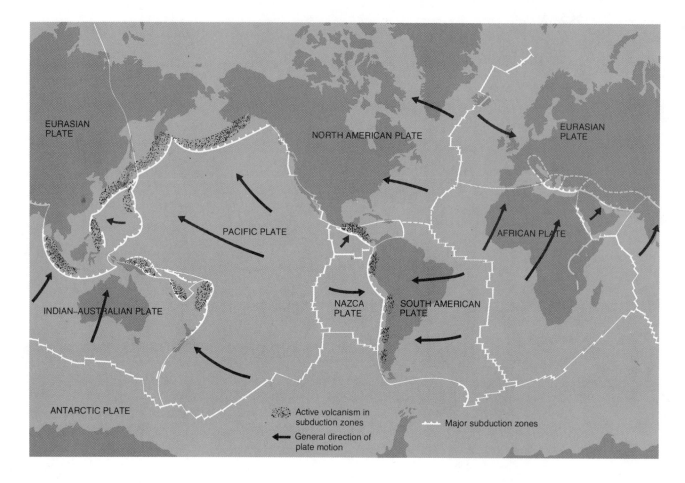

Fig. 23.10 The major subduction zones of the world and the associated areas of active volcanism. Note the relationship between subduction and island arcs on the northern and eastern sides of the Pacific Ocean. (Compiled from a variety of sources.)

tion, initially a single line of volcanoes is formed, but as the magma body is dragged down along the contact and more diapirs develop, a broader zone of volcanism is created, leading to more islands. As the volcanoes grow in size and number, they may eventually coalesce to form arc-shaped island archipelagos, such as the Japanese Islands and the Philippines (Fig. 23.10).

In the advanced stages of subduction, the plunging slab of lithosphere extends through the asthenosphere and into the underlying mantle (Fig. 23.11). Because the slab is much colder than the mantle rock, it extends intact as deep as 700 km in some areas, where it is physically assimilated into the mantle. Although the seismic zone is extended to a corresponding depth, the magnitude and frequency of faulting decline, apparently because the slab is heating up and is not as brittle as it was earlier.

How does the subduction process come to a halt? There are probably many contributing factors. The forces driving the plates may stop or change direction, but there is really no way of knowing when this happens. There is evidence, however, to indicate that when continental blocks are drawn into each other, subduction slowly grinds to a halt. This appears to have been the case when the northern edge of the Indian-Australian plate subducted under the Eurasian plate. Subduction apparently continued until the Indian subcontinent lodged against the Asia continental block. In the course of convergence, a narrow ocean basin called the Tethys Sea was formed between these blocks, and it became a depository for volcanic and sedimentary rock. As the

Fig. 23.11 The advanced stage of subduction. The slab extends into the mantle under the asthenosphere. By this time the island arc has grown into large islands, and the zone of earthquaking may reach a depth of 700 km or more. (Illustration by William M. Marsh)

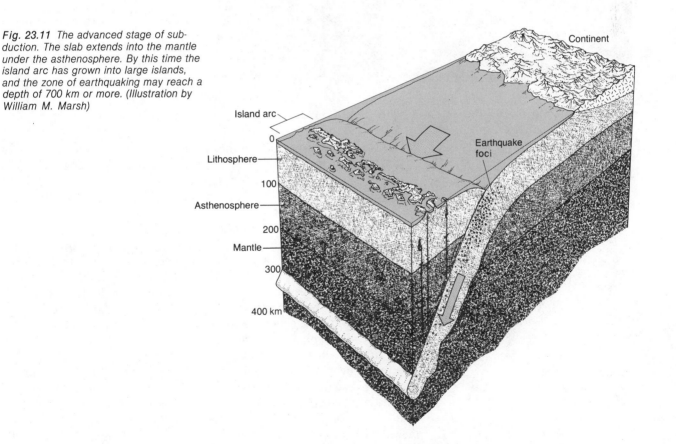

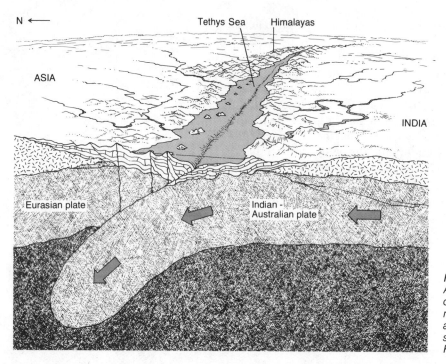

Fig. 23.12 *Subduction of the Indian-Australian plate along the southern edge of the Eurasian plate. The sedimentary rocks of the Tethys Sea were compressed and uplifted to form huge mountain ranges such as the Himalayas, Pamirs, and Hindu Kush. (Illustration by William M. Marsh)*

basin was compressed, these rocks were uplifted to form the massive mountain ranges of Tibet and neighboring lands (Fig. 23.12).

Constructive Zones

More or less opposite the subducting side of a plate is the zone of crustal spreading where magma is emerging to form new lithosphere. Here tensional motion, which is the opposite of the compressional motion on the subducting side of the plate, is set up as the two plates move apart. As a result, the crust fractures and small blocks of ocean floor downdrop between the parting plates. These faults are usually shallow and in turn generate shallow, low-magnitude earthquakes.

As the lithosphere is pulled apart, it is reduced to a relatively thin layer that is subject to intrusions of magma from the asthenosphere. The faults provide magma with ready access to the surface, where it flows between and over the parting blocks and piles up to form a rise or mountain ridge. If the rate of spreading is relatively slow, say, less than 3 or 4 cm per year, the ridge may grow to heights of 3000 m or more. Figure 23.13 shows a number of ridge profiles associated with different rates of spreading.

Although the process of spreading and emergence is known to occur on the continents, it is most prevalent in the ocean basins, where it has produced the worldwide system of oceanic ridges. Among the continents, in fact, there appear to be only two major zones of spreading—one in Africa and the other in Asia. The African zone, which is the better developed of the two, extends from the Red Sea down the East African Highlands and is marked by a system of rift valleys that form the basins for Lakes Albert, Tanganyika, and Malawi (Nyasa) as well as hundreds of smaller water bodies (Fig. 24.10). Geologists are uncertain, however, whether this zone is currently active.

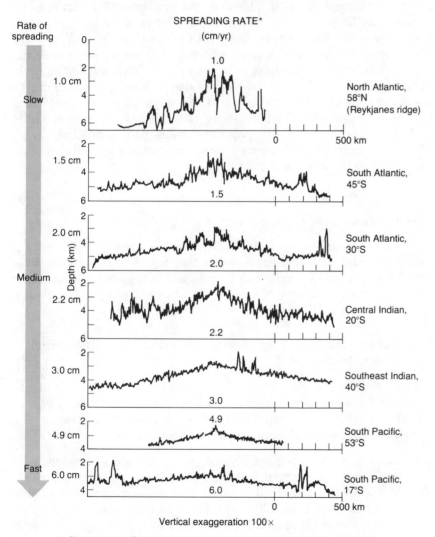

Fig. 23.13 These topographic profiles of midoceanic ridges and the accompanying rates of spreading show that where spreading is slow, ridges tend to be much higher than where spreading is fast. (From H. W. Menard, Geology, Resources, and Society, *San Francisco: Freeman,* © 1974. Used by permission.)

* (Data represent half-rates per year, i.e., the annual average rate for one side of the ridge.)

Fig. 23.14 The main faults of the San Andreas system, North America's prominent conservative zone.

Conservative Zones

In conservative zones, plate movement is mainly parallel to the boundary, and therefore crust is neither created nor destroyed. The most famous fault system in North America, the San Andreas, is a conservative zone along the California coast. It separates the northward-moving Pacific plate from the North America plate and is the source of frequent and often strong earthquakes (Fig. 23.14). However, volcanism is uncommon, which is typical of such zones.

Conservative zones are also found within active midoceanic ridges. This is related to a complex border configuration resembling a sort of interfingered pattern between the two plates (Fig. 23.15). At the ends of the fingers, the boundary is constructive, but along the sides of the fingers, the boundary is conservative. The term "transform fault" has been given to the conservative segments of the border. The origins of this border configuration are related to the rotational motion of the plates about their Euler poles. As the ridge pulls apart, lines of crustal weakness develop perpendicular to the ridge along the small-circle paths of movement. Some of these lines become transform faults, which, when interconnected with the faults that parallel the ridge, form the interfingered border pattern. This concept is portrayed schematically in the inset of Fig. 23.15.

Earthquake Intensity and Magnitude

The size of an earthquake can be described in terms of its *intensity* or *magnitude*. Intensity is a measure of an earthquake's effect on the local landscape, primarily on human structures. Geologists have developed a twelve-level intensity scale, called the Mercalli scale, and levels 1, 4, 8, and 12 are as follows:

1. Not felt except by a very few people under specially favorable circumstances.

4. During the day, felt indoors by many, outdoors by few. At night some are awakened. Dishes, windows, and doors rattle; walls make creaking sounds. Sensation like a heavy truck striking building. Standing motorcars rocked noticeably.

8. Damage is slight in specially designed structures; considerable in ordinary substantial buildings, which may partially collapse; and great in poorly built structures. Panel walls are thrown out of frame structures. Chimneys, factory stacks, columns, monuments, and walls may fall. Heavy furniture is overturned. Sand and mud are ejected from the ground in small amounts, and

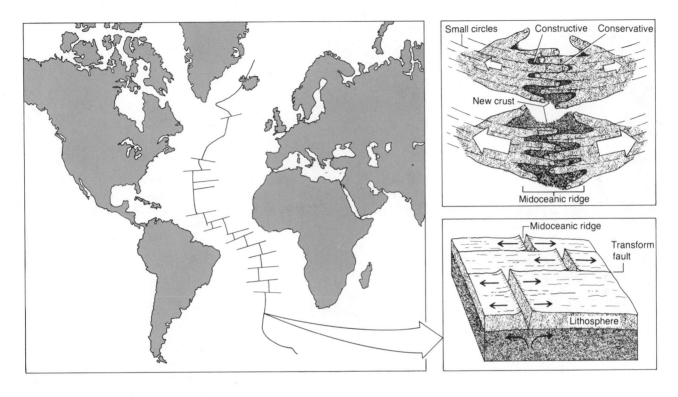

Labels in figure: Small circles, Constructive, Conservative, New crust, Midoceanic ridge

Midoceanic ridge, Transform fault, Lithosphere

there are noticeable changes in well water. Persons driving motorcars are disturbed.

12. Damage is virtually total. Waves are seen on ground surfaces, and objects are thrown upward into the air.

Fig. 23.15 The interfingered configuration of the plate border along the mid-Atlantic ridge. The east-west–trending fault lines represent conservative borders. (Illustration by William M. Marsh)

As you can see, the intensity of an earthquake can depend as much on the nature of the landscape and where people live as on the size of the quake itself. Adobe houses, such as the one shown in Authors' Note 3.2, are especially susceptible to destruction by earthquakes because of the brittleness of the material.

Magnitude is a measure of the actual seismic energy released in an earthquake. It represents the energy released at the point of displacement, called the earthquake focus. The "Richter scale," developed by Charles F. Richter of the California Institute of Technology, is the most widely used magnitude scale. To interpret this scale properly, we must be aware that it is not arithmetic, but logarithmic; for each unit on the scale, magnitude is multiplied by about 32. For example, an increase on the scale from 4 to 6 represents a 1000-fold increase in earthquake magnitude.

(U.S. Geological Survey photograph)

The most destructive earthquakes combine the following:

1. high magnitude, generally 7.5 or higher on the Richter scale;

2. shallow foci;

3. heavily populated areas;

4. mountainous topography.

The last is especially important, because the tremors often trigger landslides and avalanches that can bury populated areas. Finally, it is worthwhile to note that an earthquake is almost always occurring somewhere in the world; by way of example, those with a magnitude of 3.0 to 3.9 take place every eleven minutes, on the average.

THE EARTH'S INTERNAL ENERGY

So far in this chapter we have described the tectonic plates and the nature of their movement. But one major question remains: What forces drive the plates? To answer this, we must examine the internal energy of the earth and how it is released from the interior of the planet.

Heat is the primary form of energy emitted from the earth's interior. In the asthenosphere and lithosphere this thermal energy drives massive rock movements that produce breaking, slipping, and bending in the crust. But the energy dissipated through these processes amounts to only a fraction of the total annual heat supply, leaving the bulk of it to be dissipated by flowing through the crust into the soil and the oceans and finally into the atmosphere.

We have long known that heat is emitted from the earth's crust. Hot springs and volcanoes, in particular, are indisputable evidence of this fact. What was not known until this century, however, is that heat is continuously being emitted from every square centimeter of the crust's surface. The rate of emission, however, is so small that the presence of this heat in soil, water, and surface rock is masked out by much larger amounts of heat derived from solar radiation.

Distribution of Heat at Depth

Measurements made in mine shafts and bore holes reveal that temperature increases with depth in the upper crust at a rate of 2–3 degrees Celsius per 100 meters. Beyond a depth of 10 km or so, the rate decreases slightly, and at a depth of 100 kilometers the temperature reaches about 1200°C. Although this temperature is sufficient to melt surface rock, it is not high enough to melt the same rock at such depth. The reason for this is that the melting point of rock increases with pressure; under the massive pressure exerted by 100 km of rock overburden, the melting point of the lithospheric rock is pushed beyond 1200°C. Farther down, however, a curious thing happens. The melting-point temperature begins to level out, whereas the rock temperature continues to rise; at a depth between 100 and 200 km, a rock temperature close to melting point is reached, and the early phases of melting are initiated. We know that this corresponds to the asthenosphere. Farther into the mantle, the trend changes, and the rock is mainly solid to a depth of 2900 km (1800 miles). At this point wholesale melting marks the outer portion of the core.

Geographic Distribution of Earth Heat

From the geographer's standpoint, it is interesting to explore the spatial distribution of the earth's heat emissions. As with so many subsurface phenomena, though, accurate measurements of heat outflow are difficult to obtain. This is because they must be made at depths greater than 50 meters or so in order to avoid the effects of surface heat. Consequently, there is a lack of geothermal data for some areas of the earth; nevertheless, geophysicists have been able to piece together a reasonably detailed picture of the earth's heat outflow.

The average heat flux across the earth's surface is about 0.06 $J/m^2 \cdot s$, which amounts to 47 $cal/cm^2 \cdot year$, or 2×10^{20} calories per year for the entire planet. Although this quantity is negligible compared with the amount of heat generated by solar radiation, it is the single greatest source of terrestrial energy on the planet. For comparison, it exceeds by ten times the total energy used by human beings.

The difference between the continents and ocean basins in the average heat flow appears to be insignificant. But within each, the differences between the major divisions are very significant. In the ocean basins the averages are as follows: oceanic ridges, more than 0.08 $J/m^2 \cdot s$; abyssal plains, about 0.05 $J/m^2 \cdot s$; and trenches, less than 0.05. On the continents, the shields average less than 0.04 $J/m^2 \cdot s$, whereas major mountain regions, such as the Rocky Mountains and the Alps, average about 0.08 $J/m^2 \cdot s$ (Fig. 23.16). Generally speaking, then, it seems that the geologically active regions of the world are about 25 percent warmer than most of the rest of the planet (Fig. 23.17).

Narrowing our focus to the mountain regions, it is possible to identify individual ranges where heat emissions are several times higher than the average. Probe tests carried out in the Atlantic, for example, showed that the main range of the midoceanic ridge in the area between the Caribbean and Saharan Africa has an emission rate more than four times above the earth average. Compared with that of the

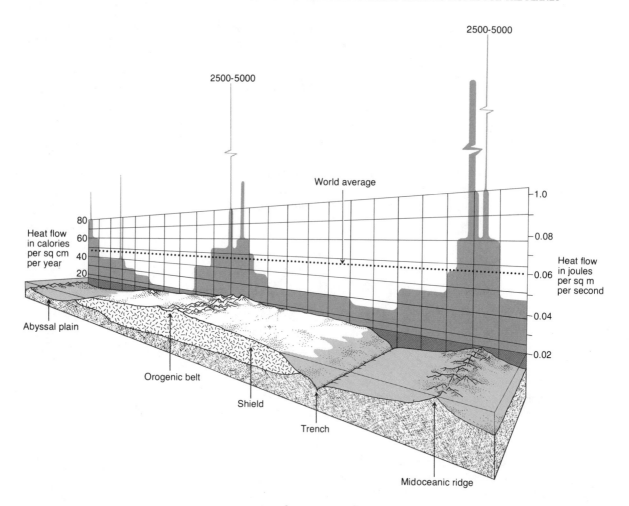

Fig. 23.16 *The general change in the rates of heat flow associated with the major features of the continents and ocean basins. Although geothermal hot spots may exceed the earth average (0.06 J/m² • s) by more than a hundredfold, mountainous regions as a whole are about 25 percent warmer than the average. (Illustration by William M. Marsh)*

surrounding ocean floor, this rate is ten to fifteen times greater (see enlargement in Fig. 23.17).

In this age of energy shortages, we are very interested in the spots of concentrated heat outflow on the continents, called *geothermal areas*. Typically these are small areas, usually about 100 to 1000 square kilometers, situated in or near mountain ranges or volcanoes. Heat emissions may be 100 times greater than the average, and part of these emissions may be released through hot springs and geysers. Geothermal areas located close to cities and industry can be tapped as sources of steam to drive electrical generators. The map in Fig. 23.18 shows the locations of geothermal power sites presently in operation, and Authors' Note 23.2 provides some additional information on geothermal power. The relationship between the rates of heat flow and the major features of the crust are summarized in Fig. 23.16.

*Fig. 23.17 Map of global heat flow in milljoules/m²s. Enlargement represents a heat-flow profile across the mid-Atlantic ridge. (Map used by permission from David S. Chapman and Henry N. Pollack, "Global Heat Flow: A New Look," Earth and Planetary Science Letter **23**, 1975. Used by permission.)*

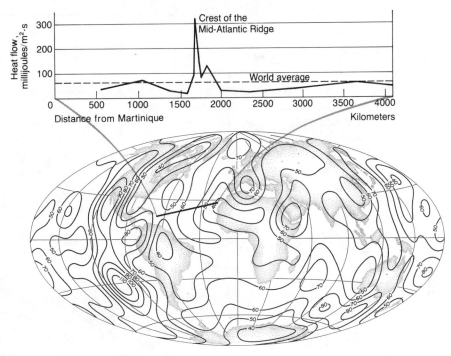

Sources and Mechanisms of Heat Flow

What is the source of earth heat? Is it heat that was trapped during the formation of the earth, or is it somehow generated continuously by an internal energy system? Actually, both answers appear to be correct. It is thought that the interior of the earth received a great charge of thermal energy with the conversion of gravitational energy into heat when the core formed about four billion years ago. To this source is added the heat produced by the radioactive disintegration of the ele-

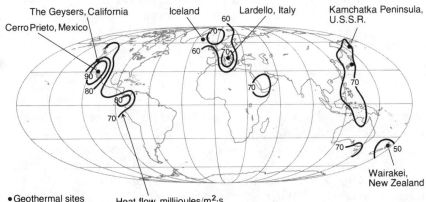

Fig. 23.18 Geothermal power sites known to be in operation, related heat-flow zones, and other terrestrial zones with high heat flows. (After Cargo and Mallory, 1977; Chapman and Pollack, 1975)

Map projection by Waldo Tobler.

AUTHORS' NOTE 23.2
Geothermal Energy

In order to utilize geothermal energy, it is necessary to not only find a shallow source of concentrated heat, but also gain some means to transfer the heat rapidly to the surface. The most efficient natural means of heat transfer is provided by groundwater that seeps into the source area, is heated, and is then discharged to the surface through cracks. This process is best developed where groundwater has access to the heat reservoir, but is restricted from ready release to the surface by impermeable rock overlying it. This produces a pressure-cooker condition, which if tapped can yield superheated steam to drive turbines for generating electrical power.

Geothermal areas can be classed into three groups. The weakest is the low-temperature field, which consists largely of water in the range of 50–82°C (120 to 180°F). Its uses are limited to heating homes, factories, greenhouses, and so on. The second class is the wet-steam field, which contains superheated water at temperatures between 180 and 370°C (350 to 700°F). As it is brought to the surface, the pressure on the water declines, and about 10 to 20 percent of it bursts into steam. Dry-steam fields, the third type, are high-temperature sources that are comprised mainly of steam. Although they are less abundant than wet-steam fields, dry-steam sources are currently the most widely developed of the world's geothermal areas. Both the wet- and the dry-steam sources are well suited for electrical-power generation.

The demands for new energy sources have led many nations to seriously consider the potentialities of

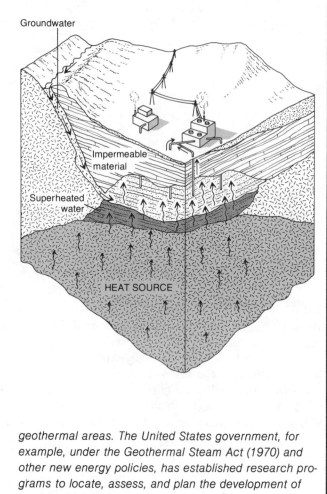

geothermal areas. The United States government, for example, under the Geothermal Steam Act (1970) and other new energy policies, has established research programs to locate, assess, and plan the development of geothermal areas.

ments uranium, thorium, and potassium, which are contained in the rocks of the mantle and crust.

The heaviest concentrations of these elements are found in granite, the rock of the continents, whereas significantly lighter concentrations are found in basalt, the rock of the ocean basins, and only a trace is thought to exist in the rocks of the mantle (Table 23.1). Thus the continents are radioactively "hot," with about 35 percent of their over-

Table 23.1 Heat from radioactivity in rocks.

ROCK GROUP	URANIUM (PPM)	THORIUM (PPM)	POTASSIUM (% WGT)
Sialic	3.0	10.0	2.2
Basaltic	0.7	3.0	1.1
Mantle	0.013	0.05	0.001

all heat flow contributed by the granitic rock. In contrast, the ocean basins are radioactively "cool," and only about 10 percent of their heat is provided by the basaltic rock. The rest of the heat emitted from both oceans and continents comes from the mantle. The reason that the oceans, with their lower radioactivity, are not markedly cooler than the continents is because the oceanic lithosphere is so much thinner than the continental lithosphere (Fig. 23.19). Thus heat must travel across a relatively short distance of rock before reaching the ocean floor.

How does the heat get out of the mantle? Although conduction is virtually the only mechanism of heat transfer through the lithosphere, by itself it is insufficient to account for the rate of heat flow into the base of the lithosphere, because solid rock is such a poor conductor of heat. Therefore, in the massive mantle, flow by conduction cannot possibly bring enough heat to the crust to establish the present rates of flow. Geophysicist Frank Press and geologist Raymond Siever underscore this point:

Heat entering one side of a plate of rock 250 miles thick should take about 5 billion years to flow out the other side. In other words, if the earth were to have cooled by conduction only, heat from depths greater than about 250 miles would not yet have reached the surface!

(From *Earth*, 2d. ed., San Francisco: Freeman, 1978.)

What are the alternatives? One is transfer by radiation, but tests show that the minerals in the mantle are too opaque to transmit enough infrared radiation to account for the rate of heat flow. The other, and more probable, candidate is convection.

Convection is a highly efficient means of heat flow because it involves movement of the heat-charged material. The greater the mobility of the material, the faster the rate of transfer. Examples of convection are common in the day-to-day routine of life: boiling water, warm air rising in a room, and thunderstorms on a hot summer day. The principle of convection is based on the differential expansion and contraction of a mobile substance. As heat is increased, molecular vibrations increase and the substance expands. Since the total mass of

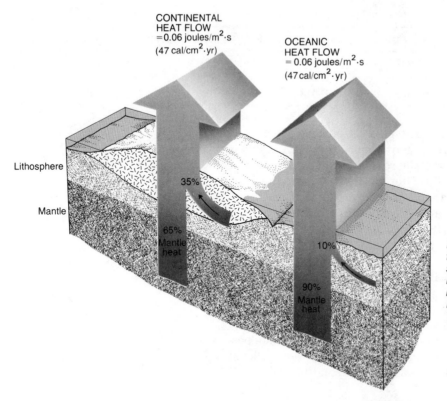

CONTINENTAL
HEAT FLOW
$= 0.06$ joules/m^2·s
(47 cal/cm^2·yr)

OCEANIC
HEAT FLOW
$= 0.06$ joules/m^2·s
(47 cal/cm^2·yr)

Lithosphere

Mantle

35%

65%
Mantle
heat

10%

90%
Mantle
heat

Fig. 23.19 The contributions to continental and oceanic heat flow by the lithosphere and mantle. Note that considerably more of the heat emitted from the continents is generated in the lithosphere than is the case in the oceans. (Illustration by William M. Marsh)

the substance is unchanged, its density decreases with expansion. If a low-density zone is situated under a higher-density (i.e., low heat) zone, the substance is gravitationally unstable. In gases and fluids, this produces a turbulent motion as the substance seeks gravitational equilibrium. This motion is convection, and it is the most effective known means of heat redistribution through substances of large volume, such as air, water, and molten rock.

The rate of convection in a substance is maximized under three conditions: (1) if the temperature difference is great between hot and cold points; (2) if the coefficient of thermal expansion is high, i.e., if the substance undergoes much expansion with the application of heat; and (3) if the viscosity (i.e., resistance to flow) is low. These conditions are frequently met in the atmosphere, and the incidence and intensity of thunderstorms are vivid testimony of this fact. But in the rock materials of the earth? On first thought it seems preposterous. However, when we recall that seismic findings show conclusively that the rock of the asthenosphere exists in a plastic state and thus may be capable of flow-type movement, the idea sounds more plausible. Furthermore, tests show that a rate of rock flow of only 2 to 3 cm per century is enough to produce a heat flow greater than that possible by conduction through stable rock. Thus it appears that under the long-term strain of convective forces, heat is transferred through the upper mantle by an upward flow that is perhaps best described as a creeping motion. Although the nature of this motion in terms of rates and direction of movement, zones of origin, and continuity of flow are unknown, modern scientists are convinced of its existence.

The theory of convection in the mantle represents one of the great advances in the history of earth science because it not only accounts for the rate of heat flow from the earth, but also establishes the presence of a force to move the huge tectonic plates. Since the early decades of this century, the debates over the meaning of the large body of evidence supporting plate tectonics often stalled on the question of the driving force responsible for the movement. Because of the size of the crustal blocks involved, it was difficult to identify a mechanism powerful enough to produce the movement. Convective flows in the mantle provided a plausible answer. Moving upward through the plastic layer, they not only transfer heat, but, as they spread out at the base of the lithosphere, also exert differential stress on the lithosphere, causing it to weaken and break. When these flows cool with the loss of heat to the overlying rock, they sink back into the asthenosphere. Thus the lithosphere can be pulled apart, pushed together, and pulled downward, depending on the pattern of flow (Fig. 23.20).

Fig. 23.20 The convection of the mantle. Hot currents probably form convective chimneys that rise through the asthenosphere and generate three types of movement: upward, downward, and lateral. As a result, zones of divergence and convergence are formed near the base of the dithosphere, which in turn appears to produce corresponding motions in the lithosphere.

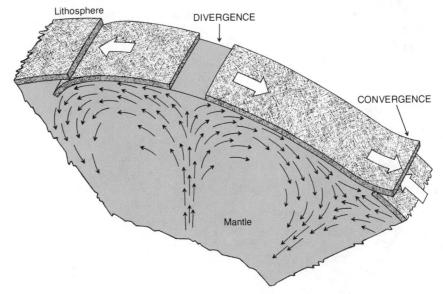

SUMMARY

Never in the history of earth science have so many geological mysteries been unraveled so quickly as in the past two decades. The door was cracked open by certain geologists, geophysicists, and seismologists who, in the first half of this century, produced data on earth heat flow and the physical properties of the mantle. This work led to the deduction of convective movement in the mantle. Coupled with Wegener's ideas on the past movements of the continents, the corridor was identified leading to the theory of plate tectonics.

The truly giant stride, however, did not come until the rocks of the ocean floor could be explored. This required sophisticated apparatus that would allow scientists to measure water depth, bottom temperature, and water chemistry and to collect samples of rocks and sediments. Although many projects have been launched with the support of many nations, the Deep Sea Drilling Project is particularly noteworthy. This project was implemented in 1969 as a joint effort of the National Science Foundation, several American oceanographic institutes, and some foreign nations. The principal thrust of the project centers on a specially designed ship, the *Glomar Challenger*, named for the *HMS Challenger*, the ship of the first oceanographic survey held in 1873. From the *Glomar*, the ocean floor can be drilled and rock samples extracted.

The results of the modern oceanographic projects have been impressive, to say the least. They show that the ocean basins, which earth scientists once thought to be the enduring, stable parts of the crust, are in fact just the opposite. They are very young, nowhere more than 200 million years old, whereas the interiors of the continents are several billion years old. Moreover, there are striking age differentials in the oceanic rock, especially along the midoceanic ridges, where the youngest rock is found near the ridge axis and the older rock at increasing distance from the ridge. This complements the geographic patterns of magnetic reversals in the rock of the ocean floor, which indicate that the crust is emerging from the midoceanic ridges.

If the crust is emerging in certain zones, it must be consumed in others. The groundwork for this problem was prepared by Hugo Benioff, who in the 1930s demonstrated that earthquake foci along the west coast of the Americas are concentrated along large slippage planes that extend from the trenches into the mantle at angles of 30 to 60 degrees. Seismic data from other parts of the world showed similar patterns near trenches, and the idea of plate subduction eventually became fact in the minds of geoscientists.

CHAPTER 24

GEOLOGIC STRUCTURES: ARCHITECTURE IN ROCK

INTRODUCTION

The plate-tectonics model give us more insight into the gross geography of the planet than we would realistically have hoped for only two or three decades ago. Among other things, it shows us why most mountain chains form where they do. Simply put, plate tectonics tells us that the best place to look for a mountain chain is along the edge of the plates. This is where island arcs, midoceanic ridges, and orogenic belts are found, and together they constitute the bulk of the world's mountains.

In general, the major mountain chains of the earth fall into two main categories: volcanic and folded. Those of the ocean basins, including both arcs and ridges, are almost exclusively volcanic, whereas the orogenic belts of the continents are mainly folded mountains. It is generally agreed that the orogenic belts result from the compression of sedimentary rocks on the margins of the continents. Where neighboring continental plates move against each other, these rocks are folded into massive structures that are pushed upward thousands of meters. The folding is usually accompanied by faulting and volcanism. The ultimate result of all this deformation is long chains of complex mountains such as the Andes (Fig. 24.1), Alps, and Rockies.

Our objective in this chapter is to examine the rock structures that make up the orogenic belts. The scale of study must be somewhat finer than the one we used to study plate tectonics, because our goal here is to learn about individual structures or groups of structures and how they influence the landscape, especially the shapes of mountains and valleys. We will examine three basic types of geologic structures: fault, folds, and volcanoes.

FAULTS AND FOLDS

On the night of March 26, 1872, the American naturalist John Muir, living in his beloved Yosemite Valley, was awakened by a violent earthquake. His description of the event was extraordinary:

At half past two o'clock of a moonlit morning in March, I was awakened by a tremendous earthquake, and though I had never

383

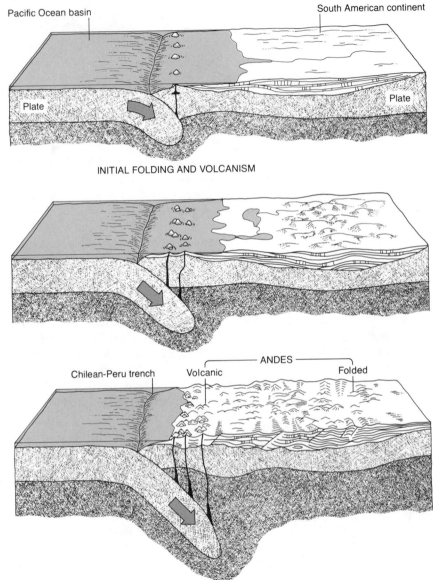

Pacific Ocean basin

South American continent

Plate

Plate

INITIAL FOLDING AND VOLCANISM

Chilean-Peru trench

ANDES

Volcanic

Folded

Fig. 24.1 A generalized interpretation of the formation of the Andes. The thick wedge of sedimentary rock on the Pacific side of South America was compressed as the Nazca and South American plates converged. At the same time, volcanic mountains were built off shore. The volcanic and folded mountains coalesced to form the modern Andes.

before enjoyed a storm of this sort, the strange thrilling motion could not be mistaken, and I ran out of my cabin, both glad and frightened, shouting, "A noble earthquake! A noble earthquake!" feeling sure I was going to learn something. The shocks were so violent and varied, and succeeded one another so closely, that I had to balance myself carefully in walking as if on the deck of a ship among waves, and it seemed impossible that the high cliffs of the Valley could escape be-

ing shattered. In particular, I feared that the sheer-fronted Sentinel Rock, towering above my cabin, would be shaken down, and I took shelter back of a large yellow pine, hoping that it might protect me from at least the smaller outbounding boulders. For a minute or two the shocks became more and more violent—flashing horizontal thrusts mixed with a few twists and battering, explosive, upheaving jolts,—as if Nature were wrecking her Yosemite temple, and getting ready to build a still better one.

Edwin W. Teale, *The Wilderness World of John Muir*, Boston, Houghton-Mifflin, 1954.

The cause of this earthquake was a sudden movement of rock along a fault in the Owens Valley, situated on the east side of the Sierra Nevada from Yosemite. Despite the spactacle created by this event, one of the strongest earth movements in California in the past two centuries, little deformation of the land was actually produced in the Owens Valley (Fig. 24.2). But this is typically the case with a single movement along a fault; therefore, we should appreciate that it takes a long time to produce mountains the size of the Sierra Nevada. Folding, tilting, and uplift may be even slower, so the orogenic belts must take millions of years to form (Authors' Note 24.1).

The Nature of Rock Deformation

In order to understand the processes of rock deformation, we should begin with a look at the behavioral characteristics of rocks when

Fig. 24.2 Surface displacement in the Owens Valley, produced by the 1872 earthquake, one of the strongest earthquakes in California in the past two centuries. The foothills of Sierra Nevada are on the left. (Photograph and diagram by William M. Marsh)

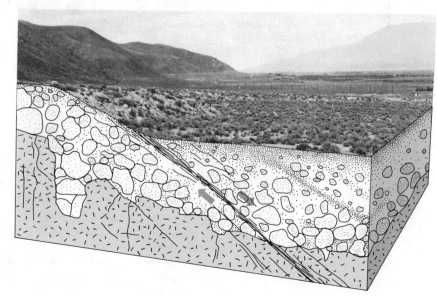

AUTHORS' NOTE 24.1

Detection of Contemporary Tectonic Movement

"Recent land uplift of as much as about 10 inches (25 centimeters) has been discovered astride a large section of California's San Adreas fault about 40 miles north of Los Angeles, according to scientists of the U.S. Geological Survey, Department of the Interior. The land swelling, in the shape of a huge kidney, has a 120-mile axis oriented roughly east-west and extending from the Pacific Ocean into the Mojave Desert. This uplift is receiving close attention among USGS earthquake specialists because similar swelling has occurred prior to some earthquakes in California and elsewhere. The scientists emphasize, however, that such uplifts also have occurred without subsequent earthquakes. Discovery of the uplift was made by R. O. Castle, J. P. Church, and M. P. Elliott, scientists at the USGS Menlo Park, California field center. Centered north of Los Angeles near Palmdale in the western Mojave Desert, the swelling apparently began about 1960 near the junction of the San Andreas and Garlock faults. Since then, it has grown east-southeastward to include an area of about 4,500 square miles (12,000 square kilometers). The uplift discovery resulted from analyses of repeated measurements taken over a number of years along precisely surveyed elevation lines crossing the southern California region. The measurements were made by various organizations, including the USGS, the National Geodetic Survey, and several southern California municipalities and counties. The significance of the uplift is not fully understood, according to USGS scientists; they are concerned, however, because it occurs astride a sector of the San Andreas fault that has remained locked since a great earthquake in 1857. Thus, the scientists explain, considerable strain could be building up in this area."

(U.S. Department of the Interior, *News Release*, February 1976)

Some of the immediate effects of earthquakes, ground cracks: left, San Francisco, 1906; right, Madison Valley, Montana, 1959.

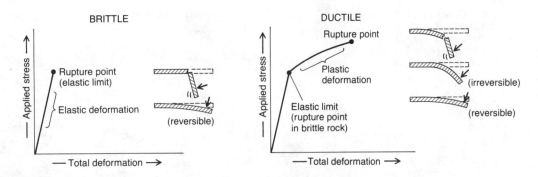

Fig. 24.3 The responses of brittle and ductile rocks to stress. With the initial application of stress, all rocks behave like an elastic and deform relatively little (represented by the lower part of both graph lines). The difference between brittle and ductile rocks occurs at the elastic limit. At this point the brittle rock ruptures, whereas the ductile bends further, but as a plastic rather than as an elastic. Ultimately, of course, even the ductile rock ruptures.

placed under stress. *Stress* is defined as a force that is acting on a body. If the force directed on a body is not equal in all directions, it is referred to as *differential stress*. When rock is subjected to differential stress, such as that produced by the motion of tectonic plates, it may undergo three stages of deformation: elastic, plastic, and rupture. Under *elastic* deformation, the rock is able to return to its original shape and size if the stress is withdrawn. For each type of rock there is an *elastic limit,* beyond which the rock does not return to its original shape if the stress is released. If stress exceeds this limit, *plastic* deformation is the result. Plastic deformation of rock is called folding, and it is irreversible. If stress is increased still further, exceeding the plastic limit, and fractures develop, the rock is said to fail by *rupture.*

How different rocks behave under differential stress depends on their internal properties, such as mineral composition and crystal structure, as well as on the rate at which stress is applied. By applying equal amounts of stress to many different rocks, one can compare their deformational characteristics. Such tests show that rocks tend to fall into two main groups: brittle and ductile. *Brittle* rocks rupture before any plastic deformation occurs; *ductile* rocks show a large range between the elastic limit and the rupture point (Fig. 24.3). Table 24.1 gives the compressive rupture strengths of ten common rocks. Although the crystalline rocks are higher on the average, it is noteworthy that most rock types show a wide range in compressive strengths.

Table 24.1 Compressive rupture strength of rocks (in kilograms per square centimeter).

ROCK	AVERAGE	RANGE
Basalt	2750	2000–3500
Quartzite	2020	260–3200
Diorite	1960	960–2600
Gabbro	1800	460–4700
Gneiss	1560	810–3270
Granite	1480	370–3790
Slate	1480	600–3130
Marble	1020	310–2620
Limestone	960	60–3600
Sandstone	740	110–2520

Marland P. Billings, *Structural Geology,* Englewood Cliffs, N.J.: Prentice-Hall, 1954. Used by permission.

Properties of Faults and Folds

Faults are fractures or ruptures along which differential displacement of rock has taken place. Relative to the fracture itself, the displacement in a fault may be up, down, back and forth, in or out, or any combination of these. Faults range in size from only a few centimeters long to tens of kilometers through the lithosphere. The largest faults

Fig. 24.4 A segment of the San Andreas fault north of San Francisco. This fault can be traced for about 750 km across California. (Photograph by R. E. Wallace, U.S. Geological Survey)

are formed along the edges of the tectonic plates, where networks of faults such as the San Andreas can be traced hundreds of kilometers along the surface (Fig. 24.4).

A standard set of terminology is used to describe faults. The faces of the blocks on either side of the fault are called the *walls*; the surface separating the walls is the *fault plane*. If the fault plane is inclined, which is normally the case, the upper face is called the *hanging wall*, and the lower face is called the *footwall* (Fig. 24.5). The trend of the fault along the earth's surface, as it would appear on a map, for example, is termed the *fault line*. The part of a wall exposed as a result of displacement is the *fault scarp*.

Folds are structures that resemble warps or wrinkles in rock. They are the most common geologic structure in the orogenic belts and are often elegantly displayed in exposures of sedimentary and metamorphic rocks along canyons and cliffs (Fig. 24.6). Folds are also common in nonmountainous areas, e.g., the midwestern United States, northern France, and southern England, where they appear as broad upward and downward flexures called *domes* and *basins* and may be hundreds of kilometers in diameter.

The basic terms used to describe folds are as follows. The two sides of a fold are the *limbs*. An imaginary plane drawn between the limbs, which divides the fold in half, is called the *axial plane*, and the crest of the fold is the *axis*. If the axis is inclined from the horizontal,

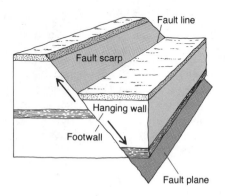

Fig. 24.5 The basic features of a fault: fault scarp, fault line, footwall, hanging wall, and fault plane.

Fig. 24.6 Folds in sedimentary rocks exposed on a mountain side. (Photograph by
Jeff Dozier)

the fold is said to *plunge*. Imagine a plunging fold to be somewhat like a
submarine emerging on the ocean surface (Fig. 24.7). Authors' Note
24.2 describes the method used by scientists to map the directional patterns of folds and faults in the field.

Fig. 24.7 The basic features of a fold:
axial plane, axis, limbs, and angle of
plunge. If the axis is vertical, the fold is
symmetrical; if tilted, it is asymmetrical.
(Illustration by Peter Van Dusen)

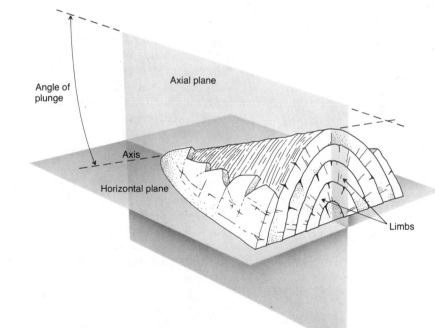

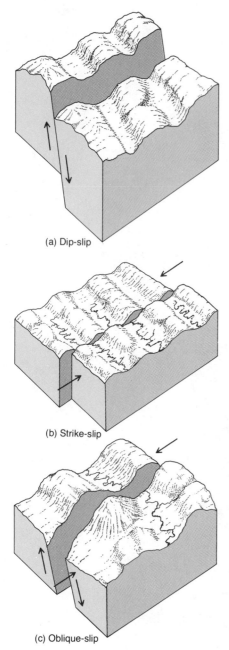

Fig. 24.8 *The three principal types of displacement in faults: (a) dip-slip; (b) strike-slip; (c) oblique-slip. (Illustration by William M. Marsh)*

(a) Dip-slip

(b) Strike-slip

(c) Oblique-slip

Types of Faults

In Chapter 23 we discussed faults in connection with the motion of tectonic plates. Three basic types of motion were described: tensional (pulling apart) in constructive zones, compressional (pushing together) in destructive zones, and lateral in conservative zones where the crust is displaced horizontally. The same types of motion also produce faulting on a smaller scale, and these faults are classed according to the direction of displacement of one wall relative to the other (Fig. 24.8). A displacement that is up or down along the fault plane is called a *dip-slip* fault; one that is parallel to the fault line, as in the transform faults on the ocean floor, is termed a *strike-slip* fault. Displacements that combine strike- and dip-slip are *oblique-slip* faults.

Several kinds of dip-slip faults are recognized: normal faults, reverse faults, and thrust faults (Fig. 24.9). In a *normal fault* the hanging wall is displaced downward relative to the footwall, exposing the upper part of the footwall in the form of a fault scarp. Many mountain ranges in the American West, including the Wasatch of Utah and the Sierra Nevada of California, represent normal faults. In the Sierras the fault scarp faces east; in the Wasatch the fault scarp faces west (Fig. 24.12). Between the Sierra Nevada and the Wasatch lies an extensive area called the Great Basin, in which range after range of mountains has been formed by normal faulting.

The opposite displacement produces a *reverse fault,* in which the hanging wall moves up relative to the footwall. Under great compressional force, reverse faults with gently inclined fault planes may produce mostly horizontal movement. These *thrust faults* are known to drive slabs of rock laterally for tens of kilometers over the surface. In some parts of the Rocky Mountains rock formations were thrust great distances to form prominent front ranges along the Great Plains. The Canadian Rockies in Alberta are a case in point; great slabs of rock were thrust eastward and piled against one another in an overlapping fashion (Fig. 24.12).

Finally, some faults involve two fault planes. A rift, or *graben,* forms where a block is displaced downward between two normal faults. The opposite type of structure, called a *horst,* results in the ele-

Geologists have traditionally mapped the distribution of folds and faults in an attempt to work out the broad deformational patterns in the crust. Until the development of seismic techniques, however, little of the underground parts of the features could be measured, and so mapping was based on the directional characteristics of rock outcrops, i.e., surface exposures of bedrock. Upon locating a fault line or the eroded limb of a fold, the geologist first measures its directional trend. Next, if beds such as sedimentary formations can be discerned along the line, the geologist measures the amount of inclination of the beds and the direction of the inclination. These two directional properties of an outcrop are known as strike and dip, and from many such recordings, we can learn much about the form of faults and folds. Strike is the directional trend of a formation along the surface; dip is the angle between the formation and

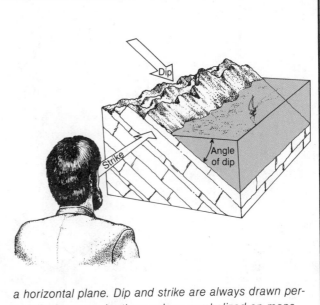

a horizontal plane. Dip and strike are always drawn perpendicular to each other and are symbolized on maps by T-shaped figures.

vation of a block between parallel fault planes. Rifts and horsts are common features in zones of tensional motion. Today the most extensive system of rifts on earth begins in the Gulf of Aqaba on the east side of the Sinai Peninsula and extends through the basin of the Red Sea and into the West African Highlands, where it forms the basins for

Fig. 24.9 The main types of dip-slip faults: (a) normal; (b) reverse; (c) thrust; (d) graben; (e) horst.

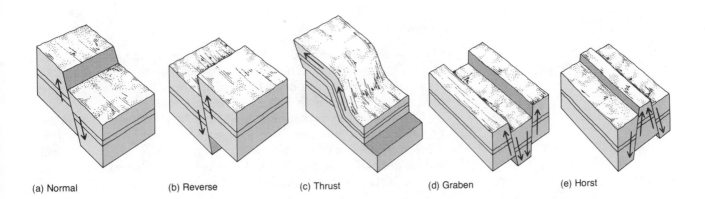

(a) Normal   (b) Reverse   (c) Thrust   (d) Graben   (e) Horst

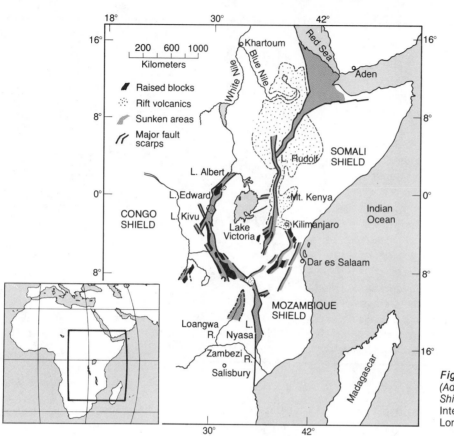

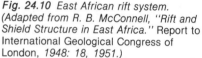

Fig. 24.10 *East African rift system. (Adapted from R. B. McConnell, "Rift and Shield Structure in East Africa."* Report to International Geological Congress of London, *1948: 18, 1951.)*

large inland lakes such as Lake Tanganyika and Lake Malawi (Nyasa) (Fig. 24.10).

Types of Folds

When rock undergoes deformation, folding usually accompanies faulting, and in most orogenic belts folding is actually the more prevalent form of deformation. Although folds in deformed rock occur in every size and shape, four or five shapes are most common.

The simplest folds can be described in two-dimensional terminology: monoclines, anticlines, synclines, and overturned folds (Fig. 24.11). A *monocline* is a single bend in an otherwise horizontal formation; an *anticline* is a double bend upward in the shape of an "A"; a *syncline* is the downward counterpart of an anticline. Either may be

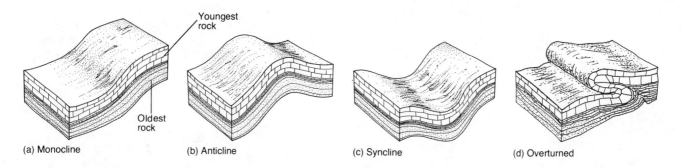

Youngest
rock

Oldest
rock

(a) Monocline   (b) Anticline   (c) Syncline   (d) Overturned

Fig. 24.11 Four basic types of folds: (a) monocline; (b) anticline; (c) syncline; and (d) overturned.

symmetrical or asymmetrical. An *overturned* fold represents such extreme bending that the formation on the lower part of the fold is turned upside down.

The Ridge and Valley section of the Appalachian Mountains contains spectacular examples of folding in sedimentary rocks (Fig. 24.12). The ridges and valleys trend north and south; when viewed in cross-section, as in Fig. 24.12, they resemble a washboard terrain. On closer examination, however, it is apparent that the topography produced by the folding deviates from the structural pattern of anticlines and synclines because valleys are formed from both types of folds. Synclinal valleys are easy to figure out; but in order to figure out the origin of anticlinal valleys, you must infer that the tops of the anticlines have been eroded away, exposing the underlying, often weaker, formations to erosion. The ridges are formed from the limbs of pairs of anticlines and synclines.

VOLCANISM

Volcanoes are emissions of magma, rock fragments (called *pyroclastics*), and gases, released through openings in the earth's crust. Most volcanoes appear to be explosive in nature and produce more pyroclastics than lava. Although volcanism is common on the continents, the bulk of it occurs in the ocean basins along midoceanic ridges and subduction zones (Figs. 23.10 and 23.15). With a few exceptions, volcanism on the continents is associated with orogenic environments.

Types of Volcanoes

Although we tend to think of volcanoes as conical-shaped mountains, they actually take on a great variety of forms. In fact, some of the largest volcanoes are not mountains at all, but are more like plateaus.

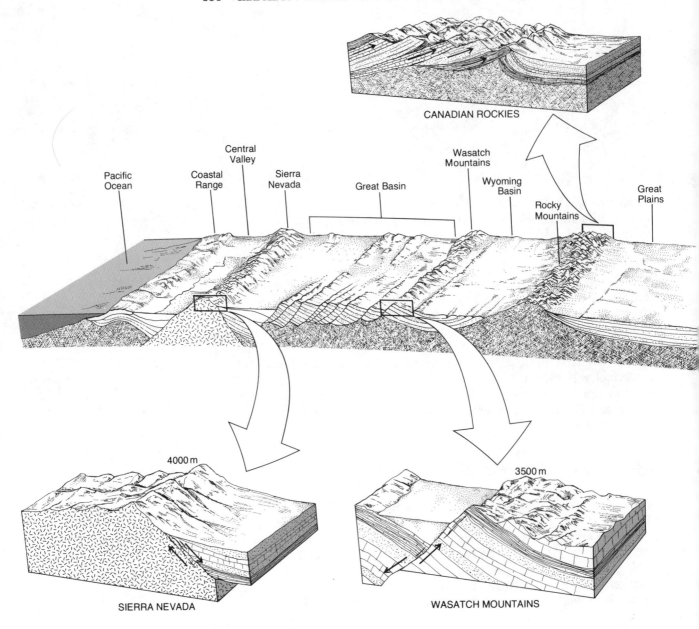

CANADIAN ROCKIES

Pacific
Ocean

Coastal
Range

Central
Valley

Sierra
Nevada

Great Basin

Wasatch
Mountains

Wyoming
Basin

Rocky
Mountains

Great
Plains

4000 m

SIERRA NEVADA

3500 m

WASATCH MOUNTAINS

These *basalt floods* are the result of massive outflows of lava from long, narrow openings, or fissures, in the crust. The lava spills over the landscape in a relatively thin sheet, typically covering hundreds of square kilometers with each eruption. The fluid behavior of the lava is attrib-

Fig. 24.12 A geological diagram across
North America from California to Mary-
land. Enlargements show the Sierra
Nevada, the Wasatch Mountains, the
Appalachians, and the Canadian Rockies.
(Illustration by William M. Marsh)

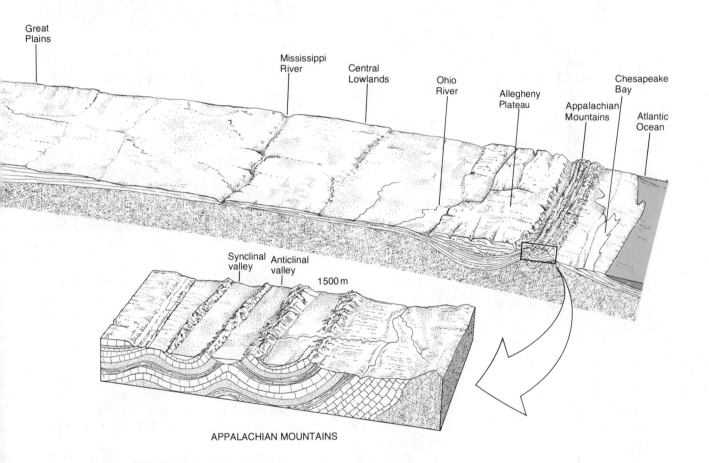

APPALACHIAN MOUNTAINS

uted to its high temperature, often up to 1200°C. On sloping surfaces it
can flow at velocities exceeding 50 km per hour. The largest continuous area of flood basalts in the United States is the Columbia Plateau,
which covers more than 400,000 km^2 (Fig. 22.12).

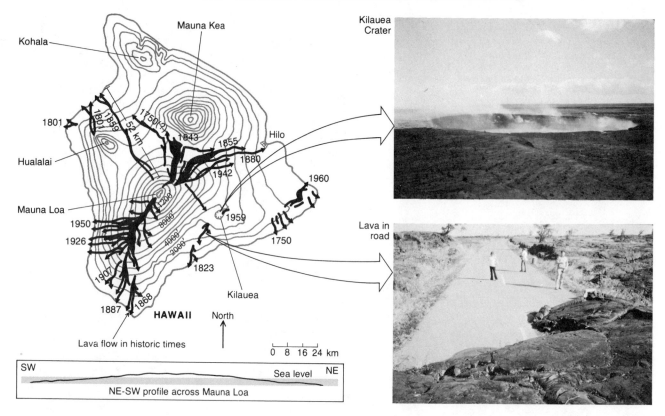

Fig. 24.13 Contour map of Hawaii, showing the five shield volcanoes that form the island. The heavy lines represent lava flows during historic times. (Illustration after Stearns and Macdonald, 1946; photographs by Bruce D. Marsh)

If a fluid magma is released from a group of tunnels, called vents, which are chronically active, the lava tends to pile up, forming a broad mound called a *shield volcano*. Although shield volcanoes may grow to heights of several thousand meters, most of their growth is lateral and is characterized by many *flank eruptions*. Side slopes are gentle, generally less than 5°, and are laced with lava flows of various sizes. The Hawaiian Islands are excellent examples of shield volcanoes. The main island, Hawaii, was built from five shield volcanoes which coalesced into a single mass as they emerged from the sea (Fig. 24.13). Like Hawaii, most shield volcanoes tend to form near the interiors of tectonic plates.

Composite volcanoes (also known as *stratocones*) are generally conical in shape and represent the popular stereotype of the "classical" volcano. Smaller and steeper than shield volcanoes, they also differ in composition; shield volcanoes are almost entirely basaltic lava, whereas composite volcanoes are made up of both lava and pyroclastic rocks. Pyroclastic material, which is described as ash and all sizes of rock particles, is commonly ejected into the air in the opening

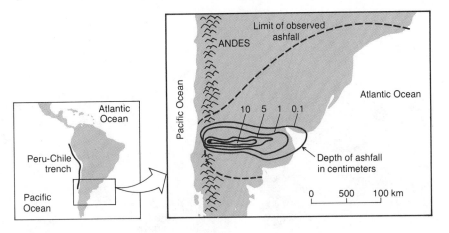

Fig. 24.14 *The distribution of ash from the eruption of Volcán Quizapú in 1932. Westerly winds carried the ash rapidly eastward, reaching Buenes Aires, located 1000 km away, in eighteen hours. Ashfall was observed in southern Brazil, 4000 km from the volcano. (After Larsson, 1937)*

stage of an eruption. The coarse particles rain back to the surface of the volcano to form a layer of some thickness; the fine particles are carried off in the atmosphere and eventually fall out elsewhere (Fig. 24.14). Later in the eruption or in a subsequent eruption, lava flows may bury the pyroclastic layer; thus the lava and the pyroclastics become interbedded in a layer-cake fashion. Many of the famous volcanoes belong to the composite class, including Vesuvius and Etna of Italy, Fuji-san of Japan, and Ararat of Turkey. Virtually all of the volcanoes in subduction zones are composites, which accounts for their abundance around the Pacific plate.

Composite and shield volcanoes often develop similar anatomies, characterized by a main passageway (called the *central vent*), multiple secondary vents (which lead from the central vent to the sides), and a *crater* at the summit (Fig. 24.15). The crater forms when magma held in the central vent is released through a lower vent, thereby causing the neck to retract. Such shifts in material can also break the hull of the volcano and create fissures which later serve as passageways for lava. A much larger depression, called a *caldera*, also forms in some volcanoes. Calderas form when a volcano collapses due to drainage of a large mass of magma through a lower vent or the expulsion of a massive amount of material in an explosive eruption. Such eruptions have been known to destroy the entire superstructure of a volcano.

Basic Mechanics of Volcanism

The underground mechanics of volcanism are in general poorly understood. Most likely, the heat that creates the magma is supplied mainly by the asthenosphere. When a pocket of magma develops at the base of the lithosphere, it begins to rise through the overlying rock like a parcel of hot air rising through the atmosphere, displacing its way upward

SHIELD VOLCANO   COMPOSITE VOLCANO

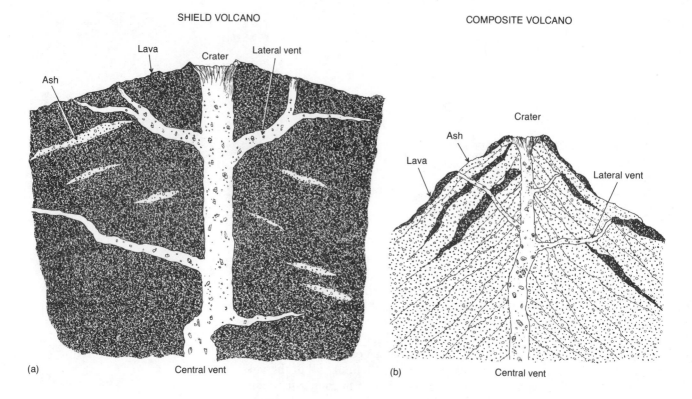

(a)   (b)

Fig. 24.15 *The basic structure and circulation features of a shield volcano (a) and a composite cone (b).*

into colder zones. It is unlikely, however, that the first pocket to rise at some point is able to reach the earth's surface. Actually, calculations show that a large pocket of magma, on the order of 1 km in diameter, must rise at a rate of 1 km per year in order not to solidify in the lithosphere. Thus it appears that many pockets of magma must rise in succession if magma is to reach the surface and burst forth or spill out there. Once the hot conduit (diapir) is created by the ascending magma, the lithosphere is not quick to destroy it. Therefore, subsequent missiles of magma may reach the surface with comparative ease, which is one of the reasons why individual volcanoes tend to remain active for a long time after they are born (Fig. 23.9).

Lava Composition and Distribution

The mineral composition of the magma depends on where it originates and what type of rock it travels through on its way toward the surface. Three types of lava are recognized, based on the percentage of silcon dioxide and ferromagnesian minerals: silicic, intermediate, and mafic.

Silicic lavas are highest in silicon dioxide, which usually accounts for 70 percent or more of the rock. *Mafic* lavas, on the other hand, contain about 50 percent silicon dioxide, but are relatively high in the ferromagnesian minerals. *Intermediate* lavas are about 60 percent silicon dioxide. As you would expect, the geography of lavas corresponds to the major divisions in the earth's crust, with the mafic lavas concentrated in the ocean basins, where, you will recall, ferromagnesian minerals are abundant and the silicic lavas concentrated on the continents (where alummosilicate minerals are abundant) (see Fig. 22.8 and Table 22.1). Intermediate lavas tend also to be geographically intermediate, forming along the edges of the continents, where both sources of minerals are available. In some places geologists have used the seaward limit of this zone, called the *andesite line*, to delimit the border of the continents.

Volcanic Events and the Landscape

Volcanoes, like earthquakes, are able to render sudden change in the landscape. Accordingly, volcanic events have received a good deal of attention, and the human drama associated with volcanic disasters has given rise to some of our most interesting history.

Andean volcano (U.S. Geological Survey photograph)

Vesuvius, Italy, A.D. 79. Ranking high among these is the burial of the Roman city of Pompeii by the eruption of Vesuvius in A.D. 79. Vesuvius was, and still is, notorious for its misbehavior. One morning in the summer of A.D. 79 it erupted with a modest burst of ash, a common sight to the residents of Pompeii. Late the following day it became clear that this eruption was larger than usual, and an evacuation of Pompeii was begun. Before everyone could escape, however, Vesuvius suddenly released a huge mass of hot ash that literally buried people and animals in their tracks. Within several days most of the city was buried to such a depth that excavation and reuse in later years were considered all but impossible. On the positive side, though, archeologists were left with an excellent sample of life in Roman times.

Mount Pelée, Martinique, 1902. The Mount Pelée eruption of 1902 destroyed a city also, but in a fashion much different from that of Vesuvius. Pelée is located on the heavily populated island of Martinique in the Caribbean. After a half century of dormancy, the volcano erupted with a series of mild explosions in the spring of 1902. Then, on the morning of May 8, it released a tremendous blast that sent a mass of incandescent gas and ash out the side of the volcano and toward the city of St. Pierre. The mass, called a *nuée ardente* (for glowing cloud), moved at a velocity approaching 200 km/hr, overwhelming the city and

Fig. 24.16 A nuée ardente, showing how the disasterous explosion on Martinique may have looked. (Illustration by William M. Marsh)

killing its 28,000 inhabitants instantly (Fig. 24.16). Only one resident survived, a prisoner in a basement cell of the jail, but a number of sailors aboard ships in the harbor watched the event. One offered this description:

The mountain was blown in pieces. There was no warning. The side of the volcano was ripped out and there was hurled straight toward us a solid wall of flame. It sounded like a thousand cannon . . . .

The wave of fire was on us and over us like a flash of lightning. It was like a hurricane of fire. I saw it strike the cable steamship Grappler broadside on, and capsize her. From end to end she burst into flames and then sank. The fire rolled in mass straight down upon St. Pierre and the shipping. The town vanished before our eyes.

The air grew stifling hot and we were in the thick of it. Wherever the mass of fire struck the sea, the water boiled and sent up vast columns of steam. The sea was torn into huge whirlpools that careened toward the open sea.

. . . The blast of fire from the volcano lasted only a few minutes. It shrivelled and set fire to everything it touched. Thousands of casks of rum were stored in St. Pierre, and these were exploded by the terrific heat. The burning rum ran in streams down every street and out into the sea.

. . . Before the volcano burst, the landings of St. Pierre were covered with people. After the explosion, not one living soul was seen on the

land. Only twenty-five of those on board were left after the first blast.

K. H. Wilcoxon, *Chains of Fire—The Story of Volcanoes*, Philadelphia: Chilton, 1966.

Parícutin, Mexico, 1943. Parícutin was born in a farmer's field about 300 miles from Mexico City in the winter of 1943. After a two-week period of earthquakes, the eruption began when a narrow fissure opened, emitting a column of gray smoke. Within twelve hours the volcano became violent, ejecting incandescent rocks and a huge eruptive column of ash and gas. In less than twenty-four hours, a cone 40 m high had developed. The eruption was now in full force, with lava bombs being hurled several kilometers into the air and ashfall reaching distances of several kilometers from the cone. The following day a lava flow spilled from the base of the cone, and the volcano began its lateral expansion. Parícutin continued to be active until 1952. In nine years it grew to a height of 250 m above its base, buried two villages with lava, and caused the abandonment of more than 50 km² of farmland (Fig. 24.17).

Krakatoa, Indonesia, 1883. Krakatoa was a small, uninhabited volcanic island situated in an ancient caldera between Java and Sumatra. In May 1883 it began to emit ash, and on August 27 it exploded in what

Fig. 24.17 *The distribution of ash and lava from Parícutin in the late 1940s. Abandoned cropland, which at that time amounted to more than 50 km², is also shown. (Adapted from Kenneth Segerstrom, "Erosion Studies of Parícutin, State of Michoacan, Mexico," U.S. Geological Survey Bulletin 965–A, 1950)*

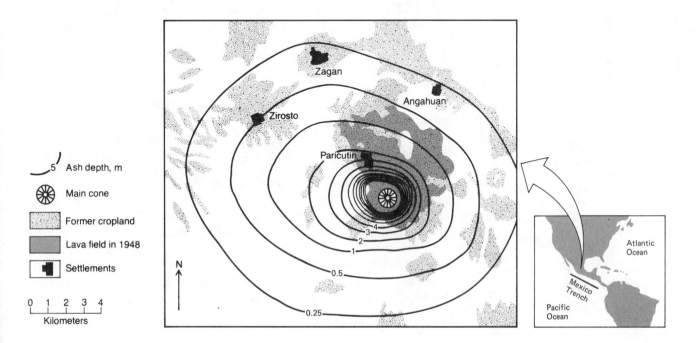

Table 24.2 Major effects of the Krakatoa blast, 1883.

TYPE	DESCRIPTION
Atmospheric	Sound of explosion heard in Australia, more than 2000 km away
	Created air pressure waves recorded on other side of earth
	Airborne debris created almost total darkness during midday at Jakarta, 150 km away
	Dust that entered higher atmosphere increased backscattering of solar radiation, resulting in temperatures several degrees cooler over much of the earth
Oceanic	Generated giant wave nearly 40 m high which overwashed near-by coastal lands, killing 30,000 people
	Sea waves produced by explosion were recorded on tidal gauges as far away as the English Channel
Geomorphic	Discharged about 20 cubic km of debris into atmosphere
	Created a hole 300 m deep in the caldera

is generally considered the greatest natural blast ever witnessed. The energy involved exceeded that of the atomic blast on Hiroshima by at least five times. The explosion was of a boiler steam-type origin, caused by the superheating of groundwater locked in the volcano. The effects of the blast were superlative and are summarized in Table 24.2.

SUMMARY

Orogenic belts are produced by massive deformation of rock near the edges of tectonic plates. In most mountain ranges folding is the principal type of deformation, and it is usually accompanied by extensive faulting and volcanism. Folds and faults are classified according to the attitude of the deformed rock and the relative directions of displacement. The Ridge and Valley section of the Appalachian Mountains is a classic example of terrain produced by folding. The rift valleys of East Africa and the mountain ranges of the Great Basin of the United States are examples of terrain created largely by faulting.

Volcanoes are classified mainly according to overall structure. The island arcs of the Pacific Ocean, such as the Aleutian and Japanese Islands, contain many composite volcanoes, whereas the Hawaiian Islands are shield volcanoes.

UNIT VI SUMMARY

- The earth's crust is the rock platform on which the landscape rests. Because of its diverse composition and irregular configuration at the earth's surface, the crust has an important influence on the formation of the landscape. This influence extends to soils, vegetation, climate, and drainage.

- The crust is also dynamic, often changing suddenly with earthquakes and volcanism. Both of these processes are related to tectonic plates and their motion.

- The earth's lithosphere, which extends to a depth of about 100 km, is subdivided into seven major tectonic plates and many smaller ones. With the exception of the Pacific plate, each major plate contains a continent. The plates move in different directions at rates ranging from 0–1 cm to 6–10 cm per year.

- Depending on the direction of movement of adjacent plates, the crust is pulled apart, pushed together, or dragged parallel to itself. These motions result in the formation of the earth's major geographical features, including orogenic belts, midoceanic ridges, and ocean trenches. In fact, the growth of the continents themselves appears to have resulted from many episodes of compression and uplift of sedimentary, volcanic, and metamorphic rocks.

- The three main types of contacts between plates are called *destructive,* where subduction is taking place and lithosphere is being destroyed; *constructive,* where spreading is taking place and new crust is formed; and *conservative,* where plates are moving past each other and crust is being neither destroyed nor formed.

- The energy that drives the movement of tectonic plates is earth heat in the mantle. The heat flux from the crust averages about 0.06 J/m$^2 \cdot$s, and for the whole earth for a year, this is more than ten times the energy used annually by all humanity.

- The *theory of plate tectonics* marks one of the great advances in the history of earth science. Although much remains to be learned, especially with respect to the mechanics of the driving force, it does provide a unified model of the planet's geology. This model gives us insight into past environments, the distribution of many plants and animals, and the geographical arrangement of major land and water features.

- Mountains are comprised of three types of geologic structures: folds, faults, and volcanoes. Folds, the most prevalent structure, are also found in nonmountainous regions where the crust has been gently warped by susidence or uplift.

- Volcanism has produced some of the more dramatic landscape changes known to humans. In addition to the direct effects of explosive eruptions and lava flows, airborne volcanic debris may influence weather and climate, and the fallout of ash may alter vegetation and land use.

FURTHER READING

Elder, John, *The Bowels of the Earth,* Oxford, England: Oxford University Press, 1978. *Despite the title and chapter titles (e.g., ''Global Jam-Pot.'' ''Crustal Retreading,'' Second-Hand Rock''), a serious and creative effort. Requires advanced work in physical science.*

Phillips, Owen M., *The Heart of the Earth,* San Francisco: Freeman, Cooper, 1968. *A concise and readable discussion of the principles of solid and fluid mechanics of the earth's interior.*

Wegener, Alfred, *The Origin of the Continents and Oceans* (translated by John Biram), New York: Dover, 1966. *Wegener's original proposition on continental drift, first published in 1915. He draws heavily on geographical and biological evidence in building the case for continent drift.*

Wylie, Peter J., *The Way the Earth Works: An Introduction to the New Global Geology and Its Revolutionary Development,* New York: Wiley, 1976. *A good description of the findings and problems in the unfolding story of plate tectonics. Very readable.*

REFERENCES AND BIBLIOGRAPHY

Bird, J., and I. Bryan, ed., *Plate Tectonics,* Washington, D.C.: American Geographical Union, 1972.

Cargo, D., and R. Mallory, *Man and His Geologic Environment*, 2d ed., Reading, Mass.: Addison-Wesley, 1977.

Chapman, D. S., and H. N. Pollack, ''Global Heat Flow: A New Look,'' *Earth and Planetary Science Letters* **23** (1975): 23–32.

Coffman, J. L., and C. A. von Hake, ed., *Earthquake History of the United States,* Washington, D.C.: U.S. Department of Commerce (NOAA), 1973.

Cox, A., G. B. Dalrymple, and R. R. Doell, ''Reversals of the Earth's Magnetic Field,'' *Scientific American* **216** (1967): 44–54.

Dalrymple, G. B., E. A. Silver, and E. D. Jackson, ''Origin of the Hawaiian Islands,'' *American Scientist* **61,** 3 (1973): 294–308.

Fridriksson, S., *Surtsey: Evolution of Life on a Volcanic Island,* New York: Halsted Press, 1975.

Green, J., and N. M. Short, *Volcanic Landforms and Surface Features*, New York: Springer-Verlag, 1972.

Hallam, A. *A Revolution in the Earth Sciences*, Oxford, England: Clarendon Press, 1973.

Holmes, A., *Principles of Physical Geology*, New York: Ronald Press, 1965.

Judson, S., K. S. Deffeyes, and R. B. Hargraves, *Physical Geology*, Englewood Cliffs, N.J.: Prentice-Hall, 1976.

Kummel, B., *History of the Earth,* San Francisco: Freeman, 1970.

Larsson, W., "Volcanic Ashes from the Eruption of the Volcano Quizapú (1932), Collected in Argentina: A Study of Aeolian Differentiation," *Geological Institute Bulletin* **26** (1937).

Marsh, B. D., "Island-Arc Volcanism," *American Scientist* **67** (1979): 161–172.

Marsh, B. D., and I. S. E. Carmichael, "Benioff Zone Magnetism," *Journal Geophysical Research* **81** (1974): 975–984.

Morgan, W. J., "Deep Mantle Convection and Plate Motions," *Bulletin American Society Petroleum Geologists* **56** (1972): 203–213.

Murphy, R. E., "Landforms of the World," Map Supplement No. 9, *Annals of the Association American Geographers* **58,**1 (1968).

Press, F., and R. Siever, *Earth*, 2d ed., San Francisco: Freeman, 1978.

Richter, C. F. *Elementary Seismology*, San Francisco: Freeman, 1958.

Stearns, H. T., and G. A. MacDonald, "Geology and Ground Water Resources of the Island of Hawaii," *Hawaii Division Hydrography Bulletin* **9**, 1946.

U.S. Geological Survey, *Atlas of Volcanic Phenomena*, Washington, D.C.: Government Printing Office, n.d.

Verhoogen, J., et al., *The Earth: An Introduction to Physical Geology,* New York: Holt, Rinehart and Winston, 1970.

Wilson, J. T., ed., *Continents Adrift—Readings from "Scientific American,"* San Francisco: Freeman, 1970.

WEATHERING
AND
HILLSLOPE
PROCESSES

UNIT VII

KEY CONCEPTS   OVERVIEW

geomorphology
erosional agents
chemical weathering
chemical stability
mechanical weathering
weathering and landforms
rainwash
sapping
piping
erosion and climate
mass movement
law of plastic deformation
mudflow
soil creep
solifluction
pore-water pressure
angle of repose
denudation
slope retreat
pediment
landforms and climate

Many processes are involved in the lowering of land masses and the shaping of landforms. The science of geomorphology focuses on those processes and the forms they produce in the landscape. Geomorphology begins with an examination of rock weathering, which involves the chemical decay and physical fragmentation of bedrock. Weathering rates tend to vary with climatic conditions, primarily heat and moisture, and with rock type. Overall, weathering is most rapid where water is abundant and temperatures are high.

The debris produced by weathering is moved downhill by hillslope processes such as rainsplash, overland flow, mudflows, and landslides. The rates at which these processes move debris vary with climate, topography, soil type, vegetative cover, and land-use activities. From the foot of a mountain or hill, the debris may be picked up by streams which carry it slowly toward the sea, where most of it emerges in the form of clay, sand, and soluble ions. The rate at which continents are lowered by this chain of processes appears to be about several centimeters per millenium.

Chapter 25 begins with a brief discussion on the nature of geomorphology and follows with a description of chemical and mechanical weathering. Hillslope processes, including rainsplash, soil creep, and landslides, are examined in Chapter 26.

CHAPTER 25 ROCK WEATHERING

INTRODUCTION TO GEOMORPHOLOGY:
THE SCIENCE OF LANDFORMS

One of the most fascinating aspects of the landscape is the form of the land. For centuries, hills, mountains, valleys, and coastlines have aroused curiosity and have provided a source of inspiration for exploration, science, and the arts. A survey of the literature of the Romantic movement illustrates the latter, as the following excerpt from Longfellow's "Sunrise on the Hills" suggests:

I stood upon the hills, when heaven's wide arch
Was glorious with the sun's returning march,
And woods were brightened, and soft gales
Went forth to kiss the sun-clad vales.
The clouds were far beneath me:—bathed in light,
They gathered mid-way round the wooden height,
And, in their fading glory, shone
Like hosts in battle overthrown,
As many a pinnacle, with shifting glance,
Through the gray mist thrust up its shattered lance,
And rocking on the cliff was left
The dark pine blasted, bare, and cleft.
The veil of cloud was lifted, and below
Glowed the rich valley, and the river's flow
Was darkened by the forest's shade,
Or glistened in the white cascade;
Where upward, in the mellow blush of day,
The noisy bittern wheeled his spiral way.

Indeed, it is interesting to note that one group of Romantic poets came to be identified with a specific landform region of England, the Lake District.

In the twentieth century the American preoccupation with land and landforms crystallized in the form of various conservationist and preservationist movements. Dedicated to the maintenance of mountain lands, streams, and coastlands, groups such as the Sierra Club and Friends of the Earth are present-day testimony of this trend. At the   **409** formal policy level, federal law has provided for public ownership, restricted use, and general preservation of a system of national parks, seashores, and wilderness areas. In most cases the designated areas are comprised of a distinctive set of landforms, as exemplified by Yosemite National Park, Grand Canyon National Park, and the Pictured Rocks National Lakeshore. The exploration urge, so pervasive in the nineteenth-century American movement westward, has come to be expressed in the activities of hiking, camping, mountaineering, boating, and motoring. Directly or indirectly, each of these is oriented toward landforms or the processes, such as streams and glaciers, associated with them.

The formal study of landforms and the processes that erode the land is called *geomorphology*. The key geomorphic processes are referred to as erosional agents: streams, glaciers, waves, currents, and wind. Geomorphology is studied by both geographers and geologists; in fact, it represents a traditionally powerful tie between these two fields of science.

The Essential Questions of Geomorphology

Three essential questions lie at the heart of geomorphology. First, at what rate is material eroded from the continents, and how does that rate compare with the rate of uplift of the continents by geophysical forces? This is a mass-balance problem, and it is of great interest to earth scientists because the results can reveal which continents are gaining and losing mass over geologic time. Unfortunately, we do not have enough data on erosion and uplift over a long enough span of time to figure out long-term trends for most continents. Over the short term, however, we are able to compute and to estimate erosion for some continents, based on field measurements of the sediment loads carried to sea by major rivers.

Second, into what forms is the land sculptured as it is being eroded? This question has received a good deal of attention from geographers, and for years one of the major issues has been whether major landforms are controlled mainly by geologic structures or by certain combinations of weathering and erosional processes that are specific to different climatic conditions. Both are important, of course, but the relative roles of geologic and climatic factors can vary from region to region. For instance, glaciers are a result of climatic conditions, and the shapes of the valleys they erode are distinctly different from those of valleys eroded by streams in areas too warm or dry to support glaciers.

Third is the question of how the erosional processes actually do their work on the land. Efforts to answer this question involve the application of physical and engineering principles to streams, glaciers, wind, waves, and currents. An important part of the search has centered on laboratory experiments and field measurements of these processes, leading to computation of the stress exerted by the moving water, air, or ice on the underlying earth materials. The amount of the stress exerted by these processes, however, is not everywhere proportional to the erosion they produce, because of the variable resistance of earth materials to erosional stress.

WEATHERING: THE FIRST PHASE

In order to appreciate how the land is eroded and how landforms are shaped, it is first necessary to understand the nature of the disintegration and breakdown of rock materials, termed *weathering*. Weathering involves a multitude of physical, chemical, and biological processes that cause physical fragmentation and chemical decomposition of rock. By definition, weathering is different from erosion; weathering involves only the breakdown of rock, whereas erosion involves the removal of the debris produced by the breakdown. This distinction is purely a matter of scientific convention, and in reality weathering and erosion are usually part of a chain of related processes, the end result of which is the loss of rock materials from the land.

Weathering is most intensive at or near the earth's surface. There are two basic reasons for this: (1) many rocks, particularly the igneous rocks, formed under conditions of heat, pressure, and chemistry so different from conditions at the surface that they tend to be unstable upon contact with surface environments; (2) the energy that drives the weathering processes is most concentrated on the earth's surface. This energy is represented by water, heat, and organic matter, as well as mechanical activity. Water is the most effective natural solvent on the planet, and heat generally increases the solubility of materials. The remains of plants contribute ions to water, which may increase its efficiency to disintegrate rock chemically. And mechanical factors, such as the expansion of ice in the cracks of rocks, tend to weaken and break solid rock.

For purposes of our discussion, we can treat weathering in two main categories: chemical and mechanical. (Some authors identify biological weathering as a third category, but most of the evidence indicates that biological factors play either a chemical or a mechanical role, contributing significantly to chemical weathering.) Most geomorphologists think that chemical weathering is the most effective type of weathering in terms of the total amount of rock breakdown. Although few data are available for comparison, it seems safe to say that for the earth as a whole, chemical weathering far exceeds mechanical weathering.

CHEMICAL WEATHERING

Throughout this book we have mentioned the principle of physical stability in reference to various parts of the environment. In the atmosphere, for instance, air is stable or unstable depending on its temperature structure. If cold, dense air overlies warm, light air, the two bodies will overturn, and the situation is thus said to be gravitationally unstable. This concept is the basis for understanding how wind is produced. By the same token, the concept of chemical stability is necessary in order to understand chemical weathering.

Chemical stability refers to the tendency of minerals to change into other states—from one mineral to another. Minerals usually form under conditions very different from those at the earth's surface; thus when they come into contact with water containing atmospheric gases and chemicals from terrestrial sources, a reaction takes place. Whether a mineral tends to be stable or unstable is related to the conditions under which it originally formed. The least stable minerals are those which formed at the highest temperatures and pressures and, in the presence of water or air, readily react to change into some other material. Calcic feldspar, for example, is transformed by chemical action into clays and soluble ions, neither of which can be reconstituted into the original mineral. On the other hand, quartz crystallizes at lower temperatures and is stable compared with calcic feldspar. Figure 25.1 lists ten minerals according to their chemical stability at the earth's surface.

Most chemical reactions take place with greatest intensity in the presence of liquid water, because of the structure of the water molecule. The hydrogen atoms are located on one side of a large oxygen atom, giving the molecule a strong negative charge on one side and a positive charge on the other (Fig. 25.2). The water molecule therefore acts like a magnet, loosening and drawing to it the ions of minerals. Most minerals react very slowly in the presence of moisture and remain in an intermediate state between stability and instability, referred to as a *metastable* state, for long periods, but this varies with bioclimatic conditions.

The power of water to affect chemical weathering is controlled by three factors: (1) the availability of water; (2) the temperature of the water; and (3) the chemicals in the water. The first is readily apparent when we examine rocks in deserts and wet environments. In wet areas rocks such as granite appear "rotten," because the feldspar and the

MOST STABLE

Quartz
Clay minerals
Muscovite
Potassic feldspar
Biotite
Sodic feldspar
Amphibole
Pyroxene
Calcic feldspar
Olivine

LEAST STABLE

Fig. 25.1 Stability of ten common minerals under various conditions at the earth's surface.

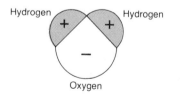

Hydrogen   Hydrogen

Oxygen

Fig. 25.2 The structure of the water molecule. Two hydrogen atoms with positive electrical charges are attached to an oxygen atom with a negative charge. Thus the water molecule is a form of an electric dipole, or magnet.

Fig. 25.3 "Rotten" granite, the result of chemical weathering in the presence of moisture. The black object is a lens cap about 5 cm (2 in.) in diameter. (Photograph by William M. Marsh and Jeff Dozier)

biotite mica, both relatively unstable, break down around the quartz crystals (Fig. 25.3). In deserts granite often looks unaltered, because the more reactive minerals, with little water to act on them, remain intact for long periods of time.

The role of heat in chemical weathering is to influence the rate of chemical reactions. Most reactions double with each 10°C rise in temperature. Therefore, we could expect that in a tropical environment with an average temperature around 27°C, chemical weathering can decompose a rock as much as four times faster than in the subarctic at an average temperature of 7°C or so (Fig. 25.4).

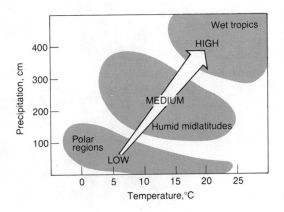

Fig. 25.4 The combinations of atmosphere moisture and heat that are conducive to high, medium, and low rates of chemical weathering, respectively.

Rain and snow are not chemically pure water. For example, when water vapor condenses in a cloud, it assimilates carbon dioxide (CO_2) from the air. Through this process the CO_2 content of rain or snow can be raised by ten to thirty times over the amount of CO_2 normally in the atmosphere (0.03 percent). Once dissolved in water, the carbon dioxide combines to form a weak carbonic acid. On the ground, additional carbonic acid can be formed if the rain infiltrates decomposing organic material. Organic matter in the topsoil is the food for bacteria, and through respiration they in turn produce carbon dioxide. In moist, aerated soils, bacterial activity may raise the CO_2 content of soil air to as much as 10 percent by volume. Added to soil water and groundwater, the CO_2 increases the concentration of carbonic acid.

Types of Chemical Weathering

When carbonic acid is introduced to limestone, it reacts with calcium (calcite mineral). From this reaction, a bicarbonate soluble in water is formed. Thus as acidic groundwater percolates through the limestone, the calcium, which makes up more than 90 percent of this rock, may be leached away (Fig. 25.5). In humid areas this process, which is known as *solution weathering*, is highly effective in the decomposition of limestone, especially if the rock is riddled with cracks. Left in place of the dissolved limestone are small residues of detrital sediments which were originally deposited with the limestone on the sea bottom. In time, these particles, which are mainly clay, accumulate along with organic material and deposits from runoff and wind to form a mantle of soil.

Fig. 25.5 Disintegration of limestone as a result of the reaction of natural carbonic acid with the calcium. The picture was taken on an exposure in the Niagara formation, Manitoulin Island, Ontario. (Photograph by Charles Schlinger)

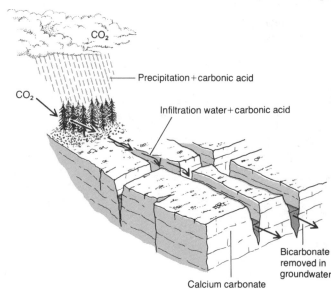

CO_2

Precipitation + carbonic acid

CO_2

Infiltration water + carbonic acid

Bicarbonate removed in groundwater

Calcium carbonate

An example of heavily weathered slopes in the Andes of Peru. Chemical-weathering rates in such environments are among the highest in the world. (Photograph by Jon Fobes)

The dissolved calcium, on the other hand, enters the groundwater and may eventually be released into a stream that ultimately discharges into the sea.

Another weathering process involving organic matter is *chelation* (pronounced key-la-tion). It is characterized by the bonding of mineral ions to large organic molecules which are excreted from vegetation. Our knowledge of this process has developed in connection with research on agricultural fertilizers, and although chelation is not well understood under natural conditions, it is considered by some researchers to be an important weathering process. Lichens, for example, excrete chelation agents, and studies have shown that lichen-covered basalt is often more deeply weathered than is lichen-free basalt.

Igneous rock is also weathered by other chemical processes involving water carrying minerals. It often starts when water penetrates the rock along the contacts between crystals and is absorbed by certain minerals. This process, called *hydration*, produces no chemical change itself, but sets up a sequence of chemical reactions which alter the minerals irreversibly. One set of these reactions is *hydrolysis*, which usually involves the reaction of water and an acid on a mineral. Hydrolysis is considered to be the most effective process in weathering granite and related rocks.

Acting on potash feldspar, a major constituent of granite, hydrolysis produces the clay mineral kaolinite as well as ions of silica in solution. Under the hot, wet conditions of the tropics, the alteration of both potash and plagioclase feldspars yields bauxite, an oxide of aluminum. In some parts of the tropics, bauxite has accumulated in such quantities that it is commercially mined for aluminum ore. Bauxite is also the main constituent of laterite, the hard layer that forms in tropical soils.

Another type of chemical weathering, *oxidation*, usually accompanies hydrolysis. It involves a number of silicate minerals, but is most apparent in rocks containing iron. The iron in olivine and pyroxene, for instance, is altered in the presence of oxygen to form ferric iron, which in turn is transformed into limonite, a mineral resembling rust.

Because of the diversified mineral composition of the igneous rocks, decomposition tends to take place at very uneven rates. In the case of granite, biotite and feldspar alter rapidly, whereas quartz is

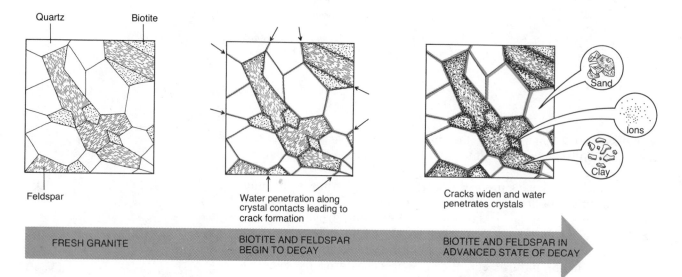

Quartz

Biotite

Feldspar

Water penetration along crystal contacts leading to crack formation

Cracks widen and water penetrates crystals

Sand

Ions

Clay

FRESH GRANITE

BIOTITE AND FELDSPAR BEGIN TO DECAY

BIOTITE AND FELDSPAR IN ADVANCED STATE OF DECAY

Fig. 25.6 The opening stages in the decay of granite. With the decay of biotite and feldspar, more cracks are opened, leading to faster and faster rates of decomposition.

slow to change (Fig. 25.6). Practically all igneous rocks, however, weather to yield four types of end products: (1) silica particles, which often end up as sand; (2) clay, which often turns out to be kaolinite; (3) soluble ions, which enter the water system; and (4) oxides of iron or aluminum, which impart reddish or brownish coloration to the soil (Fig. 25.7).

Fig. 25.7 The end products of chemical weathering: soluble ions, sand (silica), and clay.

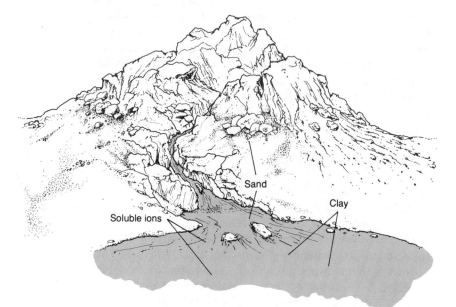

Sand

Clay

Soluble ions

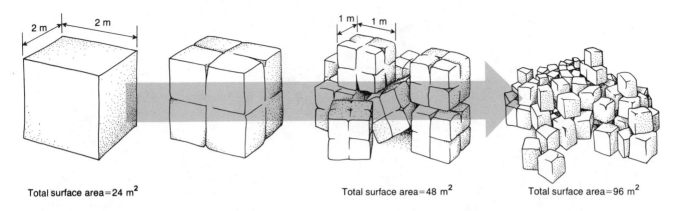

Total surface area=24 m^2 Total surface area=48 m^2 Total surface area=96 m^2

The soluble ions produced in weathering are carried off in ground-water and streamflow, eventually reaching the sea. Clay and quartz are left behind to form the soil cover. In time even the quartz is corroded by organic acids, and it too is leached from the soil, but this takes millions of years. This appears to have happened in many areas of the tropics, because soils there that have been derived from granitic rocks are poor in quartz.

Fig. 25.8 As mechanical weathering breaks rock down into smaller fragments, the amount of surface area exposed to chemical weathering increases. Here the rock is sectioned into eight pieces, and the surface area doubles with each step.

MECHANICAL WEATHERING

Mechanical weathering, the second major category, produces physical fragmentation of rock. Virtually everywhere mechanical weathering operates hand in hand with chemical processes, and in most places it is impossible to ascertain how much work should be ascribed to each. By fragmenting rock, mechanical weathering increases the total surface area over which chemical weathering occurs (Fig. 25.8). Mechanical weathering is apparently more effective in cold and dry environments, whereas chemical weathering tends to be more effective in warm, wet environments.

Mechanical weathering begins with the formation of cracks in bedrock. Crack formation is initiated in basically three ways: (1) differential expansion of rock masses; (2) chemical decomposition along bedding planes or contacts between different rock types; (3) expansion within bedrock by freezing water. When cracks widen and deepen, they are called *joint* lines (Fig. 25.5). Joint lines may extend tens or even hundreds of meters into bedrock and develop lateral offshoots. When horizontal and vertical joint lines intersect, blocks of rock are freed from the solid earth.

(a)

(b)

Fig. 25.9 Exfoliation: (a) high on the granite cliffs of Half Dome in Yosemite Valley. (Photograph by F. E. Matthes, U.S. Geological Survey, 1913); (b) an exfoliation surface in the Sierra Nevada, California, and the resultant accumulation of debris. (Photograph by G. K. Gilbert, U.S. Geological Survey, 1901.)

Types of Mechanical Weathering

Where massive bodies of igneous rock are exposed, a type of mechanical weathering called *exfoliation* tends to develop. Exfoliation is thought to be caused mainly by the differential expansion of a rock body as it decompresses with the loss of heavy rock overburden. The expansion produces a system of joint lines that yield scalelike sheets or slabs of rock (Fig. 25.9a). As the sheets form, they slide downslope or disintegrate into smaller particles which are subsequently eroded away (Fig. 25.9b). On a microscale, by comparison, individual boulders may weather in a manner that is physically similar to exfoliation. This is called *spheroidal weathering*, and although it looks like a miniature version of exfoliation, it is apparently due to chemical weathering (Fig. 25.10).

Frost wedging appears to be a very effective weathering process in polar and high-mountain environments. Unlike exfoliation, which is impossible to simulate in the laboratory, the physical process of frost wedging is easy to demonstrate. Experiments show that when water freezes, the force of crystallization can be tremendous. At a temperature of −20°C—a common surface temperature during winter in cold environments—the force exerted by the expansion of ice in a closed container is more than enough to burst steel pipes. We know from the principles of ground-heat flow that freezing in bedrock progresses from the top down (see Fig. 3.7). Therefore, the first ice to crystallize forms a cap over the water in a crack. The cap resists some of the upward expansion produced by subsequent freezing, thereby causing a

Fig. 25.10 *Spheriodal weathering is evident on the basalt boulders near the center of the photograph, taken in the Snake River Plain, Idaho. (Photograph by H. E. Malde, U.S. Geological Survey)*

large proportion of the resultant force to be directed downward and outward against the crack (Fig. 25.11). As a result, when the water is truly confined, cracks and joint lines may enlarge, and boulders may even be split open or shattered (Fig. 25.12). Freeze-thaw expansion is also effective in breaking down weathered material, such as boulder, cobble, and pebble-sized particles.

Fig. 25.11 *Ice formation in a water-filled crack, resulting in a form of mechanical weathering known as frost wedging. The force exerted by ice expansion can deepen and widen cracks and even split boulders.*

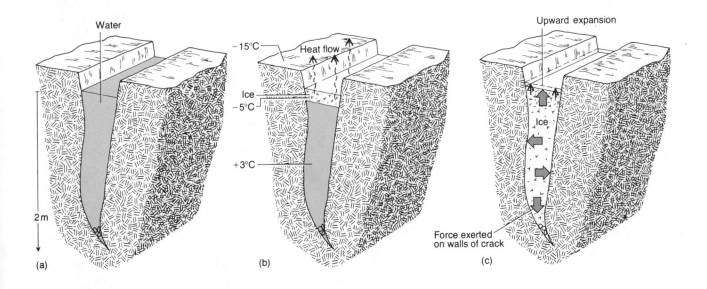

Fig. 25.12 Boulder fields, such as this one about 3500 m above sea level in the White Mountains of California, are considered to be strong evidence of mechanical weathering by frost action. (Photograph by William M. Marsh and Jeff Dozier)

Little is known about wedging caused by other substances. Crystallization of salt- or quartz-rich solutions can produce a wedging effect, although we have no information on how much work it may do. Wedging by plant roots is often cited as an example of mechanical weathering, but there is little evidence as to how effective roots are in fragmenting rock. They do cause damage to sidewalks and surely play an important role in chemical weathering, because they help to hold moisture and organic matter in joint lines and also contribute ions to soil water.

Mechanical weathering of rock is also produced by a variety of other processes; however, we know little about them in terms of their relative effectiveness. Some of these processes include rock shattering and fracturing from the impact of a long fall, such as from the Yosemite Valley cliff in Fig. 25.9(a), rock plucking by ice falling from a cliff face, and breakage from lightning strikes and the intense heat of forest fires.

WEATHERING AND LANDFORMS

Finally, it is important to ask what influence weathering has on the development of landforms. Intuitively, one would expect the least resistant rocks to form the lowest parts of the terrain. But we must bear in mind that the resistance of rock varies with climatic conditions, so any comparison must be limited to a single climatic zone. Some light was

Fig. 25.13 Comparison of tombstone weathering (a) and land elevation (b) near West Wilmington, Connecticut. Note that the resistance to weathering of these four rocks corresponds to the elevation of the terrain they underlie. (From P. H. Rahn, ''The Weathering of Tombstones and Its Relationship to the Topography of New England,'' Journal of Geological Education **19** *(1971). Used by permission.)*

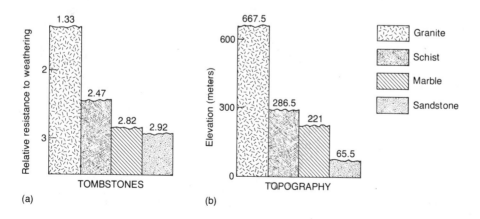

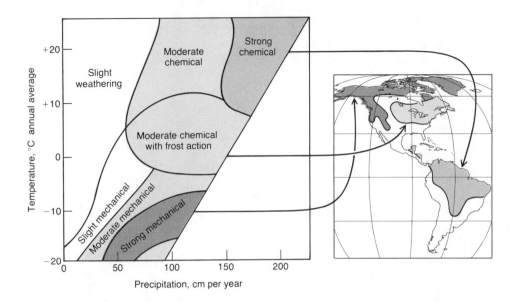

cast on this problem by P. H. Rahn and his students, who gauged the susceptibility of different rock types to weathering by comparing the amounts of disintegration in century-old tombstones of different compositions. Of the four rock types examined in a cemetery at West Wilmington, Connecticut, granite proved to be the most resistant; sandstone, the least resistant (Fig. 25.13). Compared with areas of the same rock types in the New England terrain, land elevations showed corresponding variations, with granite forming the high features and sandstone the low features (Fig. 25.13b). A comparison of terrain developed in sandstone and limestone in central Kentucky, on the other hand, shows sandstone to be more resistant (see Fig. 22.17). This can be attributed to the fact that under the bioclimatic conditions of Kentucky, this limestone is measurably weaker than sandstone.

Fig. 25.14 The general relationship between climate, represented in terms of average annual precipitation and temperature, and the types and relative rates of weathering. Geographical areas that exemplify the three major weathering environments are shown in the maps. (Graph after Peltier, 1950)

Map projection by Waldo Tobler.

SUMMARY

Water is the most essential ingredient in weathering. Both chemical weathering and frost action, the two most effective types of weathering, are water-dependent. Dry climates, whether warm or cold, generally produce low rates of weathering. In moist environments, by contrast, the collective rates of weathering tend to be high under both warm and cold temperatures. Figure 25.14 summarizes the relative rates of chemical and mechanical weathering as a function of average annual temperature and precipitation.

HILLSLOPE PROCESSES CHAPTER 26

INTRODUCTION

Weathering is only the first step in the long series of processes that culminate in the deposition of detritus in the sea. The second step is the downhill movement of rock fragments from the sites where they were freed from the bedrock. The processes responsible for this movement are collectively referred to as *hillslope processes*. There are two types of hillslope processes: *erosion* by various forms of runoff, including overland flow and flow in small channels, and *mass movement*, a variety of gravitationally induced motions, such as rockfalls, landslides, and avalanches.

Fig. 26.1 The splash created by a raindrop hitting a slope. Note the relative lengths of the downhill and uphill trajectories. (Illustration by Peter Van Dusen)

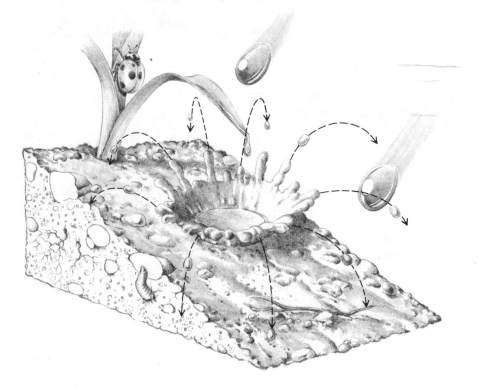

SLOPE EROSION BY RUNOFF AND RELATED PROCESSES

Rainsplash and Rainwash

Slope erosion by water begins with the impact of raindrops on the soil. When it hits a wetted surface, a raindrop sends out a circular splash of water and soil particles. If this occurs on a slope, the downhill side of the splash travels farther than the uphill side of the splash does (Fig. 26.1). The difference in the lengths of these trajectories can account for appreciable downhill movement of soil on barren or sparsely vegetated slopes.

When the intensity of rainfall exceeds the infiltration capacity of a surface, overland flow is produced. Near the crest of a hill, overland flow is slight because of the smallness of the water-collection area there; this has led some scientists to call this zone the "belt of no erosion." But as overland flow moves downslope, it increases in both volume and velocity until it is able to displace small particles, usually sand-sized and smaller. This process is referred to as *wash* or *rainwash*, and it may erode barren slopes as much as 3 to 4 cm per year (Fig. 26.2). If slopes are vegetated, however, erosion by wash is reduced substantially. In fact, studies show that in most areas rainwash is negligible on permanently forested slopes.

Gullying

As overland flow moves down the slope, it begins to coalesce into rivulets that etch small channels into the slope. Farther downslope, the rivulets may join to form ephemeral streams, which are capable of eroding gullies several meters deep into the slope face (Fig. 26.4). Upon reaching the lower gradient of the footslope, a stream may deposit part of its sediment load there. Under certain conditions, a stream may lose its flow while crossing the footslope and break down into distributaries similar to those in a river delta. This is common in arid regions, where sizable flows are lost to evaporation, or in any region where they are rapidly lost to infiltration (Fig. 26.3). The deposits resulting from this process are called *alluvial fans*, and although they are found throughout the world, they are most common in dry, mountainous regions.

The flow in ephemeral streams is not limited to the period of overland flow, because seepage from interflow may continue to drain into these channels for hours or even days after rainfall has ended. Despite the miniscule discharge of such flows, they are able to carry away fine particles and thereby advance gully formation. As with overland flow, erosion by ephemeral streams is minimized where slopes are permanently covered by vegetation.

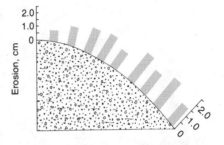
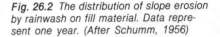

Fig. 26.2 *The distribution of slope erosion by rainwash on fill material. Data represent one year. (After Schumm, 1956)*

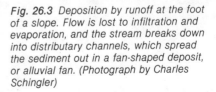

Fig. 26.3 *Deposition by runoff at the foot of a slope. Flow is lost to infiltration and evaporation, and the stream breaks down into distributary channels, which spread the sediment out in a fan-shaped deposit, or alluvial fan. (Photograph by Charles Schingler)*

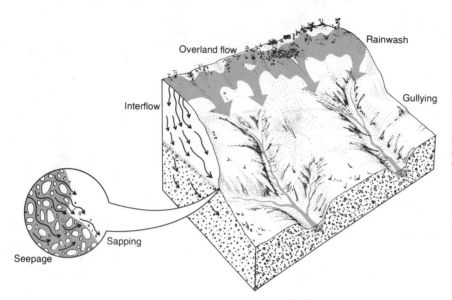

Fig. 26.4 *Hillslope erosion by runoff. The upper slope is eroded mainly by rainwash, whereas the midslope and lower slope are eroded by channel flow and sapping by seepage from interflow water. (Illustration by Wiliam M. Marsh)*

Sapping and Piping

Two additional processes help to form gullies. Sapping is a form of erosion that occurs in the walls of gullies as a result of the outflow of interflow water. As the water trickles out of the soil, it erodes fine particles, leading to undercutting of the gully wall along the seepage line (Fig. 26.4). If the overhanging material is weak, it will break off and fall into the gully. But if this material is resistant to cave-in, the seepage may erode small tunnels into the slope. This process is known as piping, and it can produce miniature caverns, or natural pipes, which extend many meters into a slope. The pipes tend to increase interflow discharge to gullies and may weaken the slope, leading to collapse of large sections of soil.

Variations in Slope Erosion Related to Climate

From field studies conducted in different parts of the world, we are able to sketch a brief picture of the relationship between bioclimatic conditions and sediment yield from slope erosion. Geomorphologists of the U.S. Geological Survey found that in North America, erosion is relatively low in arid and humid regions, but is high in semiarid regions where precipitation averages around 30 cm a year. Semiarid lands are dry enough to limit the establishment of continuous plant covers, yet moist enough to produce substantial overland flow several times a year. Where there is more rainfall, vegetation limits erosion; where there is less rainfall, low runoff limits erosion (Fig. 26.5a).

On other continents, erosion is high under not only semiarid conditions, but also those with wet and dry seasons, such as the monsoon climate of India. The combination of a dry season that severely limits vegetation and a wet season that yields many intensive rainstorms appears to be effective in producing high rates of erosion. Agriculture activity in these areas (India and Southeast Asia) undoubtedly contributes to this as well.

In those humid areas where crop agriculture is widespread, particularly where cropping is seasonal and fields are left barren for several months a year, soil erosion is also high. This is the case throughout much of the midwestern United States, for example, where it is not uncommon for farmers to leave fields cropless as much as six months per year. Without a plant cover to mitigate runoff, such fields are often severely eroded, despite outward efforts at soil conservation in most countries. With the inevitable expansion of agriculture throughout the world, we may see the balance of erosion rates shift to humid areas, if it has not already.

Finally, we can consider soil erosion related to urbanization. Urban development in North America generally followed a sequence of land use beginning in the 1800s with land clearing for agriculture. As urban areas expanded in the twentieth century, the agricultural lands gave way to suburban development. Both of these changes resulted in increases in soil erosion and stream sedimentation (Fig. 26.5b). The highest sediment yields have been associated with the construction phase of suburban development, when the land is devegetated and disturbed by earth-moving machines. Following suburbanization, sediment losses decline substantially as more and more of the land becomes urbanized and covered with hard-surface materials.

Fig. 26.5 Erosion in relationship to climate and urbanization: (a) A study conducted in the United States indicated that erosion is highest in semiarid areas, whereas a study using data from several continents showed erosion to be high in both the semiarid and the wet/dry (monsoon) climates. (After Douglas, 1967; and Langbein and Schumm, 1958); (b) Changes in sediment yield to streams in response to land-use change. (After Wolman, 1967)

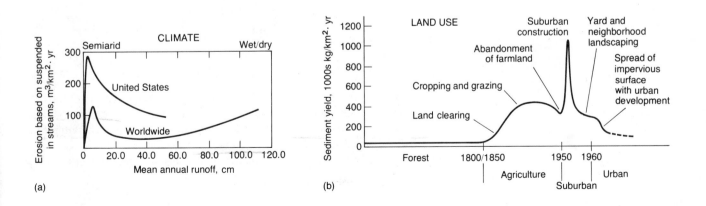

MASS MOVEMENT

When materials move by mass movement, they are drawn downslope under the influence of gravity rather than by a transporting agent such as a stream or a glacier. Mass movement may be rapid, as in a landslide or avalanche, or so slow that it is barely detectable with sensitive instruments. Moreover, mass movement may involve gigantic amounts of debris, easily enough to bury towns and villages, or just individual grains of sand rolling off an anthill.

Mechanics of Slopes

In any body of material situated on an incline, two sets of opposing forces determine whether it will move or remain in place. The forces that tend to induce movement generate *shear stress*, the magnitude of which is primarily a function of the steepness of the slope. The steeper the slope, the greater the tendency for gravity to pull objects downhill (Fig. 26.6). The opposing forces impart *shear strength* to the material. The reason the block in Fig. 26.6 does not slide downslope is because the shear strength at the contact between the block and the slope is greater than the shear stress. Shear strength is governed by factors such as the frictional resistance between adjacent particles, the cohesiveness of clays, and the binding strength of plant roots. The balance between shear stress and shear strength determines whether a slope is stable or unstable. Mass movement occurs when shear stress exceeds shear strength. When the two are equal, the slope is said to be at *critical threshold*; when shear strength is greater than shear stress, the slope is stable. The safety factor of any slope is equal to the shear strength divided by shear stress:

$$\text{Safety factor} = \frac{\text{Shear strength}}{\text{Shear stress}}.$$

When the safety factor is 1.0, the slope is at critical threshold; when it is less than 1.0, failure is imminent.

Types of Movement

Several different schemes have been devised to classify mass movements. The respected geological engineer Karl Terzaghi, for example, suggested a classification according to the rate of downhill movement. He identified landslides as rapid movements, some moving as fast as several meters per second, but most moving on the order of a meter or

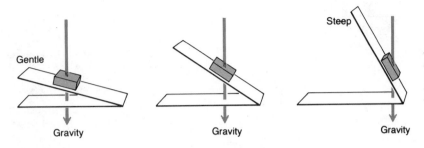

Fig. 26.6 On gentle slopes gravity pulls against the slope and therefore tends to hold objects firmly to it. On steep slopes the force of gravity tends to parallel the slope surface, thus tending to pull objects downhill.

less per day. For the slowest mass movement, Terzaghi cited soil creep, which moves at rates of only several cm per year.

Mass movement can also be classed on the basis of the type of motion involved. Five basic types of motion seem to cover most movements: fall, slump, slide, flow, and creep (Fig. 26.7). The development of a particular motion in an unstable slope depends on the physical properties of the materials in the slope. Brittle materials, such as bedrock, rock rubble, and sand, tend to rupture when they fail and to move in slides, falls, or slumps. Slumps are characterized by a back rotational movement like that of someone slouching in a chair. Materials such as clayey soils, which behave as plastics, may deform without rupturing. They bend gradually downslope in a motion that might be described as weak flow. Finally, some materials develop flowing motion when they fail. Flow is possible only in saturated materials having a liquid or near-liquid consistency. Such consistencies are found in wet clayey materials or in sand or silt within which there is a flow of pressurized water.

The relationship between stress and deformation in earth materials is described by the law of plastic deformation (Fig. 26.8). With the initial application of stress, no deformation takes place, but as stress is increased, a level is reached at which the material begins to deform. The value of this *critical threshold* in soil materials depends on the consistency of the material. Beyond the critical threshold, the material enters a phase of plastic deformation, followed by a phase approximating fluid flow.

Factors Influencing Slope Stability

Soil water and groundwater. In most areas both shear stress and shear strength vary over time, especially seasonally. During spring and winter, for example, shear stress tends to increase with the infiltration

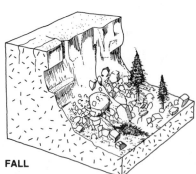

FALL

FEATURES

- Cliff exposure with fractured bedrock
- Angular blocks and slabs piled up at footslope
- Scarred and broken trees

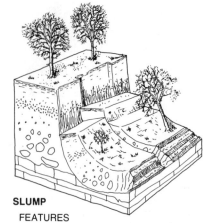

SLUMP

FEATURES

- Bank or cliff exposure in soil or bedrock
- Movement often back rotational motion
- Trees uprooted and tipped back toward slope

SLIDE

FEATURES

- Concave scar on slope
- Displaced mass resting anywhere from one meter to a kilometer downslope
- Broken and bent trees
- Buried soil

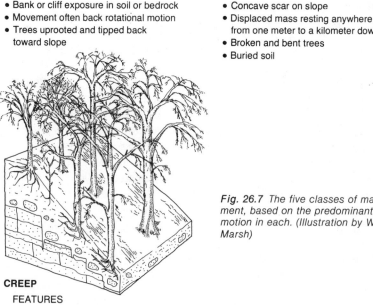

FLOW

FEATURES

- Lobate forms
- Trees bent or upright
- Buried ground vegetation and soil

CREEP

FEATURES

- Absence of deformation in trees or soil surface (contrary to widespread belief that creep causes such deformation)
- If soil is excavated, tree roots may reveal downslope bending

Fig. 26.7 The five classes of mass movement, based on the predominant type of motion in each. (Illustration by William M. Marsh)

of water from the surface. This water fills the interparticle voids, thereby increasing the mass of the soil. At the same time, shear strength decreases because the consistency of soil materials tends to become more liquidlike with the addition of water. When the critical threshold is surpassed, the material fails and moves downslope. Once

Fig. 26.8 The law of plastic deformation. With the application of initial stress, no movement occurs. At the critical threshold, shear stress exceeds shear strength, and deformation begins. With increasing stress, deformation assumes the properties of fluid flow.

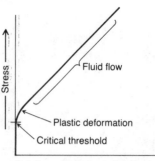

set into motion, the saturated material tends to flow downslope and, depending on its composition, forms either a *mudflow* or a *debris flow*. Mudflows consist of primarily clay and silt, whereas debris flows consist of clay and silt as well as a wide range of larger particles. Such flows are the source of serious property damage in the Los Angeles area, especially during wet winters (Fig. 26.9).

The addition of water to certain materials can also cause landslides. Groundwater may dissolve (leach) minerals that cement a mass together, resulting in loss of shear strength. Slope failures produced in this way often occur without a hint of warning and have produced some frightening disasters in Europe and Asia. In stratified materials, if groundwater is added to a clay layer that underlies a stable mass, the clay may liquefy, and the entire mass may slide on the clay layer. According to the law of plastic deformation (Fig. 26.8), this change would lower the critical threshold of the clay. Such slides are often caused by natural increases in groundwater, but in recent decades a growing number have been brought about by human-caused increases

Fig. 26.9 A localized example of destructive mass movement in Los Angeles that involved both flow- and slide-type movement. (Photograph courtesy of the Department of Building and Safety, Los Angeles)

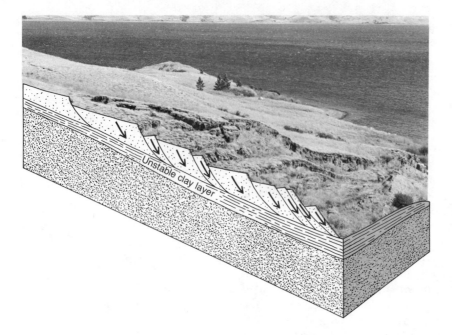

Fig. 26.10 Shallow slumps and landslides over an unstable clay layer. It appears that the clay layer liquified because of an increase in groundwater caused by the raising of the reservoir in the background. (Photograph by William M. Marsh)

in groundwater, e.g., the raising of reservoirs and seepage from irrigation and septic drain fields (Fig. 26.10).

The raising of a water table may also produce another source of slope instability: *pore-water pressure*. Within stable slopes the water pressure on soil particles created by an elevated mass of groundwater is offset by the confining pressure exerted by the overlying soil mass. If the confining pressure is reduced by a loss of soil overburden, or if the water pressure is increased by a rise in groundwater, pore-water pressure may drive the soil particles apart, causing them to float. This reduces the shear strength of the soil, because interparticle friction and cohesion are lost, and can cause failure, even in low-angle slopes. Pore-water pressure, by the way, is the cause of quicksand. Unless pore water is pressurized, sand is very stable when saturated.

Ground frost. Several kinds of mass movements are caused by ground frost. One of these is rockfall produced by ice wedging on cliffs and steep hillslopes. Although data are limited, records of rock debris on highways and railroads suggest that most rockfalls do not occur with freezeup in the fall, but rather in spring, when ground ice expands and melts (Fig. 26.11a). Rockfall often leads to the buildup of piles of rock rubble, called *talus*, which may bury as much as half of the slope (Fig. 26.11b).

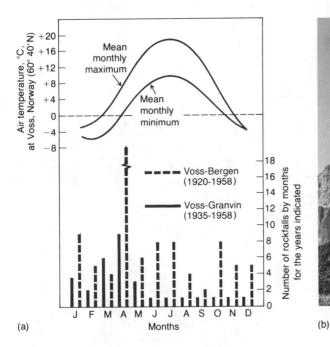

Fig. 26.11 (a) The general relationship of rockfalls and monthly temperature along two railroads in the mountains of Norway. The period of maximum rockfalls does not occur upon freezeup, but is delayed until ground ice expands and thaws in spring. (Data from Rapp, 1960); (b) A large talus slope in Glacier National Park. (Photograph by H. E. Malde, U.S. Geological Survey)

Frost penetration into moist soil is one of the most widespread causes of mass movement in mid- and high-latitude regions. As the moisture in a soil freezes, it expands. Initially the expansion is taken up in empty interparticle spaces, a fact discovered by Stephen Taber in experiments conducted nearly fifty years ago. But as freezing progresses, Taber found, additional soil water is drawn by capillary forces up to the frozen layer to form more ice. When the interparticle spaces are filled with ice, the frozen layer itself expands in the direction of least resistance, which is outward, or perpendicular to the slope face. When the ground thaws, the soil shrinks, but not along the line of expansion. Instead, it settles a little downhill, in the direction of the gravitational force, which is perpendicular to the plane of the earth's surface (Fig. 26.12). The total downslope displacement caused by freeze-thaw activity may amount to only a few millimeters with each freeze-thaw episode, but many such episodes each winter and spring can produce appreciable mass movement.

In general, slow movements of this type produce *soil creep*. Creep is also known to be caused by contraction and expansion of clay with wetting and drying, although field data suggest that freeze-thaw is a more effective creep mechanism. Many geomorphologists think that creep is the dominant slope process along the crests of hills because rainwash is so slight there (Fig. 26.4). Table 26.1 gives some repre-

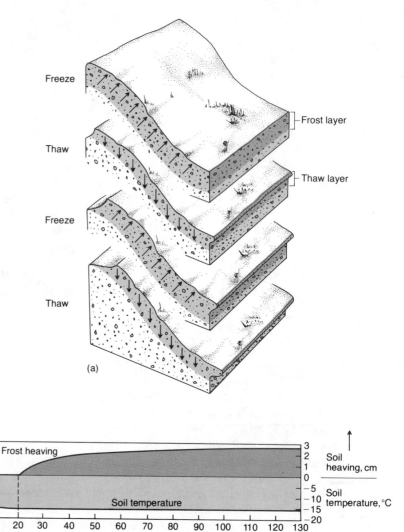

(a)

Fig. 26.12 *(a) How freeze-thaw activity can produce soil creep. The amount of downhill movement suggested is greatly exaggerated. (b) Heaving in clay does not begin until after twenty hours of freezing ground temperatures. During this time, soil water migrates to the frost layer, where it fills soil voids, ultimately causing heaving. (Graph from Stephen Taber, "The Mechanics of Frost Heaving," Journal of Geology, 1930. Reprinted by permission.)*

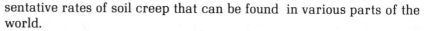

(b)

sentative rates of soil creep that can be found in various parts of the world.

In polar and high-mountain environments—often referred to as *periglacial environments* because of the predominance of frost-related processes—freeze-thaw activity and wet soil conditions combine to produce a form of mass movement called *solifluction*. Solifluction usually takes place in the active layer above the permafrost and is dis-

Table 26.1   Soil-creep rates.

LOCATION	ENVIRONMENT	CREEP RATE	SOURCE
Southern Alps, New Zealand	Mountain valley slope	1.1 cm/yr	Owens, 1969
Lake Superior	Backshore slope, herb-covered	0.68 cm/yr	Marsh and Koerner, 1972
England	Grassed and wooded hillslopes	0.025 cm/yr	Young, 1960
Tien-Shan (southern Siberia), U.S.S.R.	Mountain valley slope	Negligible	Iveronova, 1963
Southwest Scotland	Hillslope	1.28 cm/yr	Kirkby, 1967
New Mexico	Gully slopes	0.84 cm/yr	Leopold, Wolman, and Miller, 1964

tinctive because it produces lobes (tongues) of material that look as though they are flowing downslope. But despite the flow-type form, solifluction lobes move at slow rates, generally on the order of 5–15 cm per year.

Although frost action appears to be the chief driving force for solifluction, field studies indicate that movement may be promoted by the addition of liquid water as well. If meltwater builds up in a solifluction lobe, the pore-water pressure may increase, thereby reducing interparticle friction and cohesion and in turn reducing the shear strength of the mass. Some scientists reason that lobe movement in spring may be related to rises in pore-water pressure with the melting of ground ice and snow.

Solifluction is extremely widespread in tundra-type environments, even on low-angle slopes of only four to five degrees. For more than a century, its curious forms have captured the attention of scores of explorers and scientists in many parts of the world. Human fascination notwithstanding, we are beginning to learn that solifluction may be of second-order importance as a hillslope process in terms of total debris movement in cold regions. Based on several seasons of field measurements in the cold mountain valleys of northern Norway, for instance, the Swedish geomorphologist Anders Rapp found that the principal means of denudation was the loss of minerals in solution with runoff. Other interesting surface features of periglacial environments are

AUTHORS' NOTE 26.1
Stones and Patterned Ground in Periglacial Environments

Concentrations of stones are one of the most conspicuous surface features in periglacial environments. The concentrations originate in various ways. Some are derived from mechanical weathering of surface rock or deposits laid down by rivers, waves, or glaciers. Others, on the other hand, appear to have been worked out of the underlying ground by frost action. One explanation holds that these stones are lifted up with the surface layer of soil as it expands upon freezing each fall. When the layer thaws in spring, the stones are unable to settle back into their original positions because small soil particles have fallen into the cavity created when the stone was lifted. Another explanation holds that because stones have higher thermal conductivity than the surrounding soil material, frost penetrates through them rapidly, inducing the buildup of a small pocket of ice at their base, which forces the stones upward. When the ice melts in spring, the fine soil particles adjacent to it fall into the cavity, thereby limiting the depth of settlement of the stone.

Often the stones are arranged in curious geometric patterns such as circles and polygons. The origin of such patterned ground is poorly understood, but appears to be related to frost action involving heaving (upward movement) and thrusting (lateral movement) of stones. Evidence also shows that expansion, contraction, and cracking of the soil mass because of changes in soil moisture contribute to stone sorting and the formation of patterned ground. Other forms of patterned ground are manifested in the distribution of vegetation or water features rather than stones. Patterned ground can also be found outside permafrost regions, but rarely does it take the form of sorted stones. In those cases where stone patterns are found in a nonperiglacial setting, it is considered to be good evidence that periglacial conditions once existed there.

Stone polygons, Cambridge Bay, Victoria Island, Northwest Territory, Canada. (Photograph by A. L. Washburn)

Large polygons marked by vegetation and ponded water, near Prudhoe Bay, Alaska. (Photograph by A. L. Washburn)

large concentrations of stones and patterned ground (Authors' Note 26.1).

Undercutting. Many causes of slope failure result in slide-type movements, and among the most important is slope undercutting by streams, waves, or other erosional agents, including humans. Undercutting is actually a form of excavation of the slope base that produces an oversteepened slope angle. A slope angle that is oversteepened exceeds the *angle of repose,* or the maximum incline that materials such as sand, clay, or boulders can maintain without failing. If loose, dry sand, for example, is steepened to an angle greater than 33°, failure will result because the critical threshold, represented by a balance between shear stress and shear strength, has been surpassed. For larger or more angular particles, the angle of repose is steeper, exceeding 40° for talus.

Earthquakes. The more dramatic types of mass movement, such as large landslides and avalanches, are limited mainly to large, steep slopes in mountainous regions where earthquakes are frequently the major cause of failure. As seismic waves pass through a mass of unconsolidated material, particles tend to move differentially and even rotate somewhat, thereby breaking ingranular bonds formed by minerals, ice, and clay. As a result, shear strength can be drastically reduced, and the material often fails instantaneously (Fig. 26.13). The landslides and avalanches of May 31, 1970, which slid and sailed thousands of meters downslope and across a valley to bury a city of 18,000 people in Peru, are a prime example of slides triggered by an earthquake. The total volume of this mass of rock, soil, and ice was estimated at 50 million to 100 million cubic meters.

The destructiveness of such events is due to not only the great mass of material involved, but also the great distance over which it moves. The distance of horizontal movement, as across the floor of a valley, can be so great that slides appear to defy the law of gravity. We now understand, however, that such movement is facilitated by a layer of compressed air that forms under the advancing mass. This allows the slide to move with little or no friction, as in the movement of a hovercraft.

Lowering of the Continents

It is important to underscore the role of hillslope processes in lowering the land masses of the earth. These processes move weathered debris to a position in the landscape where it can be picked up by streams and glaciers and then transported to even lower elevations. As the material

Fig. 26.13 One of many rockfalls triggered by the Madison Canyon earthquake of 1959, which also triggered a major landslide that dammed the valley of the Madison River in Montana. (Photograph by J. R. Stacy, U. S. Geological Survey)

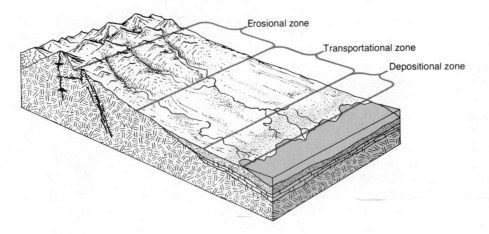

Erosional zone

Transportational zone

Depositional zone

Fig. 26.14 *The three main zones of activity associated with the lowering of the continents. Hillslope processes are most intensive in the erosional zone. In the transportational zone, material is moved by streams to the coast, where it is deposited. (Illustration by William M. Marsh)*

is transported downvalley, it undergoes further weathering, and most of it ultimately emerges on the sea coast as sand, silt, clay, or dissolved ions (Fig. 26.14). For example, the Mississippi River each year dumps nearly 500 million tons of sediment into the Gulf of Mexico.

The rate at which sediments are brought to the sea should, if we could measure it, tell us how rapidly the continents are being unloaded, or denuded. Understandably, the rate of denudation for an entire continent is difficult to measure accurately, and thus our estimates are crude at best. For smaller areas, though, somewhat more accurate estimates are possible, and measurements of sediment loads of large rivers in the United States suggest that the coterminous United States is being lowered at an overall rate between 2.5 and 7.5 cm per 1000 years. As you would expect, the rate is much higher in dry, mountainous areas, such as the Southwest, where denudation may be more than 100 cm per millenium.

Lowering of the land is partially offset, however, because the continents uplift as they are being unloaded. This process, termed *isostatic uplift*, is a response to the release of mass (weight) from the continent and represents a process likened to a ship rising in the water as it is being unloaded. Through isostatic uplift, as much as 80 percent of the elevation lost in erosion can be recovered. In other words, for every 1000 meters of land elevation lost, it appears that about 800 meters are recovered through isostatic uplift.

FORMATION OF HILLSLOPES

Mass Balance of Slopes

In a general way the rate at which hillslope processes move debris over a given surface can be treated as a function of slope steepness. The American geomorphologist Grove K. Gilbert observed this many years ago, noting that for a given volume of runoff, transporting power increases with slope inclination. If we extend this concept to include mass movements, we should find, other things being equal, that more material should be moved on steep slopes. For a particular region we can thus refer to steep slopes as high-energy slopes. Gentle slopes, on the other hand, can be thought of as low-energy slopes, and the rate of movement of debris on them is correspondingly lower. Bear in mind, though, that for any area, change in the resistance of ground materials, as through deforestation, can radically alter the relative meaning of high- and low-energy slopes.

With the idea of slope energy in mind, let us examine what happens on a slope of very irregular configuration, one similar to the walls of the Grand Canyon, with alternating steep and gentle segments (Fig. 26.15a). Where a steep segment is situated above a gentle one, the debris brought downhill piles up on the gentle segment because the energy available there is inadequate to move the debris as fast as it is received (Fig. 26.15b). This is evident on the upper parts of the Grand Canyon. But this condition cannot be permanent, for as the pile of debris grows, the slope is steepened, thereby increasing the energy to move additional debris. At the same time, the steep segments are eroded down and partially buried under debris piles. The outcome of all this is that the slope trends toward a more uniform configuration over which the distribution of energy for debris transport is less variable.

Actually, Grove K. Gilbert first presented this argument for stream gradients, but it is also applicable to slopes, especially those on which runoff is the dominant agent. Gilbert observed that other things being equal, steep slopes erode faster than gentle ones and that when the slope reaches an overall gradient such that the energy of each slope segment is just adequate to move the debris brought to it and produced by it, the slope has assumed an equilibrium profile (Fig. 26.15c).

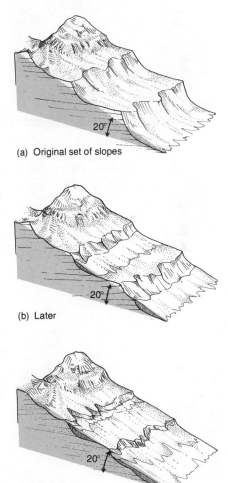

(a) Original set of slopes

(b) Later

(c) Much later

Fig. 26.15 This sequence of illustrations is based on one of G. K. Gilbert's concepts of slope, or grade. The original set of slopes (a) are largely unaffected by slope processes. Both (b) and (c) show the form changes resulting from hillslope erosion and deposition. Note the trend toward a less irregular profile over the entire slope, although the overall angle of the slope remains the same. Gilbert also argued that resistant rock formations maintain steeper slopes. (Illustration by William M. Marsh)

Grove K. Gilbert (1843–1918), keen observer and scientist whose work forms much of the basis for modern geomorphology. (Photography courtesy U.S. Geological Survey)

When a slope reaches this stage, it is in *mass balance*; that is, the amount of debris moved downslope is proportional to the amount of debris produced by weathering.

Slope Retreat

The mass-balance concept enables us to understand how slopes retreat, that is, how they are worked back over long periods of time. The problem of slope retreat has long been a topic of lively debate among geomorphologists, and the main thrust of the debate has developed around the question of whether slopes wear down, i.e., grow gentler, or wear back, i.e., maintain a fixed angle as they retreat. These two modes of slope retreat are known as *parallel* and *nonparallel* retreat. Either mode is possible, depending on the mass balance of a slope, especially the footslope.

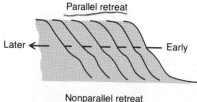

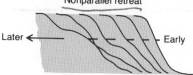

If debris is brought downslope at some rate and is removed from the footslope at the same rate by erosional processes, retreat will tend to be parallel. If the balance is altered, say, by reducing the rate of debris erosion from the footslope, the slope will grow gentler as the debris accumulates at its foot. But this only leads to a lesser slope angle and in turn a reduced rate of downslope transport of debris by hillslope processes. When the downslope transport and footslope removal are back in equilibrium, parallel retreat ensues again. So parallel retreat tends to be the rule where the slope is in equilibrium mass balance.

But what conditions are necessary in order to maintain mass balance on a slope? One condition is essential: The slope must be part of an *open geomorphic system,* a system characterized by a through-transport of rock materials (as sediments) from slopes to the edge of the continent. Through-transport is possible only if the slope is part of a drainage basin that empties into the sea. This ensures that debris released from the slope is carried away by streams.

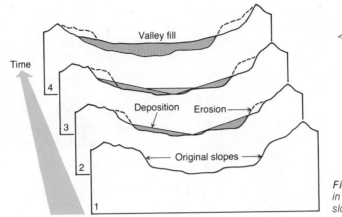
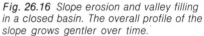

Fig. 26.16 *Slope erosion and valley filling in a closed basin. The overall profile of the slope grows gentler over time.*

Where drainage is closed, on the other hand, rock debris cannot be transported out of its source area. Therefore, it builds up in valley bottoms and on footslopes, producing gentler angles on the lower slopes (Fig. 26.16). Farther upslope, however, on the erosional segments, parallel slope retreat is still possible, but as slopes shift back, they grow shorter and eventually disappear at the topographic divides. Standing back from this scene over a long span of time, one would see the gross profile undergoing nonparallel retreat, as the base of the slopes built out into the valley and the upper slopes retreated toward the surrounding divides.

A surprising number of areas in the world qualify as closed geomorphic systems; most are situated in arid, mountainous regions. Death Valley and the Dead Sea Basin are rather famous examples of small basins where, as Fig. 26.17 suggests, basin filling and lowering of

Fig. 26.17 *Death Valley, a closed basin being filled by sediments eroded from side slopes. Compare with Fig. 26.16. (Photograph by John S. Shelton)*

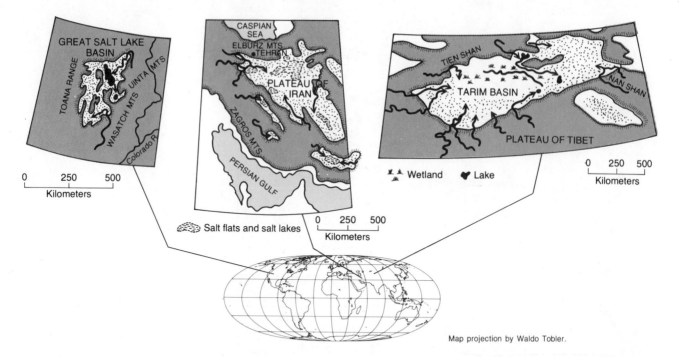

Map projection by Waldo Tobler.

Fig. 26.18 Examples of large, closed basins: Great Salt Lake Basin, the Plateau of Iran, and Tarim Basin.

footslopes are not difficult to envision. But there are large areas of closed drainage, too. In particular, the Tarim Basin of northeast China, the Plateau of Iran, and the basin of Great Salt Lake are examples of closed systems of such dimensions that basin filling and slope reduction are not as readily apparent as in smaller basins, although the trend is nonetheless present in each (Fig. 26.18).

Trends in Slope Development

It is probably rare that a particular trend in slope development would continue undisturbed for millions of years. The earth is too dynamic in most places for this to happen. Areas of active mountain building are a case in point. From observations in the American West, we know that faulting can reestablish slopes that were originally worn down by erosion and thus trigger a new episode of erosion and retreat (Fig. 26.19).

Another source of alteration in slope development is climatic change. Increased aridity in the American Southwest in the past 10,000 years, for example, caused the reduction of many lakes and streams and thus the loss of an erosional agent with sufficient power to remove debris from footslopes (Fig. 26.20). As a result, former wave and river-cut slopes in many areas have tended to grow gentler in the

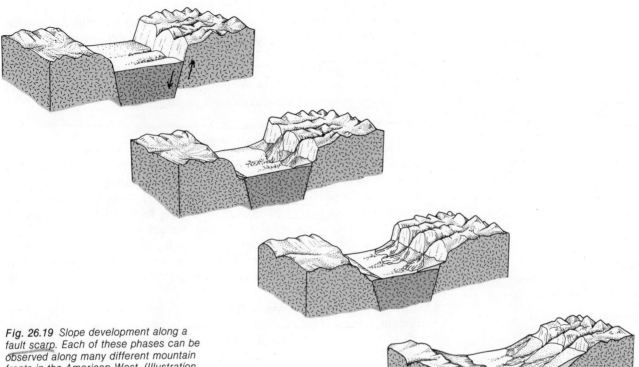

Fig. 26.19 *Slope development along a fault scarp. Each of these phases can be observed along many different mountain fronts in the American West. (Illustration by William M. Marsh)*

Fig. 26.20 *Typical slope changes in response to a drop in lake level or streamflow. (Illustration by William M. Marsh)*

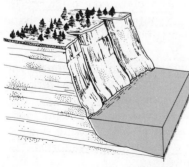

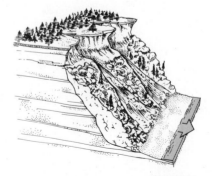

High water

Low water

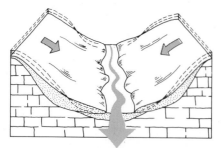

Small flow

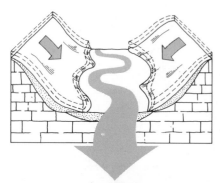

Large flow

Fig. 26.21 Slope steepening resulting from an increase in streamflow and lateral erosion. (Illustration by Peter Van Dusen)

past several thousand years. The opposite is also possible, of course; Fig. 26.21 shows steepening in the walls of a stream valley as a result of increased erosion from greater discharge. Sometimes we see examples of increased streamflow and erosion where humans have diverted water into a stream or where storm discharge has been increased by urbanization or agricultural development.

Common Slope Forms and Their Origins

If we examine the landforms in most localities, we are sure to notice some distinct similarities in the shapes of different hillslopes. Similarly, slopes of a particular form tend to be common to different geographical regions as well, and geomorphologists have diligently sought to provide explanations for this. The best we can do at the present, however, is to say that the causes appear to be multiple; in some regions appearing to be related to rock structure and type; in others, to vegetation and hillslope processes; and in yet others, to erosion and deposition by certain agents.

Gentle S-shape slopes are very common in nonmountainous regions where there is a thick soil overburden and heavy vegetative cover. Such areas include much of the United States east of the Mississippi River, northwestern Europe, and the wet tropics of South America, Africa, and southeastern Asia. The geometry of this slope form falls into three parts: (1) convex crest slope, (2) straight midslope, and (3) concave footslope. The overall angle of the slope may vary from less than 10° to as much as 45°. Beyond 45° most soil materials are unstable, even with a vegetative cover, so very steep S-shaped slopes are uncommon.

The origin of the S form appears to be tied to a particular combination of hillslope processes, notably runoff and soil creep, working on an erodible soil held in place by a heavy plant cover. The role of plants is very important and is especially evident when these slopes are devegetated for land development or farming, and accelerated erosion sets in. The slope is quickly transformed from the S shape into a concave shape as gullies are cut into it and large chunks of material are removed by slumps and slides (Fig. 26.22).

Concave slopes also develop under natural conditions, though typically in areas of rough terrain. This slope form is usually the work of a powerful erosional agent focused on the footslope or midslope. Glaciers, for example, are able to scour the sides of mountains into deep concavities called *cirques*, and a mountain glacier flowing down a stream valley carves both side slopes into broadly concave forms. The inclination of such slopes increases upward and may approach 90° near the midslope (Fig. 26.23).

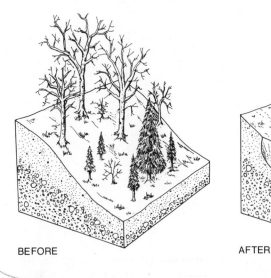

Fig. 26.22 The transformation of an S-shaped slope to a concave slope due to accelerated erosion and mass movement following devegetation. (Illustration by William M. Marsh)

BEFORE   AFTER

Convex slopes, by contrast, are usually related to resistant rock formations or to uplift of the slope itself and can be found in any climatic region. In a slope composed of sedimentary rock, for example, resistant formations in the midslope may induce the development of a convexity or a "snout." Occasionally, convexities may also be produced by mass movements such as landslides or solifluction, which produce a bulge in the midslope.

Convex slopes may also result from the uplift of the mountains by tectonic forces. Walther Penck, a European geomorphologist who theorized about the origin of landforms, argued that a trend toward convex-

Fig. 26.23 Concave slopes are often the work of a powerful erosional agent such as a glacier. (Illustration by William M. Marsh)

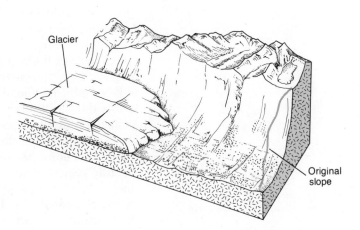

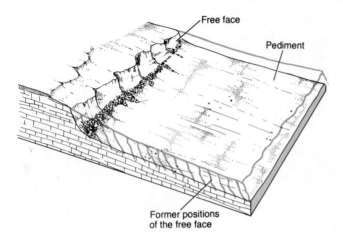

Free face

Pediment

Former positions
of the free face

*Fig. 26.24 Broad, gentle slopes called
pediments form as the free face retreats.
Pediments are prevalent in arid regions.
(Illustration by William M. Marsh)*

ity is the norm in mountain slopes because the slope is uplifted as it
forms. Though Penck's idea is not given much credance today, we can
see a trend toward convexity in rock domes which uplift as they are unloaded by exfoliation and associated processes (Fig. 25.9b).

Finally, there are slopes called *pediments*, originally defined by
G. K. Gilbert. These are broad, gentle slopes, usually only several degrees inclination, which lie at the foot of cliffs, called *free faces*. Pediments are composed of bedrock over which is strewn a light veneer of
gravelly rock debris. These slopes are thought to form in the wake of a
retreating free face. Because they are widespread and very apparent
in arid lands, many geographers and geologists feel that pediments are
unique to dry climates (Fig. 26.24).

Landforms and Climate

There is a widespread belief among students of the landscape that different climates give rise to different landforms. Surely most Americans
who have thought about the geography of their nation would be inclined to think that way, if only because the landforms of the arid West
look so different from those of the humid East. And indeed certain kinds
of landforms are found almost exclusively in the West (Fig. 26.25). But
is this due to differences in climate or to differences in the original
geology of the two regions?

The problem of documenting the relationship of landforms to climate is not easy, because of the difficulty of finding areas in different
climatic regions where the rock type, rock structure, and geologic history are similar enough to allow reliable comparison. Moreover, there
is the problem of climatic change. In general, landforms respond to climatic conditions less acutely than do vegetation and soils. Thus it is

likely that the existing landforms in most regions are products of past as well as recent climatic conditions. Finally, there is such a range of factors involved in slope formation—rainfall intensity, vegetation, rock type, rock structure, and soil formation—that some geomorphologists argue that it is possible for virtually any kind of landform to develop in any climatic region. The only exception is glacial areas, where certain landforms are uniquely associated with glaciated or frozen ground.

Barring glaciated areas, there are, nevertheless, some observable differences in landforms of contrasting climatic regions, but they tend to be more relative than absolute. For example, the landform called a *mesa* is a common feature in parts of southwestern United States. Mesas are flat-topped "islands" of relatively resistant, flat-lying sedimentary rock. The sides are usually steep free faces at the base of which are debris deposits leading onto the gentler pediments, the lowermost slopes (Fig. 26.24).

Are mesa forms found in humid areas of comparable geology? Not exactly, but similar features are there if we look through the forests and heavy mantles of soil. The humid-region versions are not so bold, because the soil and vegetation tend to soften the angular form of the bedrock. If we had to put our finger on one factor to explain the difference, it would have to be vegetation. Plants hold weathered debris in place and minimize erosion by runoff. As a result, a heavy soil cover of-

Fig. 26.25 Rock towers such as the Captains of the Canyon in the Canyon de Chelly National Monument, Arizona, are examples of landforms often associated with the American West rather than the East. (Photograph by John Hillers, U.S. Geological Survey, circa 1890)

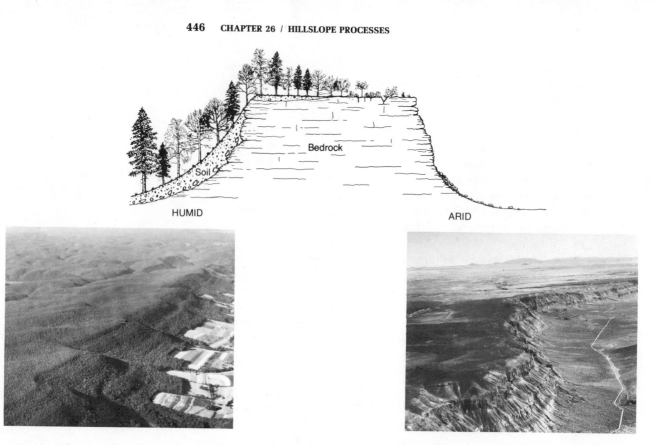

HUMID

ARID

Fig. 26.26 A mesa in which the left side
has developed under humid conditions
and the right side under arid conditions.
Below left, the Appalachian Front, an
escarpment in a humid region; below
right, an escarpment on the edge of the
Grand Mesa in Arizona. (Photographs by
John S. Shelton)

ten forms on the side slopes, which may make them be a little gentler than their nonvegetated counterparts along the arid mesa (Fig. 26.26).

SUMMARY

Hillslope processes entail the downslope movement of rock debris and soil by mass movement and erosion associated with runoff. Mass movements are gravitationally induced displacements of materials in the form of processes such as soil creep, solifluction, rockfalls, and landslides. Erosion on hillslopes involves a variety of processes, including rainsplash, sapping, and gullying. The rate at which hillslope processes perform work is related in a general way to climate, inasmuch as the magnitude and frequency of rainfall and the density of vegetation are climatically controlled. Overall, the highest rates of denudation are found in semiarid and seasonally dry climates. The manner in which slopes retreat and the forms they assume are related to many factors, including slope mass balance, rock structure, and vegetation.

UNIT VII SUMMARY

- Geomorphology is the study of landforms and the processes that sculpture them. It begins with the examination of rock weathering and goes on to include hillslope processes and the work of the major geomorphic agents: streams, glaciers, winds, and shoreline processes.

- Weathering is most effective at or near the earth's surface because many minerals are unstable in the presence of the moisture, heat, and chemicals of the landscape.

- Of the two principal types of weathering, chemical weathering is the more effective worldwide. Chemical weathering involves the breakdown of minerals by many complex processes, including dissolution, hydrolysis, and oxidation. The principal end products of chemical weathering are ions, oxides, and clays. Chemical weathering is most intensive in warm, moist climates such as the tropical rainforest.

- Mechanical weathering produces disintegration of rock and usually works hand in hand with chemical weathering. Expansion of rock masses and freeze/thaw activity are important mechanisms of mechanical weathering, and among the products are rock slabs and boulders. Mechanical weathering appears to be pronounced in cold or dry climates, although the rates at which it works are decidedly less than those of chemical weathering in warm, moist climates.

- Once particles of rock are freed by weathering, they are susceptible to downhill movement by hillslope processes. These fall into two categories: erosion by runoff and mass movement.

- Slope erosion takes place in several different ways and appears to be most effective in semiarid lands, where rainstorms can be heavy, yet the vegetative cover may not be substantial enough to protect the soil against rapid runoff.

- Mass movement involves the downslope displacement of materials under the force of gravity. Displacement may be triggered by an increase in shear stress, a decrease in shear strength, or both. In areas where rainfall or ground frost is seasonal, mass movement also tends to be seasonal.

- The mechanisms that commonly induce mass movements are groundwater, ground frost, undercutting, and earthquakes. The type of mass movement that results is related to the type of material involved, its moisture content, and its internal structure.

- Slopes may retreat in either a parallel or nonparallel fashion, depending on the balance between debris production and removal. Slope form depends on many factors, including rock type and structure, vegetation, erosional processes, and geologic uplift.

- In general, slope forms tend to be more angular in arid lands than in humid lands. Mountainous terrain generally exhibits the greatest variety in slope forms, ranging from strongly concave to convex forms.

FURTHER READING

Carson, M. A., and M. J. Kirkby, *Hillslope Form and Process,* Cambridge, England: University Press, 1972, 475 pp. *An advanced treatment of hillslope processes, including the influences of different climatic regimes.*

Cooke, R. U., and J. C. Doornkamp, *Geomorphology in Environmental Management,* London: Oxford University Press, 1974, 431 pp. *Application of geomorphology to problems of land management, emphasizing processes and measurement.*

Terzaghi, K., and R. B. Peck, *Soil Mechanics in Engineering Practice,* 2d ed., New York: Wiley, 1967, 729 pp. *Standard civil engineering text on soil behavior and related phenomena.*

Washburn, A. L., *Periglacial Processes and Environments,* New York: St. Martin's Press, 1973, 320 pp. *A comprehensive survey of research findings on the forces, processes, and geomorphic features of cold environments.*

REFERENCES AND BIBLIOGRAPHY

Bryan, K., "Cryopedology—The Study of Frozen Ground and Intensive Frost-Action with Suggestions on Nomenclature," *American Journal of Science* **244** (1946): 622–642.

Douglas, I., "Man, Vegetation and the Sediment Yield of Rivers," *Nature* **215** (1967): 925–928.

Gilbert, G. K., "The Convexity of Hill Tops," *Journal of Geology* **17** (1909): 344–350.

———, *Report on the Geology of the Henry Mountains,* Washington, D.C.: U.S. Government, 1877.

Ivernova, M. F., "Stationary Studies of the Recent Denudation Processes on the Slopes of the R. Tchon-Kizilsu Bazin, Tersky Alatau Ridge," U.S.S.R. Academy of Sciences (1963): 207–212.

Jackson, T. A., and W. D. Keller, "A Comparative Study of the Role of Lichens and 'Inorganic' Processes in the Chemical Weathering of Recent Hawaiian Lava Flows," *American Journal of Science* **269** (1970): 446–466.

Kirkby, M. J., "Measurement and Theory of Soil Creep," *Journal of Geology* **75**, 4 (1967): 359–378.

Langbien, W. B., and S. A. Schumm, "Yield of Sediment in Relation to Mean Annual Precipitation," *Transactions of the American Geophysical Union* **30** (1958): 1076–1084.

Leopold, L. B., M. G. Wolman, and J. P. Miller, *Fluvial Processes in Geomorphology,* San Francisco: Freeman, 1964.

Marsh, W. M., and J. M. Koerner, "Role of Moss in Slope Formation," *Ecology* **53,** 3 (1972): 489–493.

Owens, I. F., "Causes and Rates of Soil Creep in the Chilton Valley, Cass, New Zealand," *Arctic and Alpine Research* **1**, 3 (1969): 213–220.

Parizek, E. J., and J. F. Woodruff, "Mass-Wasting and the Deformation of Trees," *American Journal of Science* **225** (1957): 63–70.

Peltier, L., ''The Geographical Cycle in Periglacial Regions as it is Related to Climatic Geomorphology,'' *Annals of the Association of American Geographers* **40** (1950): 214–236.

Rapp, A., ''Recent Developments of Mountain Slopes in Karkevaage and Surroundings, Northern Scandinavia,'' *Geografiska Annaler* **42,** 2–3 (1960): 65–200.

Schumm, S. A., ''Evolution of Drainage Systems and Slopes in Badlands at Perth Amboy, New Jersey,'' *Geologic Society of America Bulletin* **67** (1956): 597–646.

Taber, S., ''Perennially Frozen Ground in Alaska: Its History and Origin,'' *Bulletin of the Geological Society of America* **54** (1943): 1433–1548.

Ter-Stepanian, G., ''On the Long Term Stability of Slopes,'' *Norwegian Geotechnical Institute Publication* **52** (1963): 1–15.

Terzaghi, K., ''Mechanism of Landslides,'' *Berkey Volume,* Geological Society of America (1950): 83–123.

Varnes, D. J., ''Landslide Types and Processes,'' in E. B. Eckels, ed., *Landslide and Engineering Practice,* Washington, D.C.: U.S. Highway Board, Special Report 29 (1958), pp. 20–47.

Washburn, A. L. ''Classification of Patterned Ground and Review of Suggested Origins,'' *Geological Society of America Bulletin* **67** (1956): 823–866.

Williams, P. J., ''An Investigation into Processes Occurring in Solifluction,'' *American Journal of Science* **257** (1959): 481–490.

Wolman, M. G., ''A Cycle of Sedimentation and Erosion in Urban River Channels,'' *Geografiska Annaler* **49A** (1967): 385–395.

Young, A., ''Soil Movement by Denudational Processes on Slopes,'' *Nature* **188** (1960): 120–130.

SCULPTURING
THE LAND
BY EROSION
AND
DEPOSITION

UNIT VIII

KEY CONCEPTS   OVERVIEW

bed shear stress
sediment load
channel dynamics
meanders
floodplain
geographic cycle
graded profile
dynamic equilibrium
wave generation
wave refraction
littoral transport
net sediment transport
progradation
shore protection
glaciation
Pleistocene Epoch
névé
glacier mass budget
moraine
drift
boundary layer
wind stress
deflation
saltation
loess
sand dunes

Streams, glaciers, wind, and waves and currents are the principal geomorphic agents of the landscape. Streams are the most effective of the four in moving rock debris off the continents. The ability of a stream to erode and transport particles on its bed increases with flow velocity and water depth. Both velocity and depth are generally greater in large rivers, such as the Amazon and Missouri, than in small streams. The valleys in which rivers flow are carved by the rivers themselves, and over long periods of time, the valleys are often widened as the rivers shift laterally. Grandiose models that describe the way in which entire land masses are lowered by river systems are generally inadequate, owing to the difficulty of determining the effects of long-term changes in isotatic, climatic, and tectonic conditions on landform development.

Wind waves produce the bulk of erosion and sediment transportation along coasts. Wind waves are generated when the momentum of surface airflow is transferred to the water surface. Wave motion is characterized by rotational movement of the water particles beneath the wave form. In shallow water this movement exerts force against the bottom, thereby setting up the potential for erosion and sediment transport. The flow of sediment along a coast is driven by both waves and currents. Sediments are moved from source areas, such as river deltas and eroded shorelines, and are deposited in various forms both along the shore and in deeper water offshore.

Currently covering less than 5 percent of the earth, glaciers are found mainly in high mountain and polar regions. During the Ice Age, glaciers extended well into midlatitude lands, covering nearly 10 percent of the earth. Erosion by the continental glaciers was concentrated in mountainous areas, in shield areas, and in certain basins such as those of the Great Lakes, which were enlarged by glacial scouring. As the glaciers melted back, deposits were laid down in the form of irregular hills, sandy flats, and related landforms. These deposits have a pronounced influence on the development of soils, drainage, and vegetation.

Wind is most effective as an erosional agent on nonvegetated surfaces, such as deserts, farmfields, and fresh beach and glacial deposits. Although wind at ground level can attain very high velocities, it is rarely capable of moving particles larger than coarse sand or small pebbles. Where materials of mixed particle sizes are eroded by wind, the silt and sand are the sizes winnowed away, leaving a residue of coarser particles. The sand and silt in turn are segregated from each other and are often deposited in the form of sand dunes and blankets of loess.

Chapter 27 opens with principles of streamflow and goes on to examine the movement of debris by streams and rivers and the sculpturing of landforms by rivers. Chapter 28, concerned with the geomorphology of shorelines, describes wave motion, sediment transport, and coastal landforms. Glaciation is taken up in Chapter 29, and wind erosion and deposition are examined in Chapter 30.

STREAMS, CHANNELS, AND VALLEYS

John Wesley Powell (1834–1902), scientist, explorer of the Grand Canyon, and founder of the United States Geological Survey.

INTRODUCTION

When the forerunners of the fields of physical geography and geology were emerging in the eighteenth century, there was a great deal of debate among scholars over the origin of the earth and its landforms, with much of the argument centered on how the land was sculptured. There were two major schools of thought on the topic: One held that the earth is comparatively young (6000 years or so) and that the flood cited in the Old Testament was the chief cause of the erosion and in turn of many of the earth's present landforms. According to the other view, the earth is comparatively old (millions of years), and the processes that are eroding the land today are the same ones that were active in the past, having sculpted the landforms over long periods of time. This concept became known as *uniformitarianism* and by the phrase "the present is the key to the past."

The British geologist Charles Lyell (pronounced Lye-ELL) wrote a convincing treatise in the 1830s that largely eliminated the biblical flood from serious intellectual consideration. The argument was now reduced to which of the erosional agencies—the sea (waves and currents) or runoff (streams or rivers)—is the earth's primary erosional force. Lyell argued that the sea at higher levels was mostly responsible, a view supported by his colleague Charles Darwin. But other scientists were observing and measuring runoff and sediment transport by streams, and the results were pointing to runoff as the primary erosional agent of the land. In particular, geographers and geologists sent into the American West by the federal government to explore and map the new territory found evidence that runoff was the principal means of erosion even in arid environments. In the late 1800s John Wesley Powell, explorer of the Grand Canyon, showed that the Colorado River had eroded this huge cataract nearly 2000 m deep in places. Grove K. Gilbert and other scientists demonstrated that runoff in the forms of rainwash, rivulets, and gully flow eroded massive amounts of sediment from uplands, plateaus, and valleys alike. When combined with the sediment moved by streams and rivers, it became clear to these scientists that total sediment production by runoff in lowering the land far outweighed the combined production of waves, currents, wind, and glaciers.

This chapter is concerned with the work of streams and rivers. It picks up on the thread of the discussion in the previous chapter by tracing the movement of sediment down the river valley to the ocean. We open the discussion with an examination of flow in open channels and the principles governing stream velocity, erosive power, and sediment transport. This is followed by a description of the various features built by rivers as they move their sediment loads and sculpture their valleys.

HYDRAULIC BEHAVIOR OF STREAMS

In order to understand how streams erode and deposit material in the landscape, we shall first examine a few principles about running water and the associated forces in channels. In addition to velocity, we need to examine types of flow, stream energy, and the stress exerted by the river on its channel bed.

Velocity

The velocity of the streamflow is clearly important, as the water must itself move if it is to move any particles on the streambed. In stream channels the flow velocity is apparently related to three factors: the slope of the channel, the depth of the water, and the roughness of the channel. Increases in slope or depth cause increases in velocity, whereas an increase in the channel roughness causes a decrease in velocity. The effect of an increase in depth is slightly greater than that of an increase in slope of equal proportion. Hence the fastest velocities may be found not in the mountainous headwaters of a river, but farther downstream, where the depth and discharge are greater. For example, average flow velocity near the mouth of the Amazon, where slope is only a few inches per mile and the depth around 50 m, is about 2.5 m per sec, whereas average velocity in the Grand Canyon of the Yellowstone, where slope is 200 feet per mile and the depth is 1–2 m, is only about 1 m per sec.

Within a stream channel the velocity is lowest near the bottom and the sides of the channel and highest near the top in the middle. The highest velocity is generally slightly below the surface because of friction between the water and the overlying air. At channel bends, the faster-moving water responds more to centrifugal force than slower water does and slides toward the outside of the bend (Fig. 27.1). In addition, this may also produce *superelevation* of the stream, in which the water surface near the outside of the bend is higher than the water surface near the inside.

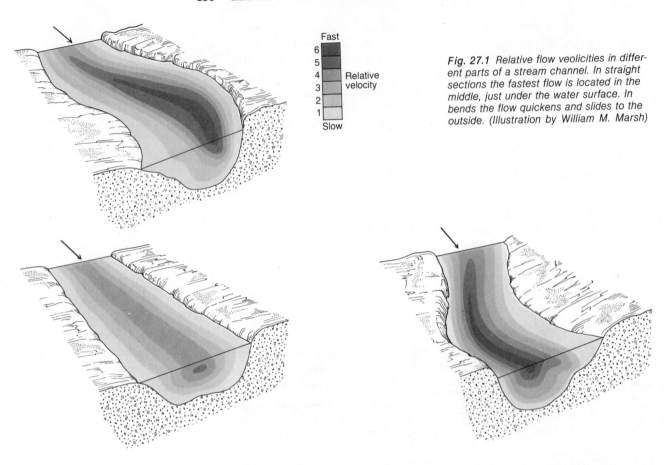

Fast
6
5
4   Relative
3   velocity
2
1
Slow

Fig. 27.1 Relative flow veolicities in different parts of a stream channel. In straight sections the fastest flow is located in the middle, just under the water surface. In bends the flow quickens and slides to the outside. (Illustration by William M. Marsh)

Flow Types

Water flow in streams is always *turbulent*. This means that the primary source of flow resistance is the mixing between the slower-moving water molecules on the streambed and the faster-moving water above the bed. In *laminar* flow, which occurs in streams only within a very thin layer near the bed, but which prevails in groundwater flow and which occurs in sheetwash, one thin layer of water simply slides over another, without any more mixing than would occur if the water were still (Fig. 27.2). The source of the flow resistance in laminar flow is just the intermolecular friction, called the *viscosity*, of the fluid. Since increased temperature of water results in lower viscosity (i.e., less intermolecular friction), the velocity of laminar flow increases with higher water temperature. The velocity of turbulent flow, how-

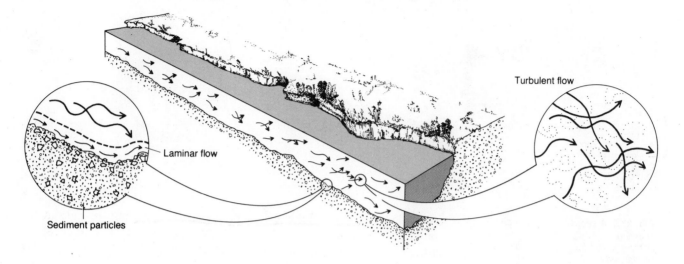

Laminar flow

Sediment particles

Turbulent flow

ever, is not affected by water temperature. The intensity of turbulence tends to increase with flow velocity.

Fig. 27.2 Turbulent flow is the predominant type of flow in streams. Laminar flow is found in a thin layer immediately on the streambed. (Illustration by William M. Marsh)

Conservation-of-Energy Principle

Any interpretations about stream behavior that we can infer from average flow velocities are limited by an important observation: Flowing water very quickly establishes an equilibrium with the prevailing frictional environment of the channel. Unless there is a change in channel shape or roughness, or unless the discharge is increased or decreased, the water does not accelerate or decelerate. Therefore, in a given stretch of channel, called a *reach*, the kinetic energy of the water (the portion of the total energy due to the motion) does not change.

We often approach problems in the physical sciences through the conservation-of-energy principle. Applied to a stream's erosive power, we would reason that if the amount of energy in the water is reduced, it must have gone somewhere, and a likely possibility is expenditure of energy at the streambed. But the velocity itself gives us no clues about the amount of energy involved.

Bed Shear Stress

The key observation we use to help us approach this problem is that the water in a channel of given shape does not accelerate. Yet a force—gravity—is always on this water, and one of the basic concepts of physics is that force causes acceleration. The water in a channel

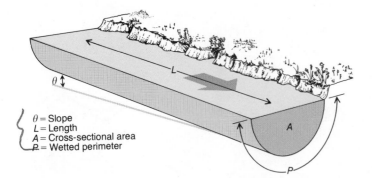

θ = Slope
L = Length
A = Cross-sectional area
P = Wetted perimeter

Fig. 27.3 A section of stream channel, showing the key parameters needed to determine the force exerted by the stream on its bed.

reach is losing potential energy as it loses elevation, but this loss in potential energy does not result in a gain in kinetic energy. It must be otherwise dissipated, and one place it can be dissipated is at the streambed.

We can easily calculate the *bed shear stress*, which is the amount of energy loss per unit area of streambed. Consider Fig. 27.3, which shows a portion of a stream channel, a uniform reach of length L. The distance from one side of the stream to the other, measured along the bottom, is the *wetted perimeter*, P. Theta (θ) is the angle between the channel floor and the horizontal, and A is the cross-sectional area. The volume of water involved is $L \times A$, and the total surface area of streambed affected by the reach is $L \times P$. The force pushing the water downhill is equal to the mass (m) times the gravitational acceleration (g) times the sine of the slope of the channel:

$$F = mg \sin \theta.$$

But since mass is the product of density (ρ) and volume, and since for small angles (<10°) $\sin \theta$ is equal to slope S (vertical drop divided by horizontal distance), the downhill force is equal to water density times gravitational acceleration times reach length times channel cross-sectional area times channel slope. In notational form, downhill force (F_d) is:

$$F_d = \rho\, g\, L\, A\, S.$$

Because this downhill force does not result in any acceleration, there must be an equal and opposite force holding the water back. This opposing force is the bed shear stress (per unit area) times the area over which it is applied. Since it is equal to the downhill force, we can say that bed shear stress for a given length of reach and wetted perimeter is equal to $\rho \cdot g \cdot L \cdot A \cdot S$, or:

$$\text{Bed shear stress} = \rho\, g\, \frac{A}{p}\, S.$$

We call the ratio of cross-sectional area to wetted perimeter, A/p, the *hydraulic radius*. Because most streams are much wider than they are deep, the hydraulic radius is approximately equal to the depth (D); hence bed shear stress can generally be given as a quantity equal to water density times gravitational acceleration times water depth times channel slope, or:

Bed shear stress = $\rho\, g\, D\, S$.

The density (ρ) of water varies slightly with temperature, but for our purposes we can consider it constant, at 1000 kg/m³. Similarly, gravitational acceleration varies slightly with latitude and altitude, but we can consider it constant, at 9.8 m/sec². Therefore, the only variables are depth and slope, and the mean bed shear stress is proportional to the product of depth and slope. The units for bed shear stress are N/m² (newtons per square meter), a force per unit area. As noted in Appendix 1, the energy unit, a joule (J), is equal to a newton applied over a one-meter length. Hence, one N/m² is equal to one J/m². If we multiply this unit by stream discharge (m³/sec), which is equal to stream velocity times its cross-sectional area, the product is joules per second, or watts, a *power* term. The product of bed shear stress and discharge, therefore, tells us the rate at which the stream is losing energy. Variations in energy, in turn, can be related to changes in either discharge or slope and water depth.

What happens to the energy dissipated by the stream? Most of it is converted to heat, but a small amount is also devoted to the work of moving sediment. The generation of heat is the principal reason streams do not freeze out in winter and why a stream can melt its way into the surface of a glacier (see Fig. 27.13).

EROSION AND DEPOSITION

In the previous section we derived an expression for the rate at which a stream loses energy in the form of bed shear stress as it flows down its channel. The next problem is to relate bed stress to erosion. In trying to formulate this relationship, however, we encounter two problems. First, only part of the energy dissipation takes place at the bed itself; some is lost near the surface in fluid turbulence. Second, actual stresses at the bed may vary quite a bit from the mean value, both from place to place in the channel and at different times. Our understanding of these two problems is such that our calculations of erosion and sediment transport are only approximate, although we can explain the principles involved.

Fig. 27.4 Suspended sediment being released into Flathead Lake by the Flathead River of Montana. (Photograph by John S. Shelton)

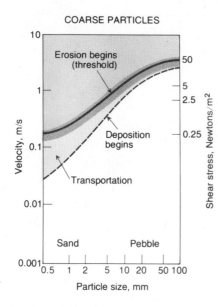

Fig. 27.5 Erosion, transportation, and deposition of coarse particles in relation to shear stress. The threshold of erosion increases with particle size. (After Hjulström, 1939)

Sediment Transport and Load Types

Most of the work of streams is devoted to the movement of sediment brought to the channel by hillslope processes and tributary streams. Sediment movement by streams occurs in two modes: as *bed load* and as *suspended load*. The bed load consists of particles that roll along in virtually continuous contact with the streambed. The suspended load consists of small-size particles that are held aloft in the stream by the turbulence of the flow. The size distribution and rate of movement of the bed load are determined by the level of bed shear stress and the sizes of the particles available on the streambed. In contrast, the characteristics of the suspended load are determined by supply. Almost all stream reaches are capable of carrying a large suspended load if such a load is produced by erosion farther upstream or in a tributary (Fig. 27.4).

Bed-load sizes range from sands to boulders. As water flows over particles on the bed, it exerts *drag* force (along the stream) and a *lift* force (upward). Both of these forces can be estimated from the bed shear stress. The lift force is of primary importance in initiating motion; laboratory flume experiments have demonstrated that a lift force of about 70 percent of the submerged weight of a particle is sufficient to pivot it upward on its axis, whereupon the drag force can push it downstream. For any size class of particles, therefore, there is a *threshold*, or critical, level of bed shear stress, below which no movement takes place. Above this threshold value, the rate of sediment movement increases as some power of the shear stress.

For sand and larger sediments transported as bed load, the same curve defines erosion and deposition (Fig. 27.5). For a given particle size, if the bed shear stress exceeds the threshold value, erosion and movement of sediment will take place. When the bed shear stress drops below that value, deposition occurs. For the smaller sizes, silts and clays, which are carried as suspended load, the story is different. Because silts and clays are cohesive, they are held in place in a streambed not principally by gravity (as are sands and larger sizes), but by molecular attraction. A large bed shear stress is required to overcome those cohesive forces, and the smaller the particle, the greater the stress required. However, once dislodged, the particles will remain in suspension unless the water becomes relatively quiescent. Thus the curves for erosion and deposition of fine particles are quite different (Fig. 27.6).

A third type of load is that carried in solution. Ions of minerals produced in weathering are released into streams through groundwater and runoff. Total ionic concentrations in stream water are generally on the order of 200–300 milligrams per liter (same as parts per million), but in areas with low relief and high rates of soil leaching, as in parts of the American east and south, concentrations may reach several thousand mg per liter. In dry areas dissolved load may be much lower than the general level of concentration, typically less than 10 percent of the total stream load. This is explained on the basis of the low rates of chemical weathering and is in contrast to the heavy suspended sediment loads in the streams of dry areas (Table 27.1).

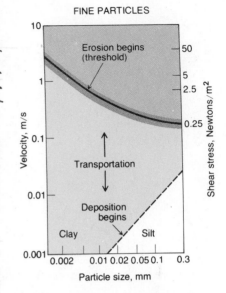

Fig. 27.6 Erosion, transportation, and deposition of fine particles in relation to shear stress. Note that greater stress is needed to erode finer particles, because cohesiveness is greater with smaller sizes. (After Hjulström, 1939)

Table 27.1 Total and dissolved loads representative of rivers in dry and humid climates.

RIVER	CLIMATE	AVERAGE DISCHARGE, CFS	TOTAL LOAD (tons/yr·mi²)	DISSOLVED LOAD AS PERCENT OF TOTAL LOAD
Little Colorado (Arizona)	Arid	63	199	1.2
Green River (Utah)	Arid/ semiarid	6,737	530	12.0
Iowa River (Iowa)	Humid	1,517	510	29.0
Delaware River (New Jersey)	Humid	11,730	270	45.0
Juniata River (Pennsylvania)	Humid	4,329	265	64.0

Adapted by permission from L. B. Leopold, M. G. Wolman, and J. P. Miller, *Fluvial Processes in Geomorphology*, San Francisco: Freeman, copyright © 1964.

Bedrock Erosion by Streams

Erosion of bedrock channels occurs through three mechanisms: (1) *chemical reactions* between minerals in the water and in rock; (2) *cavitation*, whereby sudden losses in pressure as water flows around a sharp rock corner cause formation of small bubbles, which in turn explode and release large localized pressures; (3) *scouring by material* carried in the bed load, whereby heavy particles bump and skid along the bottom, loosening and freeing material. The effectiveness of this last mechanism is evidenced by potholes, which are small pits etched into the bedrock on the streambed by stones swished around and around by currents (Fig. 27.7). Although streams can be important in the erosion of rock, especially in their upper reaches, most of their work involves the transportation of material already in the valley.

Karst Processes

One of the most dramatic effects of running water on bedrock is found in areas of carbonate rocks (limestone and dolomite), where large volumes of rock are lost in solution to groundwater. What distinguishes karst processes from solution weathering in general is that the loss of rock is concentrated along cracks and bedding planes and leads to the formation of bedrock caverns and cavities (see Fig. 12.13c). The bulk of runoff, which would otherwise be carried in surface channels, is piped underground through these voids. As the voids grow, the overlying rock may sag and then collapse, forming a depression called a doline, sink, or sinkhole. In Fig. 22.17 the southern part of the area shown is an example of karst topography characterized by abundant sinkholes. As sinkholes grow, they may coalesce, eventually revealing the pattern of underground drainage and even the stream channels themselves.

Karst processes take place in all climatic regions, but are most pronounced in midlatitude, subtropical, and tropical environments where rainfall and vegetation are abundant. The atmosphere and the organic processes in the soil contribute carbon dioxide to precipitation and infiltration water, forming a weak carbonic acid which reacts with the calcium carbonate. A bicarbonate is produced from the reaction and is soluble in water (see Fig. 25.5). In order for karst topography to form, however, certain bedrock conditions are also necessary: The

Fig. 27.7 These potholes illustrate the influence of scouring on the streambed. (Photograph by G. K. Gilbert, U.S. Geological Survey)

rock should be relatively pure limestone in massive, heavily fractured strata sufficiently elevated in the terrain to allow for free circulation of groundwater.

FEATURES ASSOCIATED WITH STREAM CHANNELS AND THEIR VALLEYS

In trying to interpret the features we see in and around stream channels in terms of the processes described above, we must keep certain facts in clear perspective. First, the stream is flowing in a channel and valley that it has created, as contrasted with a valley created by faulting, glaciation, or some other means and into which the stream merely found its way. Second, the types and magnitudes of stream processes that can be observed on most days may have little to do with the origins of many channel and valley features. Rather, it is the processes associated with a limited number of flows of particular magnitudes or selected combinations of flows that determine many features. Third, through erosion and deposition the stream is continually adjusting its channel and in turn its slope, depth, and bed shear stress. If, for example, the slope and stress of the bed shear stress are low at one point in the stream profile, deposition will take place there. But this in turn will increase the slope and thus the bed shear stress, eventually leading to a restoration of equilibrium, whereby the sediment brought to that point will be transported on through.

Pools and Riffles

One feature whose partial explanation gives us an idea of the complexity of the processes involved in channel formation is the pool-riffle sequence. During low flow in some streams, the water is distributed in a sequence of quiescent places (pools) linked together by steeper reaches of more rapid flow (riffles). Further investigation reveals that the pools contain fine sediment, whereas the riffles contain gravels and larger particles. Moreover, the spacing is surprisingly regular, generally from five to seven times the channel width from midpool to midriffle (Fig. 27.8). (Riffles fulfill an important ecological role as well: Trout eggs are laid in the riffles, as the newly hatched fish can escape through the gravels into the water.)

Through the concept of bed shear stress, we can at least explain how the pool-riffle sequence is maintained. We hasten to add, though, that the questions as to how it is formed and why the spacing is regular are more difficult to answer. The key to the maintenance question lies in the high flows rather than the low flows. If we examine a pool-riffle

Fig. 27.8 Riffles in the bend of a stream near Jackson Hole, Wyoming. (Photograph by William M. Marsh)

sequence during the low-flow period, we find, first, that depth is greatest over the pools, but that the slope is so low there that the bed shear stress is greatest over the riffles. Second, we find that bed shear stresses over the riffles are *not* great enough to dislodge the larger particles; hence the major consequence of the bed shear stress distribution is to cleanse the fine particles from the riffles and deposit them in the pools.

If we return to the stream at high flow, we immediately notice something different: The pool-riffle sequence is drowned, and the distinction between pools and riffles is not apparent. Closer inspection reveals that the pools and riffles are still there, but that the slope of the water surface is virtually constant. Thus in contrast to the low-flow situation, slope is almost the same over both pool and riffles: On the other hand, the water depth is still greater over the pools. This indicates two things: (1) bed shear stresses are greater everywhere than at low flow; and (2) bed shear stresses are maximized over the pools. Hence at high flow the pools are scoured, and the larger particles are deposited on the riffles.

Although this description explains only part of what we would like to know, it does illustrate an important principle. This particular feature is a product of the range of discharges that flow through the stream channel. It cannot be explained solely by what happens during the average flow, the most frequent flow, the dominant discharge, or any other meaningful flow you may choose.

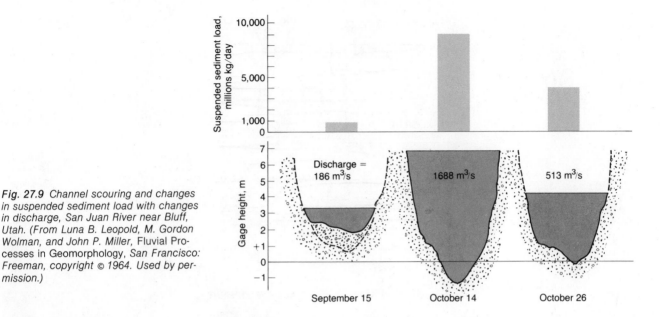

Fig. 27.9 Channel scouring and changes in suspended sediment load with changes in discharge, San Juan River near Bluff, Utah. (From Luna B. Leopold, M. Gordon Wolman, and John P. Miller, Fluvial Processes in Geomorphology, *San Francisco: Freeman, copyright © 1964. Used by permission.)*

Channel Geometry

Studies reveal that streams can undergo substantial changes in geometry in response to changes in discharge and sediment supply. When discharge rises, both velocity and water depth increase, producing widespread scouring of the streambed (Fig. 27.9). Conversely, much of the channel bed fills in as the discharge falls. These channel changes, referred to as degradation and aggradation, respectively, are also associated with land-use changes. When land is cleared for farming, soil erosion and sediment input to streams increase, resulting in channel sedimentation. When farmland gives way to suburban development, the rate of soil erosion increases substantially for a short time, and stream channels aggrade (Fig. 27.10). If urbanization follows, sediment yield to streams declines because soils are now covered with hard-surface materials such as concrete and asphalt. At the same time, however, these materials help to increase runoff, resulting in greater magnitudes and frequencies of flows. Together, the increased flows and reduced sediment supplies tend to produce channel degradation in the urban streams.

Channel Patterns and Dynamics

Two distinctive channel patterns occur in natural streams: *braided* and *single-thread*. Single-thread channels are seldom straight, almost

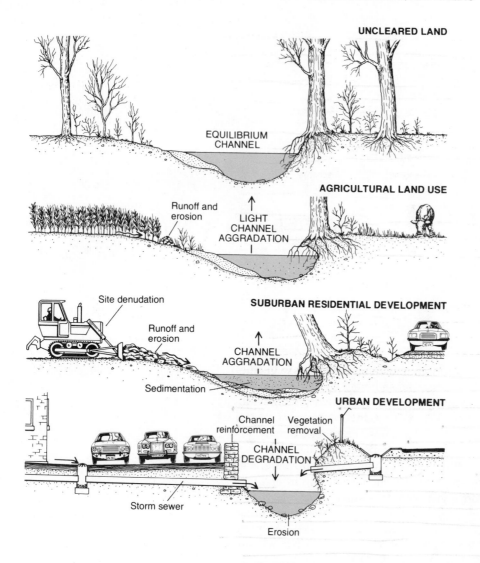

Fig. 27.10 *Sequence of channel changes related to land-use changes. Heavy aggradation is associated with land clearance for residential development. Degradation is often brought on by urbanization. (Illustration by William M. Marsh)*

always displaying some sinuosity. If this sinuosity is marked enough, we say that the stream is "meandering." Although many rivers are clearly braided or clearly meandering, we often find both patterns in the same river. Figure 27.11 shows various channel patterns.

Braided channels are often found in highly active environments, such as near the front of a glacier where massive amounts of sediment are being moved and the vegetative cover is weak. Braided channels usually form in coarse, noncohesive bed material under conditions of sharply fluctuating discharge. While the discharge is decreasing,

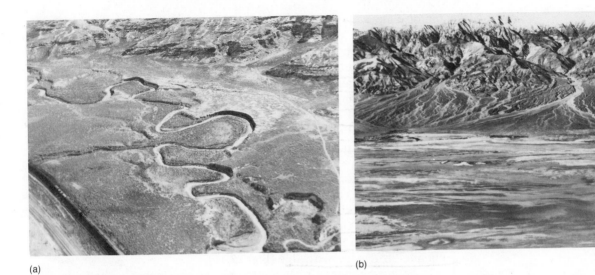

(a)

(b)

coarse debris is deposited on the channel bed in the form of gravel bars. When the discharge declines further, the bar forms a barrier and splits the flow. Subsequent high discharges may erode the gravel bar, completely modifying the appearance of the channel. If a bar survives several seasons, plants may become established on it, adding to its stability (Fig. 27.12).

Meandering channels occur in a wide variety of environments—in arid and humid regions, on all kinds of surface materials, and even on

Fig. 27.11 Channel patterns: (a) meandering single-thread; (b) braided. (Photographs by J. R. Balsley (a) and H. E. Malde (b), U.S. Geological Survey)

Fig. 27.12 According to Lewis and Clark, no islands existed at this location in the Missouri River in 1805. This 1903 photograph shows a heavily vegetated island in midstream. In 1964, after having been overwashed many times by flood waters, much of the island was still intact, testimony to the stabilizing effect of vegetation. (Photograph by T. W. Stanton, U.S. Geological Survey)

top of glaciers (Fig. 27.13). Numerous explanations for this phenomenon have been put forth, along with intricate descriptions of the meander geometry. The width of the meander belt varies with the size of the stream and appears to be related to the river's mean annual discharge; the larger the discharge, the greater the width (Fig. 27.14).

Where pools and riffles occur, we often find the riffles spaced on opposite sides of the stream. In a meander sequence, the pools are generally found at the bends, with the riffles between them. Hence there are two sets of pool-riffle sequences in a full meander.

Detailed investigations of the distribution of bed shear stresses along meandering and straight reaches of the same rivers show that the meandering reaches have less variable distribution. This fact indicates that meandering flow represents a more stable state than does flow in a straight channel. It also suggests that as a river trends toward a meandering course, the distribution of energy along the channel grows less variable (Fig. 27.15).

Active meanders in most streams are continuously changing. Erosion is concentrated on the outsides of bends and slightly downstream from them. Deposition occurs on the insides of bends, forming features called *points bars*. Each year or so a new increment is added to the point bar while the river erodes away a comparable amount on the opposite bank. In this way the river shifts laterally, gradually changing

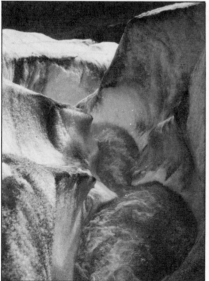

Fig. 27.13 Channel meanders in a meltwater stream flowing on the surface of a glacier. (Photograph by Jeff Dozier)

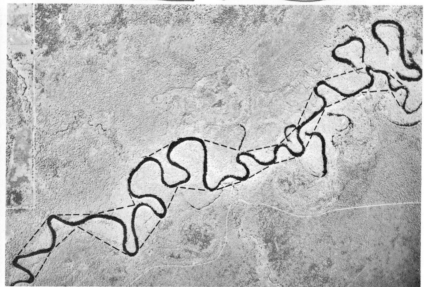

Fig. 27.14 River meanders. The meander belt is the area between the broken lines; the detached segments on the right part of the photograph are oxbows; the positions of some old oxbows can be detected from the vegetation patterns.

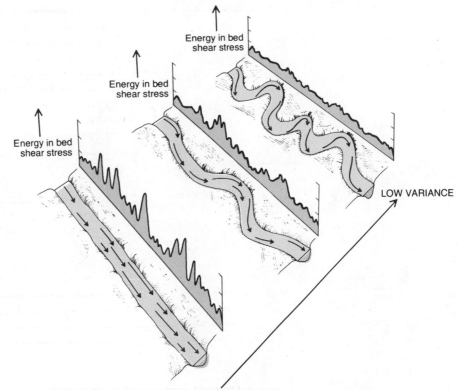

Energy in bed shear stress

Energy in bed shear stress

Energy in bed shear stress

LOW VARIANCE

HIGH VARIANCE

its location in the valley. But the river can also undergo sudden changes in location when it erodes new segments of channel and abandons old ones. This is especially commonplace where a meander forms a large loop and the river erodes toward itself from opposite sides of the loop, eventually breeching the meander. The old channel is abandoned because the new route is steeper and thus more efficient. The old channel forms a small lake, called an *oxbow*, but in time it fills with sediments and organic debris, becoming a wetland. On aerial photographs of floodplains, such features are often easily identified on the basis of the swamp or marsh vegetation over them (see Fig. 27.14).

Fig. 27.15 The relationship between sinuousity and the variance in energy distribution along a reach of stream channel. (Illustration by William M. Marsh)

Floodplains

Formation. The flat, low ground through which most rivers flow is the floodplain. Floodplains form mainly as a result of the lateral shifting by the river. The process works as follows. When the river flows against the high ground at the edge of its valley, called the *valley wall*, it undercuts the wall, which fails and thereby retreats. At the same time, new

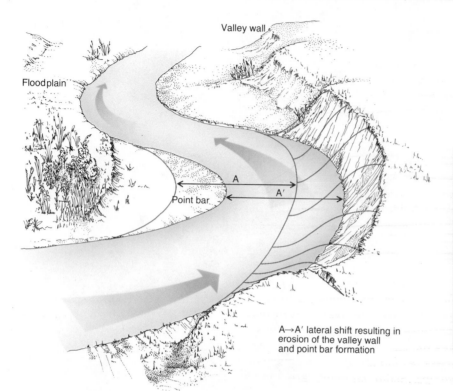

A→A' lateral shift resulting in erosion of the valley wall and point bar formation

Fig. 27.16 Floodplain formation involving undercutting of the valley wall and simultaneously building of a point bar in a meander. (Illustration by William M. Marsh)

ground is being formed on the opposite bank in the form of a point bar (Fig. 27.16). The new ground forms at a low elevation, near that of the river. In time the valley walls are cut back so far that a continuous ribbon of low ground is formed along the valley floor. This is the floodplain. When the river floods, the floodplain receives the overflow; hence the basis for the term. Although floods alter the surface of the floodplain by eroding it and leaving deposits on it, they are clearly not the main cause of its formation, and in this regard the term is a little misleading.

When the river is not in contact with the sides of the valley, hillslope processes continue to work on the valley walls, bringing new material to the floodplain. Studies of floodplain materials reveal that material brought down by mass movements, such as slumping and soil creep, may comprise an appreciable part of a small floodplain. Added to this are deposits for overland flow and small streams that drain the sides of the valley. Coupled with the channel and flood deposits, these materials help to produce a very diverse composition in floodplains.

Features. Meander scars, oxbows, and point bars (described earlier) are present in virtually every floodplain. In addition, natural levees,

Fig. 27.17 Terraces in the Madison River Valley, Montana. (Photograph by John R. Stacy, U.S. Geological Survey)

scour channels, backswamps, and terraces are common features. Levees are mounds of sediment deposited along the river bank by floodwaters. They occur on the bank because this is where flow velocity declines sharply as the water leaves the channel and part of the sediment load is dropped. In the low areas behind levees, water may pond for long periods, forming *backswamps*. *Scour channels* are shallow channels etched into the floodplain by floodwaters. They often form across the neck of a meander loop and carry flow only when floodwaters are available.

Terraces are elevated parts of a floodplain that form when a river downcuts and begins to establish a new floodplain elevation. Downcutting may be induced by many factors, including increased flow, loss of sediment load, and uplift of the land. The terraces in Fig. 27.17 resulted from the uplift of a portion of the Rocky Mountains.

THE DRAINAGE BASIN AS A GEOMORPHIC UNIT

Drainage Networks

The complex net of streams carrying water and sediment from a drainage basin may be analyzed in a variety of ways. From a geographic standpoint, three aspects of the stream networks are important: drainage density, drainage pattern, and stream order.

The *drainage density* is the total length of all channels in a unit area; typically, it is expressed in units such as miles of channel per square mile. A high drainage density is indicative of more fluvial erosion and more finely textured topography than is a low drainage density. High densities occur where the surface material is easily eroded

High drainage density on a devegetated slope.

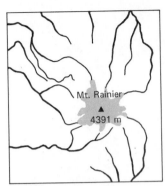

RADIAL
Mt. Rainier, Washington

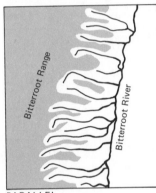

PARALLEL
Bitterroot River, Montana

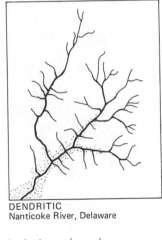

DENDRITIC
Nanticoke River, Delaware

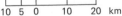

10 5 0   10   20 km

Fig. 27.18 Three types of drainage patterns: radial, parallel, and dendritic.

and the infiltration capacity is low; that is, where the coefficient of runoff is relatively high. Semiarid areas with light plant covers and fine-grained soils often give rise to very high drainage densities. In some semiarid areas south-facing slopes have higher drainage densities because the plant cover, under severe moisture stress due to intensive solar heating, is weaker there than on flat ground or north-facing slopes (see Fig. 2.5). Forested areas with low coefficients of runoff typically produce low densities. In North America drainage densities range from 2 to 500 miles of channel per square mile.

The *drainage pattern* is a description of the configuration of a network of stream valleys. Dendritic drainage, a pattern similar to the branching arrangement on an apple tree, is thought to form where control on terrain by geologic structures is not significant. Trellis drainage is generally associated with parallel-folded mountains such as the Ridge and Valley Section of the Appalachians (Fig. 24.12). Rectangular drainage is found in areas of massive igneous rock with right-angle fractures, such as on the Columbia Plateau of the United States Northwest. Radial drainage is often associated with volcanoes (Fig. 27.18).

The *stream order*, described in Chapter 13, is a ranking scheme based on where a stream fits into the drainage hierarchy. A first-order stream has no tributaries; second-order segments have only first-order tributaries, and so on. In a given basin the ratio between the number of streams of a given order and the number of streams of the next highest order is called the *bifurcation ratio* and is usually around 3 or 4, but this curious observation has not been related to any mechanism that helps to explain it. In assigning stream orders, the choice of what to call a first-order stream is arbitrary and in fact will vary with the scale of the map. In the field it is not constant in time either. During rainy periods the drainage network expands as small, otherwise dry channels become first-order streams (see Chapter 13).

FORMATION OF RIVER-ERODED LANDSCAPES

We began this chapter with some reflections on the development of thought in geomorphology, noting that by the end of the nineteenth century it was clear that running water is the principal agent of erosion of the continents. But just how that erosion takes place and what kinds of landscapes result from it were not so clear at that time, and in the following decades geomorphologists pursued these problems through various lines of research. Although considerable progress has been made, especially with respect to how erosion takes place, the answers to questions about the landscapes created by running water are still open for the most part. But there have been plenty of ideas over the years.

YOUTH

MATURITY

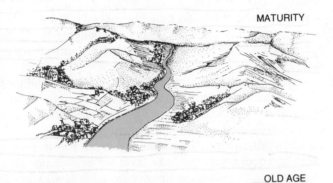

OLD AGE

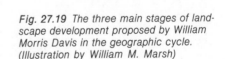

Fig. 27.19 The three main stages of land-scape development proposed by William Morris Davis in the geographic cycle. (Illustration by William M. Marsh)

The first big idea on the formation of landscapes by streams and rivers was formulated before the turn of the century by a prominent physical geographer named William Morris Davis. He proposed that the landscape evolves through a series of developmental stages as rivers deepen and widen their valleys. He envisioned three main stages of development, which, following biological nomenclature, he named youth, maturity, and old age (Fig. 27.19). During the old-age stage, the

William Morris Davis (1850–1934), prominent physical geographer. (Photograph courtesy of the Harvard University Archives)

landscape is reduced to a low, rolling plain, called a peneplain. Davis called his idea the "geographic cycle," because he thought that when the old-age stage was reached, the land mass would be uplifted and valley formation would be "rejuvenated," i.e., begin over again.

The geographic cycle was immensely popular among educators and scientists alike in the first half of the twentieth century. Modern geoscientists, however, do not view it very seriously, for several reasons. First, it is highly unlikely that land masses uplift in response to the end of the cycle. Rather, geophysical evidence indicates that uplift can take place at any time and that it is more a response to tectonic forces than to unloading of the land per se. Thus where uplift is sporadic, an orderly sequence of landscape stages is quite unlikely; where uplift is rather steady, the landscape could remain for long periods of time in one "stage" while in a state of continual change. Second, Davis thought that as river valleys widened, the side slopes would grow gentler, but field studies indicate that parallel retreat of valley slopes is probably more common. Third, the geographic cycle gave little consideration to the influence of rock type and structure on landscape formation.

Although the geographic cycle has lost most of its support, several of the concepts on which it was based are still valid. One of these is *base level*, a concept introduced by John Wesley Powell in the late 1800s. Base level is the lowest elevation to which a river can downcut its channel. Since a river cannot deepen its channel much below sea level, the ocean sets the base level for most major rivers. For streams that do not terminate in the ocean, base level may be set by a lake, another stream, or even a resistant rock formation that intersects the streambed. Essentially all base levels, whether set by the ocean or features on the continents, are temporary because their elevations are continually changing with tectonic, isostatic, erosional, and water-level fluctuations.

Another concept Davis used was *grade* or *graded profile*. If we plot an elevation profile of a stream from headwaters to mouth, we find that the overall slope of the channel flattens downstream (Fig. 27.20). Although the profiles of individual streams vary in detail, all tend to assume this long concave shape. Grove K. Gilbert termed this the *graded profile*, suggesting that it represents an equilibrium condition to which a stream adjusts itself in response to its discharge and sediment load. Like base level, however, the graded profile is a concept, not an absolute condition, because the stream is continually adjusting to changes in discharge and sediment load. In addition, base-level changes and changes in the inclination of the land due to tectonic action can affect the graded profile. It seems, then, that streams *trend* toward a graded profile, and although most approximate it, they never truly achieve it.

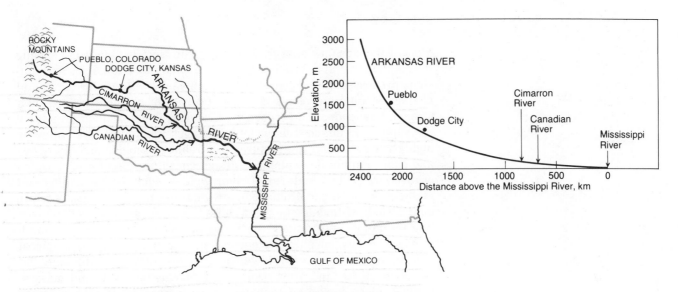

Fig. 27.20 *Profile of the Arkansas River from its headwaters in the Rocky Mountains to its terminus in the Mississippi River. The gross shape of the curve approximates a graded profile. (Data from H. Gannett, "Profiles of Rivers in the U.S.," U.S. Geological Survey, Water Supply Paper 44, 1901)*

The observation that streams trend toward an equilibrium was put to use in another major idea about river-sculptured landscapes. Around 1950, John T. Hack, a geomorphologist with the United States Geological Survey, offered a serious counterview of Davis's geographic cycle. Hack argued that the landscape is in a state of "dynamic equilibrium," meaning that it is continually trending toward an equilibrium, or steady-state condition, but because of the changeable nature of the stream's energy system (driven by climate, runoff, and uplift), rarely achieves it. This is in sharp contrast to the assumptions underlying Davis's stagewise development, in which the energy system winds down as the cycle passes approaching, in the language of systems theory, a state of entropy. Entropy is the state of minimum available energy in a system.

According to the dynamic-equilibrium concept, should the climate remain steady while the land uplifts in response to denudation, the stream system would maintain itself, because the water available for runoff (mass) and slope (energy gradient) would hold steady with the passage of time. Further, if the rock types and geologic structures also remained constant as the stream system cut its way into the land, the valley forms should remain constant as well. Thus stagewise development of stream valleys is untenable according to this concept.

SUMMARY

Most of the stream's energy for geomorphic work is devoted to moving the sediments and debris brought to the valley by tributaries and hillslope processes. Bed shear stress, which varies with water depth and channel slope, determines the size and rate of bed load moved by a stream. Suspended load transport, however, is determined largely by the supply of sediment input by tributaries upstream. Streamflow is highly variable, and the features of stream channels, such as riffles and pools, cannot be explained without consideration of a wide range of discharges. The features of river valleys, such as floodplains and terraces, are also the products of a wide range of discharges, but over long periods of time as the river shifts laterally and downcuts its way into the landscape. Among the major ideas about river-eroded landscapes, the geomorphic cycle was, and probably still is, the most widely known; however, it is generally inadequate for explaining the development of landforms over broad regions on most land masses. Modern ideas based on systems theory appear to provide a better framework for understanding river-eroded landscapes.

CHAPTER 28 WAVES, CURRENTS, AND COASTLINES

INTRODUCTION

When the sea is set into motion, the moving water exerts stress against its container, the ocean basin, thereby setting up the potential for erosion. The motion of the sea, however, is limited to the surface layer of water; therefore, erosion is restricted to the shallower parts of the basin, mainly the continental shelves. Indeed, the growth of the continental shelves themselves can be attributed in large part to the work of the sea, inasmuch as waves and currents erode material from the shallow water near shore and deposit it in deeper water farther off shore. While this is occurring, the configuration of the line of contact between land and water, the *shoreline*, is altered, being cut back at some points and filled in at others. Over the short run, the overall trend of shoreline change is toward a smoother configuration geographically (Fig. 28.1). Over the long run, however, changes in elevations of land or water and tectonic activity such as volcanism can reverse this trend.

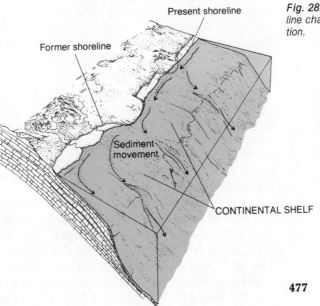

Fig. 28.1 The short-term trend of shoreline change toward a smoother configuration.

477

The sea can be remarkably effective in altering the land: A powerful storm can cut a shoreline back twenty to thirty meters overnight; the waves of a hurricane can destroy a coastal settlement in a matter of hours; and over a year waves and currents move more than 1,000,000 cubic meters of sediment along many coastlines of the world. Nevertheless, the sea is not as effective as is runoff in lowering the land. But the sea is deceiving, even to the insightful observer, for scientists no less prominent than Charles Lyell and Charles Darwin believed the sea, at higher levels, to be responsible for most of the erosion which led to the present landforms on the continents.

Because water movement is necessary for the sea to do work, we will begin this chapter by describing how water is set into motion. This will be followed by a discussion of waves and currents and the geomorphic work attributed to each. The chapter will conclude with a description of coastal landforms.

WATER IN MOTION

Causes and Types of Wave Motion

Waves are the principal form of motion in bodies of standing water. Most waves are generated by wind, but some are also generated by geophysical forces, such as landslides and earthquakes, by large air masses, such as hurricanes, and by the gravitational attraction of the moon and sun. The most common geophysical waves are produced by faulting and volcanic eruptions in the ocean floor and are called *tsumanis*. These large waves commonly reach heights of 10 meters and lengths of 150 km, but they are infrequent. Actually, the largest known wave was one created when an earthquake dislodged a mass of rock that fell 1000 m into the head of Lituya Bay, Alaska. The resulting wave reached a height of 530 m (1700 feet) above sea level.

Surges are large waves caused by atmospheric pressure and strong storm winds pushing up the water surface. Hurricanes are the most common cause of surges, although they can be produced by any storm system that exerts great stress on a water surface. Surges often reach 10 m or more above sea level and in low-lying coastal areas can produce extensive flooding.

The broadest or longest waves are actually *tides*, which are great bulges formed in the sea in response to lunar and solar gravitational forces. The most frequent tide is associated with the moon's revolution about the earth. The period of this tide is governed by the rates of lunar revolution and earth rotation; on the average, it rises and falls every 12 hours and 26 minutes. The effect of the moon is about twice that of the sun; the largest tides, called *spring tides*, form when both are aligned

Tidal flat, the area of beach covered and uncovered in each tidal cycle. (Photograph by William M. Marsh)

LOCATION	Tidal ranges	
	Daily	Spring
Englishman Bay, Maine	3.7 m	4.3 m
Belfast, Maine	3.0 m	3.35 m
Provincetown, Mass.	2.7 m	3.35 m
Chatham, Mass.	2.1 m	2.4 m
Saybrook, Conn.	1.2 m	1.2 m
Sandy Hook, New Jersey	1.5 m	1.8 m
Cape May, New Jersey	1.2 m	1.5 m
Cape Henry, Virginia	0.9 m	0.9 m
Charleston, South Carolina	1.5 m	1.8 m
Mayport, Florida	1.5 m	1.5 m

with the earth. Spring tides occur during full or new moons and, like all tidal fluxes, tend to be greatest near the equator. When the moon and sun are positioned at a right angle relative to the earth, the effect is the opposite, and the tides, called *neap tides,* tend to be lowest. The wave motion created by tides is very complex, owing to the obstructions posed by the land masses and irregular configurations of the ocean basins in shallow water. The result is a very irregular pattern of tides worldwide, ranging from massive fluctuations of 10 m or more in certain bays to modest fluxes of only a meter or so in polar regions. Along the East Coast of the United States, the tidal ranges actually decrease from north to south (Fig. 28.2). Where fluctuations are great, large flows of water, called *tidal currents,* may surge in and out of bays, estuaries, and river mouths, moving large quantities of sediment in the process.

Fig. 28.2 Ranges of the daily and spring tides along the East Coast of the United States. Note that the range decreases southward. (Data from the U.S. Army Corps of Engineers)

Generation of Wind Waves

The mechanisms involved in wave generation by wind are not well understood. They involve the transfer of momentum from moving air to the water surface, but we are uncertain how the wavey surface is produced. It is probably related to air-pressure variations associated with wind gusts which differentially depress the sea surface. In any case, where waves are being generated, the water surface is characterized by a very choppy sort of motion, called a *sea.* Ultimately, though, the drag of the wind on the water surface sets up a particular direction to the wave motion. As the wave travels from the area of generation, it grows more symmetrical and may travel great distances with little loss of size. Such waves are termed *swells.*

The size that a wave can attain is controlled by four factors: wind velocity, wind duration, fetch, and water depth. (Fetch is the distance

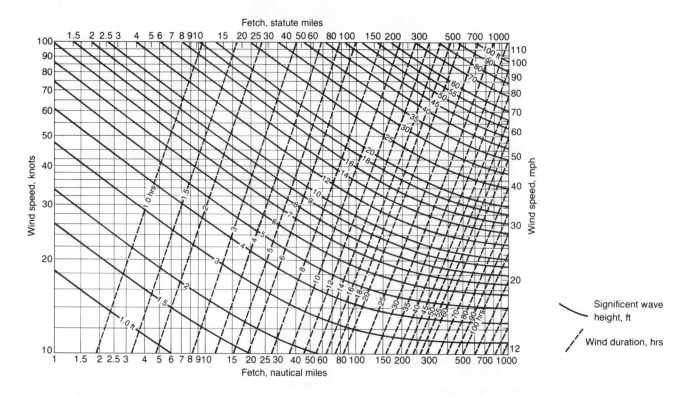

Fig. 28.3 Chart for forecasting wave size (height) in deep water. For example, given a wind speed of 20 knots, a fetch of 25 nautical miles, and a wind duration of 5 hours, the wave height will be 4 feet. (Adapted from Coastal Engineering Research Center, U.S. Army Corps of Engineers)

of open water in one direction across a waterbody.) Large values of each are necessary to generate the largest waves: in other words, fast wind blowing from one direction for a long time over a great expanse of deep water. Figure 28.3 provides a graphical solution to forecasting wave size in deep water based on wind velocity, wind duration, and fetch. For any combination of these factors, there is a maximum wave size that can be generated. The first wave of a storm to reach maximum size under a particular fetch is termed the $^t min$ wave, for "minimum time" wave. For that set of conditions, no larger wave can be generated.

Wave Forms and Motion

Waves may be simple and relatively easy to describe and analyze or so complex as to defy coherent description and analysis. The description of a wave must include both its surface form and the fluid motion beneath it. The terminology used for surface forms is as follows. The part of the wave that extends above the calm-water level is the *crest;* that below the calm-water level is the *trough.* The *wavelength* is the dis-

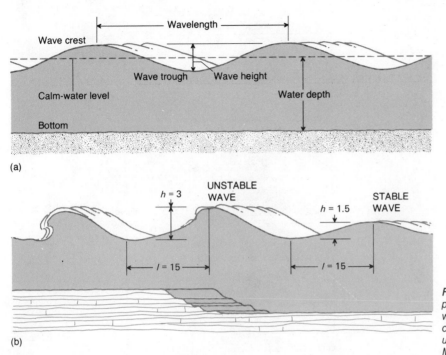

(a)

(b)

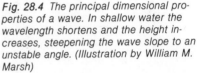

Fig. 28.4 The principal dimensional properties of a wave. In shallow water the wavelength shortens and the height increases, steepening the wave slope to an unstable angle. (Illustration by William M. Marsh)

tance from crest to crest, or trough to trough, and the *wave height* is the vertical distance between crest and trough (Fig. 28.4a). The typical slope of a wave, expressed as the ratio of wave height to wavelength, ranges from 1:25 to 1:50; above a slope of 1:7, a wave is unstable and falls over itself, or *breaks*. (Fig. 28.4b).

Unlike water in a river, the passage of a wave in deep water does not result in the transfer or flow of water. Rather, it is only the wave form that actually travels; moreover, although water particles *do* move when a wave passes, the motion is circular only under the wave form. Such waves are termed *oscillatory*, or *nearly oscillatory*, and the time it takes a wave to travel a distance of one wavelength is called the *wave period*. The velocity of a wave, which is also termed *wave celerity*, is equal to the distance traveled by a wave in one wave period:

$$\text{Velocity} = \frac{\text{Wavelength}}{\text{Wave period}}.$$

The fluid motion of an oscillatory wave is described in Fig. 28.5. As the wave trough approaches, water particles on the surface move toward the crest; as the crest passes, the particles are lifted up, car-

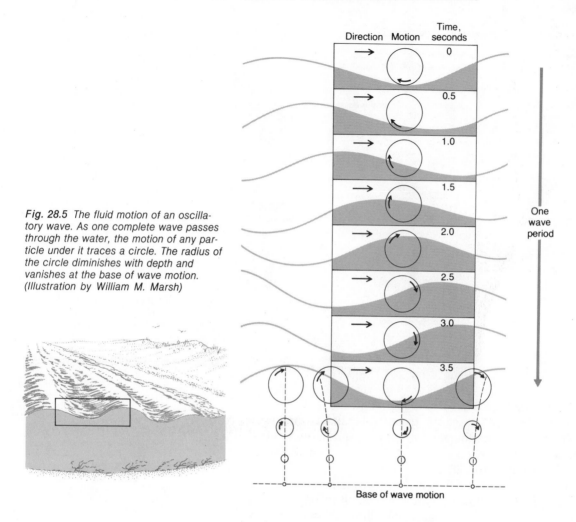

Fig. 28.5 *The fluid motion of an oscillatory wave. As one complete wave passes through the water, the motion of any particle under it traces a circle. The radius of the circle diminishes with depth and vanishes at the base of wave motion. (Illustration by William M. Marsh)*

ried ahead, and finally lowered back to their original position. This is repeated with the passage of each complete wave.

The motion of a wave also extends below the surface, but the radius of rotation decreases with depth beneath the wave. As a wave approaches the coast, it reaches a point off shore where the lowermost particles of motion touch bottom, changing the orbit of rotation from a circle to an ellipse. Closer toward shore, the elliptical motion is compressed into a linear motion; the water "slides" back and forth over the bottom with the passage of each wave. This produces friction, which slows down the base of the wave. At a depth of about 1.3 times wavelength, bottom drag becomes so great that the upper part of the wave outraces the lower part, and the wave becomes unstable and breaks. In contrast to oscillatory waves, breaking waves produce mass transport of water and are called *waves of translation*.

WAVE ENERGY AND ITS DISSIPATION

Energy Forms

The energy of a wave consists of both potential and kinetic forms. Potential energy is represented by the mass of water displaced above the wave trough; kinetic energy, by the combined velocities of the water particles associated with wave motion. The energy balance of a wave can be estimated by comparing the total energy of a group of waves entering an area of water with that of the same group leaving the area. A reduction in wave velocity and/or wave size indicates a reduction in total energy.

Waves can lose energy in a variety of ways. In polar waters, for example, floe ice increases the resistance of surface water to displacement by wave motion, which in turn reduces wave size. Wherever waves are propagated into a region of calm water, they grow smaller as potential energy is converted to kinetic energy. From our standpoint, the most important energy reduction occurs when waves enter shallow water and dissipate energy in erosion, friction against the bottom, and in the turbulent motion of wave action. In terms of energy balance, erosion represents the conversion of wave energy into work, whereas friction and turbulence result in the conversion of wave energy into heat.

Breaking waves.

Wave Energy and Circulation Near Shore

According to Douglas Inman and Birchard Brush of the Scripps Institution of Oceanography, the energy of a 3-m wave is equivalent to a row of full-sized automobiles approaching shore side by side at full throttle. How the resultant force would be exerted against the shore depends on the topography of the shoreline the wave encounters. On coasts with gently sloping offshore topography, such as most sandy coasts, most wave energy is spent as the wave crosses the shallow-water zone. The force of the wave is spread over a broad "ramp" extending from the point where the wave first "feels" bottom to the point of farthest landward penetration on the beach (Fig. 28.6a). We can see evidence for this on swimming beaches as wavelength and wave velocity grow smaller toward shore.

As the waves break, however, some of the water mass of each wave is displaced toward shore, and in time the water surface becomes elevated there, forming a seaward gradient. This sets up the

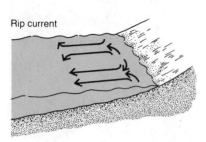

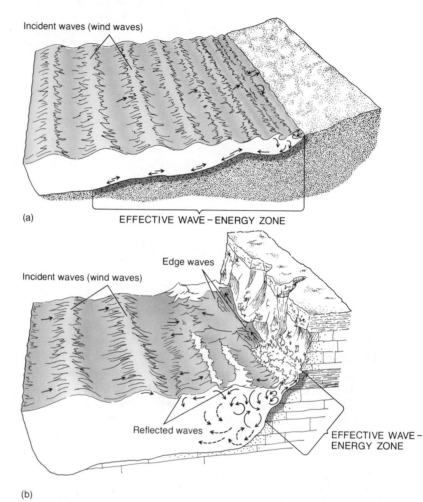

Fig. 28.6 Wave-energy distribution near shore along: (a) a shallow-water shoreline; and (b) a deep-water shoreline. The effective wave-energy zone is much broader on the shallow-water shoreline; however, the magnitude of force exerted on shore is much greater per unit area on the deep-water shoreline. (Illustration by William M. Marsh)

potential for return flows of water in the form of a current. The principal form of this current, called a *rip current,* is a narrow jet of water that shoots seaward through the base of the breaking waves (inset Fig. 28.6a). Together, the shoreward flow of surface water and the seaward flow of rip currents form near-shore circulation cells.

The picture is quite different on rocky coasts where deep water runs close to shore. Here wave energy undergoes little attenuation before reaching shore, and waves tend to smack the shore nearly full-force, somewhat like the row of automobiles striking a cliff head-on. As a result, a massive amount of energy can be focused on a relatively narrow zone along the shoreline. Because of the small surface area of this zone, it is impossible for all of a large wave's energy to be dissipated in the usual manner, and there is often a lot left over. This excess energy is redirected into new wave forms. *Reflected waves* rebound off the shore and move seaward into the path of approaching waves; the opposing forces that are set up result in an energy reduction in both waves. *Edge waves* move parallel to the shore, crossing shore-bound waves at nearly right angles. This mixture of incident waves, reflected waves, and edge waves creates a very complex zone of turbulence along rocky shores during periods of heavy wave action (Fig. 28.6b).

Energy Redistribution Through Wave Refraction

From a geographical standpoint, we need to know why certain parts of a coastline are the focus of erosion and other parts the sites of deposition. This leads us to the question of the areal distribution of wave energy in shallow water. To begin with, no coast has truly uniform offshore topography; therefore, as a wave approaches shore, some segments touch bottom before others do. The velocity of the wave is thus reduced over shallow water (where it first touches bottom), but not over nearby deep water. This produces a bend in the axis of the wave, and since a wave always travels in a direction perpendicular to its axis (crest), this results in a reorientation of wave energy relative to the shoreline (Fig. 28.7).

The process of wave bending, called *wave refraction*, is analogous to refraction in light and sound waves. To define a refraction pattern in water waves, we need to know: (1) the orientation of the wave crest; (2) the wave size; and (3) the location of the contour lines that define the underwater topography. Where offshore the wave begins to refract depends on wave size and water depth. On the basis of the relationship between wave size and wavelength, the depth at which a wave first feels bottom can be estimated from wavelength alone. At depths greater than one-half of the wavelength, the wave functions as a deep-water wave, and no refraction takes place. Where water depth is between 0.5 ($\frac{1}{2}$) and 0.04 ($\frac{1}{25}$) of the wavelength, conditions are considered to be transitional; velocity begins to decline, and the wave refracts correspondingly. Maximum wave refraction can be expected where water depths are less than 0.5 of the wavelength. Wavelength is greatly shortened in this zone; velocity is proportional to the square root of water depth, and the direction of advance is nearly perpendicular to the contours of bottom topography.

The redistribution of wave energy due to refraction can be defined by drawing *orthogonals* (lines perpendicular to the wave crest) for a group of waves traveling across the shallow zone (see Fig. 28.7). The relationship to the shape of the coastline is plainly evident: Wave

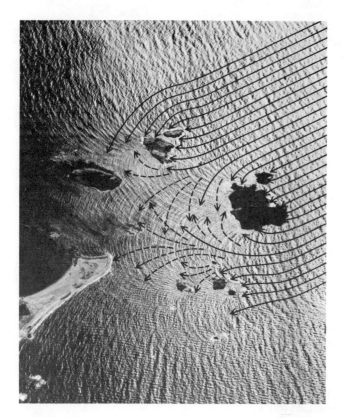

Fig. 28.7 Wave refraction and the corresponding orthogonals around some small islands along the coast of Rhode Island. (Photograph by John S. Shelton)

energy is convergent on promontories and divergent in embayments, meaning that wave energy is greater than average where the land protrudes into the sea and less than average where the sea protrudes into the land (Fig. 28.8a). Along straight coasts, wave energy is more or less evenly distributed, but oriented in the direction of the approaching waves (Fig. 28.8b). Although refraction can reduce the angle between a wave crest and the shoreline to as little as 10°, rarely does a wave approach the shore straight-on. The fact that the force of waves is exerted in a direction oblique to the shore is very important in terms of sediment transport, because when particles are lifted off bottom by wave turbulence, they tend to be carried along the shore with the flow of wave action.

Longshore Currents

Currents are also an important component of water movement and sediment transport. Currents are driven by waves, and when waves

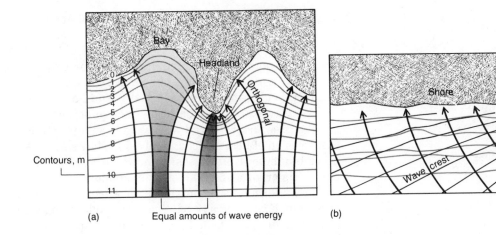

(a)   Equal amounts of wave energy   (b)

move in one direction along the coast, a gentle flow of water is set up parallel to the shoreline. This *longshore current* moves with the waves at velocities between 0.25 and 1.0 m/sec. Along coasts where winds limit waves to the one-approach direction in all seasons, the flow of longshore currents is consistently in one direction, and only velocity varies with wave energy. Where winds shift seasonally, longshore currents may reverse from summer to winter. Along some north-south–trending coasts in the midlatitudes, for example, longshore currents change from north-flowing in summer to south-flowing in winter as wind systems shift with the seasonal change in air masses. In other areas, owing to the orientation of the coast and the variable direction of winds, longshore currents are highly variable, changing over a matter of days with different weather systems.

Fig. 28.8 Distribution of wave energy: (a) Seaward of the 10-m contour, wave energy, represented by the area of the cells formed between the orthogonals and the contour lines, is equal all along the wave front. Landward, the cells compress or enlarge with wave refraction around the headland and in the bay. (b) Along a straight coast, waves refract evenly; thus wave energy is evenly distributed.

WAVE EROSION AND SEDIMENT TRANSPORT

Wave Erosion

Waves driven by wind are the primary erosional agent on shorelines. Wave erosion results in the removal of material from the shore. It may take place underwater, at the water line, or above it, and it usually results in *recession* of the shoreline. Recession is also known to be caused by a rise in water level, a subsidence of land, or both, Erosion is measured in terms of the volume of material removed, whereas recession is measured in terms of the distance of landward displacement of the shoreline.

Wave erosion is most effective in unconsolidated materials such as glacial deposits. By exerting tremendous *hydraulic pressure* on particles, storm waves can in a matter of hours erode deeply into the foot

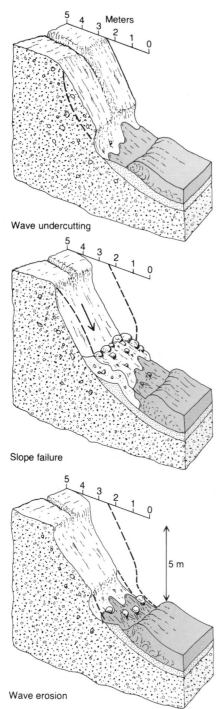

Wave undercutting

Slope failure

Wave erosion

of a bank of sand or gravel, causing it to collapse onto the beach. Subsequent waves overwash the heap of debris, quickly removing all but the heaviest particles, usually the large boulders. The fine particles are moved both downshore and offshore into deeper water (Fig. 28.9). For a 100 m of the shoreline shown in Fig. 28.9 to retreat a distance of 2 m, 1000 cubic meters of material would have to be eroded.

Wave erosion is much less effective on bedrock, because the hydraulic pressure required to break fragments free is so high. As a result, erosion of bedrock is not only slow, but also usually the product of a combination of processes. In addition to hydraulic pressure, which can be as great as 60,000 kg/m^2 in shallow water, *corrasion* and *solution weathering* also contribute to erosion. *Corrasion*, or abrasion, takes place when heavy particles are rolled, bounced, and hurled by storm waves against solid rock, and fragments of rock are broken free; solution weathering takes place when minerals are dissolved into sea water.

The shoreline features formed as a result of wave erosion are familiar to most of us (Fig. 28.10). Erosion of bedrock often results in the formation of a *sea cliff*, which may be undercut near water level to form a *wave-cut notch*. Where bedrock has variable resistance to wave erosion, an assortment of interesting features may form, including *sea caves* and *sea stacks*. The latter are pillars of resistant rock left standing offshore.

Along nonbedrock shorelines, most of the features are formed in sediments that are in transit along the coast. Only the *backshore slope*, the steep bank or bluff landward of the shore, is comprised of in situ (in place) material. The *shore*, or *beach*, stretches from the foot of the backshore slope to the shallow water just beyond shore; along erosional shorelines the beach may be only meters wide. Near the water, wave action forms a small ridge called a *berm*. On wide shores the area between the berm and the foot of the backshore slope is designated the *backshore*. Seaward of the shore is the *inshore zone*, the inner portion of which is the *surf zone*, where waves break. Beyond the inshore is the offshore zone, where waves behave like deep-water waves (Fig. 28.11). Longshore currents are strongest in the inshore zone, but do extend into the adjacent part of the offshore zone as well.

Fig. 28.9 The three phases in the erosion of a shoreline comprised of unconsolidated materials. (Illustration by William M. Marsh)

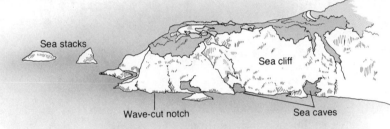

Fig. 28.10 Features of a shoreline formed by wave erosion of bedrock near the Big Sur, California. The diagram pinpoints and names the features. (Photograph and illustration by William M. Marsh)

Sediment Transport

The material eroded by waves is incorporated into a train of sediment that moves along the shore with the flow of wave and current action. The source of more than 90 percent of this sediment is the debris brought to the coast by streams and rivers. Each year the world's land masses contribute around 15 billion m³ of sediment to the coastlines. This material is moved by waves and currents, and some of it is depos-

Fig. 28.11 The principal zones and features of a coast comprised of unconsolidated materials and sediments.

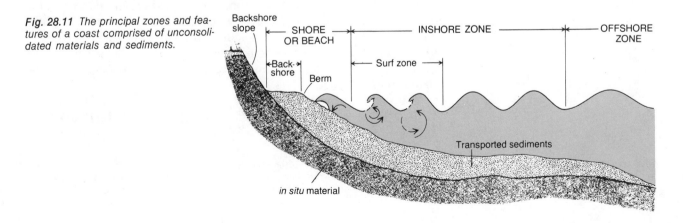

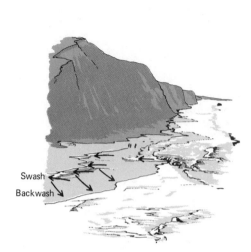

Fig. 28.12 The pattern of swash and backwash. The white wave represents the zone of breaking waves on the surf zone. (Photograph by Jeff Dozier; diagram by William M. Marsh)

ited in *sediment sinks,* environments along the coast that are favorable to massive sediment accumulations.

The movement of sediments along a coastline is called *littoral transport,* and the material that is moved is known collectively as *littoral drift.** Littoral transport is made up of two components: longshore transport and onshore-offshore transport. The longshore component consists of sediment movement associated with wave and current action mainly in the surf zone. Part of the longshore transport also takes place on the beach itself, driven by wave action in the form of *swash* and *backwash.* Swash is a thin sheet of wave water that slides up the beach face; backwash is its counterpart in return flow. Because waves strike the shore at an angle, swash flows obliquely onto the beach. Backwash, on the other hand, flows more perpendicularly to the shoreline. Together, swash and backwash produce a ratchetlike motion of water and sediment, resulting in a net downshore movement (Fig. 28.12). Much of the sediment, especially the pebbles and larger particles, moved in this fashion are rolled along the beach and are referred to as *bed load.*

Beyond the beach, longshore transport is concentrated where waves are breaking (Fig. 28.13). The turbulence created by the break-

---

*Although the term ''drift'' is often used to describe the process of sediment movement along the coast, as in the ''westward drift of sediment,'' the term ''littoral drift'' should be used in reference to the sediment itself, as in ''the littoral drift is moved by waves and currents.''

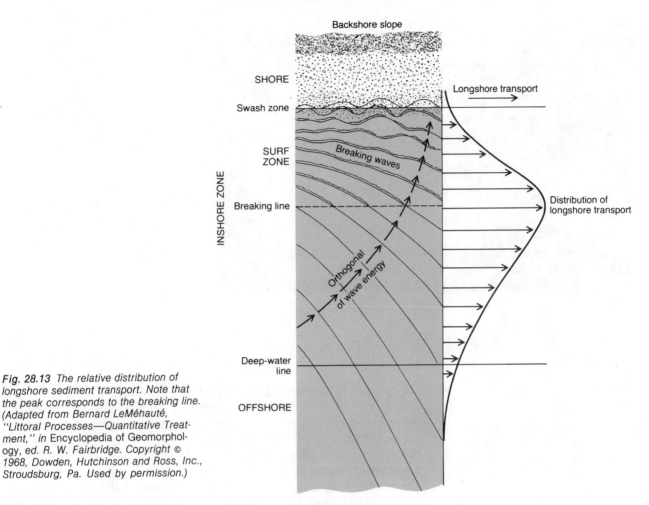

Fig. 28.13 *The relative distribution of longshore sediment transport. Note that the peak corresponds to the breaking line. (Adapted from Bernard LeMéhauté, "Littoral Processes—Quantitative Treatment," in* Encyclopedia of Geomorphology, *ed. R. W. Fairbridge. Copyright © 1968, Dowden, Hutchinson and Ross, Inc., Stroudsburg, Pa. Used by permission.)*

ers lifts sediment high into the water, whereupon it is carried downshore by longshore currents. Sediment moved in this fashion is called *suspended load*—because it is carried in suspension—and it constitutes the bulk of the littoral drift along most coasts. Most of it is sand; the fine sediments (silt and clay) raised by wave action or introduced by rivers may be carried into the offshore zone, as illustrated in Fig. 28.14.

Rates of Longshore Transport

Rates of longshore transports vary with wave energy, the angle at which waves approach the shore, the size and availability of sediments, as well as other factors such as coastal ice and vegetation.

Fig. 28.14 *A train of suspended sediment carried by longshore currents, Russian River, California. (Photograph by John S. Shelton)*

Along a sandy shoreline that is free of ice, bedrock, and other controls, longshore sediment transport is directly proportional to the flux of wave and current energy in the longshore system. In measuring annual transport rates at various places on the coasts of the United States, the Army Corps of Engineers has found, not surprisingly, considerable differences on shores located only a few hundred kilometers apart. One of the most striking differences is found in southern California; at Oxnard Plain Shore, north of Los Angeles, longshore transport is ten times greater than it is at Camp Pendleton, south of Los Angeles (Fig. 28.15). On the East Coast, rates range from 100,000–380,000 m^3/year except for sheltered locations, such as Atlantic Beach, North Carolina, where longshore transport amounts to less than 25,000 m^3 per year. In the upper Great Lakes longshore transport averages between 50,000 and 100,000 m^3 per year for most sandy shorelines. Along Arctic shorelines it is even less, on the order of 5,000–10,000 m^3, owing to the presence of grounded ice in the shallow-water zone for much of the year.

The longshore-transport figures given above represent the sum of sediment transport in two directions along the coast: one the primary direction, the other the secondary direction. This is also known as *gross sediment* transport. A more meaningful statistic in terms of coastal trends is the *net sediment transport*, which gives the balance between the two directions of movement:

$$\text{Net sediment transport} = q_p - q_s,$$

where q_p = sediment moved in the primary direction and q_s = sediment moved in the secondary direction. A large net transport indicates that sediment is being removed from a source such as a river delta and transported to a sediment sink such as a submarine canyon or the mouth of a bay. Where net sediment transport is small, but gross transport is large, sediment is merely being shifted back and forth; in other words, the same sediment mass is being reworked year after year.

Onshore-Offshore Transport

In addition to longshore transport, sediment is also transported onshore and offshore, that is, perpendicular to the shore. The amount of sediment moved in this fashion is very small compared with longshore transport, but it is important to beach topography. The onshore movement is brought about by low-energy waves, usually the summer wave regime, and is characterized by a shoreward migration of sand bars

Fig. 28.15 The general pattern of average annual longshore transport along the continental United States. Longshore transport rates for selected shorelines are represented by the heavy arrows. The river outlets represent points of sediment input to the coasts. (U.S. Army Corps of Engineers data)

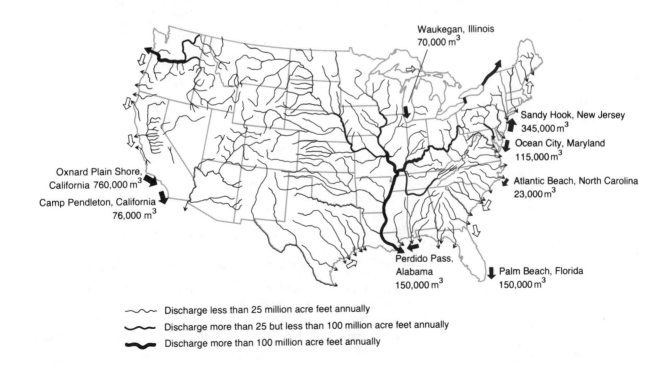

Waukegan, Illinois
70,000 m^3

Sandy Hook, New Jersey
345,000 m^3

Ocean City, Maryland
115,000 m^3

Atlantic Beach, North Carolina
23,000 m^3

Oxnard Plain Shore, California 760,000 m^3

Camp Pendleton, California
76,000 m^3

Perdido Pass, Alabama
150,000 m^3

Palm Beach, Florida
150,000 m^3

‿‿‿ Discharge less than 25 million acre feet annually

‿‿‿ Discharge more than 25 but less than 100 million acre feet annually

‿‿‿ Discharge more than 100 million acre feet annually

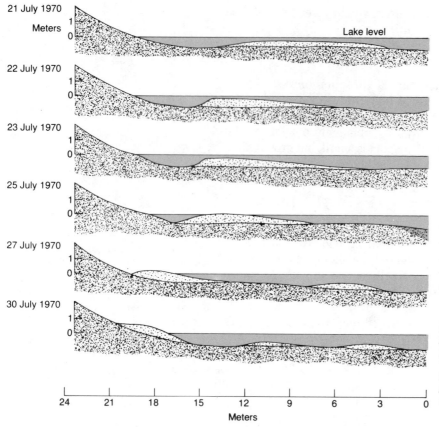

Fig. 28.16 *The onshore migration of a sand bar over nine summer days, Lake Michigan. (From R. A. Davis, Jr., et al., "Comparison of Ridge and Runnel Systems in Tidal and Non-Tidal Environments,"* Journal of Sedimentary Petrology, *1972. Used by permission.)*

(Fig. 28.16). Sand is added to the beach in the swash zone, from which it may be blown and washed farther landward (Fig. 28.17). This trend usually reverses with high-energy waves, typically the winter wave regime in the midlatitudes. Storm waves erode the beach, the beach face steepens, rip currents intensify, and sand bars migrate seaward. During both seasons, longshore transport continues.

Sediment Mass Balance and the Loss and Growth of Beaches

To help us understand how sediment transport relates to trends and rates of change in a segment of coastline, it may be helpful to discuss sediment transport in terms of mass balance. This calls for an accounting of the total sediment input and output for a parcel of coastline and

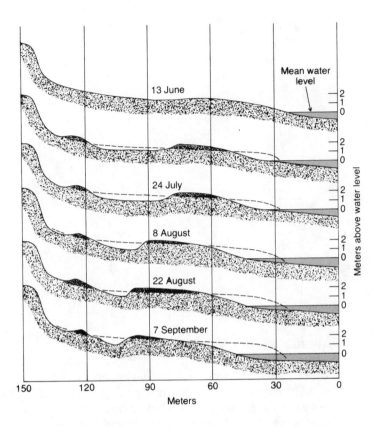

Mean water level

Meters above water level

13 June

24 July

8 August

22 August

7 September

150   120   90   60   30   0

Meters

Fig. 28.17 The growth and migration of a beach ridge over two months, Crane Beach, Massachusetts. (From M. O. Hayes, "Forms of Sediment Accumulation in the Beach Zone," Waves on Beaches and Resulting Sediment Transport, *New York: Academic Press, 1971. Used by permission.)*

then solving for net change. Basically, there are four primary inputs and outputs of sediment: longshore input (L_i), longshore output (L_o), onshore input (O_i), and offshore output (O_o). For some coasts, inputs by runoff (R_i) and outputs by wind erosion (W_o) must also be considered. These components of the mass balance are shown in Fig. 28.18; inputs are positive values and outputs are negative values:

Sediment mass balance: $L_i \; - \; L_o \; + \; O_i \; - \; O_o \; + \; R_i \; - \; W_o = 0.$

As the balance varies from season to season, year to year, or over longer intervals, the quantity of sediment stored in the parcel, which we identify with the size of the beach, also varies. If output exceeds input (negative balance), the body of stored sediment shrinks, and the

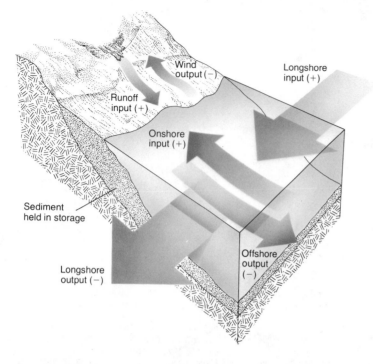

Fig. 28.18 The components of sediment transport in the mass balance of a parcel of coastline. (Illustration by William M. Marsh)

beach begins to "dry up" (Fig. 28.19). Should this condition be prolonged, the body of sediment may be drawn down so small that the *in situ* material becomes exposed, and wave energy is exerted directly on it (Fig. 28.19). In this case the backshore slope would be cut back, and a new shore profile would be established (Fig. 28.19). In time the profile would smooth out, and wave energy would become rather uniformly distributed across it, resulting in what is termed an *equilibrium profile*. Each time sea level changes or the sediment mass balance changes appreciably, the trend toward a new equilibrium profile is initiated. Figure 28.20 shows an equilibrium profile from the United States East Coast.

Erosion of the body of beach sediments can be caused by many factors. In southern California beaches began to decline when rivers were dammed for water supply, thereby trapping in reservoirs sediment that would otherwise feed the beaches. In many parts of the world, breakwaters have been constructed at harbor entrances to intercept the longshore drift. As a result, the beaches downshore dry up because the flow of the sediment train has been broken. Moreover, once the sediment is gone, wave energy erodes *in situ* material. In the Great Lakes high water levels are the most common cause of beach erosion. In the past decade lake levels have been as much as 1 m above average, and this has allowed storm waves to reach the beaches with greater force.

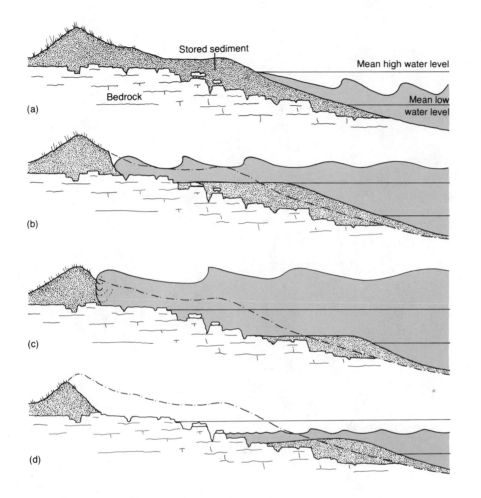

Stored sediment

Mean high water level

Bedrock

Mean low
water level

(a)

(b)

(c)

(d)

Fig. 28.19 Erosion and impoverishment of the sand supply on a beach, resulting in exposure of the underlying bedrock.

The result has been that sediment output has increased without a commensurate increase in input, and beaches have declined.

The opposite condition is one in which total sediment input exceeds total output, and the body of stored sediment grows. The beach widens, beach ridges develop, and sand dunes begin to form. Where this trend is long-term, the coastline builds seaward and is said to be *progradational*. The particular forms and causes of progradation are taken up next.

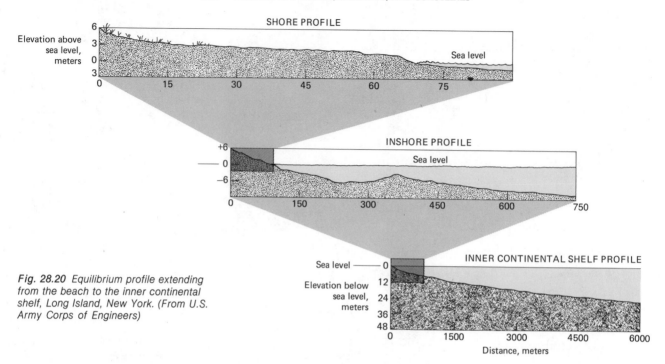

Fig. 28.20 *Equilibrium profile extending from the beach to the inner continental shelf, Long Island, New York. (From U.S. Army Corps of Engineers)*

Fig. 28.21 *A wave-energy shadow is created by this island near Gloucester, Massachusetts, resulting in the formation of a tombolo. (Photograph by John S. Shelton)*

SHORELINE DEPOSITION

Causes of Deposition

Where the stress exerted by waves and currents on the bottom undergoes a reduction, part of the train of sediment may be dropped and a deposit formed. One cause of this is a decline in wave energy, such as that related to the divergent refraction of waves at the head of a bay. Sediments transported down the sides of the bay will accumulate at the bayhead to form sand bars, a broad beach, and other depositional features. An offshore barrier, such as an island or detached breakwater, can also effect a reduction in wave energy by creating an energy shadow on its leeward side (Fig. 28.21). Sediment is dropped in this shadow, and owing to wave refraction around the barrier, deposits may build out from both the shore and the barrier. If the two are not too far apart, a neck of land called a *tombolo* may form between them.

The other major cause of deposition is a sudden increase in water depth in the path of the longshore sediment train. As the sediment train passes into deep water, sediments fall below the wave base and are deposited. This commonly occurs in three types of settings: (1) where submarine canyons run close to the shore; (2) where the sediment train is diverted into deep water by an opposing current or by a reef, pier, or breakwater; and (3) where the shoreline abruptly changes its orientation, and the momentum of the waves and currents carries the sediment train into deep water, as at the mouth of a bay (Fig. 28.22).

Depositional Features

The bulk of the sediments deposited along any coast are laid down under water and are not at all evident from the shore. This is especially so with those that slip into submarine canyons and flow down the

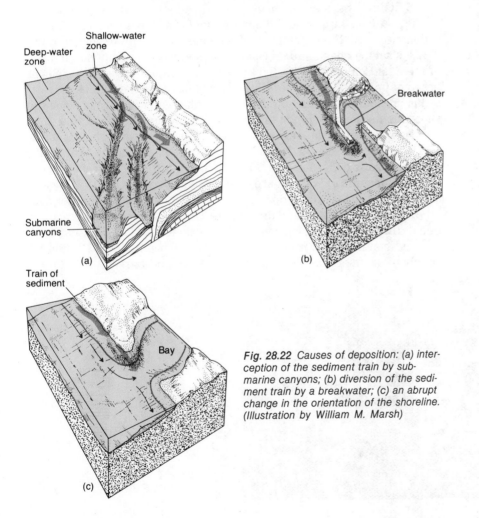

Fig. 28.22 Causes of deposition: (a) interception of the sediment train by submarine canyons; (b) diversion of the sediment train by a breakwater; (c) an abrupt change in the orientation of the shoreline. (Illustration by William M. Marsh)

Fig. 28.23 Spits and bay-mouth bars along the Massachusetts coast near Falmouth. (Photograph by John S. Shelton)

continental slope into very deep water (Fig. 28.22a). Near San Diego, California, where the continental shelf is very narrow and dissected by long canyons, much of the longshore drift is lost in this fashion.

In other settings, nearer shore, the sediments accumulate incrementally year after year and eventually form such a heap that a narrow ribbon of sand appears at the water surface. At the mouth of a bay, such a ribbon often grows from the shore itself in the form of a slender, sandy finger called a *spit*. As the spit grows, it may bend into the bay (Fig. 28.23). Depending on the directions of sediment transport and the sand supply, spits may grow from one or both sides of a bay mouth. A ribbon of sand extending across the bay mouth is called a *bay-mouth bar*. These bars are usually breached where tidal flow and/or river discharge erodes through them (Fig. 28.23).

Spits and bars are common features along coasts with irregular configurations, i.e., where there are many embayments to serve as sediment sinks. The segment of Massachusetts coastline shown in Fig. 28.23 is such a coast, and it is similar in many ways to much of the rest of the East Coast of the United States. The headlands, which are composed of unconsolidated material, are readily eroded and thus serve as sediment sources. Clearly, the trend of change is toward a smoother shoreline as the headlands are cut back and the bays are filled in.

Now examine the segment of the Maine coastline shown in Fig. 28.24. The gross configuration of this coast is similar to that of the Massachusetts coast shown in Fig. 28.23. In addition, both receive comparable amounts of wave energy and are approximately the same age, dating from the last glaciation, but the Maine coast exhibits none of the alteration from erosion and deposition. The reason for this is compositional; the Maine coast is comprised of resistant bedrock (mainly igneous and metamorphic rocks) and very little loose materials. Thus the supply of sediment is too small to build spits and bars.

Fig. 28.24 The rocky Maine coast near Brunswick. Note that very little modification by wave action is evident. (Photograph by John S. Shelton)

In certain places along coastlines that carry an ample supply of sand, deposition leads to the formation of a large triangular feature called a *cuspate foreland*. This feature often forms where two trains of longshore drift move against each other and are driven seaward. As a result, the main axis of a cuspate foreland usually points seaward. Over time the balance between the two longshore components may change, and the foreland may shift in orientation and shape. The pattern of beach ridges in a cuspate foreland often provides a clue into the history of such changes (Fig. 28.25).

PRINCIPAL TYPES OF COASTLINES

A coastline acquires its character from its geologic structure, its initial topographic configuration, and the subsequent modification of these by wave and current action. Also important in shaping the character of a coastline is the trend toward submergence, emergence, or stability. In general, *stable coasts* tend to be straight or gently curving, *submergent coasts* are heavily indented due to flooding of river valley mouths by seawater, and *emergent coasts* are often terraced because wave-cut profiles have been elevated above sea level. The following paragraphs provide brief descriptions of the principal types of coastlines (see also Fig. 28.26).

Ria Coasts

"Ria" is a Spanish term used to describe coasts marked by prominent headlands and deep reentrants (Figs. 28.23 and 28.24). Most ria coasts appear to result from partial submergence due to a rise in sea level, subsidence of the land, or both. If this includes a river mouth, flooding of the lower valley takes place, resulting in an elongated embayment called an *estuary*. Chesapeake Bay is an extreme example of valley drowning which produced a main estuary and many tributary estuaries as well. If ria coasts are composed of erodable material, they are readily modified, as illustrated in Fig. 28.23. See also Fig. 28.26.

Fiord Coasts

These coasts resemble ria coasts except that the valleys have been glaciated (Fig. 28.26). As a result, the drowned valleys are deep, steep-sided, and often very long. Norway's fiords are world-famous, but equally spectacular ones can be found in British Columbia, southern Chile, and New Zealand. Because fiords are so deep, it is virtually impossible for spits and bay-mouth bars to form at their mouths.

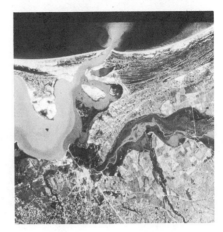

Fig. 28.25 Beach ridges, such as those evident on the right-hand side of the river mouth, reveal in part the pattern of growth of this depositional feature. (NASA photograph)

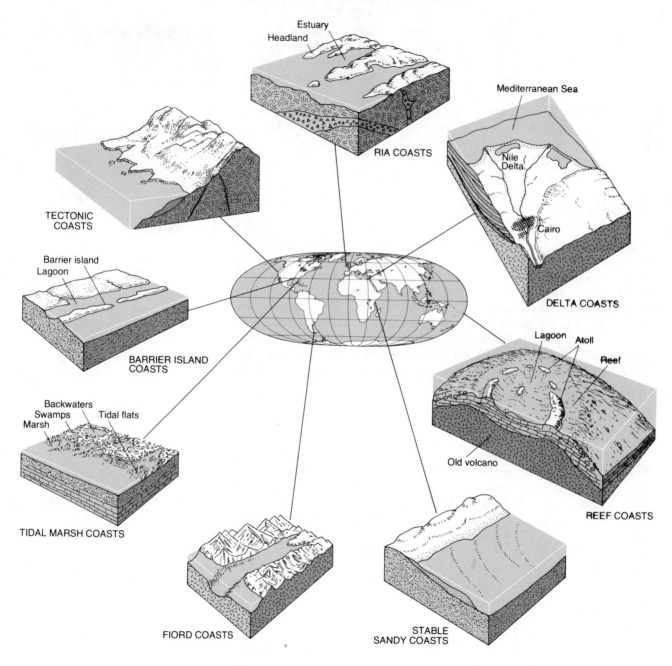

Fig. 28.26 Principal types of coastlines.

Barrier Island Coasts

Barrier islands begin as offshore bars in shallow water where there is an ample supply of sand. As the bars grow, a beach ridge emerges, small sand dunes begin to form, and vegetation becomes established. The area of water between the island and the mainland is called a *lagoon*. Owing to the shallowness and relative calmness of this environment, rates of biological productivity and sedimentation are high in the lagoon.

Barrier islands parallel much of the Texas shoreline of the Gulf Coast. The Sea Islands along the Atlantic Coast of Georgia are also barrier islands, although more segmented than the ones of the Gulf Coast. See Fig. 28.26.

Stable Sandy Coasts

These are coasts along which the input and output of longshore drift are balanced over long periods of time. As a result, net change is not appreciable, although different features and configurations may form from time to time. Large scallops and cuspate forelands are common features (see Fig. 28.26). Stable sandy coasts are often located between a sediment source and a sink.

Tidal Marsh Coasts

On shorelines where fine sediments are abundant, the influence of bedrock and topographical relief are inconsequential, and there is a sizable tidal flux, tidal marshes are often the predominant coastal type. These gently sloping environments are truly transitional between sea and land, because they are daily covered and uncovered by the sea. Vegetation plays an important role in the tidal marsh inasmuch as it secures the environment against wholesale erosion by tides and storms. (See Fig. 28.26.) In the tropics tidal marshes often give way to swamps of mangrove, a small tree that grows on stiltlike roots.

Delta Coasts

Deltas form at the mouths of rivers where the rate of river deposition exceeds the capacity of the littoral processes to carry the sediment away. Because the gradient of the delta is so slight, the river cannot maintain a single channel and thus breaks down into many distributary channels which discharge the sediment (see Fig. 28.26). Coarse sediments are deposited near the margin of the delta, whereas clays are dispersed into deeper water, often well beyond the delta environment.

The shape into which the delta is built depends on the configuration of the river mouth, the pattern of the distributaries, and the rate of

AUTHORS' NOTE 28.1
Sand Bypassing

Breakwaters are barriers constructed at harbor entrances to reduce wave energy and improve navigation safety. Because they reduce wave energy, breakwaters also interrupt the longshore sediment transport. Sediments accumulate behind or on the upshore side of breakwaters, depriving the beach downshore of its sediment supply. This often results in beach recession downshore because lacking sediment to move, wave energy is spent in the erosion of in situ material, and the breakwater area becomes clogged with sediments.

To solve this problem, a method called sand bypassing is used to maintain the flow of sediment past the harbor entrance. Three techniques may be employed. The first is hydraulic bypassing, which involves sucking up the sand and water mechanically and then pumping the slurry past the entrance to the downshore beach. Dredging and barging, the second technique, involves excavating sand by crane or hydraulic techniques, transporting it downshore, and off-loading it from the barge into shallow water. The third technique involves excavating a storage pit on the upshore side of the breakwater and piling the excavated sediment on

Pumping station

Sand is transferred hydraulically by pipe across Lake Worth Inlet, Florida, at a rate of 100,000 yd³/yr. (Photograph courtesy U.S. Army Corps of Engineers)

the downshore side of the breakwater. The pile feeds the beach downshore while the pit collects sediment upshore. The sizes of the pit and pile are scaled to the expected annual sediment transport and the desired number of years between renewal operations.

sedimentation. Four basic types of deltas are recognized: *estuarine,* which forms in the head of an estuary; *arcuate,* which forms in a bay and is fan-shaped; *cuspate,* which is pointed because of strong reshaping by waves; and *bird-foot,* which has long fingers built into the sea by the distributaries. The Mississippi delta is a bird-foot; the Nile delta is an arcuate form; and the Seine delta fits the estuarine class.

Reef Coasts

Coral reefs form in tropical seas where the water is less than 20 m deep. Reefs grow from the buildup of massive amounts of coral, an organism which secretes calcium carbonate. The calcium cements the coral and the remains of organisms together into a wave-resistant body of limestone.

Early scientists troubled over the presence of coral-reef islands in the deep ocean until it was learned that the islands had actually formed on old volcanoes, which through subsidence and erosion had been lowered to sea level. The coral builds up on the shoulders of the volcano to form a concentric ring of islands, called an *atoll*, and associated reefs, called *fringing reefs*. In the area of the crater, a small basin or lagoon forms. See Fig. 28.26.

Much larger accumulations of calcium carbonate sediments, similar to the ancient limestones found on the continents, also make up reeflike formations in tropical seas. These *carbonate platforms* form extensive shallow-water areas, such as the Great Bahama Bank southeast of Florida.

Tectonic Coasts

A tectonic coast is dominated by mountain-building processes. Not surprisingly, volcanic, faulted, and certain kinds of emergent coastlines can all be considered as tectonic coasts. Volcanic coasts are dominated by lava and ash deposits. The lava is usually fairly resistant to wave attack, but the ash gives way easily to wave action; therefore, the resultant shorelines are usually irregular. Faulted coasts, on the other hand, may be fairly straight, especially if the shoreline is formed by a fault scarp. Coasts that are elevated due to mountain building are often terraced where wave-cut platforms have been raised above sea level. See Fig. 28.26.

SAND CONSERVATION AND BEACH PROTECTION

Coastal real estate is valuable property, and people are understandably concerned about maintaining it. Unfortunately, our love for the sea side and the natural trends of change in shorelines often create incompatible arrangements. Structures are typically placed close to the water's edge; when the shore recedes, they are subject to damage and destruction. The management alternatives are to either suffer the loss and attempt to relocate the structures or reduce or stop the recession by altering the sediment mass balance. The answer usually depends on economics, and it seems that changing the sediment mass balance is the most economical solution in most areas.

Groins are the favored means of reducing the loss of beach sand in most areas. A groin is a wall built perpendicular (or nearly so) to the shore for the purpose of slowing and capturing some of the longshore drift. The drift accumulates on the upshore side of the groin and builds outward, eventually spilling around the end of the groin. A number of groins constructed in succession can slow the longshore transport sufficiently to build and maintain a beach in some locations. In other places, however, they have been known to increase erosion in beaches downshore because they interrupt the natural flow of sediment in the same manner as breakwaters often do.

Another method used to maintain beaches is *artificial nourishment*. Sand is trucked or hydraulically piped to the beach in order to sustain a body of sediment. This is an expensive method and one that must be carried out frequently to replace the sediment transported away. Where groins or breakwaters block the sediment train, it is sometimes necessary to artificially nourish the beaches downshore by a technique called *sand bypassing*. Authors' Note 28.1 offers some additional information on this shoreline-management technique.

Where erosion has reached or threatens to reach critical proportions, the beach or the backshore slope can be protected by building structures resistant to wave attack. *Riprap* (boulders or broken concrete) placed along the shore helps to dissipate wave energy and hold materials in place. Inclined walls constructed of concrete blocks called *revetment* serve the same purpose. *Seawalls* erected parallel to the shore intercept large waves and help to reduce their energy and in turn their capacity to move sediment.

SUMMARY

Wind waves are generated when the momentum of moving air is transferred to the water surface. Wave size increases with wind velocity, wind duration, and fetch. The rotational motion of water particles in a wave extends well below the water surface and in shallow water this motion exerts force against the bottom.

The geomorphic work of waves and currents includes both shore erosion and sediment transport. Longshore transport is concentrated in the surf zone, but also occurs in the deeper water as well as in the swash zone on the beach. The direction of longshore transport along the coast follows that of the transporting waves and currents, and on many coasts the prevailing direction of flow changes from season to season. For any parcel of space along a shoreline, the sediment mass balance is equal to the relative inputs and outputs of sediment by waves, currents, runoff, and wind.

Among the depositional features built by shoreline processes are spits, bars, and cusps. Land use has given rise to serious problems along changeable shorelines, and management programs have traditionally favored installation of various kinds of structures to protect against erosion and recession.

CHAPTER 29 GLACIAL PROCESSES AND LANDFORMS

INTRODUCTION

Practically every modern school child is aware of those slow-moving ice masses called glaciers. But the notion that glaciers are important erosional agents has not been acceptable in the scientific community for very long, for only 150 years ago most natural scientists themselves would have scoffed at the idea.

Much of the credit for alerting earth scientists to the nature and importance of glaciers goes to the naturalist Louis Agassiz (pronounced Egg-a see) (1807–1873). Agassiz, who first studied glaciers in the Swiss Alps, was responsible for not only drawing attention to the action and work of mountain glaciers, but also introducing the proposition that glaciers once extended over vast areas in the midlatitudes. In the 1840s he argued, and today we firmly agree, that large parts of North America, Europe, and northern Asia were once covered by massive sheets of ice called *continental glaciers*.

The Ice Age

From studies of land deposits, ocean sediments, soil features, and buried plant and animal remains throughout the world, we reason that in the past million years or so there have been at least four and as many as eighteen major episodes of continental glaciation. In North America the glaciers formed in central and eastern Canada and spread both southward and northward. The sheets that moved southward covered New England, the Great Lakes region, and the upper Great Plains, an area in which today more than 100 million people live. On the west, the ice sheets met another sheet of ice coming from the Canadian Rockies, and for a time glacial ice stretched completely across North America (Fig. 29.1).

A similar pattern of glaciation has also been documented in Europe. In both North America and Europe the periods of glaciation were interrupted by *interglacials*—relatively warm, dry periods when the bulk of the continental glaciers apparently melted away. Together, the glacials and interglacials constitute a part of earth history known as the *Pleistocene Epoch*.

Louis Agassiz (1807–1873), the Swiss natural scientist who championed the argument for continental glaciation in Europe and North America. (Photograph courtesy of Harvard Archives, Harvard University)

507

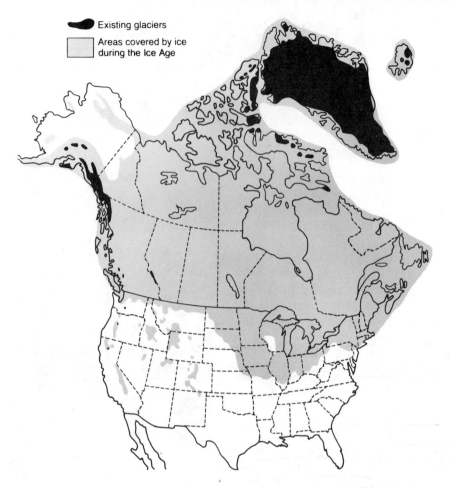

Existing glaciers

Areas covered by ice during the Ice Age

Fig. 29.1 The shaded area shows the maximum extent of glaciations in North America during the Pleistocene Epoch. The dark areas show the areas of modern glaciers. (From R. D. Goldthwait and G. W. Smith, "Glacial Geology: Introduction," in Encyclopedia of Geomorphology, *ed. R. W. Fairbridge. Copyright © 1968, Dowden, Hutchinson and Ross, Inc., Stroudsburg, Pa. Reprinted with permission.)*

The Pleistocene is of special interest to scientists because it was a time of not only the shaping of the landscape we see today, but also rapid biological change on the planet. Many large mammals, e.g., the woolly mammoth, became extinct near the close of the Pleistocene. Of paramount importance is the fact that the rise and geographic dispersal of early humans took place during the Pleistocene. Near the end of the last glaciation, coinciding with the extinctions of the large mammals, humans invented agriculture. Agriculture began 10,000 to 12,000 years ago, and in the ensuing millenia it spread across the world.

Occurrence and Types of Glaciers

Currently glacier ice covers about 10 percent of the land areas, compared with as much as 30 percent during the glacial maxima of the

Pleistocene. The most extensive glaciers are now found in Antarctica and Greenland. Elsewhere, glaciers are confined to mountain ranges, but the elevations at which they occur vary with latitude. Near the Arctic Circle alpine glaciers form as low as 500 m above sea level; at 45° latitude they form only above 3000 m; and in the tropics mountains less than 5000 m cannot support glaciers (Fig. 29.2). A brief examination of the environments that foster modern glaciers, however, tells us that in addition to cold temperatures, considerable snowfall is necessary to sustain glaciers. Indeed, there are many cold areas, especially in Siberia, that because of inadequate snowfall do not support glaciers.

Glaciers are generally classified according to size. The largest, of course, are the continental glaciers, which at their maxima covered land areas of the midlatitudes and subarctic on the order of 5,000,000 km². Although they are no longer around, the continental glaciers are not to be ignored, because they not only were important in shaping the present landscapes of North America and Eurasia, but also are likely to appear again on the planet. The polar ice sheets of Greenland and Antarctica can be considered contemporary versions of continental glaciers.

Piedmont glaciers, the next class, are formed by the merger of many mountain glaciers. Piedmont glaciers range from several thousand to several tens of thousands of km² in area and are found mainly along the lower slopes of major mountain ranges, such as the Himalayas or the Andes of southern Chile.

Mountain, or alpine, glaciers are the third class and are clearly the most widespread glaciers in the world. They may range in size from a patch of ice several hundred meters across to rivers of ice many kilometers long. Mountain glaciers are found in all latitudes, including the tropics, where they occur only on the highest mountains, such as Kilomanjaro and Mt. Kenya in East Africa or the Andes of South America.

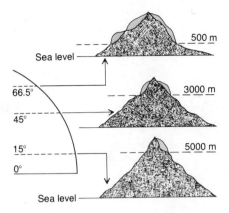

Fig. 29.2 *The elevation above sea level at which glaciers can form decreases from the tropics to the poles. Three representative elevations of glacier formation are shown here.*

GROWTH AND MOTION OF GLACIERS

Snow to Ice

Ice that forms on rivers and lakes originates directly from liquid water, whereas the ice that forms glaciers originates as snow. New-fallen snow is composed of loosely packed lacy crystals (Fig. 29.3a). The density of new snow ranges from about 50 to 300 kg/m³ (the density of pure water is 1000 kg/m³). After the snow falls, the crystals are reduced by: (1) *ablation*, which includes both melting and sublimation; and (2) physical compaction. As ablation and compaction proceed, the density of the snow increases. In the spring additional changes take place due to melting and refreezing, and the original snowflakes have now been transformed into rounded crystals. This partially melted, compacted

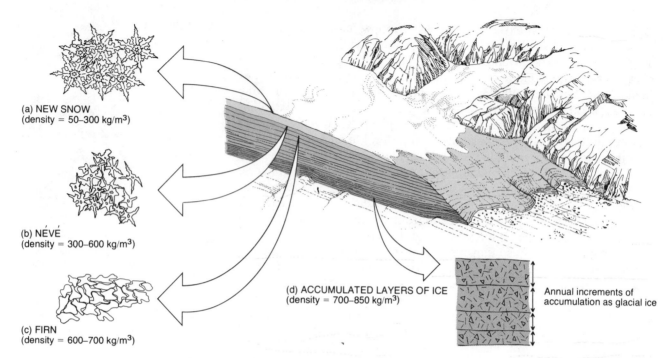

(a) NEW SNOW
(density = 50–300 kg/m³)

(b) NÉVÉ
(density = 300–600 kg/m³)

(c) FIRN
(density = 600–700 kg/m³)

(d) ACCUMULATED LAYERS OF ICE
(density = 700–850 kg/m³)

Annual increments of
accumulation as glacial ice

Fig. 29.3 New snow (a), névé (b), firn (c), and glacial ice (d) and the relative position of each in a glacier. (Illustration by William M. Marsh)

snow is called *névé* and has a density exceeding 500 kg/m³ (Fig. 29.3b). In most areas of the world where snow falls, all of the névé melts before the end of the ablation season.

In some places, though, because of either heavy winter snowfall or low ablation rates, some of the névé survives the entire ablation season. It is now called *firn* (Fig. 29.3c). Where this happens year after year, each year's layer of firn accumulates on top of the previous years' layers, and the layers below the surface are compressed. This causes a further increase in density, and the firn is changed into *glacial ice*, with typical densities of 850 kg/m³. This is somewhat lower than the normal density of ice (917 kg/m³) because some air is trapped in the glacial ice. The transformation of névé into glacial ice may take 25 to 100 years, and the exact nature of the processes involved is not fully understood. In the ice the different strata representing each year's increment of accumulation can often be seen in the crevasses of the glacier (Fig. 29.3d). Eventually all glacial ice is destroyed. In addition to ablation by melting and sublimation, glaciers can also lose ice when it breaks off into the sea. This process is known as *calving*.

Glacier Movement

In order to qualify as a glacier, a mass of ice must be capable of sustained movement. In most glaciers movement tends to be a continuous, flowing motion, in marked contrast to the erratic sliding motion that characterizes avalanche ice, for example. Movement of glacier ice begins when the downward stress produced by the weight of the ice mass exceeds the resisting strength (shear strength) within the mass, resulting in plastic deformation. In other words, at some point in the buildup of an ice mass, it grows too heavy to maintain its shape and literally squashes out in the lower portions. In mountain glaciers this begins when the ice reaches a thickness of 20 m. Deformation is most pronounced where stress is greatest and the confining pressure on the perimeter of the ice mass the least. For a mass of ice situated on a mountain slope, this is usually the lower, downslope side (Fig. 29.4).

Once set into motion, the ice develops zones of different rates of flow. If we look down on the surface of a glacier, the central zone appears to move fastest; the margins, adjacent to the valley wall, move much more slowly. Glaciologists have known this since the time of Agassiz, because surface velocity can be measured simply by installing a row of stakes across the glacier and measuring their displacement over time. The velocity of the interior of the glacier is not so easily measured, but scientists believe that the fastest zone in the glacier is located at or near the surface (Fig. 29.5). Near the bottom of the glacier, velocity drops off to rates comparable to those on the sides.

From velocity measurements made at many places on many glaciers, we are able to construct a picture of the overall pattern of ice flow in a typical valley glacier. In the upper part of the glacier, flow lines tend to converge downslope, forming a zone of compressed flow in the middle part of the glacier. At the lower end of the glacier, the ice

Fig. 29.4 The weight of a thick ice deposit exerts sufficient stress to cause the ice to deform as a plastic. Deformation is most pronounced on the downslope edge of the ice.

VELOCITY PROFILE

Fig. 29.5 The distribution of velocity in eight profiles along the Saskatchewan Glacier, Alberta, Canada. Generally, velocity decreases steadily toward the snout from a high of 117 meters per year on the surface at far left. (From Mark F. Meier, "Mode of Flow of Saskatchewan Glacier, Alberta, Canada," U.S. Geological Survey Professional Paper 351, 1960)

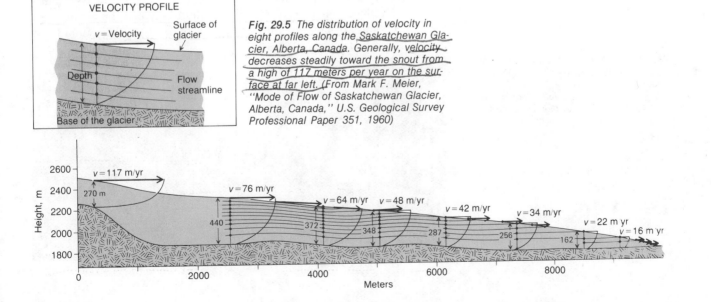

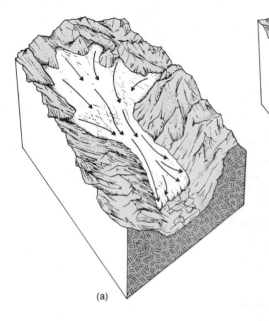

(a)

(b)

Fig. 29.6 *Pattern of flow in a valley glacier. (a) From the surface, flow lines converge in the upper region and diverge in the lower region. Flow is most constricted in the center, where flow lines closely parallel the long axis of the glacier. (b) In the interior of the glacier, flow lines tend to move downward in the upper region and upward in the lower region. (Illustration by William M. Marsh)*

decompresses, and the flow lines spread out (Fig. 29.6a). There is a corresponding flow pattern in the interior of the glacier; flow lines in the upper part move from the surface toward the center of the glacier, and in the lower part from the center toward the surface (Fig. 29.6b).

The velocity of flow of the glacier is influenced by many factors. Generally, it appears that velocity increases with the steepness of the valley floor, the thickness and temperature of the ice, and the constriction posed by the sides of the valley. The actual movement of the ice over the ground is similar in principle to the mechanism involved in ice skating. The immense pressure of the contact between the runner and ice causes sudden meltout and reduction of friction. This facilitates

Fig. 29.7 *The snout of the La Pérouse Glacier, Alaska, in 1899. Rapid advance of the ice is evidenced by the destroyed and damaged trees on the left. (Photograph by G. K. Gilbert, U.S. Geological Survey)*

slippage of the ice over the land and allows some glaciers to move as fast as 50 m per day. Most, however, are much slower, moving at rates of less than 1 m a day.

Glacier Mass Budget

In order to understand how glaciers can grow and shrink, advance and retreat, and erode and deposit, it is necessary to examine their energy balances, or what in glaciology is termed *mass budget*. In brief, the mass budget of a glacier includes three main components: (1) accumulation of snow on the upper glacier; (2) forward movement of the glacier; and (3) ablation of the lower glacier. If over a year accumulation and ablation are equal, the glacier has neither lost nor gained any mass. If accumulation, forward movement, and ablation are all equal, the glacier has gained and pushed ahead an amount of ice equal to the amount lost. Given a year in which more ice accumulates than ablates, the glacier mass grows, and the terminus of the glacier, called the snout, may advance downslope (Fig. 29.7). Although some modern glaciers are advancing, most are currently in a state of retreat (Fig. 29.8).

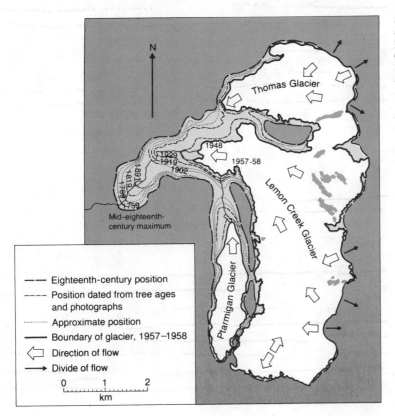

Fig. 29.8 Variations in the limits of the Lemon Creek Glacier of Alaska over 200 years. During the Little Ice Age of the eighteenth century, the ice extended 2 km farther down the Lemon Creek Valley than it did in 1957–1958. (Adapted from C. L. Hausser and M. G. Marcus, ''Historical Variations of Lemon Creek Glacier, Alaska, and Their Relationship to the Climatic Record,'' Journal of Glaciology, 1964. Used by permission of the International Glaciological Society.)

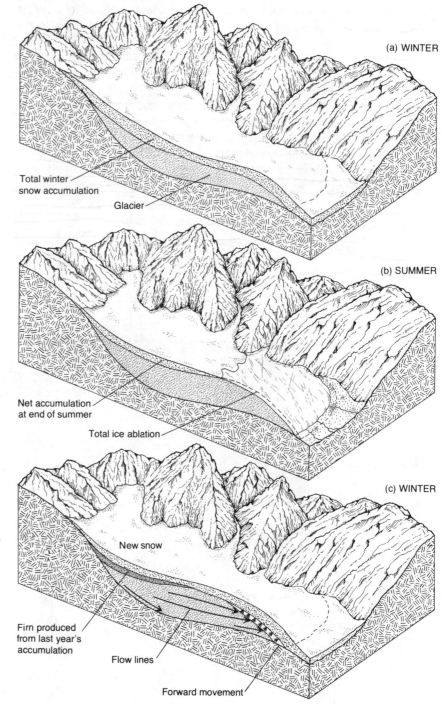

(a) WINTER

Total winter
snow accumulation

Glacier

(b) SUMMER

Net accumulation
at end of summer

Total ice ablation

(c) WINTER

New snow

Firm produced
from last year's
accumulation

Flow lines

Forward movement

Fig. 29.9 The main phases in the mass
budget of a glacier: (a) winter snow
accumulation over the entire glacier;
(b) snow remaining and total ice ablation
at the end of summer; (c) firn and forward
movement produced over the year.
(Illustration by William M. Marsh)

Fig. 29.10 *The firn line on the South Cascade Glacier, Washington, in August 1961. This line marks the lower edge of last season's snowfield, below which the glacier has completely lost its snow cover to ablation. (Photograph by Mark F. Meier, U.S. Geological Survey)*

This indicates that their mass balances are negative, which suggests that climate in this century has grown either warmer or less snowy or both. But this trend can reverse, as it did in the 1700s, when glaciers over much of the world made such strong advances that the period is referred to as the Little Ice Age (Fig. 29.8).

Let us describe the processes associated with the annual mass budget of a mountain glacier such as one we might find in the Rocky Mountains, the Alps, or the Himalayas (Fig. 29.9). The budget year begins with the accumulation of snow over the entire surface of the glacier. Snow is received from not only direct fall, but also avalanches that bring snow, névé, and ice to the glacier from mountain slopes.

The second phase of the year, the ablation period, begins in spring and usually lasts until late summer or early fall. Ablation is most intensive in the lower parts of the glacier, where the previous winter's snow cover may be lost by late spring. As the summer passes, the lower edge of the snowfield migrates farther and farther up the glacier (Fig. 29.10). While this is taking place, ablation directly consumes the ice below the accumulation zone, as is suggested by the appearance of the surface in the lower half of the glacier in Fig. 29.10. As the ablation period draws to an end, the lower edge of the accumulation zone, called the *firn line*, stabilizes, and the loss of snow subsides. The total amount of snow remaining above the firn line represents the glacier's nourishment for the year (Fig. 29.9b). We can compute this quantity of

[handwritten note: névé - partaily compacted + melted snow]

snow based on measurements giving the area of accumulation and the depth and density of the snow:

Net accumulation = Area × Depth × Density.

But to determine the mass budget, we also need to know how much ice was lost to ablation in the area below the firn line. This can be computed in the same fashion as the new accumulation, except that we are dealing with negative values. The mass budget of the entire glacier, then, represents the balance between net gain above the firn line and net loss below the firn line:

Mass budget = Net accumulation − Net loss.

While ablation and accumulation are taking place, the glacier continues to flow ahead (Fig. 29.9c). The rate of flow is mainly a function of net accumulation, but the relationship is complex because of the time lag between one season's accumulation and the corresponding movement of the glacier. The lag may amount to several decades in large glaciers, and for this reason it is not uncommon for their behavior to be out of phase with climatic trends. This may help to explain the phenomenon known as "glacial surges," in which a glacier bursts forth, driving ahead at rates of 10–20 m per day, partially in response to a massive input of energy (net accumulation) some years before.

Although forward movement is an important control on the behavior of the glacier, it is not the only control. The rate of ablation at the snout is also important; if equal to the rate of forward motion, the snout may remain stationary while the glacier continues to move. The glacier acts in this way much like a conveyor belt. However, when the ablation rate exceeds forward movement, the snout retreats; the reverse can be expected when movement exceeds ablation. These motions—stability, advance, and retreat—are critical to the following discussion concerning the erosional and depositional work of glaciers.

THE WORK OF GLACIERS

Glacial Erosion

Because of their smallness, vastness, or particular situation, certain natural phenomena are nearly impossible to observe and measure directly. Such is the case with the bottom of an active glacier. Although one can drill to the base of the ice or even build a shaft big enough to hold a person, the very presence of the hole or shaft would tend to disturb conditions there enough to cast doubt on the validity of observations. Thus an *empirical* approach, that is, one based on direct observation and measurement, to the problem of how glaciers erode is fraught with limitations.

What are the alternatives? One is to reason the problem through in reverse: in other words, to work from observable effects to causes. By examining the landforms that appear after a glacier has passed over an area, we can try to interpret what erosional processes produced them at the base of the ice. Understandably, the results obtained by this approach are subject to much guesswork, inasmuch as it is difficult to test them empirically.

Scouring and plucking. Two major erosional processes appear to occur on the bed of a glacier. One is scouring as the ice, armed with rock debris it has incorporated from various sources, skids over the land. The rocks in the ice act as an abrasive agent, literally rasping the bedrock under them. This results in a variety of features, the most notable of which are scour lines, called *striations*, and finely abraded surfaces called *glacial polish*. Polish is usually limited to well-consolidated igneous rocks, especially granite (Fig. 29.11). The abrasive action also yields a fine sediment, usually clay-sized, which often ends up in meltwater. As a result, glacial meltwater often has a light, cloudy appearance, which has led to the name *glacial milk* or *glacial flour*.

The second erosional process on the glacier bed is known as *plucking*. It appears that the ice at depth melts and refreezes in response to pressure changes and other factors; when it does, it may refreeze around blocks of jointed bedrock. When the ice moves, the block is drawn out and carried away. The plucking process occurs with greatest intensity on the down-ice side of rock knobs. Combined with scouring action occurring on the other side of a knob, the result is a curious, asymmetrical feature called a *roche moutonnée*, which is smooth on one side and steep and jagged on the other (Fig. 29.12).

Sculpturing of valleys and mountains. Glaciers flow over the land along the paths of least resistance to their movement; therefore, the ice usually concentrates in low areas, typically river valleys. Mountain glaciers illustrate this point vividly. When the ice enters a mountain valley, it takes the form of a distended tongue, called a *lobe*. As it moves down the valley, the lobe often merges with other lobes emanating from nearby mountain sides. The glacier grows with each tributary lobe; thus to a limited extent a network of mountain glaciers follows the principle of stream orders.

Fig. 29.11 *Glacial polish on a granitic dome in the Sierra Nevada. Polishing is usually most pronounced near the crests of domes and knobs composed of tightly consolidated igneous rock. (Photograph by Jeff Dozier)*

Direction of ice movement

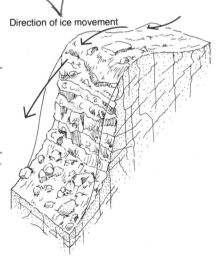

Fig. 29.12 *A roche moutonnée; the glacier moved from right to left, plucking blocks of rock from the lee side at the rock knob while scouring the iceward side.*

Owing to the deep and steep-sided nature of mountain river valleys, the glacier must flow in highly confined space. As a result, a great deal of stress is exerted against the walls of the valley. Talus and other footslope deposits which line the walls of stream valleys, such as the one shown in Fig. 29.13, are readily removed and carried downvalley in the initial advance of the ice. The glacier then begins to erode the bedrock of the valley walls and floor. The valley is both widened and deepened, eventually emerging with a beautifully symmetrical U-shape. This shape contrasts sharply with the original V-shape of the river valley, and for this reason valley shape is a good indicator of how far downvalley glaciers once extended.

Another way in which glaciated valleys differ from stream valleys is in the elevations at which small valleys join larger ones. In river networks small valleys enter large valleys at essentially the same elevations, whereas in glacier networks small valleys enter large ones at much higher elevations. The reason for this is that large glaciers have so much more erosive power than small glaciers that they are able to

Fig. 29.13 Nonglaciated stream valleys in mountainous regions are often distinguishable by their deep V-shapes. (Photograph by William H. Jackson, U.S. Geological Survey)

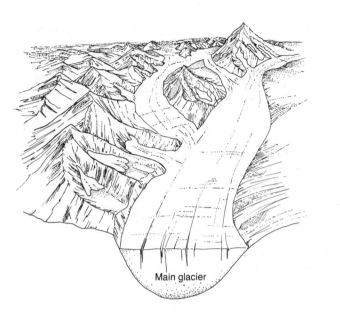

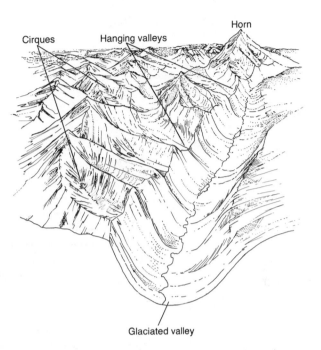

Fig. 29.14 *Typical landforms resulting from mountain glaciation. Among the most prominent features are horns, cirques, and hanging valleys. (Illustration by William M. Marsh)*

lower their valleys much faster. Although similar differences in erosive power may exist in large and small streams, the erosive power of tributary streams tends to increase as the trunk stream lowers its channel. The result is that in a river system, the lower reaches of small channels usually keep pace with large channels, whereas in glacier systems side valleys are left high above the floor of the main valley. After deglaciation, these side valleys appear to hang above the main valley and are called *hanging valleys* (Fig. 29.14). They are often the sites of spectacular waterfalls, and nowhere is this better illustrated than in the famous Yosemite Valley in the Sierra Nevada of California.

Many other interesting erosional features are formed by mountain glaciers besides distinctive valleys. Where an ice field works on the side of a mountain for a long time, it often sculpts out a small basin called a *cirque*. The formation of a cirque appears to be closely related to the development of a deep crevasse, called a *bergschrund*, at the very head of the glacier. The bergschrund cracks open in summer as the glacier slips away from the mountain side, and as it does, meltwater and rock debris fall into the chasm. Some of the meltwater freezes in contact with the bedrock, wedging blocks of rock from the mountain side (Fig. 29.15). This process, coupled with glacier plucking, steepens and deepens the valley head, producing the cirque. Many cirques that formed during the Pleistocene no longer contain glaciers, and in their bottoms small lakes, called *tarns*, have formed.

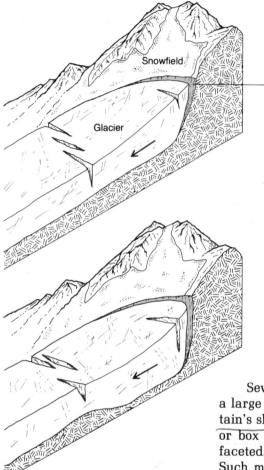

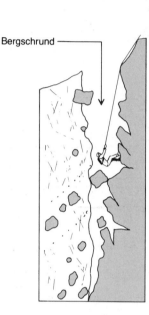

Bergschrund

Fig. 29.15 *The formation of a cirque at the head of a mountain glacier. The enlarged drawing shows the bergschrund. (After Lewis, 1938; and Gilluly et al., 1975)*

Several separate ice fields can often develop on different sides of a large mountain. As each erodes its side of the mountain, the mountain's shoulders are bevelled down, transforming it from, say, a dome or box shape into a conical shape. But since the sides are usually faceted, the shape more closely resembles that of a broad Eiffel Tower. Such mountains are called *horns*, the most famous of which is the Matterhorn of Switzerland. Each side, or facet, is called a *face*, and faces are generally separated by sharp ridges termed *arêtes*.

Sculpturing by continental glaciers. The erosional features created by the continental glaciers are in general less distinct than those created by mountain glaciers. It appears that the continental glaciers also followed river lowlands, particularly where the trends of valleys coincided with the direction of ice movement. The continental ice sheets, though, were so thick, probably 3000–4000 m, that they completely filled and overflowed most valleys. In some places, such as the Finger Lakes region of New York, U-shaped valleys were created, but they tend to be exceptions. In most areas erosion by the continental sheets produced shallow basins. It is generally agreed that the basins of the large lakes on the fringe of the Canadian Shield (including those of the Great Lakes) were sculpted from preglacial lowlands (Fig. 29.16).

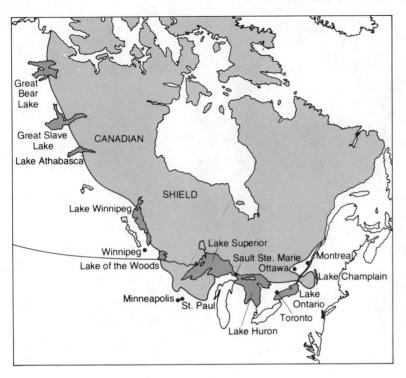

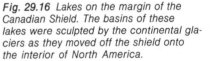

Fig. 29.16 Lakes on the margin of the Canadian Shield. The basins of these lakes were sculpted by the continental glaciers as they moved off the shield onto the interior of North America.

Glacial Deposition

Sources of glacial debris. Visitors to glaciers are often surprised to find that the ice surface is covered with a mantle of rock debris and that there may be few places where one can actually see the ice itself. This is especially common near the snout of the glacier, where most of the ice has been lost to ablation. What are the sources of the rock debris in a glacier? We have already mentioned one important source: the material picked up by plucking at the base of the ice.

A second important source of rock debris is the slopes at the margin of the ice. In mountainous areas steep slopes are subject to mass movement, such as rockfalls, landslides, and avalanches, which deposit large amounts of debris on top of the glacier (Fig. 29.17). Some of this material falls into crevasses, but most is buried by snow and eventually incorporated into the firn. Since the debris falls from side slopes, most of it is concentrated in a narrow belt along the margins of the ice. Thus when two glaciers flow together, the two debris belts on the inside merge to form an interior belt of debris, called a *medial moraine* (Fig. 29.18).

Fig. 29.17 Rock debris on the surface of the Bandaka Glacier, Hindu Kush, Afghanistan. The large pedestal in the center is more than 3 m high. (Photograph by William M. Marsh)

Fig. 29.18 *A medial moraine in the Kaska-walsh Glacier, Alaska. (Photograph by John S. Shelton)*

Fig. 29.19 *An interpretation of conditions at the front of a continental glacier, including most of the depositional landforms which typically form there. (Ilustration by William M. Marsh)*

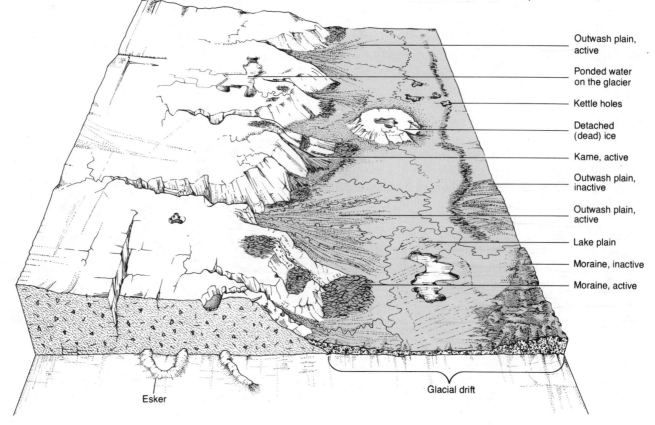

Outwash plain, active

Ponded water on the glacier

Kettle holes

Detached (dead) ice

Kame, active

Outwash plain, inactive

Outwash plain, active

Lake plain

Moraine, inactive

Moraine, active

Esker

Glacial drift

Since a glacier functions like a conveyer belt, all of the debris in the ice is eventually delivered to the terminus, where it is deposited. There are two major modes of deposition: (1) direct from the ice, or (2) meltwater flow from the ice. Meltwater deposits are called *fluvioglacial* deposits; the material deposited by the ice is called *till* or *moraine*. All glacial deposits are collectively known as *glacial drift*.

Ice deposits. Till is a heterogeneous mixture of unstratified materials ranging in size from massive boulders to clay. The actual sediment content of the till is highly variable and depends on many factors, including drainage conditions at the ice front and the nature of the debris carried by the ice. The manner in which the till is laid down is variable as well. Till deposited along the edge of the ice often forms irregular hills and mounds called *moraines* (Fig. 29.19). Moraines deposited at the point of farthest advance of the glacier are called *terminal moraines*; those deposited during halts in the retreat of the glacier are called *recessional moraines*. Other types of moraines include *lateral moraines*, which are laid down on the margins of ice, and *ground moraines*, which are deposited beneath the ice. *Till plain*, which resembles ground moraine, is formed when a sheet of ice becomes detached from the main body of the glacier and melts in place. The debris in the ice falls to the ground directly under it. Should the ice contain large boulders, such as the one shown in Fig. 29.20, they too are set down. When such boulders are deposited far from their place of origin, they are termed *erratics*.

Fig. 29.20 A glacial erratic perched on the edge of Little Yosemite Valley. Compare this boulder with the ones on the glacier in Fig. 29.17. G. K. Gilbert is standing next to the erratic. The accompanying diagram shows how the erratic was set down here. (U. S. Geological Survey photograph taken near Moraine Dome, 1908)

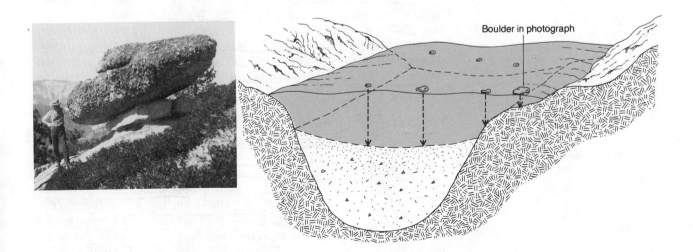

Boulder in photograph

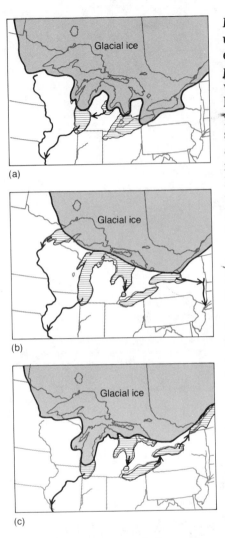

(a)

(b)

(c)

Fig. 29.21 Glacial lakes and sluiceways in the Great Lakes Basin associated with advances and retreats of the Wisconsin ice sheet: (a) 12,500 years ago; (b) 12,000 years ago; (c) 11,400 years ago. (Reprinted with permission from The Encyclopedia of Geomorphology, ed. R. W. Fairbridge. Copyright © 1968, Dowden, Hutchinson and Ross, Inc., Stroudsborg, Pa.)

Fluvioglacial deposits. In contrast to till, fluvioglacial materials are usually stratified and less diverse in particle sizes. Generally, the most extensive of the fluvioglacial deposits in a glaciated area is *outwash plain.* Outwash deposits are comprised mainly of sand eroded by meltwater from the ice and nearby till deposits. As streams of meltwater leave the ice, they cross the moraines and spread over the ground beyond them. Here the streams break down into distributaries, much as a stream does on an alluvial fan, spreading sediments in a broad fan, or apron. The outwash plain is formed when the aprons deposited by numerous streams of meltwater coalesce into a single feature (Fig. 29.19). In some places the meltwater may be directed into a stream valley near the glacier. As a result, the sediments may be laid down along the valley floor, forming what is called a *valley train* deposit instead of an outwash plain.

Fluvioglacial features are also deposited in contact with the glacier. Where sediment-laden water pours off the snout of the ice, a conical-shaped pile of sediment may build up. Such deposits, called *kames,* are often situated on or at the edge of moraines (Fig. 29.18).

Although most meltwater flows off the surface of the glacier, some also flows in ice caverns at the base of the ice. The beds of such streams often build up with sand and gravel; if the ice melts from around them, the stream beds are left as curious winding ridges called *eskers.* Eskers are often prime sources of gravel for construction.

We are led to wonder what happened to all of the water emitted from a massive continental glacier. Most of it found its way into stream systems which were taking shape on the newly formed terrain. The bulk of these drainage systems are still active, although the discharges they carry are considerably lower today. Some of the largest drainage features disappeared with the ice, however. In particular, large channels called *sluiceways* and vast areas of ponded meltwater, called *glacial lakes,* which formed along the ice front, drained away with changes in the position of the ice and the formation of new courses of drainage. This is well illustrated in the Great Lakes basin, where numerous glacial lakes rose and fell as the glacier advanced and melted back (Fig. 29.21).

Pitted topography. As the various deposits are forming, the ice disintegrates not only by melting, but also by breaking up into blocks which become detached from the main body of the glacier. These ice blocks are sometimes referred to as "dead ice," since they no longer respond to the movement of the glacier. If drift is deposited around them and the blocks melt away, a depression, called a *kettlehole,* is left in the surface. Both moraine and outwash plain may develop pitted surfaces as a result of this process. Large kettleholes that reach below the water table become flooded with groundwater, forming lakes. Shallow lakes

Table 29.1 Glacial stages and interglacials of the Pleistocene Epoch.

| CHRONOLOGICAL ORDER— YOUNGEST TO OLDEST | NAMES BY GEOGRAPHIC REGION | | |
	NORTH AMERICA	N. EUROPE	ALPS
Fourth glaciation	Wisconsin	Weichselian	Würm
Third interglacial	Sangamonian	Eemian	
Third glaciation	Illinoian	Saale	Riss
Second interglacial	Yarmouthian	Holstein	
Second glaciation	Kansan	Elster	Mindel
First interglacial	Aftonian		
First glaciation	Nebraskan	Pre-Elster	Gunz

may, in turn, fill with organic material and thereby be transformed into wetlands such as bogs, swamps, and marshes.

CONTINENTAL GLACIATION AND THE LANDSCAPE

Probably no natural event in the time of humans on earth has rendered as much change in the landscapes of the Northern Hemisphere as the continental glaciers have. We mentioned at the beginning of this chapter that the continental glaciation began about one million years ago, and based on land deposits at least four stages of glaciation are discernible (Table 29.1). Each stage in turn was characterized by a number of substages which marked episodes of advance of the ice following a period of meltback. In North America the last stage of glaciation was the Wisconsin stage, and its maximum extent in the continental United States is shown in Fig. 29.22. During the Wisconsin substages, the ice sheet melted back as far as the Canadian border and then readvanced southward. In the Great Lakes region the pattern of ice movement and retreat in each substage followed the contour of the lake basins and associated lowlands, such as Saginaw Bay in Michigan and Green Bay in Wisconsin. Not surprisingly, the landforms left behind, especially the moraines, also follow the configuration of the basins, resembling a series of collars around the Great Lakes (Fig. 29.23).

How have the landforms left by the glaciers influenced the modern landscape? The influences may, for convenience of discussion, be classed into two broad categories: direct and indirect. The direct influences, mainly those already discussed in this chapter, are the erosional and depositional alterations of the land associated with the movement and wastage of the ice sheet. The indirect influences are not so easily described and in fact are not very well understood in many respects. These influences stem from changes in climate, sea level, and land-

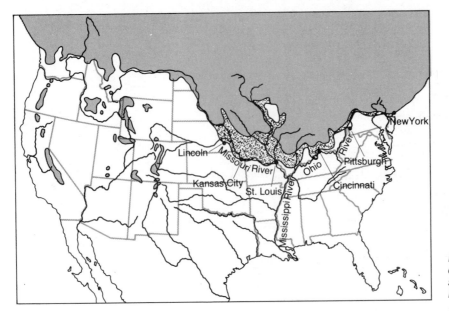

Area glaciated by Wisconsin stage

Area glaciated by earlier Pleistocene stages (not covered by Wisconsin ice)

Unglaciated area

Fig. 29.22 *The extent of Wisconsin and earlier glaciations in the United States and southern Canada. (Based on U.S. Geological Survey data.)*

Fig. 29.23 *The major moraines of the Wisconsin glaciation in the Great Lakes region. The arrows show the pattern of ice movement from the Great Lakes' basins. Compare this map with the one of the glacial lakes in Fig. 29.21. (Based on U.S. Geological Survey data.)*

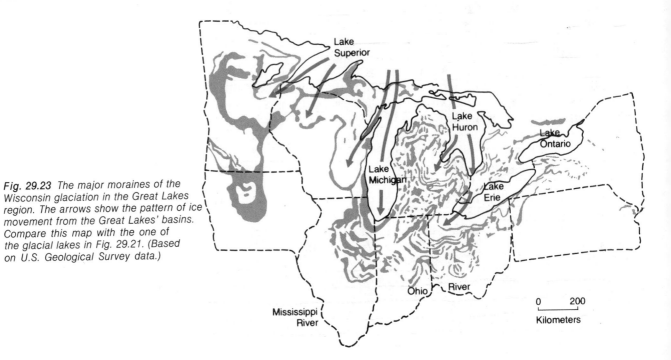

mass elevation that either accompanied or were brought on by continental glaciation. They are discussed later in this chapter.

Let us begin with a few words on some of the influences of glacial landforms on various aspects of the landscape. First, the landforms built by the continental glaciers are not overly impressive, largely because they are not very big as landforms go. Practically anywhere you can find hills and valleys of similar proportions which were formed by less spectacular processes. But despite the lack of topographic magnificence, glacial landforms generally have a pronounced influence on the development of the landscape. This influence is usually striking in freshly glaciated terrain, but declines with time and in older glaciated landscapes is quite subtle or not even evident to the casual observer.

Soils and Vegetation

Perhaps the most important influence of glacial landforms is tied to soil formation and drainage. Of course, glacial deposits vary radically in composition, ranging from those that are mainly sand and gravel to those that are practically all clay. Glaciofluvial deposits such as outwash plains and kames are usually sandy and well drained. The soils of these deposits are inherently poor in nutrients such as nitrogen and phosphorus, compared with those of till deposits. Under the cool, moist midlatitude climates of Europe and North America, these conditions are ideal for the establishment of conifer forests and the formation of Podzol soils. Indeed, so strong is the correlation among conifers, Podzols, and glaciofluvial landforms that in some areas, the occurrence of one can be taken as an indication of the presence of the other two. The great forests of white and red pine that were found in New England and the northern Midwest in the 1800s generally reflected these conditions. Today forests of jack pine are often found where the great pine forests once grew. Farther north, in the Canadian Shield, drainage of sandy soils is often poor because of the irregular configuration of the underlying bedrock. As a result, the forests on outwash plains are often dominated by spruce and tamarack instead of pines (Fig. 29.24).

Fig. 29.24 Forests and soils associated with well-drained and poorly drained outwash plains.

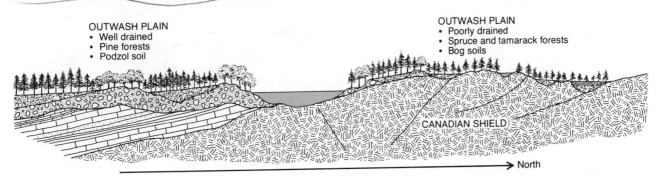

OUTWASH PLAIN
• Well drained
• Pine forests
• Podzol soil

OUTWASH PLAIN
• Poorly drained
• Spruce and tamarack forests
• Bog soils

CANADIAN SHIELD

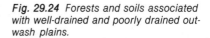 North

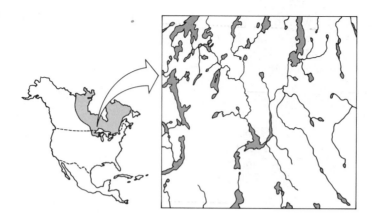

Fig. 29.25 *An example of deranged drainage from the Canadian Shield.*

Drainage

The influence of glacial landforms on drainage can be illustrated at several scales. In both North America and Europe the continental ice sheets originated in shield areas where the geology is extremely diverse. When subjected to the erosional forces of the ice, the surface of the shield yielded differentially, resulting in a highly irregular topography. The drainage lines that developed on this topography took on equally irregular patterns—so irregular, in fact, that it is referred to as *deranged* drainage (Fig. 29.25).

Outside the shields there are areas where the courses of rivers are controlled by belts of moraines. The rivers tend to flow in the lowlands between the moraines. But moraines are rarely continuous for

Fig. 29.26 *(a) Intermorainal drainage in north-central Europe. (b) The limits of continental glaciation in central and northwestern Europe.*

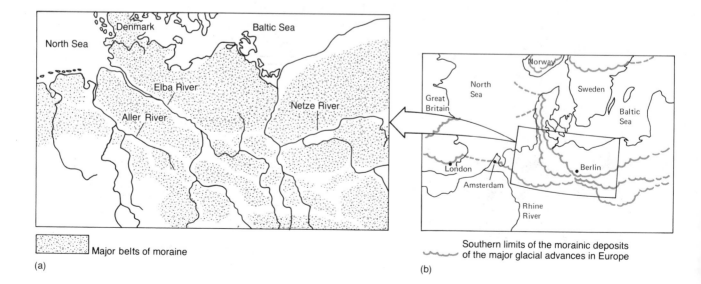

Major belts of moraine

(a)

Southern limits of the morainic deposits of the major glacial advances in Europe

(b)

great distances, and for this reason the pattern is usually broken where a river breeches a morainic belt and joins other rivers. Inter-morainal drainage is especially pronounced in northern Germany, where lengthy segments of the Aller, Elbe, and Netze rivers flow in east-west–trending lowlands between terminal moraines (Fig. 29.26).

Wetlands are also a characteristic drainage feature of glaciated terrain. They form not only in shallow kettleholes and in lakes that have filled with sediments, but also in the valleys of large drainage channels which once carried massive discharges of glacier meltwater. Today the floor of the old channel is often occupied by wetlands linked by small streams. The Minnesota River Valley, which joins the Missis-sippi River at Minneapolis, is a good example of such an area of wet-lands (Fig. 29.27).

Indirect Influences of Continental Glaciation

Since the earth has a fixed supply of water, about 98 percent of which is held in the oceans, we reason that the water that went into building the continental glaciers, which amounted to some 55 million cubic kilo-meters of ice, had to produce a lowering of sea level the world over. From underwater surveys conducted on the continental shelves, it appears that worldwide sea level during the glacial maxima was about 130 m lower than it is today. Where the continental shelf is very wide,

Fig. 29.27 Wetlands are common fea-tures of river valleys, such as the Minne-sota River, which served as a glacial drainage channel. (Illustration by William M. Marsh)

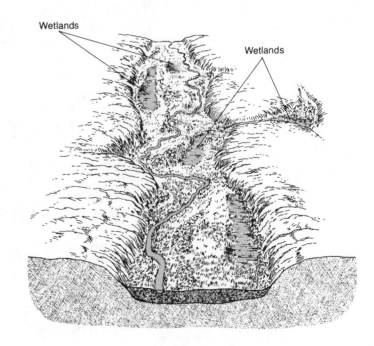

Wetlands

Wetlands

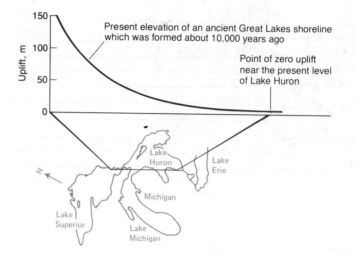

Fig. 29.28 *Profile of an unwarped shoreline in the Great Lakes Basin. The original elevation is represented by the southern end of the profile; the northern end represents the total uplift in the past 10,000 years.*

as along the east coast of North America, the ocean shoreline during the Pleistocene glaciations extended as much as 75 kilometers beyond its present location. Between Alaska and Siberia, most of the Bering Sea disappeared, creating a land bridge between North America and Asia. Some anthropologists feel that this land bridge was an important route of early human migration from Asia to North America. (By the way, if the water presently held in glacier ice—about 25 million cubic kilometers—were to melt, sea level would rise by 50 m or more, flooding all of the major coastal cities in the world.)

Another effect of the continental ice masses was to depress the earth's crust. Owing to the enormous weight of the ice masses and the elastic nature of the crust, the land actually subsided under the continental glaciers. This subsidence, known as *isostatic depression*, was greatest where the ice was thickest and lasted the longest. In North America one of these areas was Hudson Bay, where crustal depression under the ice may have exceeded 1000 m. Wherever depression occurred, however, it was not permanent, because when the ice melted away, the crust began to rebound. This is referred to as *isostatic rebound*, or glacial uplift. In the past 10,000 years *isostatic rebound* has recovered most of the depression caused by the Wisconsin ice sheet in the northern Great Lakes.

Northward from the Great Lakes, rebound is still in progress, and in the Hudson Bay region it amounts to several centimeters per year. The best measure of rebound is found in the ancestral shorelines of large water bodies, such as the Great Lakes and Hudson Bay. If we trace one of the Great Lakes' ancient shorelines from south to north, it begins to increase in elevation (Fig. 29.28). Since the entire shoreline of any water body had to have formed at a uniform elevation, such warping is solid documentation of isostatic rebound.

The Pleistocene was also a time of accelerated wind erosion, judging from the wind deposits dating from that time. On and around the shorelines of glacial lakes and on outwash plains, sand dunes formed before vegetation could stabilize new surfaces. Farther away from the glacial front, silt was deposited over extensive areas by wind. The deposits, referred to as *loess*, are composed of fairly uniform silt which was apparently winnowed from sluiceways, lake beds, and barren deposits near the ice, though no one is sure of all of its sources. In any case, vast areas of loess are found in the central part of the United States, China, and in European Russia, well beyond the southern limits of glaciation.

SUMMARY

Glaciers presently cover about 10 percent of the earth's land area, but during the Pleistocene Epoch, they covered up to three times as much land. Glaciers are fed by snow which over a number of years is transformed from névé into firn, then ice. When the thickness of the ice exceeds 20 m, it deforms as a plastic, and the glacier begins to flow. The erosional work of glaciers involves mainly scouring and plucking. The debris transported by the ice is deposited as till and fluvioglacial material in the form of till plains, moraines, outwash plains, and related landforms. Although glacial landforms are generally not large, they have a pronounced influence on the development of drainage, soils, and vegetation patterns, as is evident in the landscapes of northwestern Europe, Canada, and the Great Lakes region.

AIRFLOW, WIND EROSION, AND SAND DUNES

CHAPTER 30

INTRODUCTION

The atmosphere is an enormous ocean whose bottom is the landscape and the surface of the seas. The currents of the atmosphere, i.e., winds, are governed by the same principles of fluid motion as govern liquids, notably water. But when we examine the atmosphere's ability to erode earth materials, there are some important differences between air and water. First, the density of air (1.29 kg/m^3) is much less than that of water (1000 kg/m^3); therefore, wind cannot move as large a particle as a water current can. Second, unlike rivers, wind flows in essentially unconfined space, covering broad areas rather than narrow ribbons. Third, unlike ocean currents, which extend to a maximum depth of only a thousand meters or so into ocean, wind is everywhere capable of reaching the bottom of the atmosphere; therefore, all parts of the earth's surface can be affected by wind, but only the shallow parts of the ocean can be affected by ocean currents.

In this chapter we will examine wind as a geomorphic agent in the landscape. After discussing the motion of air as it moves over the land, we will focus on wind energy and erosive power; the latter part of the chapter describes the principal landforms created by wind action.

THE BOUNDARY LAYER OF THE ATMOSPHERE

In considering wind as an erosional agent—as opposed to a climatic agent—we need to examine only the lowermost zone of the atmosphere, called the *boundary layer*. In this transition layer between the atmosphere and the earth's surface, the motion of the atmosphere is influenced by the earth's surface. The boundary layer is about 300 m thick and more or less follows the contour of the earth's surface over both land and water (Fig. 30.1).

Recall from the discussion of mechanical convection in Chapter 3 that wind speed increases with elevation above any surface. At an elevation of 300 m or so, however, the increase ceases, and this marks the top of the boundary layer. The bottom of the boundary layer usually lies very close to the ground, but never directly on it. It is defined as the level of zero wind speed below which there is a zone of some thickness,

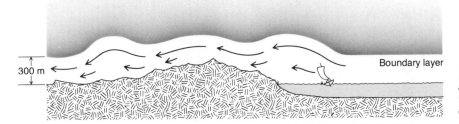

Fig. 30.1 The boundary layer of the atmosphere, a zone about 300 m deep where the atmosphere gives way to the earth's surface.

often only a few centimeters, that is essentially calm. This thin sheet of air, termed the *laminar sublayer*, is one of the key controls on wind erosion (Fig. 30.2). The thickness of this sublayer is called the *roughness length*.

THE MOTION AND VELOCITY OF AIR IN THE BOUNDARY LAYER

Types of Air Flow

Two types of flow can be detected in moving air: laminar and turbulent. In laminar flow a particle of air moves horizontally to the surface in essentially a straight-line route. The airflow in the upper part of the boundary layer approximates laminar flow, but near the ground it becomes decidedly turbulent. In the laminar sublayer flow is laminar, though movement is so faint as to be barely detectable.

Turbulent flow is characterized by an irregular, swirling motion over the ground. A particle of air may jump and eddy about in the same fashion as water moving over rapids in a stream. Unlike laminar flow, turbulent flow has a vertical as well as a horizontal component of movement, and the vertical component accounts for as much as 20 percent of the total movement. In addition, turbulent flow results in a much greater loss of wind speed than laminar flow does, owing to the resistance produced in the mixing of fast- and slow-moving molecules.

From the ground upward, the rate at which wind velocity increases varies with the type of flow. Velocity increases most rapidly in the zone of turbulent flow. Above a height of several meters or so, where turbulent flow gives way to laminar flow, the rate of increase declines and finally disappears altogether at the top of the boundary layer (Fig. 30.3). Moreover, the thickness of the turbulent zone changes with wind velocity. It thickens when velocity increases, expanding both upward and downward. This produces a decrease in the roughness length.

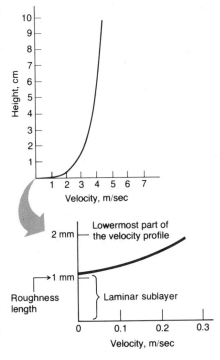

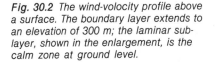

Fig. 30.2 The wind-volocity profile above a surface. The boundary layer extends to an elevation of 300 m; the laminar sublayer, shown in the enlargement, is the calm zone at ground level.

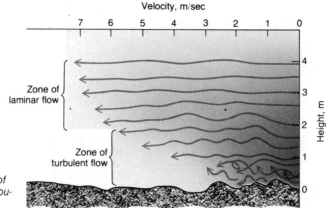

Fig. 30.3 *Wind velocity and the types of flow above the ground. The zone of turbulent flow is relatively thin and slow.*

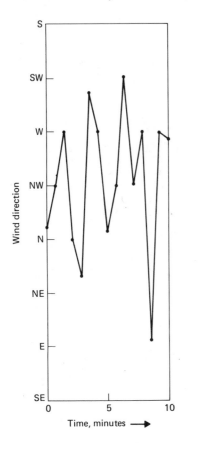

Fig. 30.4 *Variations in wind direction at a height of 5 m on a typical day in middle North America. (Data by Mark L. Hassett and William M. Marsh)*

Direction, Velocity, and Power

Wind direction and velocity at most places are highly variable and not easy to summarize in a brief statement. It is important to understand that average wind speed (based on all directions) or average wind velocity (based on single direction) tend not to be very meaningful, because the range of wind speeds and velocities is so great. This is due to the fact that near the ground, wind is very gusty, changing direction and velocity in a matter of minutes or even seconds (Figs. 30.4 and 30.5). Therefore, it is necessary to compute velocities according to various classes (e.g., 0–0.25 m/sec, 0.25–0.50 m/sec, 0.50–0.75 m/sec, and so on) in order to gain an accurate description of wind speed.*

A second, and very important, point is that wind speed is not a good indicator of wind power, because power and speed are not directly proportional. Rather, power increases according to the cube of speed or velocity. Therefore, fast winds, which on the one hand account for comparatively little total time in a sampling period, may on the other hand be the most important class in terms of power because of the power/velocity relationship. Experts are quick to point out that estimates of the electrical-generating capacity of wind based on average speed are consistently in error by more than 100 percent. The same would be true for estimates of erosion based on average speed or velocity. Figure 30.6 shows the power equivalent for the wind speed spectrum shown in Fig. 30.5. Authors' Note 30.1 describes the results of a wind-power survey of the coterminous United States.

---

*A note on the terms velocity and speed: *Velocity* is the rate of flow in one direction, whereas *speed* is the rate of flow (past a known point) in any direction. In computing wind power, data on speed are used because wind direction is irrelevant. In problems dealing with wind erosion and sediment transport, however, data on velocity are usually needed because not all wind directions are of equal importance, owing to differences in slope and exposure of the terrain.

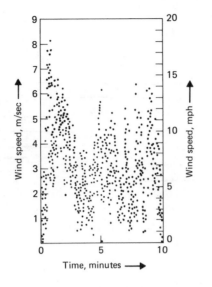

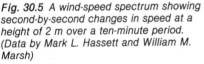

Fig. 30.5 A wind-speed spectrum showing second-by-second changes in speed at a height of 2 m over a ten-minute period. (Data by Mark L. Hassett and William M. Marsh)

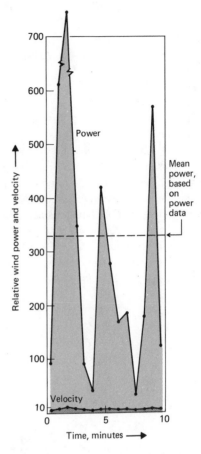

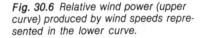

Fig. 30.6 Relative wind power (upper curve) produced by wind speeds represented in the lower curve.

SOME GEOGRAPHIC CHARACTERISTICS OF WIND

Seasonal Variations

The energy crisis has, among other things, renewed our interest in wind as a resource and improved our understanding of both its aerodynamics and geographic characteristics in North America and Europe. In the coterminous United States, for example, winter speeds are higher for the northern half of the country, where, with the exception of the Northwest, air flow is westerly (Fig. 30.7). This corresponds to the movement of air masses, especially midlatitude cyclones, across the middle of North America. In summer the overall pattern changes considerably, with the strongest winds coming off the western Gulf of Mexico and the Pacific Ocean and blowing onto the continent. This is a response to the thermal differences between land and water. As the land heats up, low pressure develops over it, which draws in air from the ocean.

Wind and Topography

If we look at specific areas, it is apparent that wind velocity is often influenced by topography (Fig. 30.8). January winds are especially strong in certain mountain settings such as Great Falls, Montana, and Casper, Wyoming, because the chinook blows down the valleys in which they are located. In other places the influence of topography is also appar-

AUTHORS' NOTE 30.1

Yes, the Great Plains Are Windy

In an effort to define the potential for energy generation from wind in the United States, a number of studies have been conducted using wind-speed data from several hundred weather stations across the country. The wind data were converted to power equivalents based on the velocity/power relationship (V^3), and the results were generalized to show regional trends. The values are in watts per square meter (W/m^2) and represent the annual average power flux for all winds at any location.

As the map shows, the region of greatest annual average power flux is the Great Plains, stretching from the Texas Panhandle into Canada. This zone also extends into adjacent parts of the Rocky Mountains in New Mexico and Wyoming (where the power flux exceeds 400 W/m^2 in one area) and the Canadian Rockies. Over both the Pacific and Atlantic oceans are large regions of great wind power; however, values decline sharply near the coast. Nevertheless, both coasts average 100–150 W/m^2 of wind power per year. The regions of lowest annual average wind power are the Southwest, the South, and the upper Great Lakes.

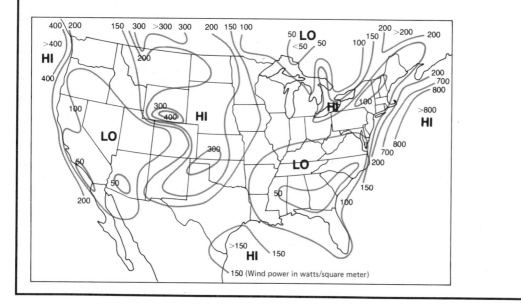

150 (Wind power in watts/square meter)

---

ent, though certainly less spectacular. Where wind moves through a constriction in a valley, an opening in a forest, or between buildings, it tends to speed up. As it moves over an isolated hill, wind at first accelerates and then decelerates on the downwind side. At this point flow can become turbulent, and large eddies may form which drift downwind in the wake of the hill. A train of such eddies, called a *Kármán*

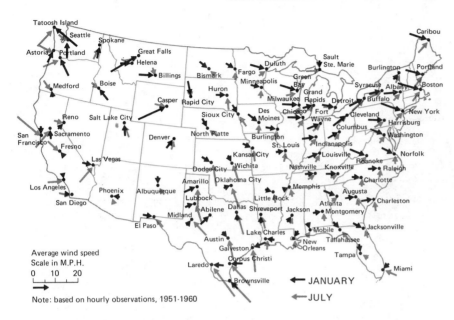

Average wind speed
Scale in M.P.H.
0   10   20

← JANUARY

← JULY

Note: based on hourly observations, 1951-1960

Fig. 30.7 Mean wind speeds and directions for January and July in the coterminous United States. The greatest seasonal changes in speed occur along the Texas Gulf and West coasts. (From Environmental Data Service, National Oceanic and Atmospheric Administration)

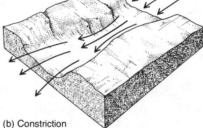

(a) Chinook

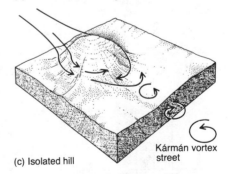

(b) Constriction

(c) Isolated hill

Kármán vortex street

Fig. 30.8 Topographic influences on airflow: (a) chinook setting, such as Casper, Wyoming; (b) constriction in a valley; (c) isolated hill with a Kármán vortex street.

vortex street (named for Theodor von Kármán, a pioneer in aerodynamic research), has been observed in the lee of some islands.

One of the most important influences of topography involves airflow over a ridge or a long hillslope. Such topographic rises form an obstacle to airflow; in order for the air in the boundary layer to maintain continuity of flow, velocity must increase as the air crosses the rise. This is portrayed in the pattern of streamlines (lines of equal wind velocity) as they approach and cross a slope. At some distance in front of the slope, the fast streamlines rise to form a large standing wave and then descend toward the crest slope (Fig. 30.9). Beneath the wave, near the foot of the slope, a low-velocity zone develops. Recent research has revealed that when a wind blows against a slope, a velocity increase of at least 100 percent occurs between the base and the crest of a slope. Surprisingly, the angle and height of the slope have relatively little influence on the maximum crest velocity relative to that of wind at the same height over flat ground; however, the minimum footslope velocity

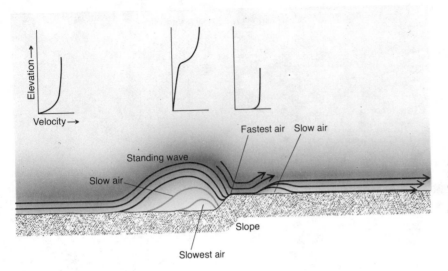

Fig. 30.9 *The approximate pattern of airflow over a slope. A large standing wave develops in front of the slope, from which fast wind descends on the upper slope. (After Bowen and Lindley, 1977)*

Fig. 30.10 *Lines of equal amplification of wind velocity over slopes of four inclinations. (From R. J. Astley et al., "The Effects of Some Hill Shapes on the Atmospheric Boundary Layer Near the Ground," New Zealand Meteorological Service Symposium on Meteorology and Energy, 1977)*

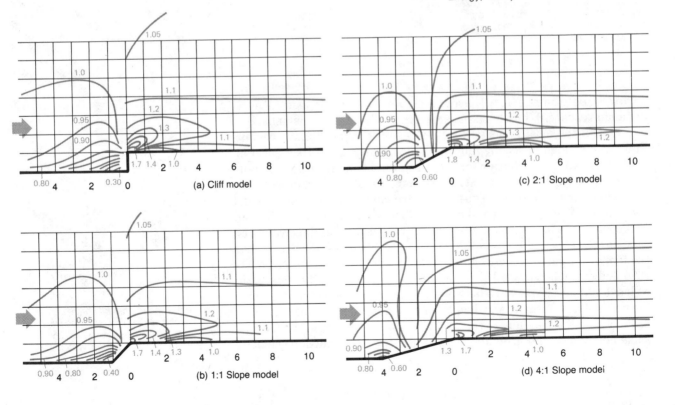

is much lower for steep slopes, especially for those steeper than 45 degrees (Fig. 30.10).

What happens to velocity in the area downwind of a slope? In general, it declines as the streamlines spread out. If the topography itself drops off sharply behind the slope, as along the back of a ridge or a sand dune, a calm zone may form. Wind-tunnel experiments show that the flow may actually undergo separation, resulting in the formation of a distinct low-velocity zone at some distance behind the crest slope.

The increase and decrease in wind velocity with topography are important to our understanding of the erosional and depositional effects of wind. They tell us that wind stress is greatest on windward slopes, particularly near the crests, and given that conditions on such slopes are conducive to erosion, that is where erosion will be greatest. By the same token, they tell us that wind stress is lightest on downwind slopes (called leeward slopes), and that is where deposition should take place. Evidence for this pattern can be seen in the formation of sand dunes, in the erosion of coastal slopes, and in the formation of snow cornices along mountain ridges (Fig. 30.11).

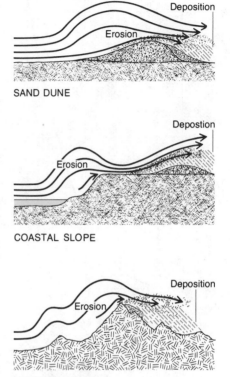

Fig. 30.11 *Examples of erosion and deposition associated with airflow over slopes.*

EROSION AND TRANSPORTATION BY WIND

The power of wind to erode is governed by two factors: air density and wind velocity. They are related in the following manner:

$$E = V^3 \rho,$$

where E = erosive power, V = velocity, and ρ = air density.

Air density has only a minor influence on the erosive force (shear stress) of wind. Moreover, it varies relatively little within several hundred meters of sea level. Velocity, on the other hand, is the principal determinant of erosive force. The point to underscore once again is that the power of moving air, which is nearly equal to the amount of shear stress exerted on the surface, varies with the third power of its velocity. Accordingly, a velocity increase from 2 m/sec to 4 m/sec produces an 8-fold increase in erosive power, whereas a velocity increase from 2 to 10 m/sec produces a 125-fold increase in erosional power. Thus fast winds are capable of doing remarkably more work than are slow winds. But let us not forget that slow winds occur much more frequently than fast ones do. Thus which winds do the most work in the long run in terms of sand movement is a more difficult question to answer. This problem is taken up in detail in Unit IX.

Sand Transport by Wind

Most of the credit for our present understanding of wind erosion and transportation of sand goes to British Brigader Ralph A. Bagnold, an

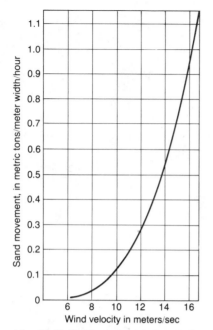

Fig. 30.12 Sand transport (across a 1-m line in one hour) relative to wind velocity. (From R. A. Bagnold, Physics of Blown Sand and Desert Dunes, *London: Chapman and Hall, 1973. Used by permission.)*

engineer and scientist stationed in the North Africa in the 1930s and 1940s. In his many experiments dealing with sand movement, Bagnold discovered most of the key principles governing sand erosion and transport. He found that the power/velocity relationship applies to the total amount of sand transported over a dune surface. Figure 30.12 shows this relationship.

The mechanisms of sand transportation by wind are very interesting and not difficult to understand. When a wind begins to rise over a sand surface, the first particle movements are discernible at a velocity of about 4.5 m/sec. The initial movement of particles is characterized by a rolling motion called *traction* (also called *creep*, which is not to be confused with soil creep). During a wind storm, sand moved in this fashion may account for 20–25 percent of the total sand transport. At the peak of a storm, traction is limited to coarser particles, which may include small pebbles.

As the wind speed picks up, grains are lifted into the air by gusts. The airborne grains travel in short trajectories, landing several centimeters downwind of their take-off points. When a grain strikes the surface, part of its momentum is transferred to another grain, and it in turn goes sailing off. This mode of transport, called *saltation*, accounts for 75–80 percent of the sand transport over sand dunes. Most sand grains travel within several centimeters of the surface, but very strong winds are able to lift grains as high as 2 m and to carry them 10 m or more downwind (Fig. 30.13).

Fig. 30.13 Sand transport by saltation over a sandy slope. (Photograph by William M. Marsh)

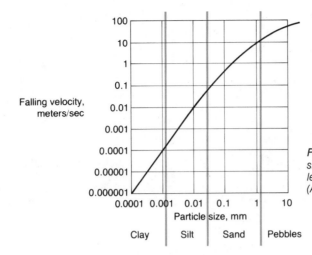

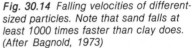

Fig. 30.14 Falling velocities of different-sized particles. Note that sand falls at least 1000 times faster than clay does. (After Bagnold, 1973)

Saltation is so effective a mode of sand transport because, first, air, which has a low density, offers little resistance to airborne missiles; second, the turbulence motion near the ground has a strong lifting component. Recall that although lifting currents occur very sporadically, on the average they constitute about one-fifth of the flow of air in terms of direction, the remainder being horizontal (Fig. 30.3). Once a particle is wafted into the air, it tends to settle back toward the surface, but can do so only if it is heavy enough to overcome the force of the uplifting currents. If a particle's terminal fall velocity is more than 20 percent of the velocity of the updraft, it cannot be held aloft. Note in Fig. 30.14 the great difference in the terminal settling velocities for sand, silt, and clay particles.

Selective Sorting of Particles

Light particles, principally silt and clay, which lack the necessary mass to settle rapidly out of moving air, are lifted well above the zone of saltation and during very strong storms may be blown thousands of meters into the air and hundreds of kilometers downwind. Therefore, on the basis of particle size (mass) alone, we can account for the separation, or sorting, of sand from pebbles and larger particles and silt and clay from sand. This helps to explain why sand dunes are made up exclusively of sand and why deposits of wind-blown silt, called *loess*, are located at great distances from areas of wind erosion (Fig. 30.15).

The magnitude of silt transportation by wind was documented in the 1930s when silt from the Dust Bowl in Kansas, Oklahoma, and

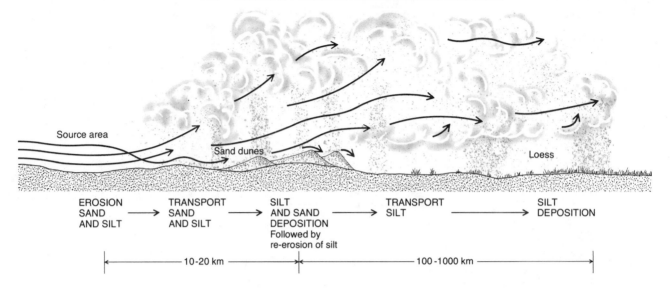

| EROSION SAND AND SILT | → | TRANSPORT SAND AND SILT | → | SILT AND SAND DEPOSITION Followed by re-erosion of silt | → | TRANSPORT SILT | → | SILT DEPOSITION |

Fig. 30.15 *The sequence of sand and silt deposition downwind from an erosion zone.*

Texas so darkened the sky that it elicited responses such as the following from a Texas schoolboy:

These storms were like rolling black smoke. We had to keep the lights on all day. We went to school with the headlights on and with dust masks on. I saw a woman who thought the world was coming to an end. She dropped down on her knees in the middle of Main Street in Amarillo and prayed outloud: "Dear Lord! Please give them a second chance."

Patrick Hughes, *American Weather Stories*, Washington, D.C.: U.S. Department of Commerce, 1976, 116 pp.

Sand Dune Formation

Sand dunes originate in aerodynamic environments that favor the deposition of sand. These are usually places where flow separates, giving rise to a zone of slow-moving air under much faster-moving air. Such zones form behind obstacles, on the leeward sides of slopes, or beneath standing waves (Fig. 30.16). As the fast air slides over the calm zone, saltating grains fall out of the air stream and accumulate on the ground.

Initially, the pile of sand that forms is stationary, but as it grows higher, it reaches faster and faster streamlines which move the sand across the top. Grains are moved from the windward to the leeward side and are piled up just over the crest. As the upper leeslope steepens to an angle around 33 degrees, the angle of repose for sand, it becomes

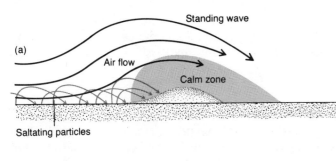

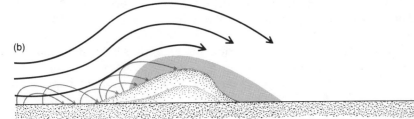

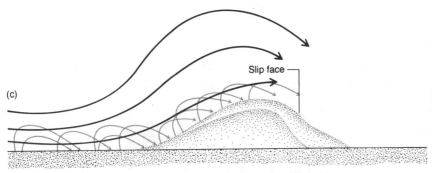

Fig. 30.16 *The formation of sand dunes: (a) saltating sand falls into low-velocity zone; (b) pile grows both vertically and horizontally, gradually filling low-velocity zone; (c) pile fills low-velocity zone, and streamlines are displaced upward, resulting in horizontal transport of sand over the crest of the pile. Low-velocity cone is not limited to lee slope.*

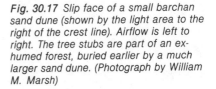

Fig. 30.17 *Slip face of a small barchan sand dune (shown by the light area to the right of the crest line). Airflow is left to right. The tree stubs are part of an exhumed forest, buried earlier by a much larger sand dune. (Photograph by William M. Marsh)*

unstable, and small slides break loose, moving sand to the lower part of the slope. The sand pile has now become a dune: a mobile heap of sand with a low-angle windward slope and a steep leeward slope, called the *slip face* (Fig. 30.17). The dune migrates over the ground as sand is eroded from one side and deposited on the other.

The height of a dune is limited by the velocity of the wind above ground level. As a dune grows, two conditions change: It reaches faster streamlines with higher elevation, and streamlines become focused near the crest with the development of the windward slope. At some point, therefore, vertical growth must cease, because wind stress becomes too great to allow further deposition. The maximum height is variable, but usually falls in the range of 10–25 m. The largest sand dunes in the world are found in Saudi Arabia and measure more than 200 m high. Actually, these are not individual dunes, but massive complexes of sand dunes which grow when smaller, fast-moving dunes migrate onto them.

Controls on Wind Erosion

Theoretically, every particle on the earth's surface is subject to wind stress, but whether the stress is sufficient to move it depends on the balance between the driving force of the wind and the resisting force of the particle. Since the driving force of wind is mainly a function of velocity, any control on velocity is a control on erosion, or the potentiality for erosion. At ground level, the roughness length is the critical control on wind velocity. The thickness of the laminar sublayer is determined largely by the roughness of the surface; the rougher the surface, the thicker the sublayer. Large boulders, trees, and buildings can raise the roughness length to 1–2 m. Shrubs produce a somewhat lower roughness length, and grasses produce one of only a few centimeters. Yet all of these materials produce virtually the same effect, because they raise effective frictional surface of the wind above the particles.

Fig. 30.18 Worldwide patterns of wind erosion.

Map projection by Waldo Tobler.

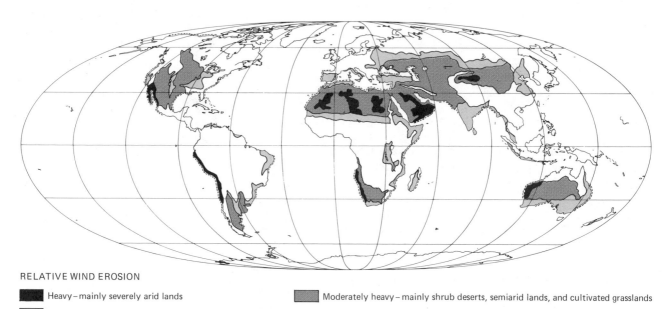

RELATIVE WIND EROSION

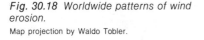 Heavy – mainly severely arid lands

Moderately heavy – mainly shrub deserts, semiarid lands, and cultivated grasslands

Moderate – mainly cultivated lands in forest, savanna, and parkland areas

Coast lines with appreciable wind erosion

The picture is quite different for smooth, unprotected ground. In particular, soils without a vegetative cover are highly susceptible to wind erosion, because the laminar sublayer is only millimeters thick, and wind stress is exerted more directly on individual particles. As a general rule, in fact, the first places to look for substantial amounts of wind erosion are areas with little or no vegetative cover. There are basically three such environments in the world: (1) the deserts; (2) environments with fresh deposits of sediments, such as beaches and alluvial fans; and (3) cultivated farmland (Fig. 30.18).

Wind erosion is not a certainty in these environments, however, because several factors other than the absence of vegetation play a part. Particle size is one of the most important factors. For any sized particle, a certain wind velocity, called the threshold velocity, is necessary to move it. The winds we experience on most days, which range between 2 and 5 m/sec (4–10 mph), are able to move only small particles, either fine sand (around 0.1 mm in diameter) or silt (0.05–0.002 mm in diameter). But for faster winds, the movable particle size is much larger. Overall, threshold velocity increases with the square root of particle size, where size is defined as particle diameter.

The influence of particle size is especially evident in some deserts where soils of mixed particle sizes have been eroded of fines, leaving only pebbles and larger particles (Fig. 30.19). These surfaces are called *desert lag* and sometimes resemble a cobblestone street when they become worn, polished, and cemented together by chemical residues. The threshold velocity required to move these lag particles is exceedingly large, probably one hundred meters per second or more. The fastest wind ever measured on the earth's surface was 103 m/sec (231 mph), recorded at the summit of Mt. Washington (2000 m elevation), New Hampshire, in April 1934.

Another factor that increases the resistance of particles to wind erosion is the cohesiveness of a soil material. Clay tends to be highly cohesive, of course, and therefore the threshold velocity required to move a clay particle is considerably greater than its small particle size suggests. In fact, a thin film of Portland cement particles requires a wind in excess of 10 m/sec to initiate movement, whereas medium-sized sand, which is not cohesive, requires a wind of only 5 m/sec to initiate movement.

Cohesiveness can also be provided by a cementing agent in a surface material. This is common to deserts, where calcium carbonate and other salts are deposited in interparticle voids and, upon hardening, bind together aggregates of particles. Ground frost can function in much the same way; unlike salts, however, crystals of ice can be sublimated away to dry winter air, thereby releasing the particles to the wind.

Fig. 30.19 A residue of pebbles left after the loss of sand to wind erosion. Such features are termed lag surfaces (Photograph by William M. Marsh)

Fig. 30.20 A rock face on which wind works in combination with runoff and various weathering processes to produce a striking microrelief. (Photograph by John R. Stacy, U.S. Geological Survey)

LANDFORMS PRODUCED BY WIND ACTION

Erosional Features

The effectiveness of wind as a primary erosional agent is difficult to ascertain. Compared with runoff, it probably produces little net erosion, but we have no idea what the relative differences between the two might be. On nonvegetated slopes, such as the one in Fig. 30.20, wind undoubtedly works in combination with weathering and mass movement to etch grooves and pits into the face of the sandstone cliff. The same can be said for many coastal slopes, where wind combines with waves, seepage, and runoff to erode and shape sea cliffs and backshore slopes. But how much work can be ascribed to wind directly is not possible with our current state of knowledge.

A small deflation hollow. (Photograph by William M. Marsh)

Unlike streams, wind can move sediment uphill as well as downhill. Where wind stress is focused on a spot in the landscape, it is possible, therefore, for wind to sculpt out a pit. In dune fields and sandy coastal areas, small cavities, called *deflation hollows*, are formed in this manner. They range from several to a hundred meters or so in diameter and may form in a matter of several days or seasons. Much larger depressions are found in the arid regions throughout the world, but most appear to be formed from faulting or from the collapse of limestone caverns. Some, however, appear to be eroded by wind.

Broad, shallow depressions (called *pans* in the Kalahari Desert of South Africa) are thought to be of aeolian origin. In the Lybian Desert of Egypt an impressive group of depressions shows strong evidence of wind erosion, although their formation was probably initiated by other processes. The largest of these, the Qattara Depression, covers around 15,000 km² (Fig. 30.21).

As we mentioned earlier, nonvegetated areas of sand- and silt-sized particles are most susceptible to wind erosion. Deserts contain the most expansive tracts of such areas, and, not surprisingly, this is where the largest surface areas of wind erosion are found. The sand eroded from such areas accumulates in large dune fields. Contrary to popular impressions, however, sand dunes are not the predominant landscape in the deserts; rather, rocky landscapes (called *hamadas* or *regs*), salt flats, and dry lake beds (called *playas*) together are more widespread. Nevertheless, there are huge tracts of sand dunes in most deserts.

Depositional Features

Sand dunes are the most prominent landforms created by wind action. The largest dune fields are found in the Middle East and North Africa and are so vast that they are called *sand seas*. One of the most spectacular sand seas, called the Rub-al-Khali, or the Empty Quarter, is located in the southern part of the Arabian Peninsula, where dunes cover about 400,000 km² of land. Another vast area of sand dunes is situated in the Lybian Desert of North Africa, about 350 km south of the Mediterranean Sea (see Fig. 30.21).

Fig. 30.21 The large depressions in the Lybian Desert of Egypt, along with the principal dune fields and the prevailing pattern of airflow. These depressions have been at least partially sculpted by wind.

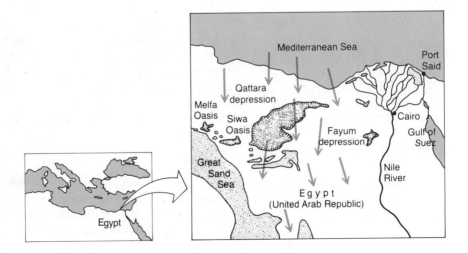

Fig. 30.22 *After a long battle against encroaching sand dunes, which involved all sorts of attempts to stop the sand, including the erection of the strange-looking board barrier on the right, the village of Biggs, Oregon, was abandoned in 1899. (Photograph by G. K. Gilbert, U.S. Geological Survey)*

Many of the large fields of sand dunes apparently function more or less as closed systems; once sand enters them, it does not leave. Although the dune fields may shift about considerably, especially with seasonal changes in wind direction, the net movement over many years is often insignificant. Thus most of the work of wind is secondary; that is, it is devoted to erosion and transportation of sand that has already been eroded by wind many times before. In some instances, however, the dune field may migrate great distances and end up at the sea or a river, where it loses part of its sand supply.

Settlements and agricultural land can be plagued by invasions of sand, and in some instances it is severe enough to destroy local land uses (Fig. 30.22). Some scholars point to dune encroachment as a contributing factor in the decline of some ancient cities in Asia and North Africa. Whether such behavior in sand dunes was related to a change in climate, such as increased aridity, is debatable, based on our present knowledge of dunes and climate.

Desert Dunes

Desert sand dunes are free-moving heaps of sand occurring in an amazing variety of forms. They are generally considered to be among the most beautiful landforms on earth and are often used as a setting or backdrop in literature and photography. The movement of desert dunes is unimpaired by vegetation, and this fact, combined with their geographic prevalence and aesthetic magnificence, suggest that the term "classical" dunes may be appropriate for them.

Although the particular conditions that give rise to desert dune forms are not well understood, it appears that sand supply and wind directions are important factors. A *barchan* is a cresent-shaped dune whose long axis is transverse to the principal sand-moving wind. The wings of the barchan are curved downwind, partially enclosing the slip face (Fig. 30.17). Barchans usually form where there is a limited supply of sand, reasonably level ground, and a fairly steady flow of wind from one direction.

As sand is added to a barchan, the wings may spawn smaller dunes, which migrate much more quickly than the parent dune does. The smaller barchans may, in turn, overtake and merge with a larger dune farther downwind. This seems to account for the tendency of barchans to migrate in schools, integrating and disintegrating into a variety of configurations as they move.

Where the sand supply is large, as in the sand seas, and formative winds come from more than one direction, sand dune forms are complex and not easy to classify. Among these are *seifs*, elongated dune masses that resemble great windrows of sand. Seifs lie parallel to the general direction of wind flow, but are clearly a product of two or more winds blowing oblique to their main axis. These winds blow sand back and forth across the axis of a dune, giving rise to a faceted topography on the flanks. Most of the gigantic dunes in the sand seas are seifs (Fig. 30.23).

Coastal Dunes

Outside of the deserts, the only appreciable areas of active sand dunes are found in coastal zones. Coastal dunes form wherever there is an ample supply of beach sand and strong onshore winds to blow it off the beach. In most cases, the beach must be broad and sufficiently agitated by wave action to keep it free of vegetation. Given these attributes, coastal sand dunes can form in any region, including the wet tropics, the humid midlatitudes, and the Arctic.

Most coastal dunes form in association with blowouts in beach ridges or berms. *Blowouts* are open-ended deflation hollows at the end of which a sand deposit builds up. As the deposit grows, it begins to migrate inland in the form of a sand dune. Because the deflation hollow serves to focus wind and sand on the middle of the dune, the flanks receive little activity and do not move much. As a result, they are suitable habitats for many beach plants, especially dune grass, sea oats, and sand cherry, which are tolerant to sand deposition. The plants add to

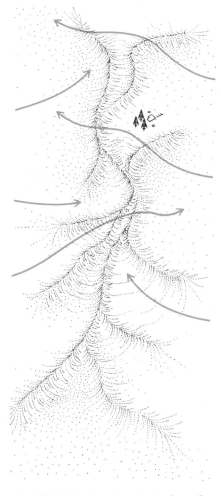

Fig. 30.23 *The formative winds of a seif flow at an angle to the main axis of the sand dune. (Illustration by William M. Marsh)*

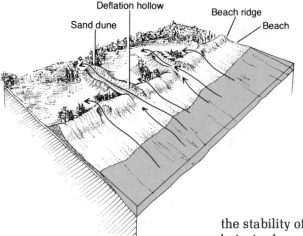

Sand dune
Deflation hollow
Beach ridge
Beach

Fig. 30.24 Setting and features associated with the formation of coastal sand dunes. The dune is the hairpin-shaped feature extending inland from the deflation hollow. (Illustration by William M. Marsh)

the stability of the flanks, and in time the dune takes on a parabolic or hairpin shape as the midsection moves farther inland (Fig. 30.24). With its stable or slow-moving wings, the coastal dune may resemble an inverted barchan. To the observer, the basic parabolic form may not be evident, though, because individual coastal dunes often coalesce to produce complex forms, as they are influenced by winds from other directions and vegetation of various types.

In addition to the differences above, coastal dunes are unlike desert dunes because they are at least partially covered by plants. Three types of vegetation are identifiable in most fields of coastal dunes: (1) vegetation being buried and killed by encroaching dunes (Fig. 30.17); (2) that such as dune grass (marram), which thrives with sand deposition and grows on leeward slopes (see Fig. 20.9); and (3) that which occupies erosional sites and helps to stabilize windward slopes (Fig. 30.25). The presence of vegetation substantially alters the dune environment by changing the patterns of airflow, raising the roughness length, and stabilizing leeslopes. Thus coastal dunes are not truly free-moving heaps of sand and in this regard are clearly in a different class from desert dunes.

Loess

Loess is the only other major deposit produced by wind. Though less prominent than sand dunes topographically, loess is very widespread and important in the landscape for its role in soil formation. Most loess appears to have been winnowed from Pleistocene glacial deposits, but the deserts also appear to have yielded appreciable amounts of the material.

Loess is principally silt, and when airborne it is carried in suspension. Observations indicate that loess can be dispersed high into the atmosphere and carried great distances by wind. This accounts for not only the widespread nature of loess deposits, but also the fact that unlike dune deposits, loess deposits tend to blanket the landscape, covering hills and valleys alike. Once deposited, however, loess is subject to erosion, especially by runoff, which may fragment the original distribution pattern.

Loess is the parent material for large areas of prairie and Chernozem soils and contributes significantly to the agricultural value of these soils. Soils formed in loess are friable, are often rich in minerals such as calcium carbonate, and possess good capillarity.

SUMMARY

Airflow over the ground is mainly turbulent, except for a thin layer, defined by the roughness length, immediately over the surface, where it is laminar. Roughness length, particle size and cohesiveness, vegetation, and wind velocity are the main controls on wind erosion. The power of wind increases with the cube of velocity; therefore, fast winds of short duration generally have a much greater capacity to do geomorphic work than do slow winds of long duration. Sediment transport by wind occurs in three modes: traction, saltation, and suspension. Sand is moved mainly by saltation, whereas silt and clay are moved in suspension. Sand dunes and loess are the principal wind deposits. The largest accumulations of sand dunes are found in the deserts, where they may form great sand seas; outside arid regions, sand dunes are most prevalent in coastal zones.

Fig. 30.25 An eroded dune ridge partially defended by vegetation. (Photograph by Charles Schlinger)

UNIT VIII SUMMARY

- Streams, waves and currents, glaciers, and wind are the principal geomorphic agents of the earth. These agents erode and transport rock debris from the continents, depositing it in the sea or in low spots on the land.

- Streams and rivers accomplish more geomorphic work than do other agents, a fact that has been generally accepted for only a century or so. Much of the credit for this knowledge goes to the geographers and geologists of American Western Surveys such as G. K. Gilbert and John Wesley Powell.

- Material is carried to streams by their tributaries and by hillslope processes. The rate of bed-load transport and the size of the particles moved increases with bed shear stress, which in turn varies with water depth and channel slope.

- The features of both river channels and valleys are products of a wide range of discharges, although those of valleys, such as floodplains, terraces, and oxbows, form over much longer time than do those of channels.

- The geographic cycle was the first of the big ideas about river-eroded landscapes; however, the notion of stagewise formation of valleys and landscapes is generally considered inadequate for most land masses. More modern ideas, such as the dynamic-equilibrium concept, appear to provide a better framework for thought on the formation of river-eroded landscapes.

- The sediment brought to the coasts by streams and rivers is redistributed by waves and currents. Most of it is deposited under water on the continental shelves and beyond, but some, particularly sand, goes into the building of beaches, bars, and spits.

- Waves are generated in several different ways, but those generated by wind do the most geomorphic work. The size of wind waves increases with wind velocity, wind duration, and fetch.

- In shallow water the motion at the wave base is exerted against the bottom, thereby setting up the potential for erosion. Near shore erosive power increases as the waves break and direct large amounts of stress against the bottom.

- Littoral transport is the movement of sediment in the coastal zone. Littoral drift is the sediment that is moved by waves and currents. Most drift is moved by longshore transport, which involves movement parallel to the coast.

- The mass of sediment that comprises any beach varies with not only net longshore transport, but also inputs and outputs by onshore-offshore transport, contributions from runoff, and losses from wind erosion.

- The erosion and retreat of shorelines constitutes one of the most serious coastal-management problems for property owners. Remedial action often involves the construction of groins which retard the rate of longshore transport and allow beaches to grow. In harbor entrances, on the other hand, sand bypassing must often be used to facilitate longshore transport.

- Glaciers are found today in mountainous regions and polar lands, but during the Pleistocene they extended far into the midlatitudes, covering large parts of North America and Eurasia.

- The growth and nourishment of glaciers depend on the mass balance between snowfall on one hand and snow and ice ablation on the other. Snowy winters and short, cool summers are conducive to the formation and maintenance of glaciers.

- Glaciers erode the land by action referred to as scouring and plucking. The debris carried in and on the ice is deposited near the glacier terminus as till or fluvioglacial material.

- Glacial landforms are diverse in both composition and morphology and generally have a strong influence on the development of soils, vegetation, and drainage, especially in recently glaciated landscapes. Continental glaciation also produced several indirect effects on the environment, including sea-level lowering, crustal depression and isostatic rebound, and loess deposition.

- The power of wind increases with the cube of velocity, but the effectiveness of wind to produce erosion is limited by roughness length, particle size and cohesiveness, and vegetation.

- Wind erosion is greatest in arid regions, cultivated lands, and environments where fresh deposits such as beaches are exposed to fast winds.

- Particle transport by wind occurs in three modes: traction, saltation, and suspension. Most sand is moved in saltation, whereas silt and clay are moved in suspension.

- Sand dunes are the most prominent landforms produced by wind action. Loess deposits, though less prominent as topographic features, are very important in the landscape, particularly for their role in soil formation.

FURTHER READING

Bagnold, R. A., *Physics of Blown Sand and Desert Dunes*, London: Chapman and Hall, 1973. *Originally published in 1941, this has become a classic in geomorphology.*

Embleton, C., and C. A. M. King, *Glacial and Periglacial Geomorphology*, London: Edward Arnold, 1968. *A synthesis of scientific findings concerning glaciers and glacial environments.*

Leopold, L. B., M. G. Wolman, and J. P. Miller, *Fluvial Processes and Geomorphology*, San Francisco: Freeman, 1964. *Generally considered the most authoritative treatment on the geomorphic effects of* running water. *Work in geomorphology and hydrology is a recommended prerequisite.*

Post, A., and E. R. LaChapelle, *Glacier Ice*, Toronto: University of Toronto Press, 1971. *A compilation of splendid photographs of glaciers and glacial environments.*

Ritter, D. F., *Process Geomorphology*, Dubuque, Iowa: William C. Brown, 1978. *A modern geomorphology text that summarizes the results of research based largely on numerical analysis of processes and landforms.*

REFERENCES AND BIBLIOGRAPHY

Bagnold, R. A., *Blown Sand and Desert Dunes*, London: Chapman and Hall, 1973.

Bascom, W. N., *Waves and Beaches*, Garden City, N.Y.: Doubleday, 1964.

Bowen, A. J., "The Generation of Longshore Currents on a Plane Beach," *Journal of Marine Research* **37** (1969): 206–215.

Bowen, A. J., and D. Lindley, "A Wind-Tunnel Investigation of the Wind Speed and Turbulence Characteristics Close to the Ground Over Various Escarpment Shapes," *Boundary-Layer Meteorology* **12** (1977): 259–771.

Brice, J. C., "Evolution of Meander Loops," *Geological Society of America Bulletin* **85** (1969): 581–586.

Chepil, W. S., and N. D. Woodruff, "The Physics of Wind Erosion and Its Control," *Advances in Agronomy* **15** (1963): 211–302.

Chorley, R. J., A. J. Dunn, and R. P. Beckinsale, *The History of the Study of Landforms*, Vol. 1, London: Methuen, 1964.

Davies, J. L., *Geographical Variation in Coastal Development*, London: Longman, 1977.

Einstein, H. A., "The Bedload Function for Sediment Transportation in Open Channel Flow," *U.S. Department of Agriculture Technical Bulletin 1026*, 1950.

Gilluly, J., A. C. Waters, and A. O. Woodford, *Principles of Geology*, San Francisco: Freeman, 1975.

Hack, J. T., "Interpretation of Erosional Topography in Humid Temperate Regions," *American Journal of Science*, Bradley Volume, 258-A (1960): 80–97.

————, "Dunes of the Western Navajo Country," *Geographical Review* **31** (1942): 240–263.

Hjulström, F., "Transportation of Detritus by Moving Water," in *Recent Marine Sediments: A Symposium*, ed. P. Trask, Tulsa, Oklahoma: American Association of Petroleum Geologists, 1939.

Inman, D. L., and B. M. Brush, "The Coastal Challenge," *Science* **181** (1973): 20–32.

Lewis, W. V., "A Melt-Water Hypothesis of Cirque Formation," *Geological Magazine* **75** (1938): 249–265.

Mackin, J. H., "Rational and Empirical Methods of Investigation in Geology," in *The Fabric of Geology*, ed. C. Albritton, Reading, Mass: Addison-Wesley, 1963.

Marcus, M. G., *Climate-Glacier Studies in the Juneau Ice Field Region, Alaska*, Chicago: University of Chicago, Department of Geography Research Paper 88, 1964.

Nye, J. F., "The Mechanics of Glacier Flow," *Journal of Glaciology* **2** (1952): 82–93.

Olson, J. S., "Lake Michigan Dune Development 1. Wind-Velocity Profiles," *Journal of Geology* (1958): 254–262.

Wahrhaftig, C., and A. Cox, "Rock Glaciers in the Alaska Range," *Geological Society of America Bulletin* **70** (1959): 383–436.

Wolman, M. G. "A Cycle of Sedimentation and Erosion in Urban River Channels," *Geografiska Annaler* **49-A** (1967): 385–395.

Wolman, M. G., and L. B. Leopold, "River Flood Plains; Some Observations on Their Formation," *U.S Geological Survey Professional Paper 282-C*, 1957.

Zenkovich, V., *Processes of Coastal Development*, trans. D. Fry, Edinburgh: Oliver and Boyd, 1967.

THE MAGNITUDE AND FREQUENCY OF LANDSCAPE PROCESSES

UNIT IX

KEY CONCEPTS   OVERVIEW

magnitude
frequency
event
recurrence interval
probability
projection
prediction
threshold force
shear stress
work
event combinations

Changes in the landscape are caused by a wide variety of physical processes: floods by streams, erosion by waves, deflation and deposition by wind, mass movements, and earthquakes. These processes can change the land only when their driving forces are stronger than the resisting forces, such as those of plant roots and sticky clay particles, which hold the land in place.

Neither the driving forces nor the resisting forces are constant over time for any place on the earth. Instead, the driving forces occasionally reach very high intensities, which in turn induce very high magnitudes in the geophysical processes. This is the case, for example, when runoff from a massive snow-melt drives a river to extraordinary flood levels. The amount of change that such levels of flow are able to render in the river valley, however, depends on more than the magnitude of flow alone, for the state of resisting forces in the valley materials is equally important. Whether the soil is frozen or thawed or barren or plant-covered at the time of the flood also controls how much soil is released in erosion.

This concept is called the magnitude-and-frequency concept, and it helps us to understand the nature of change in the landscape, including that which results in natural disasters. In Chapter 31 we shall examine how magnitudes and frequencies are determined. The problem of relating magnitude and frequency to landscape change and some of the effects of events of varying magnitude and frequencies on the earth's surface are examined in Chapter 32.

CHAPTER 31 **THE CONCEPT OF MAGNITUDE AND FREQUENCY**

INTRODUCTION

A favorite topic of folk fables concerns which kind of worker is more successful in accomplishing work: the steady plodder or the impetuous fireball. When pitted against each other, who will prove superior in total productivity? The answer is difficult for the behavioral scientist to ascertain. The parallel problem in nature is also a difficult one, but it is very important that we explore it in order to appreciate the relationship between behavioral patterns of geophysical processes and changes in the land.

ELEMENTS OF THE MAGNITUDE-AND-FREQUENCY CONCEPT

The processes that shape the earth's landscape operate over time at irregular frequencies and in highly variable magnitudes. As the magnitude of a process, such as river flow or ocean waves, increases, the frequency of its occurrence decreases. In other words, very powerful episodes of a process simply do not occur very often, whereas the gentle episodes occur often, in some cases almost continuously.

Magnitude

By *magnitude*, we mean the actual size or intensity of some aspect of a process, called a variable. For example, in a river we are usually concerned with the variable called discharge. This is a measure of the rate at which water flows past a point, such as a bridge, on a river. Discharge is expressed as the volume of water flow per time unit, such as cubic meters per second (m^3/sec) or per year (m^3/yr). In analyzing rivers we choose discharge as our significant variable because (1) the sediment-transport capability of a given river and (2) its capacity to alter vegetation, soil, and land use are related to its discharge. In the case of wind, on the other hand, we should be interested in speed and duration rather than discharge; with water waves, we would want to know wave height and velocity or period.

557

Frequency

We use the word *frequency* to express how often a given magnitude of a variable, termed an *event*, is equaled or exceeded. A high-magnitude event—such as a very high flow in a river, one that uproots trees and floods houses—occurs very seldom, perhaps once every 50 or 100 years. In contrast, a low-magnitude event might occur several times a year and produce a flow too small to create a flood. Other examples of high-magnitude low-frequency geophysical events are large, high-velocity water waves, called *tsunamis*, which are caused by earthquakes or volcanic eruptions on the ocean floor; great wind storms, including tornadoes and thunderstorms, which can damage vegetation and strip land of topsoil, silt, and sand; widespread glaze ice, which can devastate large areas of vegetation; avalanches and landslides, which may obliterate vegetation and land use; and large earthquakes, which can shake snow and soil down from slopes and collapse buildings.

The Importance of the Magnitude-and-Frequency Concept

One of the most meaningful questions facing modern physical geography and related earth sciences is: Which magnitudes and frequencies of geophysical processes* are most effective in shaping the landscape in different locales? In a river valley, for instance, is it the combined work of hundreds or thousands of small flows, or is it the work of one or two large flood flows that is most important in influencing the arrangement of vegetation, soils, and landforms in the valley? The question is compounded in both complexity and meaning when consideration is given to the effectiveness of various combinations of magnitudes and frequencies of different processes interacting in a particular landscape. In our river-valley example, if a killing forest fire or tree disease preceded a strong flood flow, the resultant changes in the valley's plant cover would likely be much different from those resulting if a strong flood flow had inundated a healthy forest.

Although a knowledge of the magnitude and frequency of a geophysical process can contribute to our understanding of the origins of many features in the landscape, it alone is insufficient to explain change, because not all magnitudes of a process everywhere produce the same results. That is, if we could isolate two events of a given energy level in two different settings, we would probably find measurable differences in the amount of work accomplished by each. Such dif-

---

*The term *geophysical* is used here in the broad sense and is meant to include processes conventionally referred to as geographic, physical geographic, geomorphic, geologic, hydrologic, atmospheric, and so on.

ferences occur because the resistance of the landscape to the force produced by the events is not equal in the two settings. Where the resistance of the landscape to a particular process is low, as is the resistance of nonvegetated soil to erosion by runoff or adobe buildings to tremors from earthquakes, a large amount of work may be accomplished, thereby rendering considerable change in the landscape. So it is not just the size of an event or how often it occurs that determines landscape change; also important is the relative resistance of the landscape to that event.

COMPUTING MAGNITUDE AND FREQUENCY: THE CASE OF RIVER FLOW

Recurrence of an Event

In order to evaluate analytically the importance of various magnitudes of events, we must express precisely their frequency of occurrence. In other words, we must determine the number of years which, *on the average*, separate events of a specific minimum magnitude. This is called the *recurrence interval*, also known as the *return period*.

A practical person might wish to know the recurrence interval of the minimum stream flow that overtops the banks of the stream near his or her house. If the recurrence interval is known, a simple probability can be calculated. For example, if the recurrence interval (t_r) of a given magnitude flood is fifty years, the *probability (p)* of occurrence is 0.02 in any given year:

$$p = \frac{1}{t_r} = \frac{1}{50} = 0.02.$$

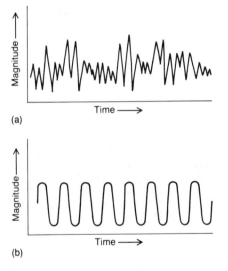

(a)

(b)

Fig. 31.1 *The nature of the frequency of noncyclical (a) and cyclical (b) processes.*

This flood, called the fifty-year flood, could ideally be expected to occur in 2 percent of the years over any period, e.g., a century. Given the opportunity to watch for the fifty-year flood over a period of, say, a thousand years, we might be surprised to find that few centuries would match the probability and produce exactly two 50-year floods. Some would produce four, others one or three, and some none. However, over many centuries the probability of two per century would be met.

It is important to realize that the term "recurrence" implies nothing cyclical about the event in question, whether a river flood or some other event. Cyclical events simply return to a given magnitude with a regular periodicity and are best exemplified in nature by biological processes such as the variations in plant respiration and growth in midlatitude environments (Fig. 31.1). River flow itself can be somewhat cyclical: High flows occur during the rainy (or snowmelt) season, which occurs around the same time every year. Except for this annual variation, however, there is no periodicity to the peak flows on a river. The magnitude of last year's peak does not influence the magnitude of this year's peak.

Accordingly, the fifty-year flood can occur in any year. The fact that it occurred last year does not mean that it will not occur again this year; the fact that a large number of dry years have passed does not mean that a big flood is more likely this year. The recurrence interval is simply the mean (statistical average) waiting time between events of a given magnitude. It can be used to determine the probability of occurrence of an event, but cannot be used for prediction. By definition, a prediction implies identification of the time and place of occurrence.

Recurrence intervals can be determined from records that give the magnitudes and times of events over a sizable time period. Without a data record of past events, there is no basis for determining the recurrence interval. Moreover, the longer the record, the better the basis for determining probabilities, especially for infrequent events.

Analysis of Magnitude and Frequency Based on Peak Annual Flow

Figure 31.2 shows the distribution of peak annual flows on the Eel River at Scotia, California, over the period 1911–1969. (Peak annual flow is the largest discharge recorded at a point on a river over the course of a year.) It is evident from Fig. 31.2, for instance, that only

Fig. 31.2 The distribution of peak annual flows on the Eel River at Scotia, California, 1911–1969. (Data from California Stream-flow Characteristics, through 1968, U.S. Geological Survey Open-File Report, 1971)

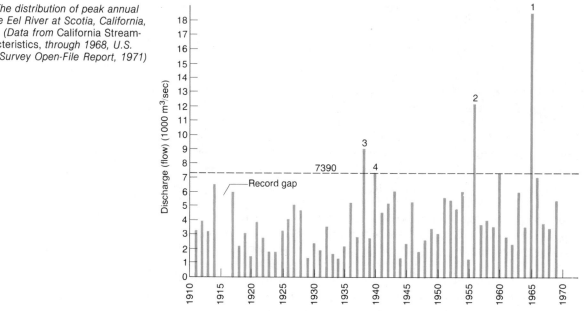

Table 31.1   Peak annual flows, Eel River at Scotia, California

YEAR	FLOW*	RANK	YEAR	FLOW	RANK	YEAR	FLOW	RANK
1911	3280	32	1932	3480	28	1951	5640	12
1912	3880	24	1933	1520	51	1952	5320	14
1913	3140	35	1934	1260	54	1953	4470	21
1914	6540	7	1935	2190	46	1954	6030	9
1917	6170	8	1936	5150	17	1955	1130	56
1918	1520	52	1937	2700	40	1956	12,300	2
1919	3060	36	1938	8950	3	1957	3680	26
1920	1470	53	1939	2730	39	1958	4930	19
1921	3910	23	1940	7390	4	1959	3310	31
1922	2830	38	1941	3510	27	1960	7389	5
1923	1960	47	1942	5210	16	1961	2680	41
1924	1760	50	1943	5890	11	1962	2320	44
1925	3430	30	1944	1230	55	1963	6000	10
1926	3990	22	1945	2450	43	1964	3450	29
1927	5070	18	1946	5270	15	1965	18,300	1
1928	4700	20	1947	1770	49	1966	7390	6
1929	1050	57	1948	2590	42	1967	3770	25
1930	2320	45	1949	3260	33	1968	3200	34
1931	1830	48	1950	2970	37	1969	5380	13

*Flow in cubic meters per second.

four flows were equal to or larger than 7390 cubic meters per second. The recurrence interval of this flow or a larger one is:

$$t_r = \frac{n + 1}{m},$$

where t_r is the recurrence interval in years, n is the total number of years or events, i.e., 57, and m is the rank of the year in question, i.e., fourth highest. Thus

$$t_r = \frac{57 + 1}{4} = \frac{58}{4} = 14.5 \text{ years.}$$

Thus a flood flow of 7390 cubic meters per second *or larger* magnitude has a recurrence interval of 14.5 years, based on records of fifty-seven years of peak annual discharge. The probability for this event's happening in any given year is:

$$p = \frac{1}{14.5} = 0.07.$$

To facilitate computation of the recurrence interval and event probability, the flow data are usually rank-ordered, that is, arranged in sequence from largest to smallest value. Table 31.1 lists the Eel River peak annual flow data, indicating ranks.

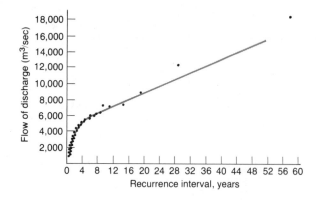

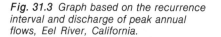

Fig. 31.3 Graph based on the recurrence interval and discharge of peak annual flows, Eel River, California.

Once rank-order is known, the recurrence interval of each event can be calculated. From these one can construct a graph, plotting recurrence intervals against their respective discharges. The curve of this graph—the generalized line drawn through these points—provides a means for estimating the flow magnitudes for various recurrence intervals or probabilities or, conversely, the recurrence intervals or probabilities for various flow magnitudes (Fig. 31.3). For example: (1) the peak annual discharge with a probability of 0.50, or two-year recurrence interval, is about 3500 m³/sec; the discharge with a probability of 0.10, or ten-year recurrence interval, is about 7000 m³/sec; (2) the recurrence interval for a peak annual discharge of 5000 m³/sec is 3.6 years, or a probability of 0.28 of occurring in any one year; the recurrence interval for a discharge of 10,000 m³/sec is twenty-five years, or a probability of 0.04 of occurring in any one year.

The Problem of Curved Distributions

In order to use the graph in Fig. 31.3 to make projections to estimate future flows and their probabilities, as well as to evaluate the meaning of the very largest flows, it is necessary to straighten the curve. This is the case with many types of magnitude-and-frequency data in physical geography. Curved graph lines indicate that the rate of increase of magnitude with recurrence interval is not arithmetically constant. If the graph line is not straightened, as you can see in Fig. 31.4, any number of lines can be projected from the curved line. But as tangents to the curve, these projections could all point in different directions, and none could be used with confidence. If, for example, the projection were made on the basis of the lower part of the curve, the estimates of very large flows (beyond the largest recorded flows) would be erroneously large. However, if the curve has been straightened, it can then be projected simply by ruling in a line beyond the largest flows.

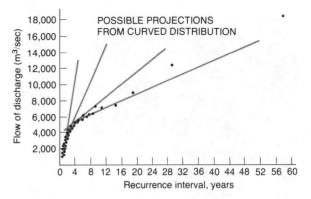

With the help of some mathematics, we can straighten the curve (Fig. 31.5). Now we are in a position to consider the problems originally posed: (1) the estimation of future flows and their probabilities; and (2) the meaning of the very largest flows. In the case of the largest flows note that the highest discharges, on the right-hand side of the graph, deviate considerably from the straight line, where the recurrence intervals are greater than twenty years. The last point, which represents the highest flood (18,300 m³/sec) recorded on the Eel River, has a

Fig. 31.4 This graph shows why it is impossible to plot one straight graph line from a curved distribution.

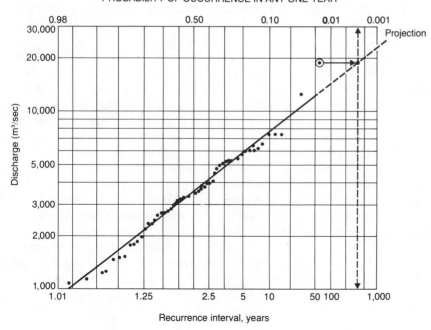

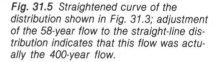

Fig. 31.5 Straightened curve of the distribution shown in Fig. 31.3; adjustment of the 58-year flow to the straight-line distribution indicates that this flow was actually the 400-year flow.

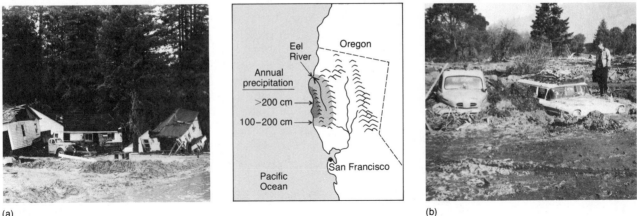

(a)   (b)

Fig. 31.6 (a) Destruction caused by the 1965 flood on the Eel River, which produced a record discharge of 18,300 m³/sec. (b) Flood damage in a neighboring river valley as a result of the same storm. (Photographs by George Porterfield and A. O. Waananen, U.S. Geological Survey)

recurrence interval of fifty-eight years and a probability of 0.017. However, since this flow deviates so markedly from the straight line of the graph, it is highly likely that it is representative not of the fifty-eight-year flood, but rather of a much larger one (Fig. 31.6). Since a flood of any magnitude can occur in any year, it appears that during the short period that records have been kept on the Eel River's flow, we recorded a flow much larger than the true fifty-eight-year flow. To place the 18,300 m³/sec flood in more accurate perspective, we may adjust its point on the graph to conform with the straight line by moving it to the right, as shown in Fig. 31.5(b). This indicates a probability of about 0.0025 for this flood, or a recurrence interval of 400 years rather than 58 years.

To check the accuracy of this interpretation, geoscientists often look for botanical evidence in the river floodplain to indicate the date of the last flow of a comparable size (Fig. 31.7). In the Blue River drainage basin in the north coastal drainage area of California (where the Eel River is located), Edward Helley and Valmore LaMarche of the U.S. Geological Survey examined an old redwood stump buried in a terrace deposit. The terrace was high enough above the river so that no recorded flood, except the flood of December 1964, had covered it. The buried stump was determined by radiocarbon dating to be about 400 years old, and the tree rings showed that the tree was 560 years old when the lower portion was buried. The indication is that a flood comparable in magnitude to the one produced by the great storms of December 1964 occurred about 400 years ago and had not occurred for at least 560 years previously. Thus the recurrence interval of the 18,300 m³/sec Eel River flow is shown by two lines of evidence to be about 400 years, not 58.

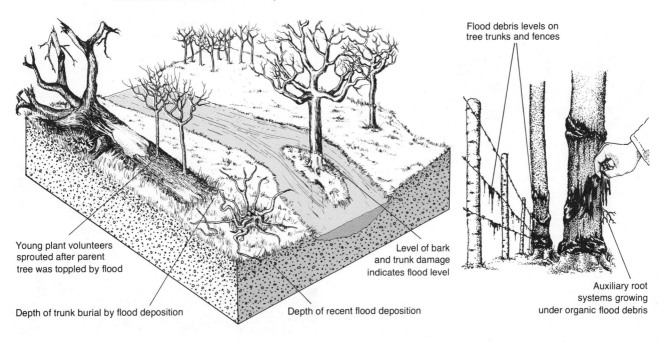

Flood debris levels on
tree trunks and fences

Young plant volunteers
sprouted after parent
tree was toppled by flood

Level of bark
and trunk damage
indicates flood level

Depth of trunk burial by flood deposition

Depth of recent flood deposition

Auxiliary root
systems growing
under organic flood debris

Projecting Future Events: Prediction versus Estimation

A projection of the curve beyond the largest recorded values shown on the graph in Fig. 31.5(b) can be useful for estimating even larger unrecorded flows. Since a projection is based on existing data representing a very short flow record compared with the total time that the river has been flowing, estimates must be interpreted with caution. Such projection estimates are *not* predictions, but only probabilities, and usually general ones at that. Although we do not mean to underrate the value of estimates based on curve projections, in this era of public bombardment by political, religious, and astrological "prediction," it is necessary to evaluate expectations of future flows as fairly as possible. This is necessary because, among other things, the climate of a river basin may change over the years, resulting in larger or smaller flows for a given frequency. In other words, the size of, say, a ten-year flow calculated on the basis of records from the early part of this century may indicate an erroneously small flow later in the century because climate has grown wetter in the basin.

A major change in the ground cover of a drainage basin can also alter flow rates. A case in point is a river basin, called Piper's Hole, in Newfoundland, Canada, whose forest cover was destroyed by fire in 1961. Both the magnitude and frequency of flood flows were greater

Fig. 31.7 Botanical indicators of former floods. (From William M. Marsh, Environmental Analysis for Land Use and Site Planning, *New York: McGraw-Hill, 1978. Used by permission.)*

AUTHORS' NOTE 31.1

Constriction of the Mississippi River and Its Effect on Flood Magnitude

"The 1973 deluge on the Mississippi River was reported by the news media as a 200-year flood, yet the discharge records show that the flow was only a 30-year event. At Saint Louis, Missouri, the flood began on 10 March and continued for seventy-seven consecutive days, exceeding the record set in 1844, when the river was in flood for fifty-eight days during the entire year. The river crested at Saint Louis on 28 April 1973 at a gage height of 13.18 m (4.03 m above flood stage) and a peak discharge of 24,100 m³/sec. This stage topped the 189-year record by 0.3 m and was 0.61 m higher than the 1844 crest; however, the discharge was about 35 percent less than the estimated peak flow in 1844. Compared to the 1908 flood, the 1973 flood had the same flow, but the peak stage was 2.51 m higher.

"Many investigators attribute higher stages of modern flows to a combination of levee confinement

The confluence of the Mississippi (right) and Missouri (upper left) rivers during the March 1973 flood. (Photograph by Charles B. Belt, Jr.)

for at least five years after the event (Fig. 31.8). Land-use change can create even more dramatic and longer-term effects. With the current trends toward both urbanization and agricultural development throughout much of the world, many river basins are producing increasingly larger discharges for a given storm as more and more land

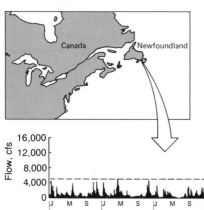

Fig. 31.8 Discharge from Piper's Hole River, Newfoundland, before and after a forest fire swept over the entire drainage basin. In the 6.5 years prior to the fire, no monthly flow exceeded 5000 cfs, whereas in the 4.5 years after the fire, eight flows exceeded 5000 cfs. (From S. I. Solomon, "Urbanization: Its Effects on the Hydrologic Regime," Urban Forum, Urban Research Council of Canada, 1975)

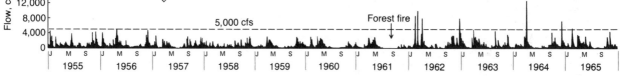

and channel constriction from navigation works such as wing dikes, side channel dikes, and revetments, which reduce the cross-sectional area of the channel. Tampering with the river started in 1837, when Lieutenant Robert E. Lee built the confinement dikes to remove sandbars threatening the Saint Louis harbor. Engineers narrowed the river from 1300 m in 1849 to 610 m in 1907 and finally to 580 m in 1969.

"A natural alluvial river generally widens its channel in response to a large flood, depending on the relative erodibility of its bed and banks. The width of an alluvial river channel over a long period of time is a function of average discharge, other hydrological and geomorphologic factors being equal. Between 1803 and 1860, the Mississippi widened itself mostly in response to four large floods. After 1881, it became more difficult for the river to widen itself, because of the navigation dikes and

the structural bank protection. The Mississippi responded by downcutting its bed, creating a deeper channel. This is reflected in the declining level of minimum and average annual discharges. But for peak annual flows, the trend has been toward higher stages.

"Since 1837, the channel at Saint Louis has lost about a third of its volume due to the installation of navigation works. As a result, during a flood on the thus modified Mississippi, the stages are higher for a given discharge. In some reaches of the river, deposition also occurs, causing a further rise in stages. Hemmed in by levees, flows that would otherwise be contained by the channel and lower floodplain are forced to higher elevations, creating more hazardous floods. Navigation works and levees make big floods out of moderate ones."

(Charles B. Belt, Jr., personal communication)

is devegetated and drained by ditches and storm sewers. In addition, some river channels have been structurally altered by the construction of navigation works and levees; as a result, the rivers no longer respond to a given discharge in the way they once did. In Authors' Note 31.1, geologist Charles B. Belt, Jr., of Saint Louis University discusses such a problem on the Mississippi River at Saint Louis, Missouri.

Despite these limitations, flood projections based on flow records are often the only source of information on which to base decisions concerning land use and engineering projects. How large must a bridge or culvert be in order not to obstruct the 50- or 100-year flow? What areas in a river valley are subject to flooding from the ten-year flow, and should they be zoned against residential development? Although magnitude-and-frequency projections cannot pinpoint the time and size of anticipated flows, they can narrow the range of possibilities and serve as the basis for regulatory policies concerning floodplain land use. The diagrams shown in Fig. 31.9 are from a government report on

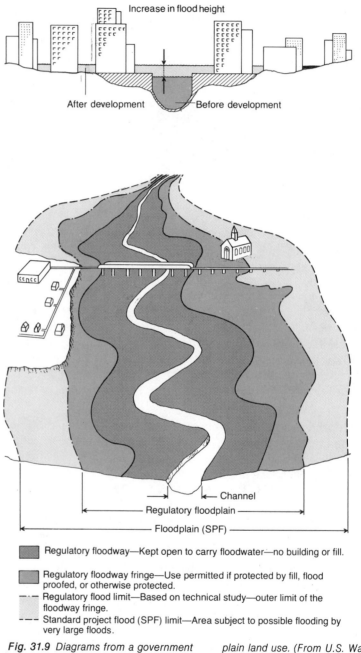

Fig. 31.9 *Diagrams from a government report showing how river-flow probabilities and projections are used as the basis for policy recommendations concerning flood-plain land use. (From U.S. Water Resources Council,* Regulation of Flood Hazard Areas, *Vol. 1; part 1, Fig. 6, 1971)*

regulation of flood hazard areas and are exemplary of the use of magnitude-and-frequency projections in today's society.

SUMMARY

The magnitude and frequency of geophysical forces are important concerns in physical geography because of the need to know which events produce the greatest work in the landscape. In general, powerful events occur infrequently, but exert great stress on the landscape and thus are capable of rendering great change. The frequency of an event, called the recurrence interval, can be determined from records of past events; in flood analysis this is based on peak annual discharge. Strictly speaking, prediction of an event is not possible, but the probability of an event can be estimated based on a projection of the curve representing the magnitude/recurrence interval relationship. Projections are normally limited to the short term because environmental change may render unreliable the records on which they are based.

MAGNITUDE AND FREQUENCY APPLIED TO THE LANDSCAPE

CHAPTER 32

THE STRESS-THRESHOLD CONCEPT

Throughout this book we have tried to show the relationship among driving forces, processes, and work in the landscape. At many points we also demonstrated that the rates at which the geophysical processes on the earth's surface perform work are not directly related to the amount of force driving them. For example, soil displacement in the form of mass movements such as landslides does not necessarily increase with shear stress until some level of stress is reached, at which point wholesale displacement takes place. The same holds for vegetation under various kinds of stress; change is negligible for stresses approaching a threshold level, but beyond that level, each increment of stress induces larger and larger amounts of change.

In the case of a river, the amount of sediment carried by a river and the erosion and deposition that take place in the river channel are not easily computed from a knowledge of the flow alone. One reason for this is that the resistance of materials over which streams flow varies greatly from place to place, and therefore the minimum force that must be reached before erosion will begin, called the *threshold force*, is not everywhere the same. This also holds true for wind, waves and currents, glaciers, and even human actions.

Figures 27.5 and 27.6 show the threshold force of moving water necessary to move particles of clay, silt, sand, and pebbles. A sand grain of a given size, for example, will not move with an initial application of force unless that force is sufficient to overcome the grain's inertial and frictional resistance. When the source of this force is a moving fluid, such as air or water, the physical relationship is very complicated, and the actual force transmitted to the grain depends on the velocity and turbulence of the fluid as well as the protective effect of surrounding grains. The amount of sediment transported by a river, then, cannot be easily expressed as an analytic function of the discharge.

We can, however, make a rough approximation of sediment transport based on discharge. Generally, sediment transport, or "work" done, is related to the force, or *shear stress*, exerted by the flowing water on the river bed. (How shear stress is measured is discussed in Chapter 27.) Shear stress is the force applied along the river channel by the moving water. Since the magnitude and frequency of shear stresses vary with the magnitude and frequency of flow, we can estimate the magnitude-frequency relationship of sediment transport from flow records.

Figure 32.1 shows the key relationships between shear stress and the magnitude and frequency of river flow. Figure 32.1(a) represents the relationship between the magnitude of an event and the shear stress it produces, but does not consider frequency. As you would expect, the bigger the event, the greater the shear stress. The opposite is considered in Fig. 32.1(b), where shear stress is expressed solely as a function of event frequency, and all events are assumed to produce the same shear stress. Since small events greatly outnumber large ones, the greatest amount of shear stress (cumulative over many years in this case) is associated with the small events. Clearly, the most realistic portrayal must combine the magnitude/stress relationship and the frequency/stress relationship, as is shown in Fig. 32.1(c). This graph shows that the moderately large events are responsible for the shear stresses that produce most of the sediment transportation in a river. Granted, the largest events usually generate the greatest stress and therefore erode and transport the greatest amount of sediment. Figure 27.9 illustrates this, inasmuch as the largest flow, at 1688 m³/sec, moved ten times more sediment per day than the smallest flow, at 186 m³/sec. However, very large events occur so infrequently that over long periods of time, the total work accomplished by them is not very great. Small events, on the other hand, occur with great frequency, but each generates so little shear stress that the combined work output is low. On balance, then, it is the moderately large events that accomplish the most work in the long run. Smaller than the largest events, but nonetheless large in comparison with an entire population of events, these events have a short enough return period to maximize their overall erosional and total work effectiveness.

The comparison of small, moderate, and large events is further illustrated through the following parable by M. Gordon Wolman of The Johns Hopkins University and the late John P. Miller of Harvard University.

A dwarf, a man and a huge giant are having a woodcutting contest. Because of metabolic peculiarities, individual chopping rates are roughly inverse to their size. The dwarf works steadily and is rarely seen to rest. However, his progress is slow, for even little trees take a long time, and there are many big ones which he cannot dent with his axe. The man is a strong fellow and a hard worker, but he takes a day off now and then. His vigorous and persistent labors are highly effective, but there are some trees that defy his best efforts. The giant is tremendously strong, but he spends most of his time sleeping.

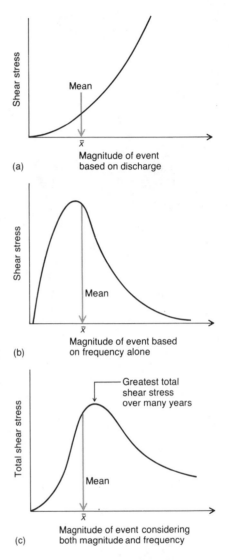

Fig. 32.1 *Three relationships between magnitude and frequency and shear stress, or work: (a) shear stress increases with the magnitude of the event; (b) if defined on the basis of frequency of occurrence alone, shear stress would be greatest for events with the greatest frequency, that is, the small flows; (c) a combination of (a) and (b), showing that total shear stress over a period of time including many flows is greatest for moderately large events.*

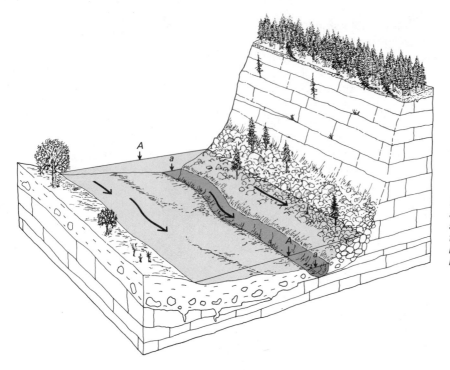

Fig. 32.2 *Stages of two magnitudes of stream flow, A and a. Note that the upper stage not only covers a larger area than the lower stage does, but also reaches the boulders at the foot of the cliff above the lower stage.*

Whenever he is on the job, his actions are frequently capricious. Sometimes he throws away his axe and dashes wildly into the woods, where he breaks the trees and pulls them up by the roots. On the rare occasions when he encounters a tree too big for him, he ominously mentions his family of brothers—all bigger, and stronger and sleepier.

From M. G. Wolman and J. P. Miller, ''Magnitude and Frequency of Forces in Geomorphic Processes,'' *Journal of Geology* **68**, 1 (1960). Reprinted by permission of the University of Chicago.

EVENT MAGNITUDE AND LANDFORMS

From the question of the relationship of magnitude and frequency to the work done by a process, we can go to the question of the relationship between the magnitude of an event and the landforms produced. Again, studies show that this relationship is not easily defined. High-magnitude flows in a river, for example, can produce landforms distinctly different from those of low-magnitude flows. This is related to two factors. First, high-magnitude flows cover a much larger sector of the river valley in terms of both elevation and area and therefore may render change in places that small events do not even reach. Second, high-magnitude flows can move large materials, such as boulders and trees, which have stress thresholds far greater than can be produced by low-magnitude flows (Fig. 32.2).

The effectiveness of a high-magnitude flow in shaping stream valleys in the Central Appalachian Mountains was dramatically illustrated in a study by John T. Hack of the U.S. Geological Survey and the late John C. Goodlett of The Johns Hopkins University. The following excerpt summarizes their observations:

The violent cloudburst flood of June 1949 that caused severe erosion in the Little River valley afforded an opportunity to study the importance of extremely low frequency floods as agents of erosion and as factors in forest ecology. During this storm, which probably lasted only a few hours, rainfall in excess of 9 inches fell on an area centered over Buck Mountain. The runoff produced dozens of debris avalanches on the upper mountain slopes, enlarged most of the channelways, and reworked the debris on the bottom lands of many larger valleys and in places removed the forest cover on the entire valley floor. The high rates of runoff were effective in eroding mountain slopes, sorting surficial debris, transporting debris, and producing terraces, alluvial fans, and cones. It is believed that such floods, though rare, occur frequently enough to exceed in importance, as erosive agents, all intervening lesser floods that do not damage the forest. The floods are also an important element in the life history of the forest; they provide open spaces for the growth of trees that require open sky, thus keeping the species composition in a state of flux.

From J. T. Hack and J. C. Goodlett, "Geomorphology and Forest Ecology of a Mountain Region in the Central Appalachians," U.S. Geological Survey Professional Paper 347, 1960.

Note that the two observers were led to believe that such floods, despite their infrequent occurrence, do occur frequently enough to exceed the work of all intervening floods in shaping the forest cover. Figure 32.3 shows the forest destroyed at one site as a result of this event.

On lakes and oceans, very large waves, resulting from landslides, volcanoes, earthquakes, or high-velocity long-duration winds may wash over a beach and erode a cliff that smaller waves never reach. As a result, the infrequent, large waves may produce distinctive coastal landforms as well as major alterations in vegetation and even land-use patterns. For example, Fig. 32.4 shows the result of a massive

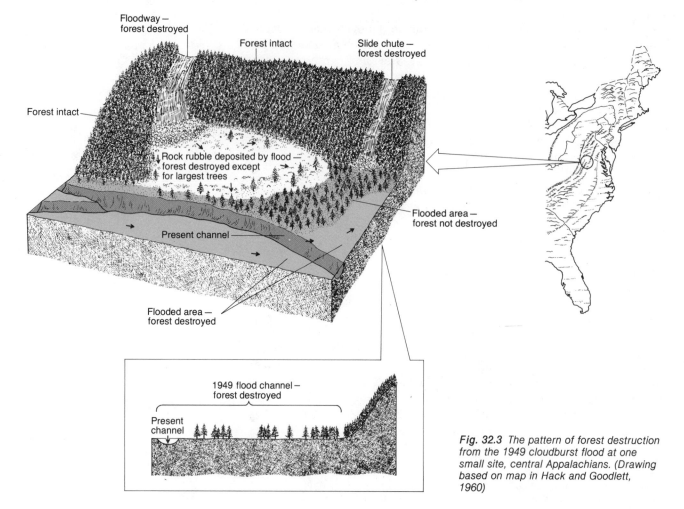

Fig. 32.3 *The pattern of forest destruction from the 1949 cloudburst flood at one small site, central Appalachians. (Drawing based on map in Hack and Goodlett, 1960)*

wave generated in Lituya Bay, Alaska, by an icefall from a glacier. Destruction of the forest cover in the broad, low areas near the mouth of the bay is a typical effect of a wave of this size.

THE MAGNITUDE-FREQUENCY CONCEPT AS APPLIED TO THE LANDSCAPE

This brings us to the magnitude-frequency concept as a means of understanding landscape and landscape change. The landforms, soils, and vegetative patterns on the earth's surface are products of processes which operate at different magnitudes and frequencies. Some of

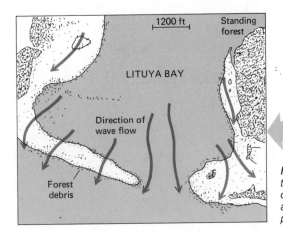

Fig. 32.4 *The area of forest destroyed by the Lituya Bay wave. Measurements indicated that this wave reached 1700 feet above sea level. (Drawn from aerial photograph)*

the most exciting questions in physical geography involve the explanation of the forms we see in terms of the magnitudes and frequencies of different processes we can observe and measure. It is important to remember that the landforms result, in part, from the variability in the rates at which the processes of river flow, waves, wind, rain, fires, mass movement, rock deformation, and vulcanism operate. A river, for example, is not like a canal carrying a regulated flow that varies little over time. Instead, a river channel and valley are a result of the entire range of flows—large ones occurring infrequently, smaller ones almost continuously—and associated events, such as landslides and mudflows, occurring on the valley sides.

Consideration of relationships between combined events such as slope failures and floods in a river valley adds another dimension to the problem. Although they may be driven or triggered by the same force, e.g., intensive rainfall, the mechanisms involved are different, and how one event influences the other is not well understood, at least analytically. In some cases one event may reduce the resistance of a part of landscape, such as vegetation, setting it up, as it were, for change by another event.

Shore Erosion and Magnitude-Frequency Relations

A coastline is the result of both small and large waves in addition to a multitude of other processes of various magnitudes and frequencies. For example, the recent high rates of erosion on the Great Lakes, although usually attributed to a single factor—namely, high water levels—are in some years actually the result of a set of processes whose magnitude-and-frequency relationships work in combination to maximize erosion. Specifically, the coincidence of high lake levels,

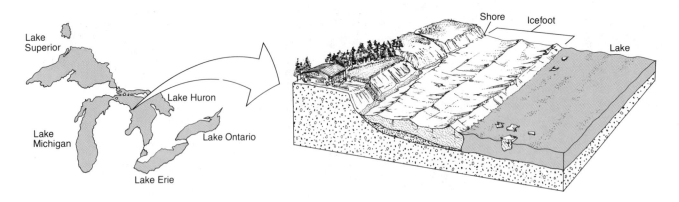

Fig. 32.5 The foot-ice formation that forms along most Great Lakes' shorelines each winter. (Illustration by William M. Marsh)

large storm waves, and weak coastal ice results in conditions conducive to a high rate of shore erosion.

Coastal ice, called foot ice, usually forms along the shorelines of most Great Lakes in late December and early January, remaining until April or May. During this time the shore is protected from the strong winter storm waves by the foot ice, as shown in Fig. 32.5. Should the foot ice form late in the season and should it have a weak and discontinuous form, the shoreline is left poorly protected and is thus susceptible to erosion by winter storm waves. This appears to have been the case throughout much of the Great Lakes in the 1972–1973 winter season, when water levels were 0.30 m or so above average. This increased the capacity of storm waves to penetrate across the shallow offshore zone, thereby further increasing their erosional effectiveness on the shoreline. Property owners reported that the 1972–1973 fall-winter period was one of the most severe periods of shoreline erosion on the Great Lakes in many decades. In sum, it appears that two high-magnitude sets of events combined with a lowmagnitude event to establish a condition in which sediment erosion and transport were maximized. This combination is shown schematically in Fig. 32.6.

Mass Movement and Magnitude-Frequency Relations

In many mountainous and hilly areas mass-movement processes account for most of the sediment delivered to stream channels. This is especially true in forested areas of undisturbed soils, where overland flow and surface erosion are generally negligible. Except for the gradual movement of material by soil creep, mass-movement processes usually involve sporadic movement of earth material, with large amounts moved over a very short time period. As these movements are a result of a complex set of environmental conditions, it will be instruc-

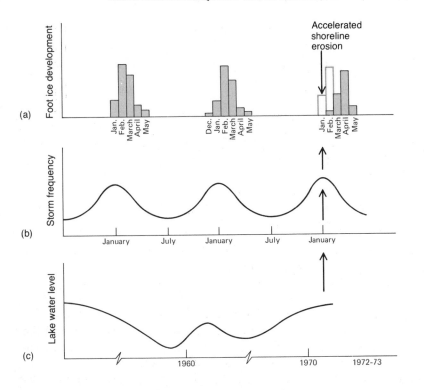

(a)

(b)

(c)

Fig. 32.6 *The combination of events that gave rise to the accelerated rates of shoreline erosion along the Great Lakes in the winter of 1972–1973. Two high-magnitude events—storm frequency (b) and lake level (c)—combined with a low-magnitude event—foot-ice development (a)—to produce the erosion.*

tive to examine what factors influence the magnitude and frequency of mass-movement events.

On any slope, there are always forces, called shear stress, which induce downslope movement of material, and opposing, or resisting forces, called shear strength, which hold the material in place. (These concepts are discussed in detail in Chapter 26.) Whenever shear stress exceeds shear strength, downslope movement will take place, and this condition can result from either an increase in shear stress or a decrease in shear strength or both.

Most landslides result from *combinations of events* which together cause the shear stress to exceed the resisting forces. For example, in forested areas two logging practices that may help to bring about this condition are clear-cutting and road construction. Clear-cutting reduces soil cohesion near the surface because the roots die soon after the trees are cut. The cuts and fills associated with road construction result in oversteepening of slopes, surface loading, and disruption of drainage, all of which can contribute to instability (Fig. 32.7).

The addition of water to a soil mass increases its susceptibility to slope failure. Water increases the shear stress on the material because it adds mass to it, but more important, it transforms clayey soils

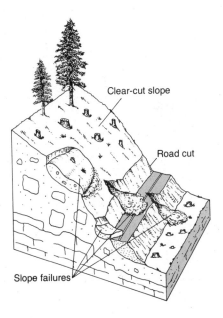

Fig. 32.7 *The types of slope failures associated with forest clear-cutting and road construction.*

Fig. 32.8 *Landslide scar on a clear-cut slope in the Cascade Mountains. A person is standing in the upper central part of the scar. (Photograph by F. J. Swanson)*

into plastics or near liquids and reduces the internal holding strength between soil particles, thus decreasing the soil's resistance to movement.

F. J. Swanson of the University of Oregon and C. T. Dyrness of the U.S. Forest Service examined the frequency of landslides in the H. J. Andrews Experimental Forest in the western Cascade Range in Oregon. Here they found that most of the slides took place in response to rainstorms; the great December 1964 rains, which caused the record flow on the Eel River, resulted in a record forty-three landslides. Swanson and Dyrness estimate that after clear-cutting, at least twelve years of vegetative growth are required to reestablish slope stability. If large rainstorms occur during this period, as is likely, much material is removed from the slopes by landslides (Fig. 32.8). Swanson and Dyrness estimated that the combined effects of clear-cutting and logging-road construction increased the amount of material removed by landslides by a factor of five over the natural state.

Wind Storms and Forest Distribution in New England

The magnitude and frequency of wind is easily observed from any location. In some places, however, the magnitudes of largest wind events are significantly greater than those in other places. In North America the southern Great Plains receive a high frequency of tornadoes, the most powerful wind storms known. Likewise, the Gulf Coast is subject to more hurricanes than are other locations.

If we trace the tracks of hurricanes, we find that many tend to follow the Atlantic coast northward from Florida. As they move along this path, the hurricanes tend to weaken, and only the strong ones reach New England with much force. Thus it appears that the magnitude and frequency of wind storms in the form of hurricanes decrease with distance northward along the Atlantic coast (Fig. 32.9).

In 1938 a very powerful hurricane reached New England and blew down large tracts of forest (Fig. 32.10). Hugh M. Raup of Harvard University, whose comments on the effects of changing lake levels on shore vegetation were presented in Authors' Note 20.2, argues that such storms have played an important role in shaping New England's forest patterns. From studies of the accounts of early settlers, Raup concludes that the precolonial forest had a patchwork distribution, characterized by patches, or stands, of forest, each comprised of essentially even-aged trees. Based on the observed patchwork pattern of forest destruction by powerful winds, there is ample reason to argue that hurricane winds have been an important agent in shaping New England forests. Coupled with forest fires, disease, and the diverse character of the New England terrain, the magnitude and frequency of

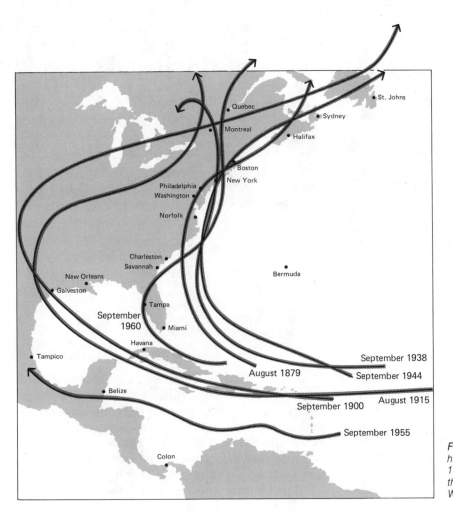

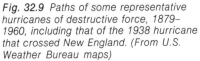

Fig. 32.9 Paths of some representative hurricanes of destructive force, 1879–1960, including that of the 1938 hurricane that crossed New England. (From U.S. Weather Bureau maps)

wind storms may have limited the New England forest from ever achieving the uniformity and stability of the climax community that some scientists think once existed.

SUMMARY

The forms and features we see in the landscape are the products of many processes, operating in different combinations and at various magnitudes and frequencies. In general, the work accomplished in-

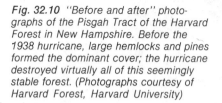

Fig. 32.10 *"Before and after" photographs of the Pisgah Tract of the Harvard Forest in New Hampshire. Before the 1938 hurricane, large hemlocks and pines formed the dominant cover; the hurricane destroyed virtually all of this seemingly stable forest. (Photographs courtesy of Harvard Forest, Harvard University)*

creases with the stress generated by a process, but the relationship may be complex and indirect. One reason for this is that the critical thresholds of the materials in the landscape vary greatly. In considering which events do the most work in the long run, it appears that the moderately large events are the most effective. The very large events occur too infrequently to do great amounts of work; however, the change they do render is notable because these events can move larger particles and reach farther into the landscape than can smaller events.

UNIT IX SUMMARY

- The processes that shape the landscape operate at uneven rates, and one of the most exciting questions in physical geography involves the relationship between the forms we see in the landscape and the magnitudes and frequencies of these processes.

- The magnitude of a process is a measure of its size or intensity; frequency is an expression of how often a given magnitude, termed an event, is equaled or exceeded.

- For a particular process, such as river flow or wave action, small events occur with high frequencies, whereas large events occur with low frequencies.

- Prediction of the magnitude and frequency of river flow or wave size, for example, is not possible based on our current knowledge of earth processes. Estimates of future events based on magnitude-and-frequency data from past events are possible, though, but they are limited to short time ranges. Such estimates are very important in engineering and land-use decisions.

- In general, it appears that intermediately large events, such as the annual or semiannual peak river flows, in the long run perform the most work in terms of sediment movement.

- The forms we see in the landscape are the result of an entire range of magnitudes, both small and large, produced by many different processes. Which magnitudes and frequencies of a process are responsible for doing the most work in shaping the land, however, is not easily answered, for two reasons: (1) total work cannot be directly related to the total stress produced by a process, because the resistance of earth materials to stress is uneven; (2) a combination of processes can produce a set of events, some of which produce high stress and some of which lower resistance to stress, and together these events can yield an exceptionally large amount of work.

- Although the very large events do not accomplish the greatest amount of work in the long run, they can produce rather distinctive changes in the landscape because they are able to move larger objects and extend farther over the landscape than smaller events can.

FURTHER READING

Burton, I., R. W. Kates, and G. F. White, *The Environment As Hazard*, New York: Oxford University Press, 1978. *A readable compilation of findings on human response to natural hazards.*

Hack, J. T., and J. C. Goodlett, "Geomorphology and Forest Ecology of a Mountain Region in the Central Appalachians," *U.S. Geological Survey Professional Paper 347*, 1960. *Results of an investigation into landforms, geomorphic processes, and vegetation which demonstrates that erosion and vegetation patterns can be explained on the basis of observable processes active today or in the recent past.*

Morgan, A. E., *Dams and Other Disasters*, Boston: Porter Sargent, 1971. *A deeply critical interpretation of the practices of the United States Army Corps of Engineers.*

Schuster, R. L., and R. J. Krizek, eds., *Landslides: Analysis and Control*, Washington, D.C.: National Academy of Sciences, 1978. *Discussion of the various aspects of landslide analysis, including detection, measurement, and stabilization.*

REFERENCES

Belt, C. B., Jr., "The 1973 Flood and Man's Constriction of the Mississippi River," *Science* **189** (1975): 681–684.

Bennett, H. H., "A Permanent Loss of New England: Soil Erosion Resulting from the Hurricane," *Geographical Review* **29** (1939): 196–204.

Bishop, D. M., and M. E. Stevens, "Landslides on Logged Areas in Southeast Alaska," *U.S. Forest Service Research Paper NOR-1* (1964), 57 pp.

Browing, J. M., "Catastrophic Rock Slides, Mount Huascaran, North-Central Peru, May 31, 1970, *American Association Petroleum Geologists Bulletin,* **57** (1973): 1335–1341.

Costa, J. E., "Response and Recovery of a Piedmont Watershed from Tropical Storm Agnes, June, 1972," *Water Resources Research* **10** (1974): 106–112.

Dyrness, C. T., "Mass-Soil Movements in the H. J. Andrews Experimental Forest," *U.S. Forest Service Research Paper PNW-42* (1967), 12 pp.

Fredrickson, R. L., "Christmas Storm Damage on the H. J. Andrews Experimental Forest," *U.S. Forest Service Research Note PNW-29* (1965), 11 pp.

Hack, J. T., and J. C. Goodlett, "Geomorphology and Forest Ecology of a Mountain Region in the Central Appalachians," *U.S. Geological Survey Professional Paper 347*, 1960.

Henry, J. D., and J. M. A. Swan, "Reconstructing Forest History from Live and Dead Plant Material—An Approach to the Study of Forest Succession in Southwest New Hampshire," *Ecology* **55**, 4 (1974): 772–783.

Marsh, W. M., B. D. Marsh, and J. Dozier, "The Geomorphic Influence of Lake Superior Icefoots," *American Journal of Science* **273** (1973): 48–64.

Rahn, P. H., "Lessons Learned from the June 9, 1972, Flood in Rapid City, South Dakota," *Bulletin of the Association of Engineering Geologists* **12** (1975): 83–97.

Raup, H. M., "Some Problems in Ecological Theory and Their Relation to Conservation," *Journal of Ecology*, British Ecological Society Jubilee Symposium **52** (supplement) (1964): 1928.

Richardson, D., "Glacial Outburst Floods in the Pacific Northwest," *U.S. Geological Survey Professional Paper 600-D* (1968): 79–86.

Sharpe, C. F. S., *Landslides and Related Phenomena*, New York: Columbia University Press, 1938.

Sigafoos, R. S., "Botanical Evidence of Floods and Floodplain Deposition," *U.S. Geological Survey Professional Paper 485-A* (1964).

Strahler, A. N., "Dynamic Basis of Geomorphology," *Bulletin of the Geological Society of America* **63** (1952): 923–938.

Swanson, F. J., and D. N. Swanston, "Complex Mass-Movement Terrains in the Western Cascade Range, Oregon," *Reviews in Engineering Geology,* Geological Society of America, Vol. III (1977): 113–124.

Trefethan, J. B., *The American Landscape: 1776–1976 Two Centuries of Change*, Washington, D.C.: Wildlife Management Institute, 1976.

Waanan, A. O., *et al.,* "Flood-Prone Areas and Land-Use Planning—Selected Examples from the San Francisco Bay Region, California," *U.S. Geological Survey Professional Paper 942* (1977).

Photograph by I.C. Russell, U.S. Geological Survey.

UNITS OF MEASUREMENT AND CONVERSION

CONVERSION FACTORS AND DECIMAL NOTATIONS

Energy, Power, Force, and Pressure

Energy Units and Their Equivalents

- *joule* (abbreviation J): 1 joule = 1 unit of force (a newton) applied over a distance of 1 meter = 0.239 calorie
- *calorie* (abbreviation cal): 1 calorie = heat needed to raise the temperature of 1 gram of water from 14.5°C to 15.5°C = 4.186 joules
- *British Thermal Unit* (abbreviation BTU): 1 BTU = heat needed to raise the temperature of 1 pound of water 1° Fahrenheit from 39.4° to 40.4°F = 252 calories = 1055 joules

Power

- *watt* (abbreviation W): 1 watt = 1 joule per second
- *horsepower* (abbreviation hp): 1 hp = 746 watts

Force and Pressure

- *newton* (abbreviation N): 1 newton = force needed to accelerate a 1-kilogram mass over a distance of 1 meter in 1 second squared
- *bar* (abbreviated b): 1 bar = pressure equivalent to 100,000 newtons on an area of 1 square meter
- *millibar* (abbreviation mb): 1 millibar = one-thousandth $(\frac{1}{1000})$ of a bar
- *pascal* (abbreviation Pa): 1 pascal = force exerted by 1 newton on an area of 1 square meter
- *atmosphere* (abbreviation Atmos.): 1 atmosphere = 14.7 pounds of pressure per square inch = 1013.2 millibars

Length, Area, and Volume

1 micrometer (μm) = 0.000001 meter = 0.0001 centimeter

1 millimeter (mm) = 0.03937 inch = 0.1 centimeter

1 centimeter (cm) = 0.39 inch = 0.01 meter

1 inch (in.) = 2.54 centimeters = 0.083 foot

1 foot (ft) = 0.3048 meter = 0.33 yard

1 yard (yd) = 0.9144 meter

1 meter (m) = 3.2808 feet = 1.0936 yards

1 kilometer (km) = 1000 meters = 0.6214 mile (statute) = 3281 feet

1 mile (statute) (mi) = 5280 feet = 1.6093 kilometers

1 mile (nautical) (mi) = 6076 feet = 1.8531 kilometers

Area

1 square centimeter (cm^2) = 0.0001 square meter = 0.15550 square inch

1 square inch (in.^{2}) = 0.0069 square foot = 6.452 square centimeters

1 square foot (ft^2) = 144 square inches = 0.0929 square meter

1 square yard (yd^2) = 9 square feet = 0.8361 square meter

1 square meter (m^2) = 1.1960 square yards = 10.764 square feet

1 acre (ac) = 43,560 square feet = 4046.95 square meters

1 hectare (ha) = 10,000 square meters = 2.471 acres

1 square kilometer (km^2) = 1,000,000 square meters = 0.3861 square mile

1 square mile (mi^2) = 640 acres = 2.590 square kilometers

Volume

1 cubic centimeter (cm^3) = 1000 cubic millimeters = 0.0610 cubic inch

1 cubic inch (in.^{3}) = 0.0069 cubic foot = 16.387 cubic centimeters

1 liter (l) = 1000 cubic centimeters = 1.0567 quarts

1 gallon (gal) = 4 quarts = 3.785 liters

1 cubic ft (ft^3) = 28.31 liters = 7.48 gallons = 0.02832 cubic meter

1 cubic yard (yd^3) = 27 cubic feet = 0.7646 cubic meter

1 cubic meter (m^3) = 35.314 cubic feet = 1.3079 cubic yards

1 acre-foot (ac-ft) = 43,560 cubic feet = 1234 cubic meters

Mass and Velocity

Mass (Weight)

1 gram (g) = 0.03527 ounce* = 15.43 grains

1 ounce (oz) = 28.3495 grams = 437.5 grains

1 pound (lb) = 16 ounces = 0.4536 kilogram

1 kilogram (kg) = 1000 grams = 2.205 pounds

1 ton* (ton) = 2000 pounds = 907 kilograms

---

*Avoirdupois, i.e., the customary system of weights and measures in most English-speaking countries.

1 tonne = 1000 kilograms = 2205 pounds

Velocity

1 meter per second (m/sec) = 2.237 miles per hour

1 km per hour (km/hr) = 27.78 centimeters per second

1 mile per hour (mph) = 0.4470 meter per second

1 knot (kt) = 1.151 miles per hour = 0.5144 meter/second

Quantities, Decimal Equivalents, and Scientific Notation

QUANTITY	DECIMAL NOTATION	SCIENTIFIC NOTATION		PREFIX
One trillion (U.S.)	1,000,000,000,000	10^{12}	T	tera-
One billion (U.S.)	1,000,000,000	10^{9}	G	giga-
One million	1,000,000	10^{6}	M	mega-
One thousand	1,000	10^{3}	k	kilo-
One hundred	100	10^{2}	h	hecto-
Ten	10	10	da	deka-
One tenth	0.1	10^{-1}	d	deci-
One hundredth	0.01	10^{-2}	c	centi-
One thousandth	0.001	10^{-3}	m	milli-
One millionth	0.000001	10^{-6}	μ	micro-
One billionth (U.S.)	0.000000001	10^{-9}	n	nano-
One trillionth (U.S.)	0.000000000001	10^{-12}	p	pico-

COMMON CONVERSION PROBLEMS

Energy

- calories to joules; e.g., 900 cal = ? joules
  solution:
    $900 \times 4.186 = 3767.4$ joules

- BTUs to joules; e.g., 252 BTUs = ? joules
  solution:
    $252 \times 1055 = 265,860$ joules

Energy Flux

- cal/cm² • min to joules/m² • s; e.g., 1.31 cal/cm² • min to ? J/cm² • sec
  solution:

$$\frac{1.31 \times 4.186}{60} \times 100^2 = 913.94 \text{ J/m}^2 \cdot \text{s}$$

- Kcal/cm² • year to joules/m² • day; e.g., 107 Kcal/cm² • year to ? J/m² • day
  solution:

$$\frac{107{,}000 \times 4.186}{365} \times 100^2 = 12{,}271{,}287 \text{ J/m}^2 \cdot \text{day}$$

- BTU/ft² • hour to cal/cm² • min; e.g., 4.7 BTU/ft² • hr to ? cal/cm² • min
  solution:

$$\left(\frac{\frac{4.7 \times 252}{60}}{6.452 \times 144}\right) = 0.02 \text{ cal/cm}^2 \cdot \text{min}$$

Length/Distance

- inches to centimeters; e.g., 12 inches = ? cm
  solution:
  12 × 2.54 = 30.5 cm

- feet to meters; e.g., 712 feet = ? meters
  solution:

$$\frac{712}{3.2808} = 217 \text{ m}$$

- centimeters to meters; e.g., 152 cm = ? meters
  solution:

$$\frac{152}{100} = 1.5 \text{ m}$$

- yards to meters; e.g., 35 yards = ? meters
  solution:
  35 × 0.9144 = 32 m

Area

- square inches to square centimeters; e.g., 10 in.² = ? square centimeters
  solution:
  10 × 6.452 = 64.5 cm²

- square centimeters to square meters; e.g., 137 cm² = ? square meters
  solution:
  137 × 0.0001 = 0.014 m²

- square feet to square meters; e.g., 427 ft² = ? square meters
  solution:
  427 × 0.0929 = 39.67 m²

- acres to square kilometers; e.g., 192 ac = ? square kilometers
  solution:

$$\frac{192 \times 4046.950}{1{,}000{,}000} = 0.78 \text{ km}^2$$

- square miles to square kilometers; e.g., 19.9 mi^2 = ? square kilometers
  solution:
    19.9 × 2.592 = 51.58 km^2

Volume

- cubic feet to cubic meters; e.g., 7100 ft^3 = ? cubic meters
  solution:
    7100 × 0.02832 = 201.07 m^3
- gallons to cubic feet; e.g., 510 gallons = ? cubic feet
  solution:
    $$\frac{510}{7.48} = 68.2 \text{ cubic feet}$$
- gallons to liters; e.g., 23 gal = ? liters
  solution:
    23 × 3.785 = 87.1 liters
- acre-feet to cubic meters; e.g., 97 ac-ft = ? cubic meters
  solution:
    97 × 4046.95 = 392,554.2 cubic meters

Mass

- ounces to grams; e.g., 31 ounces = ? grams
  solution:
    31 × 28.3495 = 878.83 grams
- pounds to kilograms; e.g., 109 pounds = ? kilograms
  solution:
    109 × 2.205 = 240.4 kilograms

Temperature

- Fahrenheit to Celsius; e.g., 52°F = ?°C
  solution:
    $$\frac{(52° - 32°)}{1.8} = 11.1°C$$
- Celsius to Fahrenheit; e.g., 19°C = ?°F
  solution:
    (1.8 × 19) + 32 = 66.2°F
- Celsius to Kelvin; e.g., 5°C = ?°K
  solution:
    5° + 273.15 = 278.15°K

COMPUTING THE SOIL-MOISTURE BALANCE, USING THE THORN-THWAITE METHOD

APPENDIX 2

In order to determine the soil-moisture balance for a location, it is necessary to know: (1) the mean monthly temperatures; and (2) the soil-moisture holding capacity, i.e., the amount of water held in the soil-root zone at field capacity. According to the Thornthwaite method, ten basic steps are involved, and each is carried out for each month of the year:

1. *Heat index:* The first step involves calculating the heat index for each month. The heat index is based on the long-term mean monthly temperature, and the monthly values (*i*) can be computed, using this formula:

$$i = (\frac{T}{5})^{1.514},$$

or it can be read from the graph in Fig. A.2.1. Once the monthly values are known, they in turn are summed for the year to obtain the annual heat index (*I*).

2. *Unadjusted potential evapotranspiration:* The second step is to estimate unadjusted PE, based on the annual heat index (*I*) and mean monthly temperature (*T*). This is figured on a monthly basis also, and for months with average temperatures between 0° and 25°C, the graph in Fig. A.2.2 can be used to approximate the unadjusted PE. For months with average temperatures of 0°C or lower, unadjusted PE is 0. For months with average temperatures above 25°C, unadjusted PE can be read from the graph in Fig. A.2.3, which is based on average monthly temperature.

3. *Adjusted potential evapotranspiration:* The third step is to correct the unadjusted PE values to account for variations in the length of daylight hours in each month. The correction factor for each month is given according to zone of latitude (Table A.2.1). The unadjusted PE is multiplied by this factor to obtain the adjusted PE. For all locations poleward of the limits given in Table A.2.1, use the values for 60°N or 50°S.

4. *Precipitation minus evapotranspiration:* Using the adjusted PE values and the mean monthly precipitation data, the fourth step is to subtract PE from P for each month.

5. *Accumulative potential water loss:* The fifth step involves summing the potential soil-moisture loss for the months with negative

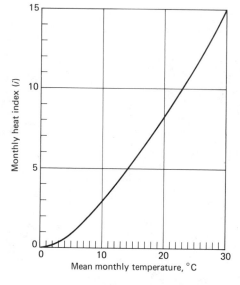

Fig. A.2.1 Graph for obtaining monthly heat index, based on mean monthly temperature (i).

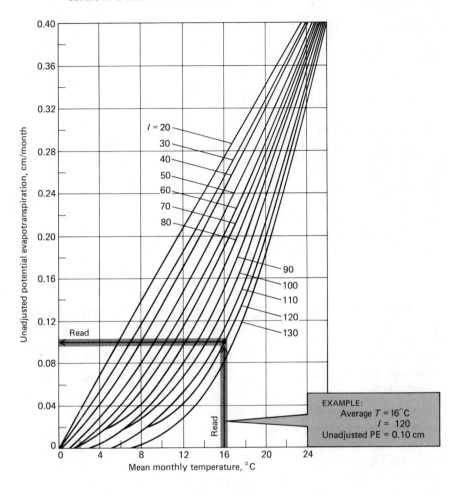

EXAMPLE:
Average $T = 16°C$
$I = 120$
Unadjusted PE = 0.10 cm

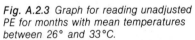

Fig. A.2.2 Curves for obtaining unadjusted PE, based on mean monthly temperature (T) and annual heat index (I).

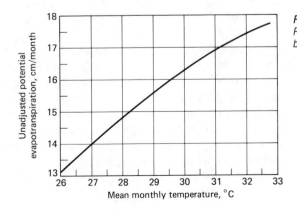

Fig. A.2.3 Graph for reading unadjusted PE for months with mean temperatures between 26° and 33°C.

Table A.2.1 Daylight correction factors to obtain adjusted PE.

LATITUDE	JAN.	FEB.	MAR.	APR.	MAY	JUNE	JULY	AUG.	SEPT.	OCT.	NOV.	DEC.
60°N	0.54	0.67	0.97	1.19	1.33	1.56	1.55	1.33	1.07	0.84	0.58	0.48
50°N	0.71	0.84	0.98	1.14	1.28	1.36	1.33	1.21	1.06	0.90	0.76	0.68
40°N	0.80	0.89	0.99	1.10	1.20	1.25	1.23	1.15	1.04	0.93	0.83	0.78
30°N	0.87	0.93	1.00	1.07	1.14	1.17	1.16	1.11	1.03	0.96	0.89	0.85
20°N	0.92	0.96	1.00	1.05	1.09	1.11	1.10	1.07	1.02	0.98	0.93	0.91
10°N	0.97	0.98	1.00	1.03	1.05	1.06	1.05	1.04	1.02	0.99	0.97	0.96
0	1.00	1.00	1.00	1.00	1.00	1.00	1.00	1.00	1.00	1.00	1.00	1.00
10°S	1.05	1.04	1.02	0.99	0.97	0.96	0.97	0.98	1.00	1.03	1.05	1.06
20°S	1.10	1.07	1.02	0.98	0.93	0.91	0.92	0.96	1.00	1.05	1.09	1.11
30°S	1.16	1.11	1.03	0.96	0.89	0.85	0.87	0.93	1.00	1.07	1.14	1.17
40°S	1.23	1.15	1.04	0.93	0.83	0.78	0.80	0.89	0.99	1.10	1.20	1.25
50°S	1.33	1.19	1.05	0.89	0.75	0.68	0.70	0.82	0.97	1.13	1.27	1.36

P − PE values. Potential water loss represents the maximum possible water loss from the soil when PE is greater than P; it is equal to the P − PE value. Beginning with the first negative month, each month's deficiency is added to the next. For dry areas, where the annual P − PE value is negative, the monthly value used to begin the summation must be approximated.

6. *Soil-moisture storage:* Since the accumulative potential water-loss values do not take into consideration the true response of the soil in giving up moisture, a more accurate value of soil-water loss must be obtained. The sixth step takes into account the lag in the transfer of capillary water from the soil-moisture reservoir to evaporation and transpiration. The curves in Fig. A.2.4 give approximate soil-moisture storage values based on the monthly accumulative potential soil-moisture loss. The curves are for soils with 10-cm, 20-cm, and 30-cm soil-moisture holding capacity. The value obtained represents the moisture remaining in the soil at the end of the month in question.

After the storage-water values have been determined for the negative months, the storage-water values for the positive months are determined by adding the P − PE value to each month's storage water. Beginning with the first positive month, this is carried out for each succeeding month, until the soil-moisture storage is brought up to the holding capacity. The time taken to reach the holding capacity, or the storage-water value closest to it, is the soil-moisture recharge period. For locations that achieve full recharge before the end of the positive months, the P − PE value goes to surplus water.

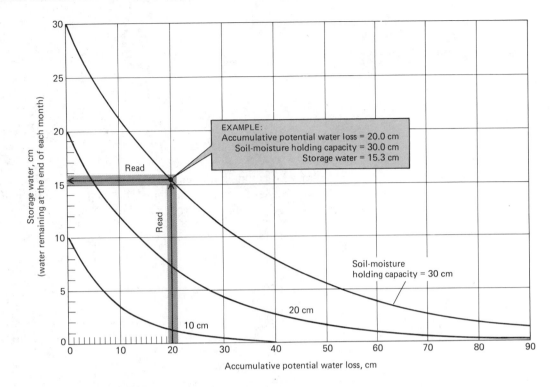

Fig. A.2.4 Curves for obtaining soil-moisture storage values for negative months.

7. *Change in soil moisture:* The seventh step, designed to set up the eighth step, calls for tabulating the change in soil moisture (storage water) from month to month and indicating whether the moisture has been gained or lost.

8. *Actual evapotranspiration:* The eighth step calls for the true amount of moisture loss, as opposed to the potential loss. Actual evapotranspiration (AE) is determined for the negative months by adding the value for change in soil moisture to the precipitation value; for the positive months, AE is equal to PE.

9. *Moisture deficit:* This represents the difference between the water demand for evaporation and transpiration and the amount actually given up. Moisture deficit is equal to PE − AE.

10. *Moisture surplus:* Following the soil-moisture recharge period, any precipitation exceeding AE (or PE) is surplus water. Moisture surplus is equal to the P − PE value for months when the soil is at moisture capacity. This water is available for runoff, although adjustments must be made for watershed detention and the storage and melting of snow.

MAPS AND ~~APPENDIX 3~~ MAP READING

INTRODUCTION

Maps are models, or miniature replicas, of the landscape. In order to use maps effectively, one must be familiar with some of their basic properties, such as direction, location, and scale.

DIRECTION

In order to be read, a map must be oriented, that is, placed in its correct relation to the earth. This is a simple matter; for essentially all maps, north is at the top of the sheet; south, at the bottom; east, at right; west, at left. Practically all of the maps in this book are oriented with north at the top. The orientation is shown by an arrow or similar symbol pointing north. On some maps two arrows are shown: one pointing to true north, and one to magnetic north; the map should be oriented to true north.

LOCATION

Location is a second important consideration in map reading. Several standard location and grid systems are used for the identification of an area covered by a map. The principal geographic coordinate system consists of a rectilinear network of orthogonally intersecting lines. This network is comprised of parallels and meridians, which are designated in degrees, minutes, and seconds, all of which are described in Chapter 5.

To describe accurately the location of a map or the places on a map, it is necessary to be able to read and interpret the coordinate system referenced on the map. Throughout much of North America, the system referenced on maps is the *township and range* system. Originally devised for the division of the landscape in the old Northwest Territories of the United States, the township and range system has since been extended to the majority of the United States as well as to northern Ontario and the western provinces of Canada. This modified rectangular grid is based on a set of selected meridians, termed *principal meridians,* and parallels, called *baselines,* which intersect at an initial point (Fig. A.3.1a).

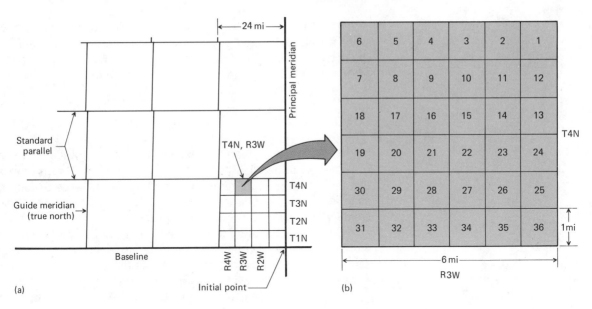

(a)

(b)

Distances are measured in the four cardinal directions from the initial point, and locations are identified at twenty-four-mile intervals along the baseline and principal meridian. However, due to the earth's shape, meridians converge toward the poles, making it impossible to fit an exact square to the earth's surface. Consequently, the ideal planimetric grid that forms the basis of the township and range system is intentionally and necessarily distorted at certain points in order to conform to the earth's curvature.

Within each set of twenty-four-mile-wide strips, six-mile strips are defined. The strips oriented east-west are defined by the parallels and are termed *townships;* those oriented north-south and bounded by the principal guide meridians are termed *ranges.* Each township and range strip is assigned a number to indicate its position vis-a-vis the initial point. Thus each small square, generally referred to simply as a township and measuring six miles on a side, is easily identified by a notation such as T4N, R3W ("township 4 north, range 3 west"). This notation identifies the township that is formed by the convergence of the fourth township strip north of the baseline and the third range strip west of the principal meridian (Fig. A.3.1a).

Every township is subdivided into thirty-six units, termed *sections,* each measuring one mile on a side. Each section within a township is given a number designation, beginning with section 1 in the northeast corner and proceeding sequentially westward to section 6, then dropping down to the next tier and proceeding back to the east, and so forth, as shown in Fig. A.3.1(b).

Fig. A.3.1(a) The basic layout of the township and range system; (b) the subdivision of a township into sections.

The errors due to the convergence of the meridians toward the poles are accumulated along the eastern and northern column and row in each township. Thus sections 1, 2, 3, 4, 5, 6, 7, 18, 19, 30, and 31 are often a fraction less than 640 acres (one square mile) in area.

SCALE

Scale is defined as the relationship between distance on the map and the corresponding distance on the earth's surface. Maps of small areas are called large-scale maps, whereas those of large areas are called small-scale maps. The level of detail on a map varies with scale; the smaller the scale, the less detail possible.

Scale is generally indicated on a map either graphically or arithmetically, and occasionally both are included as part of the map legend. The simplest scale indicator employed is the graphic, or bar, scale. This consists of an actual line or bar calibrated to indicate a precise map distance and labeled to indicate the corresponding ground distance. Any linear measurement on the map can be compared directly to the bar scale to determine the actual ground distance.

The arithmetic scale represents a ratio of units on the map to like units on the ground and is called a representative fraction. A representative fraction of 1:50,000, or 1/50,000, indicates that 1 unit on the map is equivalent to 50,000 of the same units on the earth's surface. Since the scale is expressed in terms of a ratio, the proportion between the two distances (map and ground) is constant. Thus the representative fraction is applicable to all systems and all units of measurement simultaneously. Hence 1:50,000 can be read as "1 map inch to 50,000 ground inches" or "1 map centimeter to 50,000 ground centimeters." Similarly, any other unit of measurement can be substituted, and the need for conversion factors between measurement systems, e.g., U.S. customary and metric, is thus avoided.

MAP TYPES

The symbols you see on maps depend on the type of map and the phenomena portrayed by the map. There are three basic types of maps used in geography: choropleth, dot, and isopleth. A *choropleth map* may be used to portray either numerical or nonnumerical phenomena. The key feature of this map is its patchlike appearance. Each patch (area) represents a different class or category, and any class may abut against any other. Soils (see Figs. 17.3 and 17.4) and climate (see Fig. 9.2) are usually portrayed with choropleth maps.

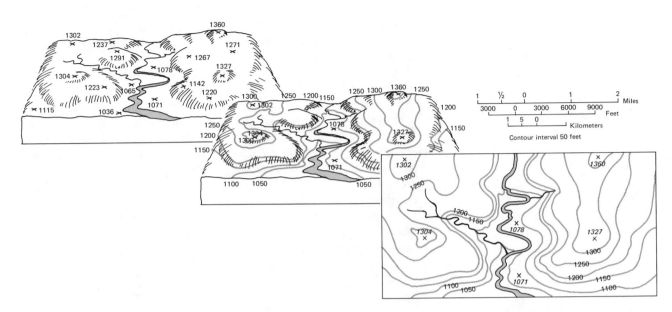

Fig. A.3.2 *The transformation of elevation data into a topographic contour map. Modern methods for building topographic contour maps use elevation data derived from both ground surveys and aerial photographs. (Illustration by William M. Marsh)*

A *dot map* is usually used to portray numerical phenomena such as population or crop production. The placement of the dots is usually intended to be representative of the location of the phenomena being portrayed. Each dot may represent a fixed value; or, the dot may be sized in proportion to different values. Groundwater withdrawal by state in the United States can be portrayed with the latter type dot map (see Fig. 12.10).

An *isopleth map* is designed to show the pattern or trend of numerical values over an area. This type of map utilizes lines, called *isolines*, to connect points (places) of equal value. If a value is not known, the location of the line is interpolated on the basis of the nearest known values. Among the rules governing isopleth maps are: (1) a given isoline must have the same value over the entire map; (2) isolines cannot cross each other; and (3) the change in value from one line to the next must not exceed the iso-interval, that is, the specified difference in value of adjacent lines in a sequence.

Isopleth maps are used extensively in physical geography, especially for climatic phenomena such as radiation (see Fig. 6.9), air pressure (see Fig. 7.11) and precipitation (see Fig. 15.6). One of their most common uses is in portraying the topography of the earth.

If isolines are used to connect points of equal elevation, they are called *contours*. The resulting topographic contour maps are constructed from elevation points measured from ground surveys and aerial photographs. The accuracy of contour maps of a given scale depends mainly on the density of data points; the more data, the greater the accuracy. Figure A.3.2 shows the basic concept behind the construction of a topographic contour map.

Topographic contour maps are perhaps the most widely used maps in the world today because they are so valuable in terrain analysis, planning, and real estate. In the United States the U.S. Geological Survey is charged with the task of preparing topographic maps for the nation. These maps, called topographic quadrangles, are prepared at a variety of scales—1:24,000, 1:62,500, and 1:250,000, for example—and are available to anyone at a relatively low cost.

In addition to contours and elevation data, the U.S. Geological Survey quadrangles provide a great deal of other information about the land. This includes drainage features, forested areas, wetlands, roads, highways, urbanized areas, and even individual structures such as homes and schools in rural areas.

GLOSSARY

Ablation The wastage of ice or snow by melting and sublimation; in the case of glaciers, it also includes calving.

Abrasion The wearing away of a substance by rasping action; for example, the scouring of bedrock by the boulders carried in the base of a glacier.

Absolute humidity An expression for the water vapor content of air; the mass (weight) of water vapor in a given volume of air (irrespective of the mass of the air); usually expressed in grams of water vapor per cubic meter of air.

Absolute zero The zero point on the Kelvin temperature scale, which represents the state at which there is no molecular vibration in a substance and hence no heat. Corresponds to − 273.15 on the Celsius scale.

Absorption The process by which incident radiation is taken up by a substance such as water vapor and converted to other forms of energy.

Abyssal plain The deep-ocean floor; the most extensive part of the ocean basin, which lies between the midoceanic ridges and the trenches, usually at an elevation of 5000 m to 7000 m below sea level.

Acidic soil A soil with a pH less than 7.0.

Active layer The surface layer in a permafrost environment, which is characterized by freezing and thawing on an annual basis.

Actual evapotranspiration The true amount of moisture given up by the soil over some time period; it is equal to soil-moisture loss plus precipitation in the period of negative moisture balance.

Adaptation A change in an organism that brings it into better harmony with its environment; two types of adaptation are acquired and genetic.

Adiabatic cooling A thermodynamic decline in a system, such as a parcel of air, in which there is no transfer of heat or mass from the system. In a rising parcel of air, decompression and expansion result in cooling.

Advection The transfer or exchange of energy in the atmosphere by the lateral movement of air as when cold air gains heat with movement over a warm surface.

Aggradation Filling in of a stream channel with sediment, usually associated with low discharges and/or heavy sediment loads.

Agronomy Agriculture science, much of which is devoted to the study of soil, water, and related phenomena.

Air pressure (see **Sea-level pressure**)

Albedo The percentage of incident radiation reflected by a material. Usage in earth science is usually limited to shortwave radiation and landscape materials.

Akaline soil A soil with a pH greater than 7.0.

Alluvial fan A fan-shaped deposit of sediment laid down by a stream at the foot of a slope; very common features in dry regions, where streams deposit their sediment load as they lose discharge downstream.

Alluvium Material deposited by a stream or river.

Alpine meadow A formation of grasses and forbs found in mountains above the treeline; similar to the tundra formation of the arctic.

Andesite An extrusive igneous rock comprised of intermediate amounts of dark minerals; the extrusive counterpart of diorite.

Andesite line The seaward extent of andesite lavas used in some places to delimit the geologic border of the continents.

Angiosperm A flowering, seed-bearing plant; the angiosperms are presently the principal vascular plants on earth.

Angle of repose The maximum angle at which a material can be inclined without failing; in civil engineering the term is used in reference to clayey materials.

Angular momentum A measure of the momentum of an object with respect to its rotation about a point; in the atmosphere the farther poleward a particle of air goes, the closer it gets to the axis of earth rotation and the faster its velocity becomes.

Anion A negatively charged ion.

Anticline A fold characterized by an upward bend, e.g., convex upward in cross-section.

Aphelion The position of the earth in its orbit when it is farthest from the sun—152 million km (94.25 million miles).

Aquiclude An impervious stratum or formation that impedes the movement of groundwater.

Aquifer Any subsurface material that holds a relatively large quantity of groundwater and is able to transmit that water readily.

Archipeligo An arc-shaped group of islands, usually of volcanic origin; most archipeligos, e.g., the Japanese Islands, are associated with subduction zones.

Arête The term given to the sharp ridges that separate cirques or faces on a glaciated mountain.

Artesian flow A pressurized flow of groundwater that reaches an elevation above the water table or even above the ground surface; groundwater that has become sandwiched into an inclined aquifer such that when the lower end is tapped, the water rises under the pressure of the water higher in the aquifer.

Artesian well (see **Artesian flow**)

Asthenosphere The layer immediately under the lithosphere, where the rock appears to be in a plastic state and capable of slow-flowing motion.

Autumnal equinox (see **Equinox**)

Azonal soil A soil order under the traditional USDA soil classification scheme; soils without horizons; those that are usually found in geomorphically active environments such as sand dunes and river valleys.

Backscattering That part of solar radiation directed back into space as a result of diffusion by particles in the atmosphere.

Backshore The zone behind the shore—between the beach berm and the backshore slope.

Backshore slope The bank or bluff landward of the shore that is comprised of *in situ* material.

Backswamps A low, wet area in the floodplain, often located behind a levee.

Backwash The counterpart to swash; the sheet of water that slides back down the beach face.

Bar A unit of force equal to 100,000 newtons per square meter; normal atmospheric pressure at sea level is slightly greater than 1 bar (1.0132 bars).

Barchan A type of sand dune that is crescent-shaped, with its long axis transverse to the wind and its wings tipped downward.

Basalt floods A form of volcanism characterized by massive outflows of lava from long fissures in the crust.

Basaltic rock A general term applied to the rocks that form the ocean basins and the lower crust; relatively high density, dark-colored igneous rock; basalt and related rocks of the sima layer.

Base level The lowest elevation to which a river can downcut its channel. For large rivers, this is controlled by sea level; for smaller ones, it may be controlled by a lake, resistant rock, or another river.

Baseflow The portion of streamflow contributed by groundwater; it is a steady flow that is slow to change even during rainless periods.

Basic soil (*see* **Alkaline soil**)

Basin A rock structure formed by a large downward flexure, often hundreds of km in diameter, e.g., Pasin Basin of France, Michigan Basin of North America.

Batholith A large accumulation of magma in a great chamber; upon cooling, it forms coarse-grained igneous rocks.

Bay-mouth bar A ribbon of sand deposited across the mouth of a bay.

Bed load The stream-carried particles (sediment) that roll along the bottom and are in nearly continuous contact with the streambed.

Bed shear stress The force exerted against a streambed by moving water; it is a function of water density, gravitational acceleration, water depth, and the slope of the channel.

Bergschrund A deep crevasse at the very head of an alpine glacier that opens as the ice pulls away from the mountain side.

Berm A low mound that forms along sandy beaches.

Biochore A major region or division of vegetation defined on the basis of the structural and compositional characteristics of the plant cover; in a general sense, it is the vegetative version of a biome; *see also* **Biome**.

Biomass The total weight of organic matter per unit area of landscape; also, the total weight of the organic matter in an ecosystem.

Biome A major division (region) of the earth's surface, defined on the basis of the plants and animals inhabiting it; geographically it may correspond to a climatic zone.

Black body A hypothetical body that is capable of absorbing all radiation incident on it and in turn is the most effective possible emitter of radiation.

Blowout (*see* **Deflation hollow**)

Boreal forest Subarctic conifer forests of North America and Eurasia; floristically homogeneous forests dominated by fir, spruce, and tamarack; in Russia, it is called *taiga*.

Boundary layer The lower layer of the atmosphere; the lower 300 m of the atmosphere where airflow is influenced by the earth's surface.

Bowen ratio The ratio of sensible-heat flux to latent-heat flux between a surface and the atmosphere.

Braided channel A stream channel characterized by multiple threads or subchannels which appear to weave in and out of one another.

British thermal unit A unit of energy used to measure heat; one BTU is equal to 1054 joules, the amount of heat required to raise the temperature of one pound of water by one degree Fahrenheit.

Brittle rock Rock that ruptures with little or no plastic deformation.

Caatinqa (*see* **Thornbush**)

Calcification A soil-forming regime of dry environments that results in the accumulation of calcium carbonate and, in grassy areas, a strong organic layer.

Caldera A large, circular depression in a volcano, resulting from an explosion or the loss of magma through a lower vent.

Caliche An accumulation of calcium carbonate at or near the soil surface in an arid environment.

Calorie A unit of energy; the amount of heat required to raise the temperature of water $1°C$, from $14.5°C$ to $15.5°C$.

Calving The process by which a glacier loses mass when ice breaks off into the sea.

Capillarity The capacity of a soil to transfer water by capillary action; capillarity is greatest in medium-textured soils.

Capillary fringe The transition between the zone of aeration and the zone of saturation, at the base of which capillary water gives way to groundwater; in fine-textured material the capillary fringe may be several meters thick, whereas in coarse-textured material it is usually only several centimeters thick.

Cation A positively charged ion.

Cation adsorption The process by which cations become attached to colloids.

Cation-exchange capacity The total exchangeable cations that a soil can adsorb; the capacity per unit volume of soil increases with finer soil textures.

Cavitation A mechanism of stream and wave erosion brought about by the sudden increase in hydraulic pressure, which causes air bubbles to burst and exert force against rock.

Celsius A temperature scale, also known as centigrade, on which 0° represents the normal freezing point of water, and 100° represents the normal boiling point of water.

Centigrade (*see* **Celsius**)

Central vent The main passageway by which magma ascends to the surface of a volcano.

Channel precipitation Rain or snow that falls directly on the stream channel and thereby contributes immediately to discharge.

Chapparal A vegetative formation of the Mediterranean climate of California, characterized by shrubs and small trees, often in shrubby thickets.

Chelation A weathering process involving the bonding of mineral ions to a large organic molecule, followed by the removal of the ions with the molecule.

Chemical stability A term referring to the tendency of a mineral to change to another form, such as another mineral; stable minerals are not readily transformed to other states and are usually those, such as quartz, that originated at relatively low temperatures.

Chemical weathering One of the two major types of weathering and generally considered to be the more effective one; it involves the chemical decomposition of rock by a variety of chemical processes, including dissolution, chelation, and hydrolysis.

Chernozem soils One of the great soil groups under the traditional USDA soil classification scheme; soil characterized by a heavy O horizon and calcium carbonate accumulation in the B horizon.

Chinook A dry, often warm, wind that descends the leeward side of a mountain range; in Germany, Austria, and Switzerland such a wind is called a *föhn.*

Circle of illumination The line dividing the illuminated half of the earth from the dark (shadow) half; the alignment of this circle changes with the seasons, thereby changing the daily length of daylight hours at different latitudes.

Climate The representative or general conditions of the atmosphere at a place on earth; it is more than the average conditions of the atmosphere, for climate may also include extreme and infrequent conditions.

Climatology The field of earth science devoted to the study of climate and climatic processes.

Climax community A group of organisms that represents an ecological equilibrium with its environment and therefore is capable of maintaining long-term stability.

Clone A genetically uniform group of plants regenerated by vegetative means (asexual) from a single parent.

Closed forest A forest structure with multiple levels of growth from the ground up; a forest in which undergrowth closes out the area between the canopy and the ground.

Coastal dune A sand dune that forms in coastal areas and is fed by sand from the beach.

Coefficient of runoff A number given to a type of ground surface representing the proportion of a rainfall converted to overland flow; it is a dimensionless number between 0 and 1.0 that varies inversely with the infiltration capacity; impervious surfaces have high coefficients of runoff.

Cold front A contact between a cold air mass and a warm air mass in which the cold air is advancing on the warm air, driving the warm air upward.

Colloid A small clay particle, less than 0.001 mm in diameter, that provides adsorption sites for ions.

Colluvium An unsorted mix of soil and mass-movement debris.

Community A group of organisms that live together in an interdependent fashion.

Community-succession concept A popular concept of vegetation change based on the idea that one plant community succeeds another in occupying a site; succession ends when a climax community is established which represents an equilibrium between vegetation and environment.

Composite volcano A cone-shaped volcano comprised mainly of pyroclastic material, e.g., Fuji-san of Japan and Vesuvius of Italy.

Compressional wave A type of seismic (elastic) wave that generates a back-and-forth motion along the line of energy propagation through a substance; also termed *primary wave.*

Concentration time The time taken for a drop of rain falling on the perimeter of a drainage basin to go through the basin to the outlet.

Condensation The physical process by which water changes from the vapor to the liquid phase.

Condensation nuclei Very small particles of dust or salt suspended in the atmosphere on which condensation takes place to initiate the formation of a precipitation droplet.

Conduction A mechanism of heat transfer involving no external motion or mass transport; instead, energy is transferred through the collision of vibrating molecules.

Cone of ascension The ascent of salt groundwater under a well in response to the development of a cone of depression in the overlying fresh groundwater.

Cone of depression A conical-shaped depression that forms in the water table or an aquifer surface immediately around a well from which groundwater is being rapidly pumped.

Conservation of angular momentum The principle that the angular momentum of a rotating mass, such as the earth or a twirling skater, will not change unless torque (stress) is applied to it; if the radius of rotation decreases (arms in), rotational velocity increases, and vice versa.

Conservation-of-energy principle The principle that energy in an isolated system can be neither destroyed nor created; thus total energy in the system remains constant.

Conservative zone A type of contact or border along a tectonic plate where lithosphere is neither destroyed nor created; the tectonic movement in conservative zones is often lateral, such as along the San Andreas fault of California.

Constructive zone A type of contact or border along a tectonic plate where new lithosphere is emerging; usually associated with midoceanic ridges.

Continental drift The term used to describe the wholesale movement of land masses over great distances on the surface of the earth; the term is generally attributed to Alfred Wegener, who advanced the first coherent theory of continental drift.

Continental shelf The seaward-sloping margin of the continents under water to a depth of 200–300 m; the shoulders of the continents where the rate sediment accumulation and sedimentary rock formation is high.

Continuity of flow A principle that describes the maintenance of flow in a system with changes in flow velocity and system capacity.

Core The innermost of the two major divisions of the solid earth, the other being the mantle; the core includes the *outer core,* which is liquid, and the *inner core,* which is solid; the core is the densest part of the earth.

Coriolis effect The effect of the earth's rotation on the path of airborne objects; winds in the Northern Hemisphere are deflected to the right and those in the Southern Hemisphere are deflected to the left of their original paths; at the equator it is negligible; the Coriolis is an effect apparent only to observers standing on the earth and as such is not a force.

Corrasion Another term for abrasion.

Convection A mechanism of heat transfer which involves mass transport (mixing) of fluid, such as occurs with turbulence in the atmosphere; any mixing motion in a fluid.

Convectional precipitation A type of precipitation resulting from the ascent of unstable air; the instability may be caused by local heating in the landscape or by frontal activity; usually short-term, intensive rainfall.

Convergent precipitation A type of precipitation that takes place when air moves into a low-pressure trough or topographic depression and escapes by moving upward; precipitation in the ITC zone is at least partially convergent.

Critical threshold The point at which shear stress (driving force) equals shear strength (resisting force) and beyond which change, such as slope failure, rock rupture, plant damage, or soil erosion, is imminent.

Crust The outermost zone of the lithosphere, which ranges from 8 km to 65 km (5–40 miles) in thickness and is bounded on the bottom by the Moho discontinuity.

Cuspate foreland A large depositional feature along a coastline, often in the form of a triangular point.

Cyclone A large low-pressure cell characterized by convergent airflow and internal instability; the two main classes of cyclones are midlatitude cyclones and tropical cyclones; in some parts of the United States, tornadoes are also called cyclones.

Cyclonic/frontal precipitation A type of precipitation that results from a large cell of low pressure and the meeting of warm and cold air masses; most of the precipitation is concentrated along the fronts; see **Cold front** and **Warm front**.

Darcy's law The principle that describes the velocity of groundwater flow; velocity is equal to permeability times the hydraulic gradient.

Debris flow A type of mass movement characterized by the downslope flow of a saturated mass of heterogeneous soil material and rock debris.

Declination of the sun The location (latitude) on earth where the sun on any day is directly overhead; declinations range from 23.27° S latitude to 23.27° N latitude.

Deerpark The term that the English use to describe the parkland vegetation of southern England, northwest France, and similar areas; see *also* **Parkland**.

Deflation hollow A topographic depression caused by wind erosion; also called a blowout.

Degradation Scouring and downcutting of a stream channel, usually associated with high discharges.

Denudation A term used to describe the erosion or wearing down of a land mass; also used to describe the process by which a site is stripped of its vegetative cover.

Denude (*see* **Denudation**)

Deranged drainage Highly irregular drainage patterns in areas of complex geology, such as the Canadian Shield, which have been heavily glaciated.

Desert lag A veneer of coarse particles left on the ground after the fine particles have been eroded away, often by wind.

Desert soils Soils characterized by a weak O horizon and salt accumulation at or near the surface; in the traditional USDA classification scheme, these soils are classed as either Red or Gray desert soils.

Destructive zone A type of contact or border along a tectonic plate where lithosphere is destroyed; see *also* **Subduction**.

Detrital rock Sedimentary rock composed of particles transported to their place of desposition by erosional processes, e.g., sandstone and shale.

Dew point The temperature at which a parcel of air is saturated.

Diapir A conduit in the lithosphere through which magma moves to the surface.

Differential stress Force, directed on a body, that is not equal in all direction.

Dike A vertical fissure in the crust that serves as a passageway for magma.

Diorite An intrusive igneous rock that is both darker and of higher density than granite.

Dip (*see* **Strike and dip**)

Dip-slip fault A fault in which the principal direction of displacement is up or down along the fault plane.

Discharge The rate of water flow in a stream channel; measured as the volume of water passing through a cross-section of a stream per unit of time, commonly expressed as cubic feet (or meters) per second.

Dissolved load Material carried in solution (ionic form) by a stream.

Disturbance Factors, other than the basic requirements, that affect a plant's well-being, e.g., floods, disease, and soil erosion.

Diurnal damping depth The maximum depth in the soil which experiences temperature change over a 24-hour (diurnal) period.

Divine Plan of Nature The idea that the earth, and indeed the entire universe, are designed according to a great plan of God's.

Drainage basin The area that contributes runoff to a stream, river, or lake.

Drainage density The number of miles (or km) of stream channels per square mile (or km²) of land.

Drainage divide The border of a drainage basin or watershed where overland separates between adjacent areas.

Drainage network A system of stream channels usually connected in a hierarchical fashion (*see also* **Principle of stream orders**).

Drift (*see* **Glacial drift** and **Littoral drift**)

Dry adiabatic lapse rate The rate of decline in the temperature of a rising parcel of air due to expansion; a thermodynamic decline in which there is no external transfer of heat or mass; equal to $-0.98°C/100$ m.

Doldrums The belt of calm and variable winds in the intertropical convergence zone; in the days of ocean sailing, a source of quiet.

Dolomite A sedimentary rock composed of chemical origin; it appears to be an altered form of limestone in which some of the calcium is replaced by magnesium.

Dome Large upward flexure of rock often hundreds of km in diameter, e.g., the Ozark Plateaus of Missouri.

Dornveld (*see* **Thornbush**)

Dowsing The practice of locating groundwater by using a stick, tree branch, or special rod; dowsing is steadfastly practiced by believers, but it has no scientific basis.

Ductile rock Rock that has a relatively large capacity for plastic deformation.

Dust Bowl The name given to the area of severe drought and wind erosion in the Great Plains during the 1930s.

Dwarfism The tendency for a plant to achieve less than full size at maturity because of environmental stress such as inadequate heat or light.

Dynamic equilibrium A term used to describe the behavior of a system, such as a river network, which is continually trending toward a state of equilibrium, but rarely reaches it. The trend may change with changes in the energy available to drive the system.

Earthquake intensity (*see* **Mercalli scale**)

Earthquake magnitude (*see* **Richter scale**)

Ecological amplitude The variation in tolerance and resource needs from member to member in a plant species.

Ecosystem A group of organisms linked together by a flow energy; also, a community of organisms and their environment.

Ecotone The transition zone between two groups, or zones, of vegetation.

Eddy The term given to a whirling or spiral motion in a fluid.

Edge wave A wave that moves parallel to the shore; usually a secondary wave of complex origins.

Effective frictional surface (see **Roughness length**)

Elastic deformation Change in the shape of a body as a result of differential stress; on the release of the stress, the body returns to its original shape.

Elastic limit The maximum level of elastic deformation of a body, beyond which it ruptures.

Elastic wave An energy wave that causes motion in a material without permanently deforming it.

Electromagnetic spectrum The classification scheme used to describe the array of electromagnetic radiation; the various categories of radiation are distinguished on the basis of wavelength.

Eluviation The removal of colloids and ions from a soil or from one level to another in a soil.

Emissivity The ratio of total radiant energy emitted at a specified wavelength and temperature by a substance to that emitted by a *black body* under the same conditions.

Empirical approach An approach to scientific investigation based on direct observation and measurement.

Energy Generally, the capacity to do work; defined as any quantity that represents force times distance. Joules and calories are energy units commonly used in science.

Energy balance The concept or model that concerns the relationship among energy input, energy storage, work, and energy output of a system such as the atmosphere or oceans.

Energy flux The rate of energy flow into, from, or through a substance; also called radiant flux density and irradiance.

Energy pyramid The attenuation of organic energy in an ecosystem; the decline of energy in an ecosystem as organic matter is passed from one level of organisms to another.

Ephemeral stream A stream without baseflow; one that flows only during or after rainstorms or snowmelt events.

Epiphyte A plant that grows in the superstructure of another plant without rooting in the soil; a nonparasitic aerial plant, e.g., Spanish moss and many orchids.

Equatorial zone The middle belt of latitude, extending 10° or so north and south of the equator.

Equilibrium profile A gently sloping topographic profile from the shore through the offshore zone across which wave energy is evenly distributed.

Equinox The dates when the declination of the sun is at the equator, March 20–21 and September 21–22, and the number of hours of dark and daylight in a 24-hour period is the same for all locations on earth. These dates are known as the autumnal equinox and vernal equinox, but which is which depends on the hemisphere.

Erosion The removal of rock debris by an agency such as moving water, wind, or glaciers; generally, the sculpting or wearing down of the land by erosional agents.

Euler's theorem The theorem that describes the movement of a plate on the surface of a sphere; such a plate moves about its own pole, called the Euler pole, along a small-circle path. This theorem helps explain the differences in the rates of sea floor spreading along the midoceanic ridges.

Evaporite A rock or mineral formed from the evaporation of mineral-rich water, e.g., rock salt and gypsum.

Evapotranspiration The loss of water from the soil through evaporation and transpiration.

Event An episode of a process defined as some quantity of a variable such as a river discharge or wind velocity.

Evolution Biological change over time that results in new or changed relationships between organisms and the environment; irreversible biological change.

Exfoliation A mechanical weathering process involving the breaking off, or "shedding," of slabs of rock in response to the differential expansion of a rock mass.

Exotic river A river, such as the Nile or Colorado, that flows through an arid region after gaining its flow elsewhere; exotic streams usually lose much of their discharge to groundwater recharge and evaporation.

Extrusive rock Igneous rock that forms at the surface and cools quickly.

Fahrenheit A temperature scale on which 32° represents the normal freezing point of water and 212° represents the normal boiling point of water.

Fault A fracture in rock along which there has been displacement of one side relative to the other.

Fault line The linear trend of a fault along the earth's surface as one would see it from the air.

Fault plane The plane representing the fracture surface in a fault.

Fault scarp The part of the fault plane exposed in a fault; in a normal fault it is the upper part of the footwall.

Feedback A return effect of a change; the consequences of a change have a feedback effect if they dampen or amplify the change or the causes of it.

Fell fields Areas of very light plant covers in polar and alpine regions; often rocky areas with scattered lichens, mosses, and small flowering plants.

Ferromagnesian minerals A subgroup of the silicate minerals; rock-forming minerals that are rich in iron and magnesium; dark-colored and relatively high-density minerals, e.g., biotite, hornblende, and olivine.

Fetch The distance of open water in one direction across a water body; it is one of the main controls on wave size.

Firn Névé on a glacier that has survived the entire ablation season; the material that is transformed into glacial ice.

Firn line The lower edge of the accumulation zone on a glacier.

Flank eruption Volcanic eruption that breaks out on the side of a volcano.

Floristic system The principal botanical classification scheme in use today; under this scheme the plant kingdom is made up of divisions, each of which is subdivided into smaller and smaller groups arranged according to the apparent evolutionary relationships among plants.

Fluvioglacial deposits Materials deposited by glacial meltwaters, including outwash plains, kames, and eskers; usually stratified.

Fold A rock structure characterized by a bend in a rock formation.

Föhn (see **Chinook**)

Foot ice An accumulation of grounded ice on and near shore in lakes, oceans, and rivers.

Footwall The lower surface of an inclined fault.

Formation A structural unit of vegetation that may be considered a subdivision of a biochore; a formation may be made up of several communities. In the traditional terminology, it is called a physiognomic unit; in geology, a major unit of rock.

Free convection Mixing motion in a fluid caused by differences in density. In the atmosphere such differences are usually caused by differential heating of air near the surface.

Free-face A steep slope or cliff formed in bedrock.

Freeze-thaw activity Weathering and mass-movement processes associated with daily and seasonal cycles of freezing and melting.

Frequency The term used to express how often a specified event is equaled or exceeded.

Friable A term used to describe the tendency of a soil to crumble or break up when plowed.

Front (see **Cold front** and **Warm front**)

Frost wedging A mechanical weathering process in which water freezes in a crack and exerts force on the rock, which may result in the breaking of the rock; a very effective weathering process in alpine and polar environments.

Fusion Another word for freezing.

Gabbro A coarse-grained, intrusive igneous rock that is dark and heavy owing to a relatively high percentage of ferromagnesian minerals.

Geodesy The science that measures the geoid and its major features.

Geographic cycle A concept developed by William Morris Davis on the formation of river-eroded landscapes. It describes three stages of landscape development (youth, maturity, and old age) and argues that rejuvenation takes place with uplift of the land, thereby renewing the cycle.

Geoid The term given to the true shape of the earth, which deviates from a perfect sphere because of a slight bulge in the equatorial zone.

Geomorphic system A physical system comprised of an assemblage of landform linked together by the flow of water, air, or ice.

Geomorphology The field of earth science that studies the origin and distribution of landforms, with special emphasis on the nature of erosional processes; traditionally, a field shared by geography and geology.

Geophysics A field of earth science devoted to the study of the earth, including the oceans and the atmosphere, through the application of models and techniques from physics.

Geostrophic winds Winds in the upper troposphere which generally flow parallel to isobars and often reach high velocities.

Geothermal energy Energy in the form of heat that is produced by the earth's interior and flows through the crust mainly by conduction.

Glacial drift A general term applied to all glacial deposits, including moraine and fluvioglacial deposits.

Glacial flour (see **Glacial milk**)

Glacial lake A natural impoundment of meltwater at the edge of a glacier.

Glacial milk Glacial meltwater of a light or cloudy appearance because of clay-sized sediment held in suspension.

Glacial polish Bedrock surfaces that have been abraded by the debris on the base of a glacier until smooth and shiny.

Glacial surge A rapid advance of the snout of a glacier.

Glacial uplift Uplift of the crust following isostatic depression under the weight of the continental glaciers.

Gleization A soil-forming regime of poorly drained areas such as bogs and swamps; it results in a heavy organic layer over a layer of blue clay.

Global coordinate system The network of east-west and north-south lines (parallels and meridians) used to measure locations on earth; the system uses degrees, minutes, and seconds as the units of measurement.

Gneiss A metamorphosed form of granite characterized by minerals arranged in bands.

Graben (see **Rift**)

Graded profile The longitudinal profile of a stream representing an equilibrium condition toward which the stream adjusts in response to changes in discharge and sediment load.

Grafting The practice of attaching additional channels to a drainage network; in agricultural areas new channels appear as drainage ditches; in urban areas, as stormsewers.

Granite An intrusive igneous rock comprised mainly of quartz and feldspar; limited in its distribution to the continents.

Granitic rock A general term applied to the rocks that comprise the continental masses; low density, light-colored igneous rock; granite and related rocks of the sial layer.

Granodiorite An intrusive igneous rock that is intermediate in composition between granite and diorite.

Graupel A frozen precipitation particle comprised of a snow crystal and a raindrop frozen together.

Gravity water Subsurface water that responds to the gravitational force (in contrast to capillary water, which responds to molecular forces); the water that percolates through the soil to become groundwater.

Great circle Any circle that circumscribes the full circumference of the earth and the plane of which passes through the center of the earth; the equator is a great circle; the shortest distance between any two points on the globe follows a great-circle route.

Greenbelt A tract of trees and associated vegetation in urbanized areas; it may be a park, nature preserve, or part of a transportation corridor.

Greenwich meridian The zero degree meridian from which east and west longitude are measured; it is named for the town of Greenwich, England, through which the line is drawn.

Groin A wall or barrier built from the beach into the surf zone for the purpose of slowing down longshore transport and holding sand.

Gross sediment transport The total quantity of sediment transported along a shoreline in some time period, usually a year.

Ground frost Frost that penetrates the ground in response to freezing surface temperatures.

Groundwater The mass of gravity water that occupies the subsoil and upper bedrock zone; the water occupying the zone of saturation below the soil-water zone.

Gulf stream A large, warm current in the Atlantic Ocean that originates in and around the Caribbean and flows northwestward to the North Atlantic and northwest Europe.

Gullying Soil erosion characterized by the formation of narrow, steep-sided channels etched by rivulets or small streams of water. Gullying can be one of the most serious forms of soil erosion of cropland.

Gymnosperm A plant that bears naked seeds; the most common group of gymnosperms are the conifers, needleleaf cone-bearing plants.

Gyre A large circular pattern of ocean currents associated with major systems of pressure and prevailing winds; the largest are the subtropical gyres.

Habitat The environment with which an organism interacts and from which it gains its resources; habitat is often variable in size, content, and location, changing with the phases in an organism's life cycle.

Habitat versatility (see **Ecological amplitude**)

Hadley cells Large atmospheric circulation cells comprised of rising air near the equator, poleward flow aloft, descending air in the subtropics, and return flow on the surface in the form of trade winds; concept first proposed by George C. Hadley in 1735.

Hair hygrometer An instrument for measuring atmospheric humidity based on the reaction of human hair to changes in vapor levels.

Hamada (see **Reg**)

Hanging valley A tributary valley that enters the main valley at an elevation well above the valley floor; most common in areas of mountain glaciation, where hanging valleys are often the sites of spectacular waterfalls.

Hanging wall The upper surface of an inclined fault.

Hardpan A hardened soil layer characterized by the accumulation of colloids and ions.

Heat island The area or patch of relatively warm air which develops over urbanized areas.

Heat transfer The flow of heat within a substance or the exchange of heat between substances by means of conduction, convection, or radiation.

High latitude The zones poleward of the Arctic and Antarctic circles.

Higher plants Generally the larger and more advanced plants; the vascular plants; pteridophytes, gymnosperms, and angiosperms.

Hillslope processes The geomorphic processes that erode and shape slopes; mainly mass movements such as soil creep and landslides and runoff processes such as rainwash and gullying.

Horizon A layer in the soil that originates from the differentiation of particles and chemicals by moisture movement within the soil column; four major horizons are recognized in a standard soil profile: O, A, B, and C.

Horst A fault characterized by a block displaced upward relative to adjacent rock formations.

Humboldt current A cold current that flows northward along the west coast of South America, contributing to the aridity there.

Humus Organic matter in the soil that has been broken down by physical, chemical, and biological processes into a granular form which is relatively stable.

Hurricane A large tropical cyclone characterized by convergent airflow, ascending air in the interior, and heavy precipitation.

Hydraulic gradient The inclination of the water table or any body of groundwater; equal to the difference in the elevation of the water table at two points divided by the distance between them.

Hydraulic pressure (*see* **Cavitation**)

Hydraulic radius The ratio of the cross-sectional area of a stream to its wetted perimeter.

Hydrograph A streamflow graph which shows the change in discharge over time, usually hours or day; *see also* **Hydrograph method**.

Hydrograph method A means of forecasting streamflow by constructing a hydrograph that shows the representative response of a drainage basin to a rainstorm; the use of a "normalized" hydrograph for flow forecasting in which the size of the individual storm is filtered out; *see also* **Hydrograph**.

Hydrologic cycle The planet's water system, described by the movement of water from the oceans to the atmosphere to the continents and back to the sea.

Hydrologic equation The amount of surface runoff (overland flow) from any parcel of ground is proportional to precipitation minus evapotranspiration loss, plus or minus changes in storage water (groundwater and soil water).

Hydrolysis A complex chemical weathering process, or series of processes, involving the reaction of water and an acid on a mineral; it is considered to be the most effective process in the decomposition of granite.

Hydrophyte Water-loving plants; aquatic plants such as water lily and water hyacinth.

Hydrostatic pressure The pressure exerted by elevated groundwater as in artesian flow.

Hygrophyte Water-tolerant plants, such as cattail, which are able to grow in saturated or lightly flooded sites.

Hygroscopic water Molecular water that resides directly on the surface of all materials; it is bound to surfaces under such great pressure that it is immobile and cannot be evaporated or used by plants.

Ice Age (*see* **Pleistocene**)

Ice wedging (*see* **Frost wedging**)

Illuviation The process of accumulation of ions and colloids in a soil.

Individualistic concept A concept of vegetation change contrary to the community-succession concept, especially the idea of a climax community; it argues that the stability of the climax community is a matter of probability because those species in greatest abundance favor regeneration of their own kind.

Infiltration capacity The rate at which a ground material takes in water through the surface; measured in inches or centimeters per minute or hour.

Inflooding Flooding caused by overland flow concentrating in a low area.

Infrared radiation Mainly longwave radiation of wavelengths between 3.0–4.0 and 100 micrometers, but also includes near infrared radiation, which occurs at wavelengths between 0.7 and 3.0–4.0 micrometers.

Interception The process by which vegetation intercepts rainfall or snow before it reaches the ground.

Interflow Infiltration water that moves laterally in the soil and seeps into stream channels; in forested areas this water is a major source of stream discharge.

Interglacial A relatively warm and dry period in the Pleistocene Epoch during which most of the continental glaciers are thought to have melted away.

Intermittent A stream with baseflow in all but the dry season when the water table drops below the streambed.

Intertropical convergence zone The belt of convergent airflow and low pressure in the equatorial zone (between the tropics) which is fed by the trade winds.

Intertropical zone The zone between the Tropic of Cancer (23.5° N latitude) and the Tropic of Capricorn (23.5° S latitude).

Intrazonal soils A soil order under the traditional USDA soil classification scheme: soils that form under conditions of impeded drainage.

Intrusive rock Igneous rock that forms within the earth and cools slowly; *see also* **Plutonic rock**.

In situ A term used to indicate that a substance is in place as contrasted with one, such as river sediment, that is in transit.

Instability A physical condition in which a fluid is gravitationally unstable; in the atmosphere it is one in which heavier (denser) air overlies lighter air.

Inversion (*See* **Temperature inversion**)

Ion A minute particle of a dissolved mineral; usually an atom or group of atoms that are electrically charged.

Isostatic depression Large-scale down-warping of the crust in response to an increase in mass (weight) on the surface; in areas of continental glaciation the crust was depressed by the weight of the ice.

Isostatic rebound The uplift or recovery of the earth's crust following isostatic depression; elastic recovery of the crust from large-scale depression; *see also* **Isostatic depression**.

Jet stream Zone of concentrated geostrophic winds; in the midlatitudes it is called the polar front jet stream because it often coincides in location with the polar front.

Joint line An open crack or fracture in bedrock, usually a result of weathering.

Joule A unit of energy equal to one newton (a unit of force) applied over a one-meter length; in terms of heat, 4186 joules are needed to raise the temperature of 1 kilogram of water 1°C, from 14.5°C to 15.5°C.

Kame A mound- or cone-shaped deposit of sand and gravel laid down by melting water in and around glacial ice.

Kaolinite A type of clay produced in the weathering of granite; it is especially widespread in tropical and subtropical regions.

Kármán vortex street A train of wind eddies downwind from a hill.

Karst topography Irregular topography in areas of carbonate rock, characterized by sinkholes, caverns, and underground drainage channels.

Katabatic wind Any wind blowing down a large incline such as a mountain slope; chinook winds are katabatic winds.

Kelvin A temperature scale based on absolute zero, the temperature at which a substance has no molecular vibration and thus generates no heat. Water freezes at 273.15°K and boils at 373.15°K.

Kilogram A metric unit of mass (weight) equal to 2.208 pounds.

Kinetic energy The energy represented by the motion of a substance; equal to mass times velocity squared, divided by two.

Kittlehole A pit in an outwash plain or moraine left from a buried block of glacial ice which melted away; *see also* **Pitted topography**.

Laminar flow Flow characterized by one layer of a fluid sliding over another without vertical (turbulent) mixing; the source of flow resistance is limited to intermolecular friction within the fluid.

Laminar sublayer In the atmosphere the layer of essentially calm air immediately adjacent to fixed surfaces such as vegetation and soil; in reality this air is not perfectly calm, but characterized by a faint laminar flow parallel to the surface.

Land cover The materials such as vegetation and concrete that cover the ground; *see also* **Land use**.

Landscape The composite of natural and human features that characterize the surface of the land at the base of the atmosphere; includes spatial, textural, compositional, and dynamic aspects of the land.

Landslide A type of mass movement characterized by the slippage of a body of material over a rupture plane; often a sudden and rapid movement.

Land use The human activities that characterize an area, e.g., agriculture, industry, and residential.

Latent heat The heat released or absorbed when a substance changes phase as from liquid to gas. For water at 0°C, heat is absorbed or released at a rate of 2.5 million joules per kilogram (597 calories per gram) in the liquid/vapor phase change.

Laterite A layer of iron and aluminum oxide accumulation in tropical soils, mainly the latosols.

Laterization A soil-forming regime of warm, moist environments that produces a strongly leached soil with light topsoil and heavy accumulations of iron and aluminum oxides; also called *ferratillization*.

Latosols One of the great soil groups under the traditional USDA soil classification scheme; soils characterized by a weak O horizon, heavy accumulation of laterite, and a deeply weathered profile.

Lava Molten rock that has reached the surface; *see also* **Magma**.

Law of plastic deformation A physical principle describing the deformational response of a substance to increasing shear stress.

Leaching The removal of minerals in solution from a soil; the washing out of ions from one level to another in the soil.

Levee A mound of sediment which builds up along a river bank as a result of flood deposition.

Life cycle The biological stages in the complete life of a plant.

Life form The form of individual plants or the form of the individual organs of a plant; in general, the overall structure of the vegetative cover may be thought of as life form as well.

Limb Term applied to the flanks of a fold when viewed in cross-section; in an anticline the limbs slope away from the axis.

Limestone A sedimentary rock of chemical and biological origins; calcium carbonate precipitated from seawater and deposited in the form of the shells of sea creatures.

Limiting factors (*see* **Principle of limiting factors**)

Lithosol An azonal soil comprised of large fragments of bedrock.

Lithosphere The upper layer of the mantle; the unit in which the tectonic plates are defined; it is about 100 km thick and includes the crust.

Little Ice Age A period of climatic change in the Northern Hemisphere generally from the fourteenth through the eighteenth centuries; it was marked by a cooling trend and manifested by glacial advances in the seventeenth and eighteenth centuries.

Littoral drift The material that is moved by waves and currents in coastal areas.

Littoral transport The movement of sediment along a coastline; it is comprised of two components: longshore transport and onshore-offshore transport.

Loess Silt deposits laid down by wind over extensive areas of the midlatitudes during glacial and postglacial times.

Longshore current A current that moves parallel to the shoreline; velocities generally range between 0.25 and 1 m/sec.

Longshore transport The movement of sediment parallel to the coast.

Longwave radiation Radiation at wavelengths greater than 3.0–4.0 micrometers; includes infrared (thermal), radio waves, and microwaves.

Mafic lava Lava with a low quartz content in which silicon dioxide constitutes about 50 percent of the rock; this lava is prevalent in the ocean basins.

Magma Molten rock within the lithosphere which cools to become igneous rock; magma that reaches the surface is called lava.

Magnetic polarization The polarization of magnetized materials such as iron particles in volcanic or sedimentary rock.

Magnitude and frequency The concept concerning the behavior of processes and the resultant changes they produce individually and collectively in the landscape; it involves which events render the greatest change and what kinds of change different sized events render.

Mantle One of the two major divisions of the solid earth, the other being the core; the mantle includes the lithosphere, asthenosphere, mesosphere (the upper mantle), and the lower mantle; the mantle contains about two-thirds of the earth's mass.

Maquis Shrubby vegetation of the Mediterranean lands of Europe; apparently a response to climate and long-term disturbance by various land uses; also called *macchia* and *garique: see also* **Chaparral**.

Mass balance The relative balance in a system, based on the input and output of material such as sediment or water; the state of equilibrium between the input and output of mass in a system.

Mass budget (*see* **Mass balance**)

Mass movement A type of hillslope process characterized by the downslope movement of rock debris under the force of gravity; it includes soil creep, rock fall, landslides, and mudflows; also termed *mass wasting*.

Meander A bend or loop in a stream channel.

Meander belt The width of the train of active meanders in a river valley.

Mechanical weathering One of the two major types of weathering; it produces physical fragmentation of rock by ice wedging, rock expansion, and a variety of other mechanisms.

Mercalli scale A scale for rating the intensity of an earthquake; intensity is a measure of an earthquake's destructive effect in the landscape.

Meridians The north-south-running lines of the global coordinate system; meridians converge at the North and South poles; the Prime Meridian marks 0° longitude.

Mesa A flat-topped mass of bedrock that rises sharply above the surrounding terrain; it is usually capped by a resistant formation of rock and has the general aspect of a broad table.

Mesophyll The inner tissue of a leaf where moisture is stored and from which it is released in transpiration.

Mesophyte Plants with intermediate water requirements, usually found in sites with well-drained soils but adequate soil moisture in most months.

Mesosphere A subdivision of the atmosphere that lies above the stratosphere, extending from 50 km to 90 km altitude.

Metastable state In chemical weathering the condition of a mineral when it is intermediate between stability and instability.

Meteorology The field of earth science that studies the weather, with emphasis on forecasting short-term changes and events.

Microflora Minute plant life in the soil, mainly bacteria, algae, and fungi, that consume vegetal matter and in turn help produce humus; the most effective consumers of the organic matter deposited on the soil by plants.

Middle latitude Generally the zone between the pole and the equator in both hemispheres; usually given as 35°–55° latitude.

Midoceanic ridge The volcanic mountain chain located in the interior on an ocean basin along a zone of seafloor spreading.

Millibar A unit of force (or pressure) equal to one-thousandth of a bar; normal atmospheric pressure at sea level is 1013.2 millibars (mb); *see also* **Bar**.

Mineral A naturally occurring inorganic substance with a characteristic crystal (molecular) structure that is fundamentally the same in all samples.

Mixing ratio An expression for the vapor content of air; the weight (mass) of water vapor relative to the weight of the dry air occupying the same space; usually measured in grams per kilogram.

Moho discontinuity The lower boundary of the crust, where seismic wave velocities show an appreciable increase; the exact nature of the Moho and its significance in the lithosphere is not known; also called the *M discontinuity*.

Moisture deficit A term in the soil-moisture balance; the difference between actual and potential evapotranspiration; the difference between the demand for and availability of soil moisture for evapotranspiration.

Monsoon A seasonal wind system in South Asia which blows from sea to land in summer, bringing moisture to the continent, and from land to sea in winter, bringing dry conditions to India and neighboring lands.

Montmorillonite A type of clay that is notable for its capacity to shrink and expand with wetting and drying.

Moraine The material deposited directly by a glacier; also, the material (load) carried in or on a glacier; as landforms moraines usually have hilly or rolling topography.

Mudflow A type of mass movement characterized by the downslope flow of a saturated mass of clayey material.

Neap tide (*see* **Tide**)

Near-deserts A term sometimes used to describe deserts with heavier than average vegetative covers; specifically, the American deserts with diverse plants, including large ones such as the sahuaro cactus.

Net photosynthesis The energy balance of a plant; the balance between the energy produced in photosynthesis and that used in respiration.

Net sediment transport The balance between the quantities of sediment moved in two (opposite) directions along a shoreline.

Névé Partially melted and compacted snow; it generally has a density of at least 500 kg/m^3.

Newton A measure of force; the force necessary to accelerate a 1-kilogram mass 1 meter in 1 second squared.

Nitrogen fixation The process by which gaseous nitrogen is converted by microorganisms living in association with certain plants to a form that can be stored in the soil and utilized by plants.

Nonferromagnesian minerals A subgroup of the silicate minerals; light-colored, low-density, rock-forming minerals, e.g., quartz and orthoclase feldspar; also called *aluminosilicate* minerals.

Nonparallel slope retreat A mode of slope retreat in which the slope angle grows smaller as the slope is eroded back; *see also* **Parallel slope retreat**.

Normal fault A fault in which the hanging wall is displaced downward relative to the footwall.

Nuée ardente A dense, "glowing cloud" of hot volcanic gas and ash which moves downhill at high speeds, scorching the landscape.

Oblique-slip fault A fault that combines both strike-slip and dip-slip displacements.

Occluded front A frontal condition in a midlatitude cyclone in which the cold and warm fronts have merged, forcing the warm-air sector upward.

Ocean trench A great trough in the ocean floor, between 7500 m and 11,000 m below sea level; trenches are associated with subduction and lie along island arcs or orogenic belts.

Open forest A forest structure with a strong upper one or two stories and limited undergrowth; a forest that is largely open at ground level.

Open system A system characterized by a through-flow of material and/or energy; a system to which energy or material is added and released over time.

Orogenic belt A major chain of mountains on the continents; one of the major geologic subdivisions of the continents.

Orographic precipitation A type of precipitation that results when moist air is forced to rise when passing over a mountain range; most areas of exceptionally heavy rainfall are areas of orographic precipitation.

Oscillatory wave A wave in which there is no mass transport of water; the motion of the wave is circular; thus water particles return to their original position with the passage of each wave.

Outflooding Flooding caused by a stream or river overflowing its banks.

Outwash plain A fluvioglacial deposit comprised of sand and gravel with a flat or gently sloping surface; usually found in close association with moraines.

Overland flow Runoff from surfaces on which the intensity of precipitation or snowmelt exceeds the infiltration capacity; also called Horton overland flow, for hydrologist Robert E. Horton.

Oxbow A cresent-shaped lake or pond in a river valley formed in an abandoned segment of channel.

Ozone One of the minor gases of the atmosphere; a pungent, irritating form of oxygen that performs the important function of absorbing ultraviolet radiation.

Pangaea An ancient supercontinent that was comprised of the world's major land masses packed together around Africa several hundred million years ago; the breakup of Pangaea led to the formation of today's continents.

Parallels The east-west-running lines of the global coordinate system; the equator, the Arctic Circle, and the Antarctic Circle are parallels; all parallels run parallel to one another.

Parallel slope retreat A mode of slope retreat in which the slope angle remains essentially constant as the slope is eroded back; see also **Nonparallel slope retreat**.

Parent material The particulate material in which a soil forms; the two types of parent material are *residual* and *transported*.

Parkland A savanna formation of the midlatitudes characterized by prairie or meadows, with patches and ribbons of broadleaf trees.

Patterned ground Ground in which vegetation, water features, or stones are arranged in a geometric pattern, e.g., circles or polygons; it is widespread in cold environments.

Peak annual flow The largest discharge produced by a stream or river in a given year.

Pedalfers A general class of soil characterized by accumulations of iron and aluminum; soils in areas that receive at least 60 cm of precipitation annually.

Pediment Long, gentle slope at the foot of a cliff or free-face; it is usually composed of bedrock with a light covering of rock debris; common in dry regions.

Pedocals A general class of soil characterized by calcium accumulations and found in areas that receive less than 60 cm of precipitation annually.

Pedon The smallest geographic unit of soil defined by soil scientists of the U.S. Department of Agriculture.

Percolation test A soil-permeability test performed in the field to determine the suitability of a material for wastewater disposal; the test most commonly used by sanitarians and planners to size soil-absorption systems.

Perennial stream A stream that receives inflow of groundwater all year; a stream that has a permanent baseflow.

Peridoite A coarse-grained igneous rock found at depth in the lithosphere; a very dark, high-density rock.

Periglacial environment An area where frost-related processes are a major force in shaping the landscape.

Permafrost A ground-heat condition in which the soil or subsoil is permanently frozen; long-term frozen ground in periglacial environments.

Permeability The rate at which soil or rock transmits groundwater (or gravity water in the area above the water table); measured in cubic feet (or meters) of water transmitted through a specified cross-sectional area when under a hydraulic gradient of 1 foot per 1 foot (or 1 m per 1 m).

pH (see **Soil pH**)

Phase change Reorganization of a substance at the atomic or molecular level resulting in a change in physical state as from liquid to vapor; also called *phase transition*.

Phase transition (see **Phase change**)

Phenological adaptation A form of plant adaptation in which the stages in the life cycle (e.g., flowering, pollination, seed germination) are in phase (adjusted to) the seasons and the periodicity of certain events in the year.

Phloem (see **Xylem and phloem**)

Photoperiod The duration of the daily light period when photochemical activity can take place in a plant.

Photosynthesis The process by which green plants synthesize water and carbon dioxide and, with the energy from absorbed light, convert it into plant materials in the form of sugar and carbohydrates.

Piedmont glacier A large glacier usually formed from the merger of many alpine glaciers.

Piezometric surface The theoretical elevation (datum) to which groundwater would adjust if released from the differential pressure under which it normally exists.

Pioneer One of the communities of the community-succession concept; the first community of plants to occupy a new site.

Piping The formation of horizontal tunnels in a soil due to sapping, i.e., erosion by seepage water; piping often occurs in areas where gullying is or was active and is limited to soils resistant to cave-in.

Pitted topography Glacial terrain characterized by a pocked surface; the pits result from buried ice blocks that have melted away.

Plane of the ecliptic The plane defined by one complete revolution of the earth around the sun.

Plant production The rate of output of organic material by a plant; the total amount of organic matter added to the landscape over some period of time, usually measured in grams per square meter per day or year.

Plant stress Limitations placed on photosynthesis by too much or too little of the basic requirements, namely, light, heat, water, carbon dioxide, and certain minerals.

Plastic deformation Irreversible change in the shape of a body without rupturing.

Plate tectonics The geophysical theory or model in which the lithosphere is partitioned into great plates which move laterally on the surface of the earth; plate tectonics emerged as a serious scientific proposition with the articulation of the theory of continental drift.

Playa A dry lake bed in the desert.

Pleistocene Epoch Generally considered the most recent epoch of geologic time; characterized by major episodes of continental glaciation beginning about two million years ago.

Plucking The process by which a glacier removes blocks of rock from the bedrock; an erosional process associated with melting and refreezing at the base of glacial ice.

Plunging fold A fold whose axis is inclined rather than horizontal to the earth's surface.

Plutonic rock Deep intrusive rock; igneous rock that forms in large chambers well within the crust.

Podzolization A soil-forming regime of cool, moist environments that produces a strongly leached soil with a distinctive hardpan layer.

Podzols One of the great soil groups under the traditional USDA soil-classification scheme; soils characterized by a strong O horizon, a leached A horizon, and a B horizon containing oxides of iron and aluminum.

Point bar Deposit in a stream channel on the inside of a meander or bend.

Polar front The zone or line of contact in the mid-latitudes between polar/arctic air and tropical air; it often coincides with the polar front jet stream.

Polar zone The upper high latitudes, 75°–90° latitude.

Polypedon A group of pedons having similar characteristics; also called a *soil body*.

Pools and riffles Features of stream channels; pools are quiescent places separated by riffles, or reaches of rapid flow.

Pore water pressure The pressure exerted by groundwater against the particles through which it is flowing.

Porosity The total volume of pore (void) space in a given volume of rock or soil; expressed as the percentage of void volume to the total volume of the soil or rock sample.

Potential energy The energy represented by the elevation of mass above a critical datum plane (elevation), e.g., the elevation of rainwater above sea level.

Potential evapotranspiration The projected or calculated loss of soil water in evaporation and transpiration over some time period given an inexhaustible supply of soil water.

Precipitable water vapor The total amount of water in the atmosphere; the average depth of water added to the earth's surface if all the moisture in the atmosphere were to condense and fall to earth.

Precipitation The term used for all moisture—solid and liquid—that falls from the atmosphere.

Pressure gradient The change in pressure over distance between two points; on weather maps the pressure gradient is measured along a line drawn at right angles to the isobars.

Primary consumer An organism that eats plants as its sole source of substance; it may be either a plant (e.g., bacteria or algae) or an animal (e.g., deer or buffalo).

Prime meridian (*see* **Greenwich Meridian**)

Principle of limiting factors The biological principle that the maximum obtainable rate of photosynthesis is limited by whichever basic resource of plant growth is in least supply.

Principle of stream orders The relationship between stream order and the number of streams per order; the relationship for most drainage nets is an inverse one, characterized by many low-order streams and fewer and fewer streams with increasingly higher orders; *see also* **Stream order**.

Productivity (*see* **Plant production**)

Progradation A term used to describe a shoreline that builds seaward.

Pruning In hydrology the cutting back of a drainage net by diverting or burying streams; usually associated with urbanization or agricultural development.

Psychrometer An instrument for measuring atmospheric humidity, comprised of two thermometers—a wet bulb and a dry bulb; humidity is measured by the difference in readings between the two thermometers.

Pteridophyte A low, nonwoody plant that reproduces via spores rather than seeds; the largest group of pteridophytes is the ferns.

Pyroclastic materials Fragments of volcanic rock thrown out in a volcanic explosion.

Quickflow (*see* **Stormflow**)

Quicksand Sand that is incapable of supporting overburden (added weight) because of high pore-water pressure.

Radiation The process by which radiant (electromagnetic) energy is transmitted through free space; the term used to describe electromagnetic energy, as in infrared radiation or shortwave radiation.

Radiation beam The column of solar radiation flowing into or through the atmosphere.

Rainfall intensity The rate of rainfall measured in inches or centimeters of water deposited on the surface per hour or minute.

Rainforest A forest formation dominated by a heavy cover of evergreen trees, with abundant secondary vegetation in the form of epiphytes and lianas; in addition to the equatorial and tropical rainforests, the conifer forests of the very humid portions of the marine west coast are often classed as rainforest.

Rainshadow The dry zone on the leeward side of a mountain range of orographic precipitation.

Rainsplash Soil erosion from the impact of raindrops.

Rainwash Soil erosion by overland flow; erosion by sheets of water running over a surface; usually occurs in association with rainsplash; also called *wash*.

Rating curve A graph that shows the relationship between the discharge and stage of various flow events on a river; once this relationship is established, it may be used to approximate discharge using stage data alone.

Rational method A method for computing the discharge from a small drainage basin in response to a given rainstorm; computation is based on the coefficient of runoff, rainfall intensity, and basin area.

Reach A stretch or segment of stream channel.

Recharge The replenishment of groundwater with water from the surface.

Recurrence The number of years on the average that separate events of a specific magnitude, e.g., the average number of years separating river discharges of a given magnitude or greater.

Reflected wave A wave that rebounds off the shore or an obstacle and is redirected seaward through shore-bound incident waves.

Reg A rocky desert landscape; also called a *hamada*.

Regolith The weathered material overlying the bedrock; usually coarse, unsorted.

Relative humidity An expression for the water vapor content of air at a given temperature; the vapor content of a body of air expressed as a percentage of the amount of vapor held by a parcel of air when it is saturated.

Relief The range of topographic elevation within a prescribed area.

Residual soil Soil formed in parent material derived from the underlying bedrock, i.e., from *in situ* material.

Respiration The internal cellular processes of a plant by which energy is used for biological maintenance.

Reverse fault A fault in which the hanging wall is displaced upward relative to the footwall.

Revolution The motion of a planet in its orbital path around the sun.

Ria coast A heavily indented coast marked by prominent headlands and deep reentrants.

Richter scale A scale for rating the magnitude of an earthquake, i.e., the amount of energy released at the focus of the earthquake. The Richter scale is a logarithmic scale; for each unit, magnitude increases about 32 times.

Riffles (*see* **Pools and riffles**)

Rift (graben) A fault in a zone of tensional stress characterized by a block displaced downward relative to adjacent rock formations.

Rift valley A valley formed by a rift fault, e.g., the valleys of the large lakes of East Africa.

Rimed A term used to describe snow crystals on which condensation has taken place as they fall to earth.

Rip current A relatively narrow jet of water that flows seaward through the breaking waves; it serves as a release for water that builds up near shore.

Riprap Rubble such as broken concrete and rock placed on a surface to stabilize it and reduce erosion.

Roche moutonnées An erosional feature sculpted from a rock knob or dome by a glacier; it is smooth on the side of ice advance and jagged on the down-ice side; the term is derived from *moutonnée*, a French word for wavy wigs of the eighteenth century that were pomaded with mutton tallow.

Rockfall A type of mass movement involving the fall of rock fragments from a cliff or slope face.

Root wedging A mechanical weathering process in which a root grows inside a crack, placing stress on the rock and widening the crack.

Rotation The spinning motion of a sphere, such as that of the earth about its axis.

Roughness length The height of the zone or envelope of calm air over a surface which marks the base of the zone of turbulent airflow.

Runoff In the broadest sense runoff refers to the flow of water from the land as both surface and subsurface discharge; the more restricted and common use, however, refers to runoff as surface discharge in the form of overland flow and channel flow.

Rupture Deformation of a substance by fracturing.

Salination Salt saturation of soil as a result of a rise in the water table due to irrigation.

Saltation The principal mode of transport by wind; it is characterized by particles "hopping" over the ground.

Sand bypassing A means of artificially feeding sand to the beach downshore from a barrier such as a breakwater across a bay mouth.

Sand sea A huge field of sand dunes such as the Empty Quarter of Arabia.

Sapping An erosional process that usually accompanies gullying in which soil particles are eroded by water seeping from a bank.

Saturated adiabatic lapse rate The rate of decline in the temperature of a rising parcel of air after it has reached saturation; it is variable but averages $-0.6°C/100$ m; this rate is less than the dry adiabatic lapse rate ($0.98°C/100$ m), because of the heat released in condensation.

Saturation absolute humidity The maximum mass of water vapor that can be held in a cubic meter of air at a given temperature; *see also* **Absolute humidity**.

Saturation mixing ratio The maximum mass of water vapor that can be held in a given mass of dry air at a given temperature and pressure.

Saturation vapor pressure The maximum value that vapor pressure can attain in air at a given temperature; *see also* **Vapor pressure**.

Savanna A biochore characterized by trees and shrubs scattered among a cover of grasses and forbs; the tropical savanna is the most extensive savanna formation and is found in the areas of the tropical wet/dry climate.

Scattering The process by which minute particles suspended in the atmosphere diffuse incoming solar radiation.

Sclerophyll forest Forest of the Mediterranean climate, characterized by small, widely spaced evergreen hardwood trees; generally considered the least prominent of the world's forest formations.

Scouring A mechanism of erosion by streams and glaciers in which particles carried at the bed abrade underlying rock.

Sea A term used to describe the choppy sort of waves in an area of wave generation.

Sea breeze A local wind that blows from sea to land as a result of the differential heating of land and water in the coastal zone; usually a daily occurrence.

Sea-level pressure The pressure exerted by the atmosphere on the earth's surface at sea level; measured by the height of a column of mercury in a mercurial barometer; normal (average) sea-level pressure is 29.92 inches, or 76 cm of mercury; 14.7 pounds of pressure per square inch; or 1013.2 millibars of force per square meter.

Seamount A volcanic mountain in an ocean basin whose origin is not connected with a midoceanic ridge or a subduction zone; volcanic rises in the abyssal plains such as Bermuda and Hawaii.

Secondary consumer An animal that preys on primary consumers; in the soil moles are secondary consumers.

Sediment sink A coastal environment, such as a bay mouth, where massive amounts of sediment are deposited.

Seepage The process by which groundwater or interflow water seeps from the ground.

Seepage lake A lake that gains its water principally from the seepage of groundwater into its basin.

Seif A large sand dune that is elongated in the general direction of the formative winds; a dune formed by winds from more than one direction.

Seismology A branch of geophysics devoted to the study of earthquakes and the interpretation of seismic waves.

Sensible heat Heat that raises the temperature of a substance and thus can be sensed with a thermometer. In contrast to latent heat, it is sometimes called the heat of dry air.

Septic system Specifically, a sewage system that relies on a septic tank to store and/or treat wastewater; generally, an on-site (small-scale) sewage-disposal system that depends on the soil to dispose of wastewater.

7th approximation The modern soil-classification system of the U.S. Department of Agriculture; it uses six levels of classification, beginning with orders—Entisols, Inceptisols, Aridosols, Mollisols, Spodosols, Alfisols, Ultisols, Vertisols, Oxisols, and Histosols.

Shear stress Differential stress acting on a body in which the forces are directed at angles to one another.

Shear wave A type of seismic (elastic) wave that produces motion transverse (perpendicular) to the direction of seismic energy propagation; also termed *secondary wave*.

Shield A major geologic subdivision of the continents; the relatively low elevation interior of a geologically stable continent. The term is derived from the shape of a battle shield placed handle side down.

Shield volcano A volcano comprised mainly of lava, with the overall shape of a shield or dome. The Hawaiian Islands are shield volcanoes.

Shifting agriculture An agricultural practice in tropical and equatorial areas characterized by the movement of farmers from plot to plot as soil becomes exhausted under cultivation.

Sial layer The upper part of the crust; the part that forms the continents and is comprised of relatively light, granitic rocks. The term is a contraction of *silicon* and *aluminum*.

Silicate minerals The principal rock-forming group of minerals; minerals composed of a basic ion of silicon and oxygen, called the silicon oxygen tetrahedron, combined with one or more additional elements.

Silicic lava Lava with a high quartz content in which silicon dioxide constitutes 70 percent or more of the rock; this lava is prevalent on the continents.

Sima layer The lower part of the crust; the part that forms the ocean basin and is comprised of relatively heavy, basaltic rock. The term is a contraction of *siicon* and *magnesium*.

Single-thread channel A stream channel characterized by a single course; it may be straight or meandering.

Sinkhole A pitlike depression in areas of karst topography, caused by the removal of limestone or dolomite by underground drainage; also called a *sink* or *doline*.

Slip face The lee side of a sand dune where wind-blown sand accumulates and slides downslope.

Slope failure A slope that is unable to maintain itself and fails by mass movement such as a landslide, slump, or similar movement.

Slope form The configuration of a slope, e.g., convex, concave, or straight.

Sluiceway A large drainage channel or spillway for glacier meltwater.

Slump A type of mass movement characterized by a back rotational motion along a rupture plane.

Small circle Any circle drawn on the globe that represents less than the full circumference of the earth; thus the plane of a small circle does not pass through the center of the earth. All parallels except the equator are small circles.

Soil-absorption systems The term applied to sewage-disposal systems that rely on the soil to absorb wastewater; see also **Septic system**.

Soil body (see **Polypedon**)

Soil Conservation Service An agency of the U.S. Department of Agriculture that is responsible for soil mapping and analysis in the United States.

Soil creep A type of mass movement characterized by a very slow downslope displacement of soil, generally without fracturing of the soil mass; the mechanisms of soil creep include freeze-thaw activity and wetting and drying cycles.

Soil-forming factors The major factors responsible for the formation of a soil: climate, parent material, vegetation, topography, and drainage.

Soil-heat flux The rate of heat flow into, from, or through the soil.

Soil material Any rock or organic debris in which soil formation takes place.

Soil-moisture balance A model that describes the changes in the availability of soil moisture as a product of precipitation, evapotranspiration, and storage water in the soil.

Soil-moisture deficiency The amount of water needed to raise the moisture content of a soil to field capacity.

Soil-moisture recharge A term in the soil-moisture balance; the replenishment of soil moisture following a period of soil-moisture loss, i.e., following a period of negative moisture balance.

Soil-nutrient system The system defined by the flow nutrients between the soil and the plant cover.

Soil order A major level of soil classification in the traditional USDA scheme; the three orders defined are zonal, azonal, and intrazonal.

Soil pH The degree of alkalinity or acidity of a soil; the ratio of hydrogen ions to hydroxl ions. On the pH scale 7.0 is neutral.

Soil profile The sequence of horizons, or layers, of a soil.

Soil regime A particular combinaton of soil-forming conditions, generally related to climate, that gives rise to certain soil processes and in turn a distinctive soil profile.

Soil structure The term given to the shape of the aggregates of particles that form in a soil; four main structures are recognized: blockly, platy, granular, and prismatic.

Soil texture The cumulative sizes of particles in a soil sample; defined as the percentage by weight of sand, silt, and clay-sized particles in a soil.

Soil-water balance (see **Soil-moisture balance**)

Solar constant The rate at which solar radiation is received on a surface (perpendicular to the radiation) at the edge of the atmosphere. Average strength is 1353 joules/m^2 · sec, which can also be stated as 1.94 cal/cm^2 · min.

Solifluction A type of mass movement in periglacial environments, characterized by the slow flowage of soil material and the formation of lobe-shaped features; prevalent in tundra and alpine landscapes.

Solstice The dates when the declination of the sun is at 23.27° N latitude (the Tropic of Cancer) and 23.27° S latitude (the Tropic of Capricorn)—June 21–22 and December 21–22, respectively. These dates are known as the winter and summer solstices, but which is which depends on the hemisphere.

Solum That part of soil material capable of supporting life; the true soil according to the agronomist; the upper part of the soil mass, including the topsoil and soil horizons.

Solution weathering A type of weathering in which a mineral dissolves on contact with water carrying a solvent such as carbonic acid.

Speciation The process by which new species originate.

Species A group, or taxon, of individuals able to freely interbreed among themselves, but unable to breed with other groups; the smallest taxon of the floristic system of plant classification.

Specific heat The relative increase in the temperature of a substance with the absorption of energy.

Specific humidity An expression for the vapor content of air similar to the *mixing ratio*; the weight (mass) of water vapor relative to the weight of the moist air (vapor plus dry air) to which it belongs; measured in grams per kilogram.

Spheroidal weathering A form of chemical weathering in which a boulder sheds thin plates of rock debris.

Spore A reproductive cell in plants; generally any nonsexual reproductive cell; among the vascular plants, the pteridophytes are spore-bearing.

Spring tide (see **Tide**)

Squall line The narrow zone of intensive turbulence and rainfall along a cold front.

Stage The elevation of the water surface in a river channel.

Stefan-Boltzmann equation The intensity of energy radiated from a body increases with the fourth power of its temperature times a constant.

Stemflow Precipitation water that reaches the ground by running from the vegetation canopy down the trunk of a tree, shrub, or the stem of grass; see *also* **Interception**.

Steppe Short-grass prairie of the semiarid climatic zones; widespread in Eurasia and North America.

Stomata The openings in the foliage of a plant through which moisture is released during transpiration; the stomata open and close in response to air temperature, humidity, and other factors.

Stormflow The portion of streamflow that reaches the stream relatively quickly after a rainstorm, adding a surcharge of water to baseflow.

Stratosphere The subdivision of the atmosphere that lies above the troposphere; it is characterized by stability and temperature that increases with altitude.

Stream order The relative position, or rank, of a stream in a drainage network. Streams without tributaries, usually the small ones, are first-order; streams with two or more first-order tributaries are second-order, and so on.

Stress A force acting on a body or substance; *see also* **Plant stress** or **Shear stress**.

Stress-threshold concept A concept of vegetation change based on the magnitude- and-frequency concept; vegetation changes according to the magnitude of various stresses and disturbances in the form of floods, drought, disease, and fire, for example, in the plant environment.

Striation Scour line etched into bedrock by the rock debris on the base of a glacier.

Strike and dip Directional properties of a geologic structure such as a fault. Strike is the directional trend of a formation along the surface; dip is the angle of incline of the formation measured at a right angle to strike.

Strike-slip fault A fault in which the main direction of displacement is lateral, or along the fault line.

Subarctic zone The belt of latitude between 55° and the Arctic and Antarctic circles.

Subduction The process by which a tectonic plate is consumed or destroyed as it slides into the earth along the contact with an adjacent plate; subduction zones are places of frequent earthquakes and volcanism and are usually marked on the surface by ocean trenches.

Sublimation A physical process by which a solid is changed directly to a gas (or vice versa) without passing through the liquid phase.

Subtropical high-pressure cells Large cells of high pressure, centered at 25°–30° latitude in both hemispheres, which are fed by air descending from aloft; these cells are the main cause of aridity in tropics and subtropics.

Subtropical zone The zone of latitude near the tropics in both hemispheres; between 23.5° and 35°.

Succulent habit A form of plant adaptation to arid conditions characterized by fleshy bodies and/or foliage with the capacity to store large amounts of water.

Summer solstice (see **solstice**)

Sun angle The angle formed between the beam of incoming solar radiation and a plane at the earth's surface or a plane of the same attitude anywhere in the atmosphere.

Surge A large and often destructive wave caused by intensive atmospheric pressure and strong winds.

Suspended load The particles (sediment) carried aloft in a stream of wind by turbulent flow; usually clay- and silt-sized particles.

Swash The thin sheet of water that slides up the beach face after a wave breaks.

Swell A wave with a relatively smooth form, usually found at some distance from the area of wave generation.

Syncline A fold characterized by a downward bend, i.e., concave downward in cross-section.

Système international (S.I.) The preferred system of units according to an international consensus of scientists; the basic units of energy, time, and space are the joule, second, and square meter, respectively.

Taiga (see **Boreal forest**)

Talus An accumulation of rock debris at the foot of a slope as a result of rockfall.

Taxon Any unit (category) of classification of organisms.

Tectonic plate A large sheet of lithosphere that moves as a discrete entity on the surface of the earth; the lithosphere is subdivided into seven major plates and many smaller ones; each major plate, with the exception of the Pacific plate, contains a continent.

Temperate forest A forest of the midlatitude regions that could be described as climatically temperate, e.g., broadleaf deciduous forests of Europe and North America, comprised of beeches, maples, and oaks.

Temperature inversion An atmospheric condition in which the cold air underlies warm air; inversions are highly stable conditions and thus not conducive to atmospheric mixing.

Temperature profile The change in temperature along a line or transect through an environment, usually expressed in a graphical format.

Terrace A surface formed by wave erosion or river processes and elevated above the existing level of the ocean or floodplain.

Tertiary consumer A carnivore that preys on secondary as well as primary consumers; animals, such as birds of prey, that are near the ends of the food chains.

Theory of continental drift (see **Continental drift**)

Thermal conductivity A thermal property of a substance describing its capacity to transmit heat given a thermal gradient of $1°K$ per meter (or $1°C/m$).

Thermal diffusivity A thermal property of a substance that describes the rate at which a given temperature, represented, for example, by an isotherm, passes through a substance. It is defined as the ratio of thermal conductivity to volumetric heat capacity.

Thermal gradient The change in temperature over distance in a substance; usually expressed in degrees Celsius per centimeter or meter.

Thornbush A vegetative formation of the tropical savanna regions, characterized by short, thorny trees and shrubs; called *caatinqa* in northeastern Brazil and *dornveld* in South Africa.

Threshold The level or magnitude of a process at which sudden or rapid change is initiated.

Thrust fault A fault in which the hanging wall is driven laterally over the footwall.

Thunderstorm An intensive convectional storm that produces heavy precipitation, strong local winds, as well as thunder and lightning.

Tide A large wave caused by bulges in the sea in response to the lunar and solar gravitational forces. The largest tides are *spring tides,* which occur when the moon and sun are aligned with the earth; the smallest are *neap tides,* which occur when the moon and sun are positioned at a right angle relative to the earth.

Tolerance The range of stress or disturbance a plant is able to withstand without damage or death.

Tombolo A depositional feature along some shore-lines which forms a neck of land between an island and the mainland.

Topographic relief (see **Relief**)

Topsoil The uppermost layer of the soil, characterized by a high organic content; the organic layer of the soil.

Township and range A system of land subdivision in the United States which uses a grid to classify land units; standard subdivisions include townships and sections.

Traction A mode of sediment transport by wind in which particles move in contact with the ground; also called *creep.*

Trade winds The system of prevailing easterly winds, which flow from the subtropical highs to the intertropical convergence zone (ITC) in both hemispheres; also called the *tropical easterlies.*

Transform fault A strike-slip fault; in particular, the term is applied to the faults that run transverse to the midoceanic ridges.

Transmission The lateral flow of groundwater through an aquifer; measured in terms of cubic feet (or meters) transmitted through a given cross-sectional area per hour or day.

Transpiration The flow of water through the tissue of a plant and into the atmosphere via stomatal openings in the foliage.

Transported soil Soil formed in parent material comprised of deposits laid down by water, wind, or glaciers.

Travel time (see **Concentration time**)

Tree line The upper limit of tree growth on a mountain where forest often gives way to alpine meadow.

Tropical cyclone (see **Hurricane**)

Tropics Correctly used, this term refers to the Tropic of Capricorn and the Tropic of Cancer; however, it is often used to refer to areas equatorward of the tropics.

Troposphere The lowermost subdivision of the atmosphere; the layer that contains the bulk of the atmosphere's mass and is characterized by convectional mixing and temperature that decreases with altitude.

Tsunami A large and often destructive wave caused by tectonic activity such as faulting on the ocean floor.

Tundra Landscape of cold regions, characterized by a light cover of herbaceous plants and underlain by permafrost.

Turbulent flow Flow characterized by mixing motion in which the primary source of flow resistance is the mixing action between slow-moving and faster-moving molecules in a fluid.

Turgor A term used to describe the status of water pressure in plant foliage; when a plant wilts, it loses its turgor, and the leaves become puckered and limp.

Ultraviolet radiation Electromagnetic radiation of wavelengths shorter than visible, but longer than X-rays.

Unified system A soil-classification scheme used in civil engineering, based on soil performance when it is placed under stress.

Uniformitarianism A concept attributed to sixteenth-century geologist James Hutton; the types of processes operating on the earth today are the same ones that were active in the geologic past; often condensed to "the present is the key to the past."

U.S. Geological Survey An agency of the U.S. Department of Interior that is responsible for mapping and analyzing rock types, minerals, earthquakes, riverflow, and related phenomena.

Urban climate The climate in and around urban areas, it is usually somewhat warmer, foggier, and less well lighted than the climate of the surrounding region.

Urbanization The term used to describe the process of urban development, including suburban residential and commercial development.

Valley wall The side slope of a river valley where the floodplain gives way to upland surfaces.

Vapor pressure An expression for the water vapor content of air; it is the pressure exerted by the weight of the water vapor molecules independent of the weight of the other gases in the air; expressed in millibars or newtons per square meter.

Vascular plants Plants in which cells are arranged into a pipelike system of conducting, or vascular, tissue; xylem and phloem are the two main types of vascular tissue.

Vegetative regeneration Asexual regeneration by plants in which some part of the plant, such as a root or a special organ such as a rhizome, is able to propagate new stems.

Vein Igneous rock or a deposit of minerals that form in a joint line or fracture.

Velocity The rate of movement in one direction, expressed as distance over time, e.g., m/sec, km/hr, or mph.

Vernal equinox (*see* **Equinox**)

Viscosity A measure of the resistance to flow in a fluid due to intermolecular friction. At a tempera-ture of 10°C, molasses has a higher viscosity than water does.

Visible light Electromagnetic radiation at wavelengths between 0.4 and 0.7 micrometer; the radiation that comprises the bulk of the energy emitted by the sun.

Warm front A contact between a cold air mass and a warm air mass in which the warm air is moving against the cold air, sliding upward along the contact.

Water table The upper boundary of the zone of groundwater; in fine-textured materials it is usually a transition zone rather than a boundary line. The configuration of the water table often approximates that of the overlying terrain.

Water witching (*see* **Dowsing**)

Watt A unit of power that is often used as an energy expression; equal to 1 joule per second.

Wave celerity Wave velocity, equal to wavelength divided by wave period.

Wave period The time it takes a wave to travel the distance of one wavelength.

Wave refraction The bending of a wave, which results in an approach angle more perpendicular to the shoreline.

Wave of translation A wave that produces a mass transport of water; in coastal areas it is often a breaking wave; see **Oscillatory wave**.

Weathering The breakdown and decay of earth materials, especially rock; see *also* **Chemical weathering**.

Westerlies The prevailing eastward flow of air over land and water in the midlatitudes of both hemispheres; also called the *prevailing westerlies*.

West-wind drift Ocean current, or drift current, which flows eastward in the midlatitudes and subarctic, driven by the prevailing westerly winds.

Wetland A term generally applied to an area where the ground is permanently wet or wet most of the year and is occupied by water-loving (or tolerant) vegetation such as cattails, mangrove, or cypress.

Wetted perimeter The distance from one side of a stream to the other, measured along the bottom.

Wien's law A physical law stating that the wavelength of maximum-intensity radiation grows longer as the absolute temperature of the radiating body decreases; also called *Wien's displacement law*.

Wind power The power generated by wind; proportional to the cube of speed or velocity.

Wind wave A wave generated by the transfer of momentum from wind to a water surface.

Winter solstice (*see* **Solstice**)

Work A concept closely related to energy, work is the product of force and distance and is accomplished when the application of force yields movement of an object in the direction of the force.

Xerophyte Plants capable of surviving prolonged periods of soil drought, e.g., the cacti.

Xylem and phloem Conducting tissue in vascular plants through which the plant fluids are transmitted.

Zenith For any location on earth, the point that is directly overhead to an observer. The zenith position of the sun is the one directly overhead.

Zenith angle The angle formed between a line perpendicular to the earth's surface (at any location) and the beam of incoming solar radiation (on any date).

Zonal soil A soil order under the traditional USDA soil-classification scheme; soils with well-developed horizons that reflect the climate conditions of the region in which they are found.

Zone of eluviation The level, or zone, in a soil losing materials in the form of colloids and ions; the zone of removal.

Zone of illuviation The level, or zone, in a soil where colloids and ions accumulate.

Zone of saturation (*see* **Groundwater**)

INDEX